Promoting Sustainable Practices through Energy Engineering and Asset Management

Vicente González-Prida
University of Seville, Spain

Anthony Raman
NTEC Tertiary Group, New Zealand

A volume in the Advances in Environmental Engineering and Green Technologies (AEEGT) Book Series

Managing Director:	Lindsay Johnston
Managing Editor:	Austin DeMarco
Director of Intellectual Property & Contracts:	Jan Travers
Acquisitions Editor:	Kayla Wolfe
Production Editor:	Christina Henning
Development Editor:	Brandon Carbaugh
Typesetter:	Amanda Smith
Cover Design:	Jason Mull

Published in the United States of America by
Engineering Science Reference (an imprint of IGI Global)
701 E. Chocolate Avenue
Hershey PA, USA 17033
Tel: 717-533-8845
Fax: 717-533-8661
E-mail: cust@igi-global.com
Web site: http://www.igi-global.com

Copyright © 2015 by IGI Global. All rights reserved. No part of this publication may be reproduced, stored or distributed in any form or by any means, electronic or mechanical, including photocopying, without written permission from the publisher. Product or company names used in this set are for identification purposes only. Inclusion of the names of the products or companies does not indicate a claim of ownership by IGI Global of the trademark or registered trademark.
     Library of Congress Cataloging-in-Publication Data

Promoting sustainable practices through energy engineering and asset management / Vicente Gonzalez-Prida and Anthony Raman, editors.
   pages cm
  Includes bibliographical references and index.
   ISBN 978-1-4666-8222-1 (hardcover) -- ISBN 978-1-4666-8223-8 (ebook) 1. Sustainable development reporting--Developing countries. 2. Sustainable development--Developing countries. 3. Energy development z Developing countries. 4. Assets (Accounting)--Developing countries. I. Gonzalez-Prida, Vicente, 1975- II. Raman, Anthony, 1968-
  HD60.3.P76 2015
  333.79028'6--dc23
                       2015003361

This book is published in the IGI Global book series Advances in Environmental Engineering and Green Technologies (AEEGT) (ISSN: 2326-9162; eISSN: 2326-9170)

British Cataloguing in Publication Data
A Cataloguing in Publication record for this book is available from the British Library.

All work contributed to this book is new, previously-unpublished material. The views expressed in this book are those of the authors, but not necessarily of the publisher.

For electronic access to this publication, please contact: eresources@igi-global.com.

Advances in Environmental Engineering and Green Technologies (AEEGT) Book Series

ISSN: 2326-9162
EISSN: 2326-9170

MISSION

Growing awareness and an increased focus on environmental issues such as climate change, energy use, and loss of non-renewable resources have brought about a greater need for research that provides potential solutions to these problems. Research in environmental science and engineering continues to play a vital role in uncovering new opportunities for a "green" future.

The **Advances in Environmental Engineering and Green Technologies (AEEGT)** book series is a mouthpiece for research in all aspects of environmental science, earth science, and green initiatives. This series supports the ongoing research in this field through publishing books that discuss topics within environmental engineering or that deal with the interdisciplinary field of green technologies.

COVERAGE

- Renewable energy
- Green Transportation
- Waste Management
- Industrial Waste Management and Minimization
- Sustainable Communities
- Policies Involving Green Technologies and Environmental Engineering
- Green Technology
- Air Quality
- Radioactive Waste Treatment
- Water Supply and Treatment

IGI Global is currently accepting manuscripts for publication within this series. To submit a proposal for a volume in this series, please contact our Acquisition Editors at Acquisitions@igi-global.com or visit: http://www.igi-global.com/publish/.

The Advances in Environmental Engineering and Green Technologies (AEEGT) Book Series (ISSN 2326-9162) is published by IGI Global, 701 E. Chocolate Avenue, Hershey, PA 17033-1240, USA, www.igi-global.com. This series is composed of titles available for purchase individually; each title is edited to be contextually exclusive from any other title within the series. For pricing and ordering information please visit http://www.igi-global.com/book-series/advances-environmental-engineering-green-technologies/73679. Postmaster: Send all address changes to above address. Copyright © 2015 IGI Global. All rights, including translation in other languages reserved by the publisher. No part of this series may be reproduced or used in any form or by any means – graphics, electronic, or mechanical, including photocopying, recording, taping, or information and retrieval systems – without written permission from the publisher, except for non commercial, educational use, including classroom teaching purposes. The views expressed in this series are those of the authors, but not necessarily of IGI Global.

Titles in this Series

For a list of additional titles in this series, please visit: www.igi-global.com

Handbook of Research on Advancements in Environmental Engineering
Nediljka Gaurina-Medjimurec (University of Zagreb, Croatia)
Engineering Science Reference • copyright 2015 • 660pp • H/C (ISBN: 9781466673366) • US $345.00 (our price)

Soft Computing Applications for Renewable Energy and Energy Efficiency
Maria del Socorro García Cascales (Technical University of Cartagena, Spain) Juan Miguel Sánchez Lozano (University Centre of Defence at the Spanish Air Force Academy, Technical University of Cartagena, Spain) Antonio David Masegosa Arredondo (University of Granada, Spain) and Carlos Cruz Corona (University of Granada, Spain)
Information Science Reference • copyright 2015 • 408pp • H/C (ISBN: 9781466666313) • US $235.00 (our price)

Optimum Design of Renewable Energy Systems Microgrid and Nature Grid Methods
Shin'ya Obara (Kitami Institute of Technology, Japan)
Engineering Science Reference • copyright 2014 • 430pp • H/C (ISBN: 9781466657960) • US $235.00 (our price)

Nuclear Power Plant Instrumentation and Control Systems for Safety and Security
Michael Yastrebenetsky (State Scientific and Technical Centre for Nuclear and Radiation Safety, Ukraine) and Vyacheslav Kharchenko (National Aerospace University- KhAI, Ukraine, and Centre for Safety Infrastructure-Oriented Research and Analysis, Ukraine)
Engineering Science Reference • copyright 2014 • 470pp • H/C (ISBN: 9781466651333) • US $265.00 (our price)

Green Technology Applications for Enterprise and Academic Innovation
Ezendu Ariwa (University of Bedfordshire, UK)
Information Science Reference • copyright 2014 • 335pp • H/C (ISBN: 9781466651661) • US $225.00 (our price)

Computational Intelligence in Remanufacturing
Bo Xing (University of Pretoria, South Africa) and Wen-Jing Gao (Meiyuan Mould Design and Manufacturing Co., Ltd, China)
Information Science Reference • copyright 2014 • 348pp • H/C (ISBN: 9781466649088) • US $195.00 (our price)

Risk Analysis for Prevention of Hazardous Situations in Petroleum and Natural Gas Engineering
Davorin Matanovic (University of Zagreb, Croatia) Nediljka Gaurina-Medjimurec (University of Zagreb, Croatia) and Katarina Simon (University of Zagreb, Croatia)
Engineering Science Reference • copyright 2014 • 433pp • H/C (ISBN: 9781466647770) • US $235.00 (our price)

Marine Technology and Sustainable Development Green Innovations

www.igi-global.com

701 E. Chocolate Ave., Hershey, PA 17033
Order online at www.igi-global.com or call 717-533-8845 x100
To place a standing order for titles released in this series, contact: cust@igi-global.com
Mon-Fri 8:00 am - 5:00 pm (est) or fax 24 hours a day 717-533-8661

Editorial Advisory Board

Charalampos Apostolopoulos, *Independent Researcher, UK*
Horacio Vergara Arancibia, *Independent Researcher, Chile*
Siham El-Kafafi, *University of Waikato, New Zealand*
Beatriz Sedano García, *Centre of Studies and Technical Research, Spain*
Adolfo Crespo Márquez, *University of Seville, Spain*
Warren Naylor, *Northrop Grumman Electronic Systems, USA*
François Pérès, *University of Toulouse, France*

Table of Contents

Preface xv

Section 1
Wind, Solar, and Other Renewable Energies

Chapter 1
Decision Support System for Wind Farm Installation Using Bipolar Analysis 1
    Yasmina Bouzarour-Amokrane, Université de Toulouse, France
    Ayeley P. Tchangani, Université de Toulouse, France
    François Pérès, Université de Toulouse, France

Chapter 2
Decreasing Wear of Large Vertical Axis Wind Turbines by Employing a Multi-Level Turbine
Concept 22
    Jan H. Wiśniewski, Warsaw University of Technology, Poland

Chapter 3
Assessing the Profitability of Changing a Turbine for a Hydroelectric Power Plant Based on
Long-Period Water Gauge Readings 35
    Jan H. Wiśniewski, Warsaw University of Technology, Poland
    Bartosz M. Olszański, Warsaw University of Technology, Poland

Chapter 4
An Overview to Thermal Solar Systems for Low Temperature: Outlining the European Norm
12976 54
    Vicente González-Prida, University of Seville, Spain
    Anthony Raman, NTEC Tertiary Group, New Zealand

Chapter 5
A Reliability Test Installation for Water Heating Solar Systems: Requirements and Design
According to the European Norm 12976 91
    Vicente González-Prida, University of Seville, Spain
    Anthony Raman, NTEC Tertiary Group, New Zealand

Chapter 6
Electricity Production from Small-Scale Photovoltaics in Urban Areas 124
> *Constantinos S. Psomopoulos, Piraeus University of Applied Sciences (TEI of Piraeus), Greece*
> *George Ch. Ioannidis, Piraeus University of Applied Sciences (TEI of Piraeus), Greece*
> *Stavros D. Kaminaris, Piraeus University of Applied Sciences (TEI of Piraeus), Greece*

Chapter 7
The Topicality and the Peculiarities of the Renewable Energy Sources Integration into the Ukrainian Power Grids and the Heating System 162
> *Vira Shendryk, Sumy State University, Ukraine*
> *Olha Shulyma, Sumy State University, Ukraine*
> *Yuliia Parfenenko, Sumy State University, Ukraine*

Section 2
Bordering Topics about Asset Management and Green Energies

Chapter 8
A System Safety Analysis of Renewable Energy Sources 194
> *Warren Naylor, Independent Researcher, USA*

Chapter 9
Predictive Maintenance for Quality Control in High Precision Processes 204
> *María Carmen Carnero, University of Castilla – La Mancha, Spain*
> *Carlos López-Escobar, Aluminium Company of America (ALCOA), Spain*
> *Rafael González-Palma, University of Cádiz, Spain*
> *Pedro Mayorga, Electrical Technology Institute (ITE), Spain*
> *David Almorza, University of Cádiz, Spain*

Chapter 10
Retrospection of Globalisation Process and the Sustainability of Natural Environment in Developing Countries 244
> *Shahul Hameed, Te Wananga o Aotearoa, New Zealand*

Chapter 11
Clean Technology Industry: Relevance of Patents and Related Service Providers 263
> *Liina Tonisson, Fraunhofer MOEZ, Germany*
> *Lutz Maicher, University of Jena, Germany & Fraunhofer MOEZ, Germany*

Chapter 12
Mathematical and Stochastic Models for Reliability in Repairable Industrial Physical Assets 287
> *Pablo A. Viveros Gunckel, Universidad Técnica Federico Santa María, Chile*
> *Adolfo Crespo Márquez, Universidad de Sevilla, Spain*
> *Fredy A. Kristjanpoller, Universidad Técnica Federico Santa María, Chile*
> *Rene W. Tapia, RelPro SpA, Chile*
> *Vicente González-Prida, Universidad de Sevilla, Spain*

Chapter 13
Challenges in Building a Green Supply Chain: Case of Intel Malaysia 311
    Yudi Fernando, Universiti Sains Malaysia, Malaysia
    Kurtar Kaur, Universiti Sains Malaysia, Malaysia
    Ika Sari Wahyuni-TD, Andalas University, Indonesia

Chapter 14
Low Carbon Footprint: The Supply Chain Agenda in Malaysian Manufacturing Firms 324
    Muhammad Shabir Shaharudin, Universiti Sains Malaysia, Malaysia
    Yudi Fernando, Universiti Sains Malaysia, Malaysia

Chapter 15
Review of Supply Chain Integration on Green Supply Chain Management (GSCM) 348
    Alia Nadhirah Ahmad Kamal, Universiti Sains Malaysia, Malaysia
    Yudi Fernando, Universiti Sains Malaysia, Malaysia

Compilation of References 369

About the Contributors 393

Index 400

Detailed Table of Contents

Preface xv

Section 1
Wind, Solar, and Other Renewable Energies

Chapter 1
Decision Support System for Wind Farm Installation Using Bipolar Analysis.. 1
    Yasmina Bouzarour-Amokrane, Université de Toulouse, France
    Ayeley P. Tchangani, Université de Toulouse, France
    François Pérès, Université de Toulouse, France

The necessity to control and reduce the negative impact of human activities on environment and life quality along with technology progress in renewable energy in general and wind energy in particular render it possible today to consider wind energy projects on a large scale. Developing wind energy on a large scale however raises other problems such as choosing an adequate site to settle a wind farm where many other issues such technical feasibility and performance levels, visual pollution, economic and social concerns, etc. must be addressed. Such decisions usually involve many parameters and necessitate the collaboration of many stakeholders. In this context, this chapter proposes an approach based on the concept of bipolar analysis through Benefit Opportunity Cost and Risk (BOCR) analysis, which permits one to address correctly a Group Decision-Making Problem (GDMP) to build a decision support system in order to assist the wind farm installation process.

Chapter 2
Decreasing Wear of Large Vertical Axis Wind Turbines by Employing a Multi-Level Turbine
Concept 22
    Jan H. Wiśniewski, Warsaw University of Technology, Poland

The chapter focuses on describing the author's own multi-level vertical axis wind turbine concept, putting emphasis on its specific features, the scope of conducted analyses, as well as general knowledge important to wind industry specialists, other people with an interest in wind energy, and engineers aspiring to achieve innovative results without needlessly complicating their design. Current results show a reduction of the maximum bending moment during a rotation at the bottom of a two level turbine of up to 19.7% after optimisation; at the same time an optimised turbine can achieve a reduction of maximum moment jump during a rotation at the bottom of a turbine of up to 73.4%. Further studies are currently being conducted, as both the study presented in this chapter and its continuations might have a definitive influence on the future development of the wind-energy sector.

Chapter 3
Assessing the Profitability of Changing a Turbine for a Hydroelectric Power Plant Based on
Long-Period Water Gauge Readings.. 35
    Jan H. Wiśniewski, Warsaw University of Technology, Poland
    Bartosz M. Olszański, Warsaw University of Technology, Poland

The chapter focuses on explaining the construction of author's own engineering-level model which calculates energy production based on historical water level and flow rate readings as well as economic factors such as net present value of the proposed investment on the example of a HPP on the Wkra River. The model methodology assumes the identification of location's hydrological features and translates them into a set of contingency scenarios. Various internal costs, such as maintenance or labor costs, related to normal HPP activity and taxation are discussed and incorporated into the economic part of the model. Test case results indicate that for a series of good years in terms of water flow and electricity production, full repayment of initial investment costs is possible after less than three years. Results for the chosen modernization parameters indicate that within 10 years of installing a new turbine, even the most pessimistic case would bring added value to the real estate valuation.

Chapter 4
An Overview to Thermal Solar Systems for Low Temperature: Outlining the European Norm
12976... 54
    Vicente González-Prida, University of Seville, Spain
    Anthony Raman, NTEC Tertiary Group, New Zealand

This chapter deals with those prefabricated systems with a steady state of operation (state in which the temporal variation of the thermodynamic properties is null), describing, in a brief manner, a methodology for testing the characterization of the thermal performance in accordance with the European normative. All of the previously mentioned form the justification for a foundation or base from which a testing installation is proposed in a later chapter that, at the same time, is compared to a real installation. Lastly, this chapter attempts to outline a simple mathematical methodology to analyze the future behavior of the reliability of a system (solar in this case), when it is still in an extremely early stage of its life cycle, such as the design phase.

Chapter 5
A Reliability Test Installation for Water Heating Solar Systems: Requirements and Design
According to the European Norm 12976 91
    Vicente González-Prida, University of Seville, Spain
    Anthony Raman, NTEC Tertiary Group, New Zealand

This chapter analyzes the requirements that a test installation design must comply with in order to carry out the test procedures of prefabricated systems mentioned in a previous chapter and based on the norm EN 12976. In other words, the authors consider for the test installation design, firstly, the requirements that the hydraulic circuit of such an installation must meet, followed by the specifications required by the custom-built systems; all of this has the aim of certifying that the characteristics of the prefabricated system are the applicable ones. Subsequently, the chapter then directly proceeds to design a test installation, which is to be later compared to a real installation.

Chapter 6
Electricity Production from Small-Scale Photovoltaics in Urban Areas 124
    Constantinos S. Psomopoulos, Piraeus University of Applied Sciences (TEI of Piraeus),
        Greece
    George Ch. Ioannidis, Piraeus University of Applied Sciences (TEI of Piraeus), Greece
    Stavros D. Kaminaris, Piraeus University of Applied Sciences (TEI of Piraeus), Greece

The interest in solar photovoltaic energy is growing worldwide. Today, more than 40GW of photovoltaics have been installed all over the world. Since the 1970s, the PV system price is continuously dropping. This price drop and the adaptation of feed-in tariffs at governmental or utility scale have encouraged worldwide application of small-scale photovoltaic systems. The objective of this chapter is to present the potential for electricity production focusing mainly on the benefits of small-scale installations in urban areas, along with the growth of the global photovoltaics market. The types of installation alternatives are described but the focus is on the rooftop installations due to their simplicity and relatively low cost for urban areas. Electricity production data are presented together with their technical characteristics. Furthermore, analysis of the cost reduction is attempted and the benefits gained from the implementation of small-scale systems are also presented, demonstrating the sustainability role they will play.

Chapter 7
The Topicality and the Peculiarities of the Renewable Energy Sources Integration into the
Ukrainian Power Grids and the Heating System... 162
    Vira Shendryk, Sumy State University, Ukraine
    Olha Shulyma, Sumy State University, Ukraine
    Yuliia Parfenenko, Sumy State University, Ukraine

The chapter proposes to approach the problem in the following ways: assess the problems in the existing power system, analyze the current state of the RES using, and determine the existing ways of computer modeling of the grid. The chapter also discusses the topicality of renewable energy use in the construction of distribution grids and the ability to model their work. It explores issues including current research attempts to identify the existing methods, which can be applied in the Decision Support System (DSS) for calculation and evaluation RES. The problem of making decisions for energy saving in district heating requires such measures as energy audit and planning. These activities require monitoring the current energy consumption in real time.

Section 2
Bordering Topics about Asset Management and Green Energies

Chapter 8
A System Safety Analysis of Renewable Energy Sources 194
    Warren Naylor, Independent Researcher, USA

This chapter is focused solely on whether renewable energies can be implemented safely and if they are safer than the technologies they are replacing or supplanting albeit in small quantities at the current pace of implementation. Renewable or sustainable energy sources are necessary due to the ultimate erosion of traditional energy sources and the harmful effects they introduce into the environment and negatively affect our health. Regardless of how you personally feel concerning renewable energy sources, they are here and here to stay. With that simple understanding, we should ensure these systems are safe. This chapter evaluates the hazards associated with renewable energies and compares and contrasts them to those hazards posed by the traditional or legacy fossil fuel energies. The advantages of renewable energies are palpable and discussed in great detail in the other chapters of this book. This chapter focuses specifically on the safety of the renewable energy systems.

Chapter 9
Predictive Maintenance for Quality Control in High Precision Processes 204
    María Carmen Carnero, University of Castilla – La Mancha, Spain
    Carlos López-Escobar, Aluminium Company of America (ALCOA), Spain
    Rafael González-Palma, University of Cádiz, Spain
    Pedro Mayorga, Electrical Technology Institute (ITE), Spain
    David Almorza, University of Cádiz, Spain

In external grinding processes, vibrations induced by the process itself can lead to defects that affect the quality of the parts. The literature offers models that cannot include all process variables in the analysis. This research applies theoretical models and experimental analysis to determine their suitability for predicting the chatter profile of parts in a plunge grinding process. The application of variance analysis to overall vibration value induced by grinding wheel-workpiece contact allows us to show that high frequency displacements vibration are sensitive to the process setup as well as to the quality of the products manufactured. The final statistical analysis has provided a determination of the spectral bands of the process in which the vibrations causes by grinding wheel-workpiece contact influence the existence of flaws in the workpieces. The methodology described can contribute to increasing the environmental sustainability of an industrial organization.

Chapter 10
Retrospection of Globalisation Process and the Sustainability of Natural Environment in
Developing Countries 244
    Shahul Hameed, Te Wananga o Aotearoa, New Zealand

Globalization is an inevitable integrating process and vital to the world economy but it generates many challenges towards the integration of "economic independence" of the nation states like (a) economic integration through investment/trade and capital flow, (b) initiating multilateral political interaction between the countries, and (c) diffusion of dominant cultural values and beliefs over other cultures. globalization accelerates structural change, which alters the industrial structure of host countries, for instance the excessive use of natural resources and contributes to the physical environmental deterioration. Further, globalization transmits and magnifies market failures and policy distortions if not properly addressed. The chapter attempts to (a) identify the key links between globalization and environment deterioration, (b) identify some issues in multilateral economic agreements in trade, finance, investments, and intellectual property rights that affect environmental sustainability, (c) identify and review priority policy issues affecting multilateral economic agreements on environment issues.

Chapter 11
Clean Technology Industry: Relevance of Patents and Related Service Providers 263
    Liina Tonisson, Fraunhofer MOEZ, Germany
    Lutz Maicher, University of Jena, Germany & Fraunhofer MOEZ, Germany

Many clean technology transfer barriers have been associated to Intellectual Property (IP) rights. The objective of this chapter is to give insights to the types of IP rights services the clean technology industry needs to overcome. By conducting in-depth qualitative interviews with a convenience sample of 25 clean technology companies in 2012, most outsourced intellectual property-related services were discovered. The clean technology companies specified the following top three IP services required from service providers: legal services for IP protection, legal services for IP transactions, and IP consultancy (i.e. IP portfolio analyses). The companies investigated outsource IP-related processes to service providers. That leads to the conclusion that outsourcing patent-related activities is an efficient management decision for the clean technology industry. Outsourcing tasks to competent service providers who are familiar with foreign technology and legal markets were found to be especially useful against infringement threats from developing countries.

Chapter 12
Mathematical and Stochastic Models for Reliability in Repairable Industrial Physical Assets 287
    Pablo A. Viveros Gunckel, Universidad Técnica Federico Santa María, Chile
    Adolfo Crespo Márquez, Universidad de Sevilla, Spain
    Fredy A. Kristjanpoller, Universidad Técnica Federico Santa María, Chile
    Rene W. Tapia, RelPro SpA, Chile
    Vicente González-Prida, Universidad de Sevilla, Spain

Generally, assets present a varied behaviour in their life cycle, which is related directly to the use given and consequently related to the technical assistance traditionally known as maintenance or maintenance policies. It can be of a diverse nature: perfect, minimum, imperfect, over-perfect, and destructive as appropriate. This feature requires the application of advanced techniques in order to model the behaviour of assets life, adapting ideally to each reality of use and wear out. In this chapter, the stochastic models PRP, NHPP, and GRP are explained with their conceptual, mathematic, and stochastic development. For each model, the conceptualization, parameterizing, and stochastic simulation are analysed. Additionally, complementing the analysis and resolution pattern, these models are concluded with a numeric application that allows one to show step by step the mathematic and stochastic development as appropriate.

Chapter 13
Challenges in Building a Green Supply Chain: Case of Intel Malaysia 311
    Yudi Fernando, Universiti Sains Malaysia, Malaysia
    Kurtar Kaur, Universiti Sains Malaysia, Malaysia
    Ika Sari Wahyuni-TD, Andalas University, Indonesia

Consumers today are focusing on products that are manufactured using sustainable, environmentally friendly methods. Profitability or even existence of an industry can be impacted by public opinion. Governments all over the world are also coming up with stricter regulations for industries to comply with on items like pollution, hazardous content, conflict minerals, child labor, exploitation, etc. A number of requirements have been set up by the semiconductor industry, and Intel worldwide is working on some of the current issues: (1) conflict-free minerals sourcing; (2) using green/sustainable energy; (3) reduction of water consumption/recycling of water; and (4) migrating to unleaded parts and halogen free parts. This chapter presents the Intel experiences and challenges in building a green supply chain at both the corporate and regional levels.

Chapter 14
Low Carbon Footprint: The Supply Chain Agenda in Malaysian Manufacturing Firms 324
Muhammad Shabir Shaharudin, Universiti Sains Malaysia, Malaysia
Yudi Fernando, Universiti Sains Malaysia, Malaysia

Malaysia has committed to a 40% reduction of carbon emissions by 2020. The government has encouraged industry, society, and non-government organizations to work together to achieve this objective. The government has provided incentives through several energy programmes such as energy efficiency, renewable energy, green technology, and green building. One key area that has been targeted is logistics and supply chain, which has been contributing to high carbon emissions in manufacturing industries. Scholars and practitioners have only recently begun to pay attention to creating a low carbon supply chain. Furthermore, Small Medium Enterprises (SMEs) have faced several challenges in adopting low carbon activities. SMEs are unable to take the advantage of energy initiatives because of a lack of knowledge, a shortage of funds, and inadequate facilities. Almost 90% of firms are in the service industry working with large manufacturing firms and some SMEs working in manufacturing industry are working closely with their supply chain networks; achieving low carbon targets is hampered by the readiness of the manufacturing itself. This chapter discusses the challenges and future agenda of creating low carbon supply chains in manufacturing in Malaysia. Possible solutions are provided at the end of the chapter.

Chapter 15
Review of Supply Chain Integration on Green Supply Chain Management (GSCM) 348
Alia Nadhirah Ahmad Kamal, Universiti Sains Malaysia, Malaysia
Yudi Fernando, Universiti Sains Malaysia, Malaysia

The world economy operates on a capitalist market system where more and more natural resources are strained to produce maximum profits on the basis of achieving the efficiency of the economies of scale. As corporations' awareness increases on the jeopardizing impact they have caused to the deteriorating environment, more corporations have established a more eco-friendly operation. Greening the supply chain is one significant example of such moves. Realizing the green supply chain tendency in the industry, this proposed chapter focuses on highlighting the supply chain integration with business partners (suppliers, shippers, distributors, and customers) on Green Supply Chain Management (GSCM) practices. The chapter shows the literature supporting the important integration of GSCM as it enables corporations to gather collective strength, skills, and capabilities in achieving its ecological as well as business objectives. Both practitioners in companies and corporations might find this review useful, as it outlines major lines of research in the field.

Compilation of References 369

About the Contributors 393

Index 400

Preface

OVERVIEW

Renewable energies are part of a sector that developed greatly towards the end of the 1990s, but most of all in the last few years in a practical sense, all over Europe, and right now in most of the emerging countries. In renewable energy, solar, wind, and geo-thermal energy, among others, have been in a process of increasing industrialization and marketing in recent years. Logically, until there is a sufficiently consolidated market volume in those emerging countries, these types of energies will not be able to compete in the same conditions with the so-called conventional ones. However, social awareness of these new energy forms, as well as their diffusion and establishment in the energy market is still in full growth. This will contribute towards the consolidation of the market for renewable energy in the coming years.

Over the last decade, a great number of books, projects, and doctoral theses have been developed regarding aspects on the application of green energy. In particular, some researches set out installations, mathematical approaches, and methodologies that represent an important support in the search for solutions to technical problems in the field of sustainable energy. These topics form the base from which this book has been intended to be developed. With that purpose, this publication aims to be an essential reference source, building on the available literature in the field of renewable energies in developing countries while providing for further research opportunities in this dynamic field. With contributions from leading researchers, each chapter presents a fresh look at today's current topics. Therefore, it is hoped that this text will provide the resources necessary for managers, technology developers, scientists, and engineers to adopt and implement sustainable practices in developing nations across the globe.

SUMMARY OF TOPICS

Highly experienced contributors have collaborated in the completion of this book. Their chapters have deeply analyzed topics related to the development of green techniques, promotion, implementation, and adoption mainly in developing countries. Therefore, the book covers areas including, but not exclusively, related to engineering and management. Some of these relevant topics are indicated here below:

- Energy and Sustainability;
- Thermal and Photovoltaic Solar Systems;
- Wind Turbines and Wind Farms;
- Industrial Assets Management;

- Globalization of Environmental Issues;
- Production and Consumption of Renewable Energy;
- Dependability;
- Power Consumption and Management;
- Social Implications of Green Energies;
- Renewable Energies in Developing Countries;
- Industrial Management and Reliability Analysis on Renewable Energies.

The book has been also intended to provide a practical view, trying to promote green techniques from educational point of view and mainly thinking in the development of emerging countries. Each chapter in the book focuses on specialized topics for greater understanding of the chosen subject and provides detailed discussions of emerging issues, offers cutting-edge views on new horizons, and deepens the understanding in these mentioned topics. This book aims to provide a holistic viewpoint of the topics covered from both an academic and applied angle.

TARGET AUDIENCE

Engineers, academicians, researchers, advanced-level students (both postgraduate and doctoral), technology developers, and managers who take decisions on this field will find this text useful in furthering their research exposure to pertinent topics in sustainable energy and assisting in furthering their own research efforts in this field. *Promoting Sustainable Practices through Energy Engineering and Asset Management* is aimed at the mentioned target audience worldwide and provides an in depth look at current global concerns. The book brings together a set of highly concentrated researches that will provide in depth knowledge to engineers, academicians, etc. This wide audience is appealed by this book providing with deeper knowledge on a broad set of emerging issues in the global promotion of sustainable practices.

ORGANIZATION OF THE BOOK

The book is organized into two sections. The first section refers to "Wind, Solar, and Other Renewable Energies" and is developed in seven chapters. The second section is related to bordering topics about asset management and green energies and is developed in eight chapters. A brief description of the fifteen chapters, related to their research matter and the conclusions they achieved, are gathered and summarized as follows:

Chapter 1 deals with wind farms installations and presents a decision support system based on a bipolar multicriteria analysis. Among renewable energy, wind power is an energy that does not require fuel and which does not generate greenhouse gas and toxic wastes. Thus, wind power is easily a form of "clean green energy." Since wind farm selection problem requires the involvement of multiple variables, positive and negative criteria are considered distinctly to represent potential wind farms by selectability (positive criteria) and rejectability (negative criteria) measures. Due to the complexity of the socio-economic environment and the rapid technological evolution, a post-evaluation study on wind farm planning is essential to optimize management ability and minimize losses. Considering the satisficing game theory, a final choice of each actor is represented with a satisficing equilibrium set. A post-evaluation

Preface

study on wind farm planning consists in a set of selection indicators proposed in literature in order to evaluate wind farm planning. Basically, the advantages of the model proposed by chapter 1 is that it can be adjusted for other forms of sustainable energy selection process such as hydroelectricity, solar energy, geothermal energy, etc.

Chapter 2 provides new and original solutions which are currently being tested to wind turbines by the employment of a multilevel turbine concept. The researcher is inclined to consider bending moments at the bottom of a large wind tower as a very probable reason for accelerated wind turbine wear. Moment direction oscillations would cause cracking of concrete foundations, with cascading effects, as a loosened tower can gain momentum before hitting the concrete instead of just pushing on it. As the research is ongoing and brings new input every month, this lecture should be treated as an introduction to the topic, and, as the author explains, any persons interested in the results or applications of this research are encouraged to contact him for newer and more specific results. Nevertheless, currently it is only certain, that limiting destructive forces at critical points of a wind turbine can be used to limit material costs or increase reliability. Therefore, the presented solution not only helps achieve specific goals for a wind turbine, but also simplifies the manufacturing and transportation of large-scale its components.

Chapter 3 presents new and easy implementable scheme for assessment of the profitability of changing a turbine for a hydroelectric power plant. The researchers indicate that small and privately owned hydroelectric power plants, which have been in use for more than a decade are usually equipped with specific types of turbines, which for most installation conditions give a much lower annual energy production than a Kaplan turbine installed for the same water conditions would. Due to these reasons, the presented method can be successfully used as a case study element of a business plan or in the long term fiscal forecast of currently running installation. It is the authors' belief that site specific data from everyday reading for the time period of no less than one decade, if available, can be used, by the method described, to create more reliable modernization profitability analyses for small hydroelectric power plants.

Chapter 4 analyzes the requirements that thermal solar system must fulfil in reference to components, work flow, safety systems etc. according to European Standard EN 12976. The installation of these systems is not considered in the above mentioned regulation, but it does include all requirements in the documentation for the person performing the installation and for the final user. In addition to that, this norm expresses also the distinct testing methods applied to domestic solar energy systems for the validation of the prior requirements. With that goal, throughout that chapter the basic concepts related to solar energy utilization and regarding to product quality are defined. Test procedures are also depicted for the validation of prefabricated systems, resulting finally a categorization of the thermal performance. At the end, this chapter presents a test method developed to provide a reliable answer, but at the same time, fast as possible and at a minimum cost.

In Chapter 5, a reliability test installation for water heating systems is presented, following the requirements and design indicated in the European Standard EN-12976. In this chapter, different testing methods applied to domestic solar energy systems are analyzed for the validation of the standard requirements. With this purpose, authors consider for the test installation design, firstly, the requirements that the hydraulic circuit of such an installation must meet, followed by the specifications required by the custom-built systems; all of this has the aim of certifying that the characteristics of the prefabricated system are the applicable ones. Subsequently, the chapter proceeds to design a test installation, which is also compared to a real installation. As far as the interest in thermo-solar energy is growing worldwide and it is usual the setting of new norms in order to commercialize related products in different markets, this chapter may be a good example and aid for those emerging countries who are nowadays applying new standards in this field along with the growth of the global solar market.

Chapter 6 shows how the interest in solar photovoltaic energy is growing worldwide. In particular, this chapter deals with electricity production from small scale photovoltaics in urban areas, highlighting that the major advantage of solar photovoltaic energy systems is that they generate electricity pollution-free and can be easily installed on residential and commercial buildings as grid-connected applications. As the authors indicate, photovoltaics present high interest in community due to fact that the sun is freely providing light in huge quantities and it is expected to be shining for the next hundred million years. Therefore, there is more than enough solar irradiation available to satisfy the world's energy demands. It represents a situation of energy for all. The benefits of the implementation of small-scale photovoltaic generators are, for example, the support to the low voltage distribution network, the contribution in the green of cities and urban areas and, of course the relevant effort of reducing the greenhouse gases emissions. All these data presented and discussed, support the opinion that the photovoltaic systems can help to achieve sustainability and the role of small- scale low cost installation can be an additional advantage, since they can be easily implemented in urban areas.

Chapter 7 describes the features of the integration of renewable energy sources into the Ukrainian power grid, trying to answer questions about the potential of Ukraine, the barriers of its implementation, and the legislation for the regulation of this process. According to the authors, the problem of limited energy resources and their wasteful usage becomes more important from year to year. The primary energy resources are essentially used for production of two types of energy: heat and electricity. The urgent problems in the field of energy management using the information modeling of the electrical power grids and district heating were studied. This chapter comments that, according to the analysis of existing methods that can be applied in information modeling problems, there are many ways of solving such issues. Thus, further research in the field of power grids construction with the renewable energy is aimed to simulate smart grids with different parameters of the renewable energy to achieve their optimal combinations.

In Chapter 8, authors apply existing theoretical models to determine their suitability for predicting the chatter profile of parts in a plunge grinding process. Maintenance has of course a significant impact on the environmental sustainability of an organization. Particularly, predictive maintenance is based on measuring and recording certain physical parameters associated with a working machine in order to obtain data and information through which failures can be detected, so the future state of the machine may be determined. The presented methodology therefore can be applied to predict the remaining life of a machine or equipment, providing a quality control in high precision processes. In other words, the researchers describe a methodology that can contribute to increasing the environmental sustainability of manufacturing organizations with a more efficient use of resources, as the quality of pieces can be guaranteed; therefore, the energy used in manufacture, the material used to make the work pieces and the effort involved in the process of remanufacture is avoided.

Chapter 9 refers to renewable energy sources from the point of view of a system safety analysis. The chapter suggests how the selection of operating parameters that guarantee process quality (acceptable chattering levels) is not possible using a theoretical analysis. Basically, the methodology described by the author may contribute to increasing the environmental sustainability of manufacturing organizations with a more efficient use of resources, as the quality of pieces can be guaranteed. Therefore, the energy used in manufacture, the material used to make work pieces and the effort involved in the process of remanufacture is avoided. Renewable energies are presented as a controversial topic depending on

Preface

several matters (like the political persuasion, proposed financial gains or lose etc.) which makes objective research difficult. The research considers that, given the likelihood of occurrence and the potential consequences of using sustainable energy sources (wind and solar), renewable energies should continue to evolve and displace as much of the traditional energy sources as feasible. In fact, it states there are no safety hazard that cannot be controlled to prohibit the growth of renewable energy sources as they are our future and the safer option.

Chapter 10 states how the major challenges that the host developing countries confront is to manage the process of globalization in such a way that it promotes environmental sustainability and equitable human development. According to the chapter, on the one hand, globalization accelerates structural change, thereby altering the industrial structure of countries, for instance the excessive use of natural resources and contributes to the pollution levels. However, on the other hand, globalization transmits and magnifies market failures and policy distortions that may spread and exacerbate environmental damage; it may also generate pressures for reform as policies heretofore thought of as purely domestic attracts international interest. Therefore and in reference to the globalization process, the author considers that in the presence of diversity of environmental endowments, assimilative capacities and preferences efficient environmental management requires sensitivity to local ecological and social conditions which cannot be ignored towards having viable environmental sustainability.

In Chapter 11 and addressing climate change, the development and deployment of advanced and innovative clean technology are an innovation imperative. According to the authors, the International Energy Agency (IEA) report from 2008 states that clean technology innovation must rise by a factor between two and ten to meet global climate change goals. Therefore, the roles of clean technology transfer and the related intellectual property rights are receiving much attention from industry and policy makers. The chapter states that clean technology companies specified the following top three intellectual property services required from service providers: intellectual property protection, intellectual property contracting, and Intellectual property portfolio analyses. Consequently, outsourcing industrial property matters to trusted external experts are evident in current clean technology industry and it is an accepted management practice among clean technology companies that saves money and time for companies.

Chapter 12 expresses that the reliability model is an essential aspect for the management and optimization of physical industrial assets. Particularly, model and analysis of repairable equipment are of great importance, mainly in order to increase the performance oriented to reliability and maintenance as part of the cost reduction. The wide range and variability of their behavior, demands the application of techniques of diverse complexity and depth, which allow adapting in a better way to each one of the realities. That research becomes an analytic and explicative procedure about the definition, calculation methodology and criteria that must be considered to parameterize industrial assets under certain degradation level after maintenance, complementing in addition its analysis with a numeric application that allows demonstrating step by step the mathematic and stochastic development as appropriate. The practical cases chosen were developed in the mining industry of Chile.

Chapter 13 shows how some companies nowadays are making efforts in reducing their environmental footprint, driving for sustainability and responsible use of natural resources where they operate in. Authors indicate how industry is facing a lot of pressure to stop pollution, to use electrical energy and water in a sustainable way, use ethically sourced material and usage of environmental friendly, non-hazardous (green) materials. The chapter particularizes the topic in a specific corporation, whose vision and mission is internalized by its various business units and departments. Alignment is done at all levels through virtual open forums, departmental meetings, Q&As, and through staff meetings. Employees are encouraged to submit ideas for sustainability and are provided funds to put those ideas into action. Nevertheless, authors consider that much work still needs to be done and the monitoring and publishing of environmental and social KPIs will help in persuading reluctant organization to adopt industry best practices.

Chapter 14 is focused on the current situation about the supply chain agenda in Malaysian manufacturing firms. As climate change becomes one of the top agendas of the world, firms are voluntarily submitting their carbon emissions and adopting renewable energy because of uncertainty in petroleum price and unattractive economic factors. According to the authors, Malaysia is expected to increase gross domestic products which concentrate on international trade. As a consequence, supply chain members should focus on reducing carbon emission and cost of carbon production through the redesign of their current process to latest low carbon process. In addition to this, they consider that manufacturing firms must voluntary participate in government effort to promote low carbon practices because it will benefit firms when policymakers start enacting environmental policy that is suitable for industrial firms.

Finally, Chapter 15 highlights the role of supply chain integrity in Green Supply Chain Management (GSCM) leading to business performance. According to the authors, corporate organizations have begun to have raised interest on the environmental impacts caused by the operation and have started to modify their business processes to be more eco-friendly. The chapter offers a conceptualization based on content analysis of academic journals where most of the articles highlight that integration among supply chain members are the most important factor in GSCM, while competing companies could mutually progress in GSCM should they agree to collaborate. Authors indicate that the adoption of supply chain integration is considerably quite slow especially in developing countries which value competition more than competition even along the vertical chain. This should be therefore improved as GSCM could only function and assist in achieving sustainable business performance, should the firm is confidence on the reliability and integrity of its partners.

CONCLUSION

Throughout this preface, we have observed an overview of the entire book, trying to describe how these topics regarding energy engineering and asset management fits in the world today. In addition to this, it is also indicated the target audience to whom this book is mainly focused, and a brief summary of the importance of each of the chapter included, providing a description of each one. As already commented, the target audience is constituted by professionals, researchers, and students working in the fields of energy engineering and asset management. In order to comment how this book impacts the field and contributes to the subject matter, it is important to underline that this work looks to discuss and address the difficulties and challenges that developing countries face in implementing this kind of green energy, while describing, in a brief manner, methodologies and tools to be applied in accordance with the environmental sustainability.

Preface

The book is, in few words, a repository of ways to improve the quality of energy processes by identifying sustainable practices, while applying asset management techniques and methods as well as an improvement of business processes. To a greater or lesser extent, each chapter provides a set of quality management methods, statistical tools, or just a diverse point of view to promote different sustainable practices in order to improve asset life cycle, customer satisfaction, reduction in pollution, cost reduction, profit increase, etc. Engineering and educational practices already implemented in developed countries can provide examples (to follow, adapt, or to avoid) in order to promote these renewable energies in the new emerging countries. The 15 chapters address different aspects for promotion of sustainable practices as well as the various engineering, mathematical approaches, and/or management tools that may provide a better understanding and awareness of the so-called green energies. Additionally, the book is intended to explore the impact of such practices on emerging countries in which the governments are implementing them.

Vicente González-Prida
University of Seville, Spain

Anthony Raman
NTEC Tertiary Group, New Zealand

Section 1
Wind, Solar, and Other Renewable Energies

Chapter 1
Decision Support System for Wind Farm Installation Using Bipolar Analysis

Yasmina Bouzarour-Amokrane
Université de Toulouse, France

Ayeley P. Tchangani
Université de Toulouse, France

François Pérès
Université de Toulouse, France

ABSTRACT

The necessity to control and reduce the negative impact of human activities on environment and life quality along with technology progress in renewable energy in general and wind energy in particular render it possible today to consider wind energy projects on a large scale. Developing wind energy on a large scale however raises other problems such as choosing an adequate site to settle a wind farm where many other issues such technical feasibility and performance levels, visual pollution, economic and social concerns, etc. must be addressed. Such decisions usually involve many parameters and necessitate the collaboration of many stakeholders. In this context, this chapter proposes an approach based on the concept of bipolar analysis through Benefit Opportunity Cost and Risk (BOCR) analysis, which permits one to address correctly a Group Decision-Making Problem (GDMP) to build a decision support system in order to assist the wind farm installation process.

INTRODUCTION

Renewable energy is to play a larger role in providing electricity due to its existence over wide geographical areas, in contrast to other energy sources, which are concentrated in a limited number of countries. Rapid deployment of renewable energy and energy efficiency is resulting in significant energy security, climate change mitigation, and economic benefits (Executive Summary Energy Technology Perspectives 2012. Pathways to a Clean Energy System, 2012).

DOI: 10.4018/978-1-4666-8222-1.ch001

At national level, at least 30 nations around the world already have renewable energy contributing to more than 20% of energy supply. National renewable energy markets are projected to continue to grow strongly in the coming decade and beyond (Renewables 2013: Global Status Report, 2013).

Among renewable energy, the wind power is an energy that does not require fuel and which does not generate greenhouse gas and toxic wastes. It helps to maintain the air quality without polluting the soil and without overexploiting them (only 2% of the soil is required). Potentially harmful, wind energy can however cause noise and visual disturbances that come into consideration when choosing wind farm installation. The location of a wind farm must consider several criteria including its impact on wildlife and wind speed for instance.

The Wind power is growing at the rate of 30% annually and a worldwide installed capacity had reached 254 GW, at the end of 2012 (World Wind Energy Association [WWEA]. Half-year report. Technical report, 2012). So far, 72 countries own wind power for commercial use, 22 countries have an installed capacity able to deliver more than 1 GW (Wu, Li, Ba, & Wang, 2013).

To achieve such performance it is required to simultaneously consider technological, environmental and political challenges related to the process of matching existing electricity generation capacities with wind energy (Pinson, 2013).

Due to the complexity of the socio-economic environment and the rapid technological evolution, a post-evaluation study on wind farm planning is essential to optimize management ability and minimize losses. For example, ignoring the importance of management has exposed China to severe overcapacity and overproducing power equipment problems since 2009. Even now, China produces 20GW of unrequired energy (Wu et al., 2013).

Post-evaluation study on wind farm planning consists in: evaluating the planning work, the guiding ideology and optimization of general design plan, judging of design, the feasibility of advanced technology and the accuracy of budget estimates as for example the selection of turbines for commercial-scale wind farms considering varying wind conditions (Chowdhury, Zhang, Messac, & Castillo, 2013; Montoya, Manzano-Agugliaro, López-Márquez, Hernández-Escobedo, & Gil, 2014). A set of selection indicators are proposed in literature in order to evaluate wind farm planning such as: wind resources, wind farm sites, equipment, policies, management, uncertainty factors, power demand and production, economics, impact on visual environment, connection with power grid and road, load-bearing capacity of soil, restrictions regarding the existence of prohibited places, varying wind velocities and directions, etc. (Wu et al., 2013; Rahbari, Vafaeipour, Fazelpour, Feidt, & Rosen, 2014).

Considering wind farm sites selection problems, because of the growing concerns of people regarding the (negative) impact of such projects and the great pressure exerted by the authorities in terms of environmental regulations, the implementation of a wind farm installation is done through a rigorous evaluation process taking into account different characteristics at different levels. Structured evaluation process leads to gain environmental and social agreements that will lead to the approval of the competent authorities and thus the implementation of the best project (Lee, Chen, & Kang, 2009). From this point of view, multi-criteria analysis appears to be a suitable tool to merge and analyze all potential projects by establishing a relationship between alternatives and factors that influence the decision (Abu-Taha, 2011; Wu, Geng, Xu, & Zhang, 2014).

With the aim of selecting a site for a wind farm installation, this chapter proposes to structure the problem in a multi-criteria framework and use a bipolar analysis to evaluate the negative and positive impact of each potential site on the objectives of the committee or stakeholders group.

In order to represent the social link and the potential influences that decision makers may respectively undergo, bipolar approach proposes to model the interactions between decision makers through concordance and discordance measures. These measures are used to represent respectively the positive and negative influence that each decision maker can undergo from its neighbourhood. The recommendation is then made using a process of consensus seeking.

The next section introduces wind farm installation and underlines the importance of post-evaluation study. Section 3 presents an overview of the proposed bipolar evaluation model. The developed consensus process to resolve wind farm installation problem is exposed in section 4. Eventually, the paper ends in Section 5 with a case study.

WIND FARM INSTALLATION

For wind farm resource selection (Pestana-Barros & Sequeira-Antunes, 2011; Han, Mol, Lu, & Zhang, 2009; Lee et al., 2009), some studies have demonstrated that the site selection is critical for optimization of wind farm project (Wu et al., 2013). Thomsen et al. (2001) have compared potential for site-specific design of MW (Megawatts) size wind turbines installed at different sites. The result showed that the variation in aerodynamically driven loads and energy production could be more than 50% between different sites. Fuglsang and Thomsen (2001) have presented a method for site-specific design of wind turbines and compared a 1.5 MW stall regulated wind turbine in normal onshore flat terrain with an offshore wind farm and showed a potential increase in energy production of 28%, installation cost reduced by 10.6–4.6% to offshore wind farm. Connection with electric networks, influence of wind turbines' height above ground, Wind Effect Gusting and micro sitting of WEGs are factors having a great influence on annual energy production (Joselin-Herbert, Iniyan, Sreevalsan, & Rajapandian, 2007).

The successful site selection goes through the assessment of some indicators such as: wind resource assessment, analysis on grid-connection conditions, analysis on traffic conditions, study on local terrain and geological conditions, land acquisition and impact on environment. Wu et al. listed these indicator as follow (Wu et al., 2013):

- **Wind Resource Assessment:** The following indicators can be used to estimate the wind resource potential: annual average wind speed, wind power's effective usage hours, capacity factor, etc. The larger these indicators are, the more abundant the wind power is. The high quality wind is necessary for more stable wind direction, smaller changes in wind speed, less weather disasters and smaller intensity of turbulence. The study of geographical distribution of wind speeds, characteristic parameters of the wind, topography and local wind flow and measurement of the wind speed are also crucial in wind resource assessment for successful application of wind turbines.
- **Grid-Connection Analysis:** The selected site should be as close to the grid as possible to reduce investment and circuit losses in grid engineering and also meet the voltage decrease requirement. What's more, enough capacity and good quality of power grid is also required, so as to avoid the damaging effects on the grid caused by wind farm's random output or stop running.
- **Analysis on Traffic Conditions:** Traffic and transportation conditions of the selected wind farm should be taken into consideration. It will have to be defined in particular whether the conditions are appropriate for equipment's transportation, whether the carrying burden of road is suitable for wind turbines and other transport vehicles, that is, the road used for transporting wind turbine should reach three or four levels at least.
- **Terrain and Geological Conditions Analysis:** The more complex is the terrain condition, the more serious will be the turbulence phenomenon, of course this

is adverse for the output of wind turbine. The geological conditions of selected site should also be taken into consideration. It will have to be decided whether they are suitable for deep mining, housing construction and installing wind turbines or not. The ideal foundation is rock, compact soil or clay, lower water table, small earthquake intensity; the ideal wind farm is a place with regular wind, small turbulences and such as there is no slope of hillside steeper than 30°.

- **Analysis on Land Acquisition and Impact on Environment:** Requires also to pay attention to wind farm construction procedure, costs associated with land acquisition, impact on local residents and status of their proper placement.

These issues represent different characteristics that need transversal skills and require several actors as associations, political groups and industry (Lee et al., 2009).

Indeed, installing a wind farm implies project leaders, elected representatives, public authorities, etc. The choice of potential implantation sites (alternatives selection) generally returns to elected representatives acting on behalf of the civil society; the choice may be subjected to the approval of public authorities.

To take into account of the multiplicity of characteristics involved in a group decision context, the following section provides a bipolar evaluation model to address the selection of wind farm location problem in a multi-criteria environment.

BIPOLAR EVALUATION MODEL

Formerly a group decision making problem (GDMP) is characterized by a set of decision makers noted $D = \{d_1, d_2, \ldots d_p\}$ involved in the selection process, a set of alternatives representing the potential sites to select $A = \{a_1, a_2, \ldots a_n\}$, a set of objectives for each decision maker d_k, $O^k = \{o_1^k, o_2^k, \ldots, o_{q_k}^k\}$. The evaluation of potential sites is realized using a set of criteria associated to each objective and noted $C(o_i^k) = \{c_1, c_2, \ldots, c_{m_k}\}$.

In the literature, the concept of bipolarity is focused by some works mainly in the field of Industrial Engineering (Imoussaten, Montmain, Trousset, & Labreuche, 2011; Felix, 2008, 1994) and the goal oriented engineering for management led by objectives in Software Engineering systems (Gonzales-Baixauli, Prado, & Mylopoulos, 2004; Giorgini, Mylopoulos, Nicchiarelli, & Sebastiani, 2002 ; Fleurey & Solberg, 2009).

These works propose models of bipolar and qualitative influence to assess the effect of actions on the performance of complex systems with not necessarily multicriteria formalism. In the multicriteria framework, Grabisch et al. (2008) define the concept of bipolarity through the scales of criteria or attributes characterizing alternatives. These scales can be bipolar and univariate or unipolar and bivariate. In the former case, the scale is divided into two zones by a neutral point. A positive appreciation is associated to the area located above the neutral point, and a negative appreciation to the zone positioned below this point. On unipolar bivariate scales, an alternative can receive both positive and negative evaluations, reflecting contradictory assessments. According to the nomenclature of bipolarity introduced by Grabisch et al. (2008), the model proposed in this chapter can be considered bipolar and univariate.

The elicitation of criteria is realized by considering their supporting nature (where variation is correlated positively with considered objective variation) or rejecting nature $C_r^{o_i^k}(a_i)$ (where variation is correlated negatively with considered objective variation) with regards to objective achievement. The set of criteria is distributed as follows:

$$C^{o_l^k}(a_i) = C_s^{o_l^k}(a_i) \cup C_r^{o_l^k}(a_i).$$

The alternatives are then represented for each decision maker d_k by selectability measures $\mu_{s_0}^k(a_i)$ and rejectability measures $\mu_{s_0}^k(a_i)$ aggregating respectively the results from the evaluation of the attributes of each supporting and rejecting category (Bouzarour-Amokrane, Tchangani, & Pérès, 2012; 2013a, Tchangani, Bouzarour-Amokrane, & Pérès, 2012)).

In this phase, the Analytic Hierarchy Process (AHP) can be used to structure and assess the various attribute values of alternatives. The AHP procedure is a flexible tool to solve unstructured complex decision problems considering quantitative and qualitative aspects in hierarchical structure. The procedure principle is to decompose a decision problem into clusters going from general to operational level. The AHP procedure can be associated to Benefit, Opportunity, Cost, and Risk analysis (BOCR analysis) to distinguish positive certain (benefit) and uncertain (opportunity) factors from negative certain (cost) and uncertain (risk) factors. Opportunities are usually considered as expectations about positive spin-off, future profits and revenue of future positive developments; whereas benefits represent current revenue or those profits from positive developments one is relatively certain of. Risks in BOCR analysis are supposed to stand for the expected consequences of future negative developments, whereas costs represent (current) losses and efforts and consequences of negative developments one is relatively certain of (Wijnmalen, 2007). Considering AHP-BOCR approach, each alternative can be represented by considering distinctly certain criteria (benefit/cost) and uncertain criteria (opportunity/risk) in supporting and rejecting attribute categories (see Figure 1).

The bipolar measures are initially called 'a priori measures' and do not reflect the potential impact of influence (positive or negative) of a vicinity.

In the social collaborative decision, decision makers are encouraged to collaborate in the evaluation phase. Since decision group members have usually different perceptions, attitudes, motivations and personalities, positive and / or negative influences can be exercised during the evaluation phase. When a decision maker is individualistic, he will prefer to consider only his point of view. Conversely, a 'collaborative' or 'holistic' decision maker will tend to integrate the advice of his vicinity in his assessment based on the importance to give to each vicinity member (Bouzarour-Amokrane, Tchangani, & Pérès, 2013b; Tchangani, 2013).

To model the influence related to each decision maker's opinion, the concordance and discordance degrees are defined and integrated in relatives' measures. These measures allow decision makers to express their level of agreement or disagreement with respect to 'influential' players. Considering that $V(d_k)$ is the influencing vicinity of decision maker d_k in positive or negative way, the concordance and discordance degrees noted respectively $\omega_{kk'}^c$ and $\omega_{kk'}^d$ are defined as relative degrees of concordance and discordance that d_k attaches to the opinion of the decision maker $d_{k'}$ compared to other members of his vicinity, where $\sum_{k' \in V(d_k)} \omega_{kk'}^c = 1$ and $\sum_{k' \in V(d_k)} \omega_{kk'}^d = 1$.

Based on this definition, we can consider that a decision maker will be as mush important in the community as the other makers give him a good confidence. Thus, the importance degree of d_k noted Θ_k can be defined with the following equation.

$$\Theta_k = \frac{\sum_{k'} \max\left(0, \omega_{k'k}^c - \omega_{k'k}^d\right)}{\sum_{j=1}^{p_k} \left\{\sum_l \max\left(0, \omega_{lj}^c - \omega_{lj}^d\right)\right\}} \quad (1)$$

for non-zero importance degree, the Equation (2) can be used.

$$\Theta_k = \frac{\sum_k \left(\frac{1}{1+(\exp(-\alpha(\omega_{k'k}^c - \omega_{k'k}^d)))} \right)}{\sum_{j=1}^{p_k} \left(\sum_l \left(\frac{1}{1+(\exp(-\alpha(\omega_{lj}^c - \omega_{lj}^d)))} \right) \right)} \quad (2)$$

where α is a turning parameter.

The 'Relative measures' noted $\mu_s^{k/V(d_k)}$ and $\mu_r^{k/V(d_k)}$ involving the vicinity opinions of decision maker d_k are then defined considering 'a priori' bipolar measure and influence of each member of the vicinity $V(d_k)$ as follows.

$$\mu_s^{k/V(k)}(a_i) = \frac{\sum_{k' \in V(d_k)} \left(\omega_{kk'}^c \mu_{s_0}^k(a_i) + \omega_{kk'}^d \mu_{r_0}^k(a_i) \right)}{\sum_i \left(\sum_{k' \in V(d_k)} \left(\omega_{kk'}^c \mu_{s_0}^k(a_i) + \omega_{kk'}^d \mu_{r_0}^k(a_i) \right) \right)} \quad (3)$$

$$\mu_r^{k/V(d_k)}(a_i) = \frac{\sum_{k' \in V(j)} \left(\omega_{kk'}^c \mu_{r_0}^k(a_i) + \omega_{kk'}^d \mu_{s_0}^k(a_i) \right)}{\sum_i \left(\sum_{k' \in V(d_k)} \left(\omega_{kk'}^c \mu_{r_0}^k(a_i) + \omega_{kk'}^d \mu_{s_0}^k(a_i) \right) \right)} \quad (4)$$

In relative selectability measure, a decision maker d_k considers also 'a priori' rejectability measure of vicinity member in relation of their

Figure 1. Bipolar hierarchical structure of criteria

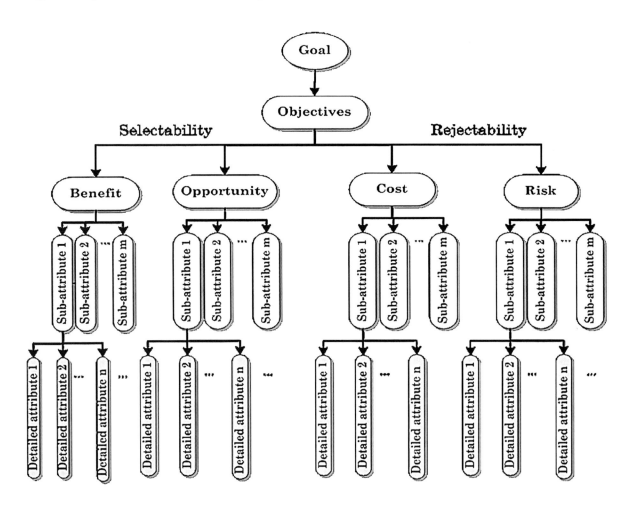

discordance degrees. Inversely, in relative rejectability measure, decision maker d_k considers 'a priori' selectability measure of vicinity member in relation of their degree of discordance.

The final bipolar measures considering the vicinity influence are then given by Equations (5) and (6).

$$\mu_s^k(a_i) = \delta^k \mu_{s_0}^k(a_i) + (1-\delta^k) \mu_s^{k/V(d_k)}(a_i) \quad (5)$$

$$\mu_r^k(a_i) = \delta^k \mu_{r_0}^k(a_i) + (1-\delta^k) \mu_r^{k/V(d_k)}(a_i) \quad (6)$$

where $\mu_{s_0}^k(a_i) / \mu_{r_0}^k(a_i)$ represent 'a priori' measures of alternative a_i.

$0 \leq \delta^k \leq 1$, is the individualism degree of decision maker d_k. When δ^k tends to 0, the decision maker is considered as 'holistic' (altruist) and gives more importance to the global opinion represented by his vicinity. When δ^k tends to 1, the decision maker is 'individualist' and considers his opinion as better than the one of his vicinity.

CONSENSUS MODEL

Once the individual responses have been obtained, a process of identifying consensus is proposed using distance evaluation to achieve the satisfactory group solution considering the individual nature and the potential interactions that may influence their choices. The individualism notion is considered to represent the attitude of each individual and its impact on the final result (Bouzarour-Amokrane et al., 2013b).

Based on the formalism of the satisficing game theory (Stirling, 2003), each decision maker represents its solutions through a satisficing equilibrium set defined as follows:

$$\varepsilon_q^{S,k} = S_q^k \cap \varepsilon^k \quad (7)$$

where S_q^k is the satisficing equilibrium set defined by Equation (8)

$$S_q^k = \left\{ a_i \in A : \mu_s^k(a_i) \geq q^k \mu_r^k(a_i) \right\} \quad (8)$$

with q^k: caution index of the decision maker d_k for adjusting his aspiration level; a low value of q^k enlarges the satisficing set and inversely, a large value of q^k reduces it.

ε^k is an equilibrium set defined by Equation (9) to avoid Pareto dominance that can exist in satisficing set.

$$\varepsilon^k = \left\{ a \in A : \mathcal{D}^k(a_i) = \phi \right\} \quad (9)$$

with $\mathcal{D}^k(a_i)$, the set of alternatives for which there are no strictly better alternatives.

The intersection result of satisficing equilibrium sets of decision maker leads to a common legitimate solution. If the intersection is non empty $\bigcap_{k=1}^{p} \varepsilon_q^{S,k} \neq \emptyset$, selection criteria are proposed to determine the final solution as follows.

1. Selection of alternative selected by the most important decision maker (considering the importance degree).
2. Selection of alternative with the highest average selectability measure.

$$\left(a^{k*} = \arg \max_{a_i \in \cup \varepsilon_q^{s,k}} \left(\sum_k \Theta_k \mu_s^k(a_i) \right) \right)$$

3. Selection of alternative providing the lowest average rejectability.

$$\left(a^{k*} = \arg \min_{a_i \in \cup \varepsilon_q^{s,k}} \left(\sum_k \Theta_k \mu_r^k(a_i) \right) \right)$$

4. Selecting the alternative having the maximum balance between selectability and rejectability for all decision makers

$$\left(a^{k*} = \max_{a_i \in \cup \varepsilon_q^{s,k}} \left(\sum_k \Theta_k \left(\mu_s^k(a_i) - \mu_s^k(a_i) \right) \right) \right).$$

5. **Selecting a Qualified Majority:** In some cases, the use of qualified majority is recommended in decision making (e.g. voting in the Council of the European Union), where, only the responses from a subset of decision makers are considered.

In the case of an empty intersection $\bigcap_{k=1}^{p} \varepsilon_q^{S,k} = \varnothing$, a consensus process is needed to find a common legitimate solution. To reach an agreement, a bipolar soft consensus process based on distance evaluation is proposed below. These measures are commonly called 'proximity measures' when a comparison of individual assessments is made with respect to the collective opinion in a 'soft' consensus process.

Soft Consensus Model

Supporting the idea that a final agreement between the decision makers is not necessary to the resolution of group decision problems, this section proposes to reach a common solution through a soft consensus process (Herrera-Viedma, Herrera, & Chiclana, 2002; Herrera, Herrera-Viedma, & Verdegay, 1996) based on proximity and bipolar measures defined from final bipolar measures.

The proposed bipolar approach based on distance analysis has to be distinguished from TOPSIS (Technique For Order Preference By Similarity To Ideal Solution) based also on distance evaluation to find the optimal solution which represents the shortest distance from the positive solution and the farthest distance from the negative solution (J. Park, I. Park, Kwun, & Tan, 2011). Indeed, the presented bipolar approach proposes more flexible method based on identification of satisficing and non-dominated individual solutions first according to caution index of each individual and reach of a consensus from distance evaluation between individual solutions with regards to convergence notions. Moreover, the proposed approach is different from TOPSIS approach which is characterized by cardinal attributes where preferences are set in advance and the best action can be selected among the poorest if all actions are bad.

The proximity measures are used to assess the gap between decision makers' evaluation regarding an alternative a_i while bipolar consensus measures evaluate the difference between decision maker d_k and the rest of the group regarding bipolar measures of alternative a_i.

Using proximity and bipolar consensus measures, the feedback process can be divided into two phases. In the identification phase, the proximity and consensus measures are used to identify alternatives with wide divergence and, decision makers showing a significant gap assessment on bipolar measures against other members. A recommendation phase is then used to send instructions to decision makers with significant differences.

Unlike soft consensus process proposed in the literature such as those presented by Herrera-Viedma et al. (2007) or Khorshid, (2010), the proposed model does not aim to converge decision maker assessments on all alternatives, but focuses on alternative with the same trend (convergent) initially. Targeted recommendations are given to decision makers with inconsistent opinions considering the rest of the group. This leads to a common solution. An iterative process is engaged with several consultation phases until an agreement is reached. In each iteration, the proximity measures and bipolar consensus are calculated to determine alternatives with strong disagreement (using proximity measures) and identify decision makers with a large divergence (using measurements of bipolar consensus).

The main problem in this situation is how to determine the convergence of individual preferences (Tapia-GarcíA, Del Moral, MartíNez, & Herrera-Viedma, 2012). To achieve this convergence, boundary conditions (tolerance threshold) are determined by the moderator and / or decision makers for each level. Alternatives and decision makers with convergent assessments are identified and targeted adjustments are recommended. This process is governed by a feedback mechanism to guide decision makers in changing their opinions. The parameters of the proposed soft consensus are developed below.

Proximity Measure

A proximity measure (alternative) is a measure used to calculate the average distance between decision makers evaluations over alternative a_i. It is obtained by Equation (10).

$$dis_i = \frac{\sum_k \sum_{k', k' \neq k} \left[\left(d_{s_i}^{kk'}\right)^2 + \left(d_{r_i}^{kk'}\right)^2\right]^{\frac{1}{2}}}{\binom{p}{2}} \quad (10)$$

with dis_i is the proximity measure of alternative a_i,

$$d_{s_i}^{kk'}(a_i) = \Theta_k \mu_s^k(a_i) - \Theta_{k'} \mu_s^{k'}(a_i) \quad ,$$

the gap between selectabilty measure of d_k and $d_{k'}$ over evaluation of alternative a_i considering the importance degree of each decision maker.

$$d_{r_i}^{kk'}(a_i) = \Theta_k \mu_r^k(a_i) - \Theta_{k'} \mu_r^{k'}(a_i),$$

the gap between rejectability measure of d_k and $d_{k'}$ over evaluation of alternative a_i considering the importance degree of each decision maker.

$$\binom{p}{2} = \frac{p!}{2!(p-2)!} \quad ,$$

the binomial coefficient taking into account the distance combinations avoiding redundancies (for example: $d_{s_i}^{12} = d_{s_i}^{21}$).

Bipolar Consensus Measure

Bipolar consensus measures allow calculating the distance between a decision maker d_k and the rest of the group over final bipolar measures of alternative a_i. Bipolar measurements consensus are given by Equations (11) and (12).

$$d_{s_i}^k = \frac{\sum_{k', k' \neq k} \left(\left|d_{s_i}^{kk'}\right|\right)}{p-1} \quad (11)$$

$$d_{s_i}^k = \frac{\sum_{k', k' \neq k} \left(\left|d_{s_i}^{kk'}\right|\right)}{p-1} \quad (12)$$

where $d_{s_i}^k, d_{r_i}^k$ represent respectively, supportability and rejectability consensus measures. These measurements are then integrated into a feedback mechanism defined below.

Feedback Mechanism

A feedback mechanism allows decision makers to change their preferences in order to achieve a tolerated degree of proximity. Literature generally represents the feedback process by an identification phase followed by a recommendation one. According to this structure, the proposed feedback process associated with the bipolar approach is given as follows.

Identification Phase

The identification phase is used to evaluate proximity degree of individual evaluations of each alternative, and compare them to tolerance threshold. Alternatives with high variation (the proximity measurements exceeding tolerance threshold) are discarded. Only converging alternatives are subsequently processed. To identify the contribution of each actor, the distance between the bipolar assessments is calculated using bipolar consensus measures to identify decision makers whose evaluations are far of the group. From these measurements, evaluations of alternatives can be modified using the following steps:

1. Identification of alternatives whose proximity measure dis_i meet the condition (1) $d_i \leq \omega$ where ω is the tolerance threshold for alternatives (average distances on the alternatives set can be considered as threshold tolerance). This allows excluding alternatives that may create conflicts and focus on alternatives already having a certain convergence.
2. Identifying decision makers with divergent opinions through the non-fulfilment of the following conditions: (2) $d_{s_i}^k \leq \omega_s$, (3) $d_{r_i}^k \leq \omega_r$, where ω_s and ω_r are respectively selectability and rejectability tolerance threshold.

Recommendation Phase

- In this recommendation phase, a discussion session allows to give targeted recommendations for divergent decision makers (i.e. those that do not fulfill conditions (2) and (3)) considering selected alternatives (that meet the condition (1)). The recommendations are based on the following rules:

- For $d_{s_i}^k > \omega_s$: decision maker d_k presents a significant difference related to its selectability measure compared to selectability measures of the rest of the group. To know the divergence direction and identify if the considered alternative has an important selectability (positive divergence) or a very low selectability (negative divergence), Equation (13) is used.

$$div_{s_i}^k = \frac{\sum_{k',k' \neq k}\left(d_{s_i}^{kk'}\right)}{p-1} \quad (13)$$

if $div_{s_i}^k > 0$, alternative a_i presents a good selectability measure, no change is required.

otherwise ( ), the selectability measure is smaller than average, an increase of considered measure is recommended.

Similar recommendations are applied for rejectability measures:

- For $d_{r_i}^k > \omega_r$: decision maker d_k presents a significant difference related to its rejectability measure compared to the rejectability measures of the rest of the group. The divergence direction indicates whether the alternative has a low rejectability (positive divergence) or a significant rejectability (negative divergence). The divergence direction of rejectability measure is given by Equation (14)

$$div_{r_i}^k = \frac{\sum_{k',k' \neq k}\left(d_{r_i}^{kk'}\right)}{p-1} \quad (14)$$

if $div_{r_i}^k > 0$, alternative a_i presents a low rejectability measure, no change is required.

otherwise ($div_{r_i}^k > 0$), the rejectability measure is bigger than average, a reduction of this measure is recommended.

Once the changes are done, the satisficing equilibrium set of each decision maker is rebuilt. The iterative process is stopped when $\bigcap_{k=1}^{p} \varepsilon_q^{Sk} \neq \emptyset$. If the solution satisfies the group, the process is stopped. Otherwise, a new iteration can be proposed.

The next section illustrates on a case study the developments of this chapter.

CASE STUDY

Deciding the key location to build a wind farm is an essential element in order to obtain optimum energy production at a cheapest cost. A strategic location is crucial to reduce costs associated with turbine foundations, access roads and construction areas. The location selection will also enable the authorities to predict the environmental impact on the wind farm surroundings.

The real world problem we have been dealing with in this section was treated initially by Lee et al. (2009) to select the most appropriate site to settle a wind farm with sustainable objectives to satisfy. The problem was reformulated to fit the developed framework with a highlight on sustainability as the main goal to reach. Three objectives have been considered: performance objective, Operational objective (business drivers), and socio-economic objectives. Performance concerns the capabilities of the conversion system for delivering the results, such as availability and efficiency, in variant processing environments. Business drivers are defined as the expectations of participants about the wind farm, such as potential, challenge, and opportunities. Socio-economic needs consider the aptitude of the envisaged solution to be competitive with respect to the other challengers (Lee et al., 2009).

Several criteria can be considered to limit the variability and disadvantages associated with the installation of a wind farm. For example, wind availability criterion takes into account the strength of the wind which can vary from zero to storm level.

The discussed method is used to evaluate the priority of criteria in selecting a location for wind farm considering bipolar environment based on BOCR analysis. The repartition of criteria on each objective in the considered context is summarized in Table 1.

Considering that a decisional committee is composed of three entities: the wind specialists, local elected officials and public authorities respectively noted d_1, d_2, d_3. The importance degree given by each decision maker, for Performance objective, Operational objective and Socio-economic objectives are respectively $\omega^{1,O} = [0.1\ 0.8\ 0.1]$, $\omega^{2,O} = [0.8\ 0.1\ 0.1]$, $\omega^{3,O} = [0.1\ 0.1\ 0.8]$

It is assumed that production department (d_1) will give more importance to the performance objective, their first goal being to establish a production site with a good yield. To manage the budget and preserve the communal heritage, elected officials (d_2) will promote socio-economics aspects and public authorities (d_3) will pay more attention to the operability of the future site.

The AHP method associated to BOCR analysis is used to evaluate alternatives over benefit, opportunity, cost and risk factors through the proposed repartition of criteria. For example, the initial results of criteria evaluation given by d_1 are summarized in Table 2.

The aggregating results of criteria evaluation considering supporting and rejecting categories are represented in Table 3 by the 'a priori' bipolar measures for each decision maker.

The social link and the potential influences between decision makers are modelled through concordance and discordance measures noted respectively $\omega_{kk'}^c / \omega_{kk'}^d$ and represented by the following matrices.

$$\omega_{kk'}^c = \begin{bmatrix} - & 0.1 & 0.9 \\ 0.7 & - & 0.3 \\ 0.2 & 0.8 & - \end{bmatrix} \omega_{kk'}^d = \begin{bmatrix} - & 0.7 & 0.3 \\ 0.2 & - & 0.8 \\ 0.6 & 0.4 & - \end{bmatrix}$$

Table 1. Repartition of criteria and sub-criteria with regards to objectives of wind farm project

Objective	Factors	Criteria	Sub-Criteria
Performance objective	Benefit	Wind availability	Geographical distribution of wind speed frequency
			Mean wind power density
			Annual mean wind speed
		Site advantage	Influence of selected height of installation
			Effect of wind gusting
			Micro-siting of WEGs
	Opportunity	Advanced technologies	Computerized supervisory
			Variable speed wind power generation
			Swept area of a turbine rotor
	Cost	Wind turbine	Design and development
			Manufacturing
			Installation, maintenance
Socio-economic objective	Opportunity	Financial schemes	Switchable tariff
			Discount of tax rate and duty rate
			Other investment and production incentives
		Policy support	Wind power concession program
			Clean development mechanisms program
			Other policy supports
	Cost	Connexion	Electric connection
			Grid connection
		Foundation	Main construction
			Peripheral construction
Operational Objective	Benefit	WEG functions	Real and technical availability
			Affordable, reliable, and maintenance free
			Power factor, capacity factor
	Risk	Human	Conflicts Entrepreneurs, policy makers, residents
		Technique	Technical complexity and difficulties
		Pertinence	Loyalty or lease agreement, geology suitability, etc.

These matrices show for example that decision maker d_2 gives good confidence to decision maker d_1 through a good degree of concordance and a low degree of discordance. Considering the concordance and discordance given above, the importance degree Θ_k of each decision maker d_k is deduced from Equation (2) and given by the following vector: $\Theta_k = [0.3334 \ 0.3285 \ 0.3381]$. The relative measures taking into account the vicinity are obtained using Equations (3) and (4) while bipolar final measures result from Equations (5) and (6). Considering that the individualism degree is medium ($\delta^k = [0.5 \ 0.5 \ 0.5]$), the results are respectively presented in Table 4 and Table 5.

Table 2. Evaluation results of criteria given by decision maker d_1

Objective	Factors	Criteria	Sub-Criteria	a_1	a_2	a_3	a_4	a_5
Performance objective	Benefit	Wind availability	Geographical distribution of wind speed frequency	0,1853	0,2265	0,1235	0,2147	0,2500
			Mean wind power density	0,1690	0,2184	0,1632	0,2431	0,2063
			Annual mean wind speed	0,1976	0,2298	0,1774	0,2137	0,1815
		Site advantage	Influence of selected height of installation	0,2196	0,2016	0,1576	0,1990	0,2222
			Effect of wind gusting	0,1698	0,1995	0,2237	0,2022	0,2049
			Micro-siting of WEGs	0,1849	0,2165	0,1825	0,2141	0,2019
	Opportunity	Advanced technologies	Computerized supervisory	0,2117	0,1964	0,1862	0,1990	0,2066
			Variable speed wind power generation	0,2100	0,1900	0,2200	0,1875	0,1925
			Swept area of a turbine rotor	0,2005	0,2112	0,1952	0,1952	0,1979
			Static reactive power compensator	0,1990	0,1914	0,2040	0,1990	0,2065
	Cost	Wind turbine	Design and development	0,1854	0,1987	0,1987	0,2053	0,2119
			Manufacturing	0,1848	0,2065	0,1957	0,1957	0,2174
			Installation, maintenance	0,1842	0,1974	0,2039	0,2039	0,2105
Socio-economic objective	Opportunity	Financial schemes	Switchable tariff	0,2000	0,2000	0,1902	0,2024	0,2073
			Discount of tax rate and duty rate	0,2101	0,1957	0,1836	0,2029	0,2077
			Other investment and production incentives	0,2174	0,1932	0,1763	0,2005	0,2126
		Policy support	Wind power concession program	0,1718	0,2154	0,1872	0,2077	0,2179
			Clean development mechanisms program	0,1963	0,2120	0,1780	0,2042	0,2094
			Other policy supports	0,1839	0,2217	0,1763	0,2141	0,2040
	Cost	Connexion	Electric connection	0,1296	0,2222	0,1111	0,2407	0,2963
			Grid connection	0,1569	0,2157	0,0980	0,1961	0,3333
		Foundation	Main construction	0,1351	0,1892	0,1892	0,2162	0,2703
			Peripheral construction	0,1290	0,1935	0,1613	0,1935	0,3226

continued on following page

Table 2. Continued

Objective	Factors	Criteria	Sub-Criteria	a_1	a_2	a_3	a_4	a_5
Operational Objective	Benefit	WEG functions	Real and technical availability	0,1740	0,2099	0,1961	0,2044	0,2155
			Affordable, reliable, and maintenance free	0,2008	0,1988	0,2008	0,1988	0,2008
			Power factor, capacity factor	0,1889	0,2111	0,1852	0,2185	0,1963
	Risk	Human	Conflicts Entrepreneurs, policy makers, residents	0,2053	0,1947	0,2105	0,2000	0,1895
		Technique	Technical complexity and difficulties	0,2078	0,1939	0,2078	0,1967	0,1939
		Pertinence	Loyalty or lease agreement, geology suitability, etc.	0,2069	0,1936	0,2202	0,1989	0,1804

The graphical representation of the evaluation results (Table 4) in the plane (μ_r, μ_s) is given in Figure 2. The index of caution q^k is assumed to be 1 for each decision maker d_k. The satisficing equilibrium sets $\left(\varepsilon_q^{S,k}\right)$ of actors are as follows:

$\varepsilon_1^{S,1} = \{a_5, a_2, a_4\}$, $\varepsilon_1^{S,2} = \{a_3\}$ and $\varepsilon_1^{S,3} = \{a_1, a_4\}$. In this case there is no common solution $\left(\bigcap_{k=1}^{3}\varepsilon_1^{S,k} = \varnothing\right)$. We propose then to use the proposed soft consensus procedure.

Table 3. 'A priori' bipolar measures for each decision maker

Decision Makers	Bipolar Measures	a_1	a_2	a_3	a_4	a_5
d_1	Selectability measure ($\mu_{s_0}^1$)	0.1943	0.2078	0.1835	0.2089	0.2055
	Rejectability measure ($\mu_{r_0}^1$)	0.2043	0.1889	0.2088	0.2226	0.1753
d_2	Selectability measure ($\mu_{s_0}^2$)	0.1944	0.1940	0.2122	0.1842	0.2152
	Rejectability measure ($\mu_{r_0}^2$)	0.1809	0.2108	0.1776	0.2055	0.2252
d_3	Selectability measure ($\mu_{s_0}^3$)	0.2090	0.1647	0.1978	0.2221	0.2064
	Rejectability measure ($\mu_{r_0}^3$)	0.2025	0.1917	0.2048	0.1933	0.2076

Decision Support System for Wind Farm Installation Using Bipolar Analysis

Table 4. Relative bipolar measures

Decision Makers	Relative Bipolar Measures	a_1	a_2	a_3	a_4	a_5
d_1	$\mu_s^{1/V(d_1)}$	0.1976	0.1861	0.1926	0.2102	0.2135
	$\mu_r^{1/V(d_1)}$	0.1996	0.1893	0.2050	0.1951	0.2109
d_2	$\mu_s^{2/V(d_2)}$	0.2008	0.1929	0.1968	0.2060	0.2035
	$\mu_r^{2/V(d_2)}$	0.2049	0.1815	0.2013	0.2166	0.1957
d_3	$\mu_s^{3/V(d_3)}$	0.1947	0.1972	0.2014	0.2026	0.2041
	$\mu_r^{3/V(d_3)}$	0.1900	0.2043	0.1894	0.2040	0.2122

Using a feedback mechanism, decision makers have the ability to change their assessments based on recommendations made by the analyst during the discussion sessions, with the aim to converge to a common solution.

- **Identification Phase:** The first phase of the feedback mechanism uses proximity measures and bipolar consensus to identify respectively, divergent alternatives and decision makers with assessments that deviate from those of other decision makers. Using Equation (10), the proximity measure is given by

$$d_i = \begin{bmatrix} 0.0077 & 0.0079 & 0.0119 & 0.0113 & 0.0090 \end{bmatrix}$$

Assuming that the average distances on the set of alternatives is the tolerance threshold, the proximity distance must not exceed 0.0096

Figure 2. Graphic representation of final bipolar measures for each decision maker

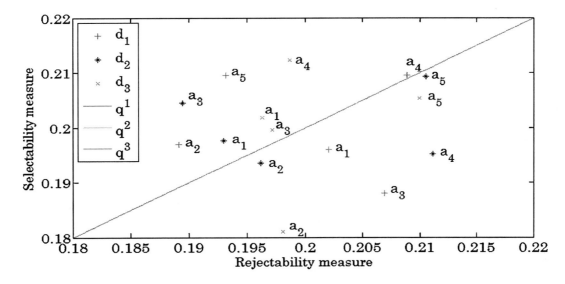

Table 5. Final bipolar measures

Decision Makers	Final Bipolar Measures	a_1	a_2	a_3	a_4	a_5
d_1	Selectability measure (μ_s^1)	0.1960	0.1969	0.1881	0.2095	0.2095
	Rejectability measure (μ_r^1)	0.2020	0.1891	0.2069	0.2089	0.1931
d_2	Selectability measure (μ_s^2)	0.1976	0.1935	0.2045	0.1951	0.2093
	Rejectability measure (μ_r^2)	0.1929	0.1961	0.1894	0.2111	0.2105
d_3	Selectability measure (μ_s^3)	0.2019	0.1809	0.1996	0.2123	0.2053
	Rejectability measure (μ_r^3)	0.1963	0.1980	0.1971	0.1987	0.2099

Table 6. Bipolar consensus measures

	$d_{s_i}^k$			$d_{r_i}^k$		
D	d_1	d_2	d_3	d_1	d_2	d_3
a_1	0,0017	0,0019	0,0031	0,0025	0,0035	0,002
a_2	0,0033	0,0022	0,0034	0,0026	0,002	0,0032
a_3	0,0046	0,0024	0,0026	0,0045	0,0056	0,0034
a_4	0,0039	0,0067	0,0048	0,0014	0,0012	0,0023
a_5	0,0007	0,0009	0,0005	0,0057	0,0033	0,0042

$(d_i \leq 0.0096)$. Consequently, alternative a_3 and a_4 (having widely divergence) should be discarded. Assuming that thresholds ω_s, ω_r were obtained from averages of bipolar distances on the set of alternatives, Table 6 shows the gaps observed at the actor level for each alternative.

- **Recommendation Phase:** The second phase allows the analyst to make targeted recommendations to divergent decision makers: Table 6 shows that:
  ○ Decision maker d_1 presents a deviation from the average concerning the selectability measure of alternative a_2. The direction of the divergence is positive ($div_{s_2}^1 = 0.003$), the selectability measure is important and cannot be modified.
  ○ Decision maker d_2 presents a divergence regarding the rejectability measures of alternatives a1 and a5. The divergence direction of the rejectability measures is given by $div_{r_1}^2 = 0.0035$ and $div_{r_5}^2 = -0.0015$. The negative

Figure 3. Graphic representation of final bipolar measures (iteration1)

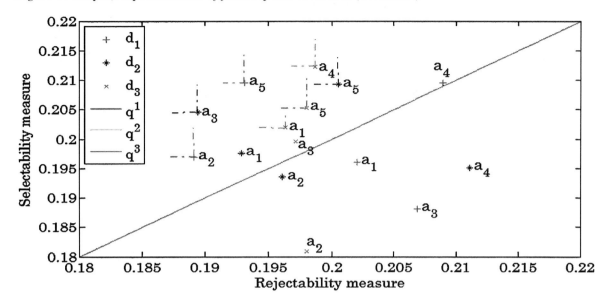

divergence of alternative a5 leads to the recommendation of reducing the rejectability measure. Alternative a1 which has a low rejectability is spared.

○ Decision maker d_3 presents a strong rejectability measure on alternative a5 compared to the rest of the group. The negative divergence direction $\left(div_{r_5}^3 = -0.0042 \right)$ implies a recommendation of reducing this measure.

The reduction of rejectability measures of alternatives a_5 by decision makers d_2 and d_3 to $\mu_r^2(a_5) = 0.2005$ and $\mu_r^3(a_5) = 0.1980$ respectively leads to the following graphical representation (see Figure 3). The satisficing equilibrium sets $\left(\varepsilon_q^{S,k} \right)$ of each decision makers d_k are deduced as follows: $\varepsilon_1^{S,1} = \{a_5, a_2, a_4\}$, $\varepsilon_1^{S,2} = \{a_3, a_5\}$ and $\varepsilon_1^{S,3} = \{a_1, a_4, a_5\}$. The solution obtained by the intersection of the sets is the alternative $a_5 \left(\bigcap_{k=1}^{3} \varepsilon_1^{S,k} = a_5 \right)$.

In the example discussed here, the integration of positive and negative influences of decision makers in the model and the relatively small number of decision makers allowed reaching a consensus quickly after a single recommendation step. The individualism average rate considered for all decision makers allows also to nuance the individual assessments and reduce differences that a high degree of individualism could make appear as shown in figure 4 for individualism degrees δ^k equal to 0.9 for each decision maker.

A sensitivity analysis can be performed by varying the caution index and/or individualism degree to test different possible scenarios and stability of recommended solutions.

Figure 4. Graphic representation of final bipolar measures for $\delta^k = \begin{bmatrix} 0.9 & 0.9 & 0.9 \end{bmatrix}$

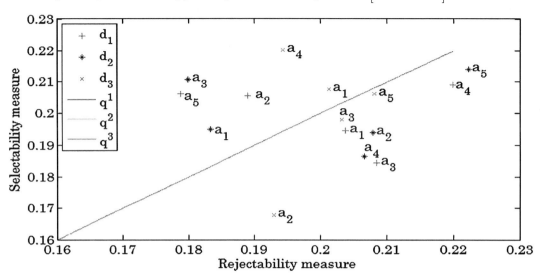

CONCLUSION

The renewable energies now is a topic of great interest for many countries. This growth is due to the depletion of perishable fossil fuels and continuous industrial development. This leads companies to greater efforts to achieve competitive energetic capacity while meeting the new environmental and ecological constraints required by the authorities. In addition, recent environmental disasters (Fukushima nuclear plant disaster, oil spill caused by the sinking of the tanker Erika, etc.) push countries to question their fossil energy sources. Wind energy has relatively safe and positive characteristics that explain the rapid growth of its exploitation in recent decades. However, in some cases, the wind farms installation is done without detailed post-evaluation study which can cause huge losses.

In this chapter, the selection of suitable wind farm problem was considered in bipolar multi-criteria context. The review of used indicators on post-evaluation study on wind farm planning was first introduced. Since wind farm selection problem requires the involvement of multiple decision makers, the second section presented the proposed bipolar evaluation model whose positive and negative criteria are considered distinctly to represent potential wind farms, for each decision maker, by selectability (positive criteria) and rejectability (negative criteria) measures. Considering group decision interactions, relative bipolar measures were defined through concordance and discordance degrees which represent respectively a positive and negative influential vicinity impact on final decision of each decision maker. Based on the satisficing game theory, a final choice of each actor was represented with a satisficing equilibrium set. If the individual solutions do not converge, a soft consensus process was proposed to lead decision makers towards a common solution based on distance evaluation. A real case problem was considered as an application of the proposed methodology. To limit the variability and disadvantages associated with the installation of a wind farm more criteria have to be considered related to its environmental and social impacts such as visual and noise pollution, the pollution level caused by the wind farm construction, the wind farm capacity considering a real need, etc., must be assessed and integrated in the evaluation process. The proposed model can be adjusted for other forms of sustainable energy selection process such as hydroelectricity, solar energy, geothermal energy, etc. This approach can also be used in the selection of used materials in the renewable energy exploitation such as the turbine selection, solar panels selection, etc.

REFERENCES

Abu-Taha, R. (2011). Multi-criteria applications in renewable energy analysis: A literature review. In *Proceedings of Technology Management in the Energy Smart World (PICMET '11)*. Portland, OR: IEEE.

Bouzarour-Amokrane, Y., Tchangani, A., & Pérès, F. (2012). *Definition and measure of risk and opportunity in the bocr analysis*. Paper presented at the 10th International Conference of Modeling and Simulation - MOSIM'12, Bordeaux, France.

Bouzarour-Amokrane, Y., Tchangani, A., & Pérès, F. (2013a). *Evaluation process in end-of-life systems management using BOCR analysis*. Paper presented at the IFAC Conference on Manufacturing modelling, Management and Control, Saint Petersburg, Russia.

Bouzarour-Amokrane, Y., Tchangani, A., & Pérès, F. (2013b). *Résolution des problèmes de décision de groupe par analyse bipolaire*. Paper presented at the 5th Doctoral Days (JDJN) GDRMACS, Strasbourg, France.

Chowdhury, S., Zhang, J., Messac, A., & Castillo, L. (2013). Optimizing the arrangement and the selection of turbines for wind farms subject to varyipreng wind conditions. *Renewable Energy*, 52, 273–282. doi:10.1016/j.renene.2012.10.017

Felix, R. (1994). Relationships between goals in multiple attribute decision making. *Fuzzy Sets and Systems*, 67(1), 47–52. doi:10.1016/0165-0114(94)90207-0

Felix, R. (2008). *Multicriteria decision making (MCDM): Management of aggregation complexity through fuzzy interactions between goals or criteria*. Paper presented at the International Conference on Information Processing and Management of Uncertainty in Knowledge-Based Systems (IPMU), Malaga, Spain.

Fleurey, F., & Solberg, A. (2009). A domain specific modeling language supporting specification, simulation and execution of dynamic adaptive systems. In *Proceedings of Model Driven Engineering Languages and Systems*. Berlin: Springer. doi:10.1007/978-3-642-04425-0_47

Fuglsang, P., & Thomsen, K. (2001). Site-specific design optimization of 1.5–2.0 mw wind turbines. *Journal of Solar Energy Engineering*, 123(4), 296–303. doi:10.1115/1.1404433

Giorgini, P., Mylopoulos, J., Nicchiarelli, E., & Sebastiani, R. (2002). Reasoning with goal models. In *Proceedings of the 21st International Conference on Conceptual Modeling*. London, UK: Springer.

Gonzales-Baixauli, B., Prado Leite, J. C. S., & Mylopoulos, J. (2004). Visual variability analysis for goal models. In *Proceedings of the 12th IEEE International Requirements Engineering Conference (RE'04)*. Kyoto, Japan: IEEE.

Grabisch, M., Greco, S., & Pirlot, M. (2008). Bipolar and bivariate models in multicriteria decision analysis: Descriptive and constructive approaches. *International Journal of Intelligent Systems*, 23(9), 930–969. doi:10.1002/int.20301

Han, J., Mol, A. P. J., Lu, Y., & Zhang, L. (2009). Onshore wind power development in China: Challenges behind a successful story. *Energy Policy*, 37(8), 2941–2951. doi:10.1016/j.enpol.2009.03.021

Herrera, F., Herrera-Viedma, E., & Verdegay, J. L. (1996). A model of consensus in group decision making under linguistic assessments. *Fuzzy Sets and Systems*, *78*(1), 73–87. doi:10.1016/0165-0114(95)00107-7

Herrera-Viedma, E., Alonso, S., Chiclana, F., & Herrera, F. (2007). A consensus model for group decision making with incomplete fuzzy preference relations. *IEEE Transactions on Fuzzy Systems*, *15*(5), 863–877. doi:10.1109/TFUZZ.2006.889952

Herrera-Viedma, E., Herrera, F., & Chiclana, F. (2002). A consensus model for multiperson decision making with different preference structures. *IEEE Transactions on Systems, Man, and Cybernetics. Part A, Systems and Humans*, *32*(3), 394–402. doi:10.1109/TSMCA.2002.802821

Imoussaten, A., Montmain, J., Trousset, F., & Labreuche, C. (2011). Multi-criteria improvement of options. In *Proceedings of the 7th conference of the European Society for Fuzzy Logic and Technology (EUSFLAT-2011) and LFA-2011*. Aix-Les-Bains, France: Atlantis Press.

International Energy Agengy (IAE). (2012). *Executive summary energy technology perspectives: Pathways to a clean energy system*. Retrieved from http://www.iea.org/Textbase/npsum/ETP-2012SUM.pdf

Joselin-Herbert, G. M., Iniyan, S., Sreevalsan, E., & Rajapandian, S. (2007). A review of wind energy technologies. *Renewable & Sustainable Energy Reviews*, *11*(6), 1117–1145. doi:10.1016/j.rser.2005.08.004

Khorshid, S. (2010). Soft consensus model based on coincidence between positive and negative ideal degrees of agreement under a group decision-making fuzzy environment. *Expert Systems with Applications*, *37*(5), 3977–3985. doi:10.1016/j.eswa.2009.11.018

Lee, A. H. I., Chen, H., & Kang, H. (2009). Multi-criteria decision making on strategic selection of wind farms. *Renewable Energy*, *34*(1), 120–126. doi:10.1016/j.renene.2008.04.013

Montoya, F. G., Manzano-Agugliaro, F., López-Márquez, S., Hernández-Escobedo, Q., & Gil, C. (2014). Wind turbine selection for wind farm layout using multi-objective evolutionary algorithms. *Expert Systems with Applications*, *41*(15), 6585–6595. doi:10.1016/j.eswa.2014.04.044

Park, J. H., Park, I. Y., Kwun, Y. C., & Tan, X. (2011). Extension of the TOPSIS method for decision making problems under interval-valued intuitionistic fuzzy environment. *Applied Mathematical Modelling*, *35*(5), 2544–2556. doi:10.1016/j.apm.2010.11.025

Pestana-Barros, C., & Sequeira-Antunes, O. (2011). Performance assessment of Portuguese wind farms: Ownership and managerial efficiency. *Energy Policy*, *39*(6), 3055–3063. doi:10.1016/j.enpol.2011.01.060

Pinson, P. (2013). Wind energy: Forecasting challenges for its operational management. *Statistical Science*, *28*(4), 564–585. doi:10.1214/13-STS445

Rahbari, O., Vafaeipour, M., Fazelpour, F., Feidt, M., & Rosen, M. A. (2014). Towards realistic designs of wind farm layouts: Application of a novel placement selector approach. *Energy Conversion and Management*, *81*, 242–254. doi:10.1016/j.enconman.2014.02.010

Renewables 2013: Global status report. (2013). *REN21 steering committee*. Retrieved from http://www.ren21.net/Portals/0/documents/Resources/GSR/2013/GSR2013_lowres.pdf

Stirling, W. C. (2003). *Satisficing games and decision making: With applications to engineering and computer science*. Cambridge, UK: Cambridge University Press. doi:10.1017/CBO9780511543456

Tapia-Garcí, A., Del Moral, J. M., Martí, M. J., Nez, M. A., & Herrera-Viedma, E. (2012). A consensus model for group decision-making problems with interval fuzzy preference relations. *International Journal of Information Technology & Decision Making*, *11*(04), 709–725. doi:10.1142/S0219622012500174

Tchangani, A. (2013). Bipolarity in decision analysis: A way to cope with human judgment. In Exploring innovative and successful applications of soft computing. Granada, Spain: IGI Global.

Tchangani, A., Bouzarour-Amokrane, Y., & Pérès, F. (2012). Evaluation model in decision analysis: Bipolar approach. *Informatica*, *23*(3), 461–485.

Thomsen, K., Schepers, G., & Fuglsang, P. (2001). Potentials for site-specific design of mw sized wind turbines. *Journal of Solar Energy Engineering*, *123*(4), 304–309. doi:10.1115/1.1408611

Wijnmalen, D. J. D. (2007). Analysis of benefits, opportunities, costs, and risks (BOCR) with the AHP-ANP: A critical validation. *Mathematical and Computer Modelling*, *46*(7-8), 892–905. doi:10.1016/j.mcm.2007.03.020

World Wind Energy Association (WWEA). (2012). *Half-year technical report*. Retrieved from http:// www.wwindea.org/webimages/Half-year report 2012.pdf

Wu, Y., Geng, S., Xu, H., & Zhang, H. (2014). Study of decision framework of wind farm project plan selection under intuitionistic fuzzy set and fuzzy measure environment. *Energy Conversion and Management*, *87*, 274–284. doi:10.1016/j.enconman.2014.07.001

Wu, Y., Li, Y., Ba, X., & Wang, H. (2013). Post-evaluation indicator framework for wind farm planning in China. *Renewable & Sustainable Energy Reviews*, *17*, 26–34. doi:10.1016/j.rser.2012.09.013

KEY TERMS AND DEFINITIONS

Bipolar Approach: Resolution approach used to evaluation a set of alternatives considering distinctly their positive and negative potential to achieve fixed objectives.

BOCR Analysis: Analysis method consisting to evaluate respectively a Benefit, Opportunity, Cost and Risk factors, related to potential solutions considered in decisional problems.

Consensus: Agreement or consent of the largest number of actors (following possible compromise) in order to response to a collective decisional problem.

Group Decision Making Problem (GDMP): Decision problem involving a set of actors (generally come from different specialties and environments) in order to solve a complex problem.

Multi-Criteria Problem: Decision problem including a set of potentially contradictory criteria used to estimate a potential of problem's alternatives.

Renewable Energy: All energy recovered from a natural resource considered inexhaustible on a human timescale (water, wind, sun).

Wind Farm Installation: Set of wind turbines installed in a suitable site or location in order to produce energy.

Chapter 2
Decreasing Wear of Large Vertical Axis Wind Turbines by Employing a Multi-Level Turbine Concept

Jan H. Wiśniewski
Warsaw University of Technology, Poland

ABSTRACT

The chapter focuses on describing the author's own multi-level vertical axis wind turbine concept, putting emphasis on its specific features, the scope of conducted analyses, as well as general knowledge important to wind industry specialists, other people with an interest in wind energy, and engineers aspiring to achieve innovative results without needlessly complicating their design. Current results show a reduction of the maximum bending moment during a rotation at the bottom of a two level turbine of up to 19.7% after optimisation; at the same time an optimised turbine can achieve a reduction of maximum moment jump during a rotation at the bottom of a turbine of up to 73.4%. Further studies are currently being conducted, as both the study presented in this chapter and its continuations might have a definitive influence on the future development of the wind-energy sector.

INTRODUCTION

Industrial wind energy production is currently dominated by Dutch-type horizontal axis wind turbines (HAVTs).

H-type vertical axis wind turbines (HAVTs), in spite of a simpler built and lower noise impacts, have yet to make a strong impact on the amount of electrical energy produced from wind, as there are so far no records of large scale H-type HAVT farms or even a single unit of over 1MW installed capacity lasting for more than a few years. Sadly, neither producers nor investors feel the urge to report the reasons for low lifetimes of large-scale VAWTs. After conducting a study of both scale and pattern of bending moments at the bottom of a large VAWT tower, I am inclined to consider them a very probable reason for accelerated wind turbine wear. Moment direction oscillations would cause cracking of concrete foundations, with

DOI: 10.4018/978-1-4666-8222-1.ch002

Figure 1. Concept drawing of a dual-rotor H-type vertical axis wind turbine

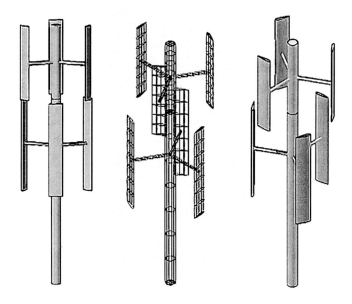

cascading effects, as a loosened tower can gain momentum before hitting the concrete instead of just pushing on it.

The proposed solution to the problem is straightforward: if bending moments and their direction shifts at the bottom of a VAWT are too great, they should, if possible, be reduced. The possibility of such a reduction due to dividing a H-type rotor into parts and shifting them in phase has been analysed with satisfying results. Current results show a reduction of the maximum bending moment during a rotation at the bottom of a two level turbine of up to 19,7% after optimisation, at the same time an optimised turbine can achieve a reduction of maximum moment jump during a rotation at the bottom of a turbine of up to 73.4%. Further studies are currently being conducted, as both the study presented in this chapter and its continuations might have a definitive influence on the future development of the wind-energy sector.

Figure 1 shows a conceptual drawing of a dual-level H-type VAWT. Proportions in it have been chosen to make the drawing easy to understand and not to provide optimal performance.

BACKGROUND

Wind turbines are an ancient technology, with traces of wind turbines in Persia dating back to around 400A.D. and claims of wind turbine use in China as far back as 12th century B.C. Wind turbines that produce electricity however are a much newer technology, with the first working examples having been built in 1887 by Prof James Blyth in Scotland (Price, 2005) and simultaneously in 1888 by faculty and students of a high-school in Jutland, Denmark(Jamison, 1997) and by Charles F. Brush and his company in Ohio, U.S.A.(Anon, 1890). It is interesting to note that the first one of those, and as such the first wind turbine used for production of electricity was a vertical axis wind turbine (VAWT). It is also very important to note that VAWT technology is still in its nascent stage, due to problems with reliability of large-scale units. If these problems could be overcome, then Vertical Axis Wind Turbines could become a very important part of the energy mix, due to a series of strong advantages over Horizontal Axis Wind Turbines (HAWTs). Rapid rotation of wing tips of a HAWT is responsible for high levels of noise pollution as compared to VAWTs. The very same quality of HAWT's, allowing for formation of strong wing-tip vortices, results in comparably high levels of turbulence behind a working turbine and the necessity to space HAWTs far apart. The fact that H-type VAWTs have a less turbulent wake (Dabiri, 2011) allows for achieving greater energy production density over an area – one of important aspects taken into consideration while conducting Strategic Environmental Assessments and Environmental Impact Assessments for energy generating investments. Finally, it is also worth noting, that the Betz limit, commonly referred to as the upper theoretical limit of the kinetic energy of an airflow that can be extracted by a wind turbine refers only to horizontal axis wind turbines. The Betz limit operates on the assumption that there is no mixing between the area swept by the rotor and the surrounding airflow.

That assumption is not only unrealistic in relation to vertical axis wind turbines, as can be seen i.e. in Figure 2, which shows contours of velocity magnitude of a 2D simulation of a VAWT rotor during operation, but would probably have devastating effects on vertical axis wind turbine efficiency had it been true. In the example shown in Figure 2, the wake behind the leading wind (the one most to the left) regains most of the initial speed of the airflow due to mixing with the flow outside the area swept by the rotor. It is very important to note that no existing VAWTs are widely known to have exceeded the Betz limit, but there is also no theoretical reason, why, for large industry scale wind turbines, they shouldn't be able to. The key could be creating a state-of-art turbine on a very large scale – as can be seen using programs like XFOIL, MSES or NASA Foilsim a foil's lift/drag ratio grows along with chord length giving the same designs very different efficiencies, based solely on scale.

BASE ASSUMPTIONS, CONSEQUENCES OF DIVIDING WINGS

For any given wind conditions e.g. speed and set turbine parameters there is an optimal rotor speed, that allows for maximizing wind energy production. Wind speed however is highly dependent on height. While within the Earth's boundary layer, air is travelling at lower speeds due to viscosity. Air particles moving closest to Earth are greatly slowed down by interacting with the immobile surface. Particles next to them are affected by the greatly slowed down particles and so on, with the wind speed increasing gradually with height (usually) – the shape of the wind speed/height curve depending on the outlay of the surrounding terrain. The most common method for determining the shape of the curve is the Hellman exponential law, based on empirical studies (Linacre, 1992) and correlating the speed of the wind at two different heights

$$V_2 = V_1 \cdot \left(\frac{h_2}{h_1}\right)^{\alpha}, \qquad (1)$$

in which V_2 is the wind speed at height h_2, V_1 is the wind speed at height h_1 and α is the friction coefficient also known as the Hellman exponent. This coefficient is dependent on the local topography of a specific site, however it can change to an extent, giving somewhat different results, depending on weather conditions.

Table 1 shows typical friction coefficient values for various landscapes. High values are expected for terrain that heavily obstructs airflow. Usually, the greater the size and number of obstacles, the greater the speed difference between different heights and therefore, the higher the friction coefficient.

A finite length wing experiences a drop in its lift coefficient, as air can move towards a point beyond the span of the wing to avoid it, and vortices can form, mostly on the outer edges of the

Figure 2. Contours of velocity magnitude of an analysed wind turbine level 2D simulation performed in ANSYS Fluent for a wind speed of 9 m/s

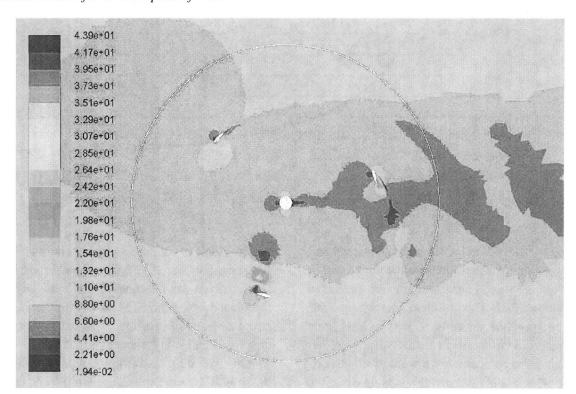

wing. Because of that, shortening wings beyond a certain point can have a strong negative influence on their usefulness.

Figure 3 shows dependence of lift coefficient value loss based on wing span for a constant chord NACA0018 wing. In case of a constant chord NACA0018 wing spanning 45 chord lengths and uniform wind speeds along the span of the wing, the lift coefficient losses would amount to slightly above 1% of the value that can be reached for an infinitely long wing. If the wing should be divided into two sections e.g. spanning respectively 30 and 15 chord lengths, the losses for the first part would amount to 1,59% and the other part 3,18% compared to wings of infinite length. Thus, in the case of a uniform wind speed, weighting in the lengths of the segments one can calculate the additional loss in lift coefficient caused by dividing the wing into parts as 1,06%.

The division of wings into segments presents one additional opportunity. As mentioned at the beginning of the section, for a given set of conditions, there is a wing movement speed that is optimal for energy production, and the optimal wing speed is dependent on wind speed. However

Table 1. Friction coefficient α values for different landscapes

Landscape Type	Friction Coefficient
Lakes, sea, ocean and smooth hard ground	0.10
Low grass	0.15
Tall crops, hedges, shrubs	0.20
Heavily forested land	0.25
Small towns with some trees and shrubs	0.30
City areas with tall buildings	0.40

(Bañuelos-Ruedas, Angeles-Camacho, & Rios-Marcuello, 2011).

Figure 3. Lift coefficient value loss based on wing span

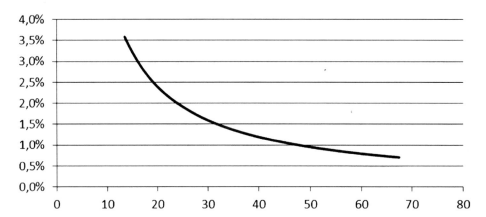

wind speed tends to grow with height, and therefore for optimal performance, the upper part of the turbine should move with greater speed than the lower part. This problem has two possible solutions – one would be to make a long, bent turbine wing with turbine diameter changing over height as to compensate for wind speed change. The other would be to divide a straight wing into segments with segments working at greater wind speeds having greater diameter, losing some lift due to shorter wing span, but working in better conditions tailored to each rotor level, and not just a point on the wing of a single level VAWT. The added benefits are shorter, straight wings, easier for production and transport, as well as obtaining the primary research goal for the project described in this chapter that is the limiting of maximum bending moments at the bottom of the turbine tower during a rotation cycle.

INPUT DATA AND 2D FLOW ANALYSIS RESULTS

Input data required for use in a study of bending moment reduction at the bottom of a VAWT tower was obtained from two dimensional models conducted in ANSYS Fluent, one of the leading programs in computational fluid dynamics simulations. The turbine diameter (22.5m) and wing chord (1.5m) were based on year 2012 assumptions on size of a Darrieus-type wind turbine for use in the Danish Deep-wind programme. The tower, affecting fluent calculations by obstructing wind and generating drag has been given a diameter of 1.5m.

The goal at this point in the analysis was to determine the wind speed, the wing speed ratio (the ratio of the speed of the wing compared to the speed of the wind) and a set wing pitch angle that would allow for optimal energy production.

The initial wind speed for Fluent calculations was determined through meta-analysis. The object of the search would be an optimal wind speed figure understood as the single wind speed responsible for the highest percentage of annual electricity production at given conditions. Parameters approximated to be a fairly good estimate of wind conditions available for a multitude of wind-farms in Poland, a mean wind speed of 6.4 m/s at a height of 50m, a terrain roughness of 0.2m and wind speed distribution fitting a Weibull curve with a shape parameter of 2.1 (Kiss, & Janosi, 2007) were used together with production curves of three different wind turbines: REPower 5MW (HAVT), Suntop 2kW (H-type VAWT) and Evolvegreen 4MW (VAWT, not constructed yet), each analysed for rotor heights of 40m, 60m and 90m.

Table 2. Optimal wind speed meta-analysis results

Rotor Height	REPower 5MW	Suntop 2kW	Evolvegreen 4MW
40 m	9 m/s	8 m/s	9 m/s
60 m	9 m/s	9 m/s	9 m/s
90m	10 m/s	9 m/s	10 m/s

Based on results presented in Table 2, a wind speed of 9 m/s was chosen as the most representative wind speed for basic level simulations in ANSYS Fluent.

The size of the mesh used for the first phase of CFD calculations and shown partially in Figure 4 has a 3 rotor diameters width and 5 rotor diameters length. A triangular mesh has been chosen, due to problems tetrahedral meshes may encounter with properly calculating flows that drastically change direction. Simulations using the Spallart-Allmaras method were carried out for 51 different turbine rotation speeds for each of 9 analysed wing pitch angles, for a total of 459 independent transient simulations of wind turbine performance. Afterwards the method was changed to k-omega SST and checks for various wind speeds, turbine diameters and mesh sizes were performed giving a total of over 1500 independent transient simulations and large amounts useful data.

The final mesh shape around the airfoils themselves was achieved by applying the bias function in ANSYS Mesher. The shape of the mesh after applying 400 divisions along the profile edge along with a bias factor of 6 has been shown in Figure 5. The results of simulations using this division were almost identical to results obtained after applying 1000 divisions along the profile edge with no bias, however there was a distinct difference in the number of time steps after simulation initialization that gave false results, that, along with the lowering computational cost by limiting the number of mesh cells, made using the bias function a much better choice.

Figure 6 shows some of the results obtained for the NACA0018 airfoil using different wing pitches and in some cases, different wind speeds and turbine diameters. The final, most favorable conditions for generating electrical power at a wind speed of 9 m/s were a wing pitch set of 3 degrees and a rotation speed of 150 degrees per second corresponding to a wing speed to wind speed ratio of 3.3.

CALCULATION MODEL, ADVANCED RESULTS

As the only purpose of a wind turbine is to generate power, optimizing turbine parameters for maximum production before proceeding to the final study was a practical necessity. The wing moment

Figure 4. Close-up of mesh used for simulations with wing pitch set at 3 degrees

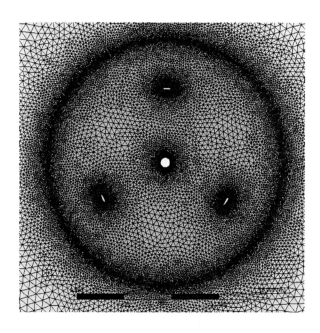

Figure 5. Close-up of mesh around a NACA0018 profile used for simulations with wing pitch set at 3 degrees

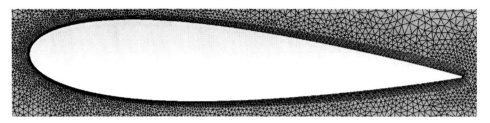

curves for optimal parameters were then used for constructing a mathematical model of a dual-level wind turbine. In the model, a moment sum vector of a single level is calculated along with its angle for 360 time steps, one for each degree the wind turbine has rotated. The second level is additionally magnified in relation to the first with a scaling factor that can be set at any value. This scaling factor depends on the relative height of both wind turbine levels, the relative wing span of levels or even the diameter of both levels. The scaling factor was used, due to the fact that it is much easier setting a single factor for scale instead of few and choosing proportions for an actual turbine project later on. Results show that the greatest bending moment and moment jump reductions at the bottom of the turbine tower are achieved for a scaling factor of 1, so to compensate for the upper turbine level moment having a greater arm it should have a smaller span, or diameter. The created model also allows for the two turbine levels to move at different speeds and even in different directions, but as the results are sub-optimal they will not be discussed in this chapter. The goal of the model was to test limiting the maximum bending moment and

Figure 6. Wind turbine power coefficient dependence on rotation speed calculated using the k-omega SST method

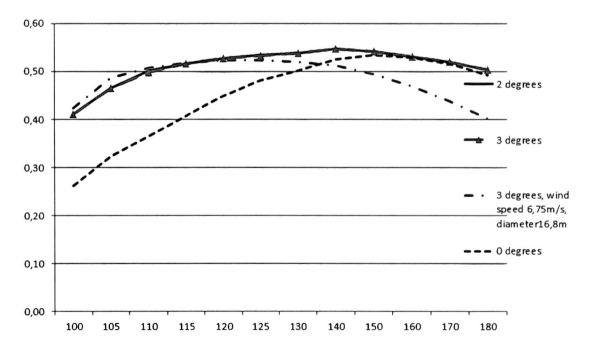

Figure 7. Values of bending moments at bottom of turbine tower from both levels, their sum and change in moment value during single time step for a scaling factor of 1.1 and starting configuration 0

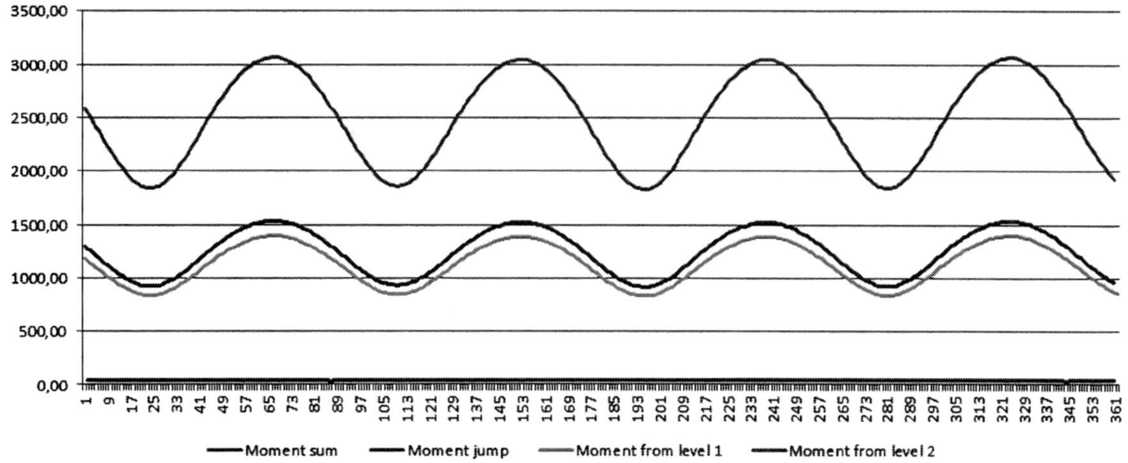

moment jump at the bottom of a turbine tower for all possible starting configurations, taking into account the shift in the second levels position in relation to the first in degrees.

In Figure 7, a curve of bending moments at the bottom of a turbine can be seen for a scaling factor of 1.1 and starting configuration 0. In this configuration the bending moment sum is the same as for a single level turbine of the same size as a dual-level turbine.

Figure 8 represents the best starting configuration with the upper and lower level wings shifted by 60 degrees, the same as can be seen in Figure 1. In this configuration the maximum bending moment sum at the bottom of the turbine tower during a cycle has been limited by 19,7%.

After receiving results for calculations with uniform wind speed it has become possible to assess the relative sizes of both levels within the turbine for a 3-dimensional model as well as

Figure 8. Values of bending moments at bottom of turbine tower from both levels; their sum and change in moment value during single time step for a scaling factor of 1 and starting configuration 60

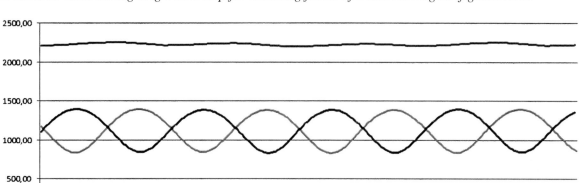

defining the relevant expected wind speeds at both levels. For this purpose, the Hellman exponential law was used, with a friction coefficient value of 0.2. Assuming that we keep the 9 m/s speed as the relevant wind speed for the first level of the wind turbine, we can, after a few iterations, calculate the sizes of both turbine levels, as well as the relevant wind speed of the second level, calculated as the wind speed having the average of the energy of the winds blowing on the second turbine level. After analyzing the created wind speed curve, a height difference of 10 meters between the ground and the bottom of the first level wings was chosen. After calculating the size of each level as to amount to the same average bending moment values at the bottom of the turbine foundation by assuming the same wind turbine power coefficient, a relevant wind speed for the second turbine level was calculated, which in turn influenced the sizes of both turbine levels.

After a few iterations, the solver arrived at a final relevant wind speed for the upper turbine level, equal to 10,4 m/s and corresponding turbine section sizes as shown in Table 3, with the lower level spanning 32.7 chord lengths and the upper level spanning 15 chord lengths. These parameters, obtained by

Table 3. Turbine section sizes calculated for 3D modelling

Height to bottom of lower turbine level	10.0m
Height of lower turbine level	49.0m
Height of upper turbine level	22.5m

using results of two-dimensional analyses, were both input into the calculation module used for determining scenarios for limiting maximum bending moments at the bottom of the turbine tower and used to create a three-dimensional model for flow analysis using ANSYS Design Modeler and ANSYS Mesher. The 3D model has the same diameter of both upper and lower rotor, and was, at the time of chapter submission, tested with uniform wind speed of 9 m/s in order to better compare results with the basic 2D analysis. An added gain of a 3D analysis will be the possibility to observe the vortices created on wing edges between both turbine levels and how these vortices interact. Including 3D flow analysis into the scope of currently conducted research has become possible thanks to a research grant at the University of Warsaw's Interdisciplinary Centre for mathematical and computational Modeling.

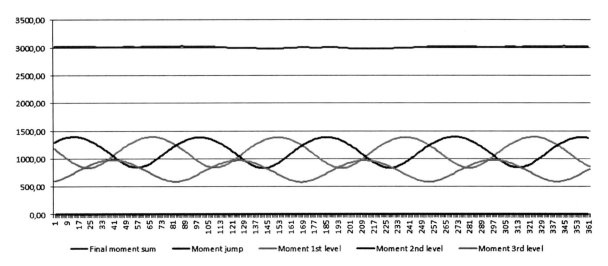

Figure 9. Values of bending moments at bottom of turbine tower from both levels, their sum and change in moment value during single time step for a second level scaling factor of 1 and starting configuration 55 and a third level scaling factor of 0.7 and starting configuration 27

Table 4. Turbine section sizes of a 3-level vertical axis wind turbine

Height to bottom of lowest turbine level	10.0m
Height of lowest turbine level	47.0m
Height of middle turbine level	23.5m
Height of upper turbine level	16.5m

In order to further test the full potential of limiting the maximum bending moments at the bottom of a turbine tower, a uniform wind speed scenario for a 3-level wind turbine was created. As with the case of a dual-level vertical axis wind turbine, a range of moment scaling factors had to be checked in order to ensure an optimal solution was reached.

The final scaling factors reached were 1 and 0.7 as seen in Figure 9, making it possible to achieve, for the best starting configuration, limiting of maximum bending moments at the bottom of a turbine tower by 19.8% and moment value jump during a single time step by 75%. The additional division allows to further take into account the growing wind speeds at greater heights, but demands further limiting the span of the upper turbine level in relation to the height of the entire turbine.

Table 4 shows an example of sizing of a wind turbine that meets the above specified conditions. As can be seen, for a 94 meter tall wind turbine, the upper level would only have a height of 16.5 meters, which is 11 chord lengths, relating to a lift loss of about 4.3% due to a finite wing span. Answering the question of whether dividing the turbine into more than two segments would prove advantageous would require wind tunnel experimentation or further extensive 3D fluid flow analyses.

CONCLUSION

The chapter provides new, original, patent pending solutions which are currently being tested. As the research is ongoing and brings new input every month, this lecture should be treated as an introduction to the topic, and any persons interested in the results or applications of this research are encouraged to contact the author for newer and more specific results. The exact effects of limiting the maximum bending moments at the bottom of the turbine tower on cracking of foundations and vertical axis wind turbine reliability are being looked into and will provide the final information on how much impact the presented research should have on the wind turbine industry. For now it is only certain, that limiting destructive forces at critical points of a wind turbine can be used to limit material costs or increase reliability. The presented solution not only helps achieve those goals for an H-type vertical axis wind turbine, but also simplifies the manufacturing and transportation of large-scale H-type VAWT's components.

REFERENCES

Anon. (1890). Mr. Brush's windmill dynamo. Scientific American, 63(25), 54.

Bañuelos-Ruedas, F., Angeles-Camacho, C., & Rios-Marcuello, S. (2011). Methodologies used in the extrapolation of wind speed data at different heights and its impact in the wind energy resource assessment in a region. In Wind farm - Technical regulations, potential estimation and siting assessment. Academic Press.

Dabiri, J. O. (2011). Potential order-of-magnitude enhancement of wind farm power density via counter-rotating vertical-axis wind turbine arrays. *Journal of Renewable and Sustainable Energy*, *3*(4), 043104. doi:10.1063/1.3608170

Jamison, A. (1997), Public participation and sustainable development: Comparing European experiences. In PESTO papers 1. Aalborg University Press.

Kiss, P., & Jánosi, I. M. (2007). Comprehensive empirical analysis of ERA-40 surface wind distribution over. *Europe*.

Linacre, E. (1992). *Climate data and resources*. London: Routledge.

Price, T. J. (2005). James Blyth - Britain's first modern wind power engineer. *Wind Engineering*, *29*(3), 191–200. doi:10.1260/030952405774354921

ADDITIONAL READING

Anaya-Lara, O., Hughes, F. M., Jenkins, N., & Strbac, G., (2006). Contribution of DFIG-based wind farms to power system short-term frequency regulation, *IEE GTD Proceedings*, 153 (2), 164–170. 238 Wind Energy Generation: Modelling and Control

Arulampalam, A., Ramtharan, G., Caliao, N., Ekanayake, J. B., & Jenkins, N. (2008). Simulated onshore-fault ride through offshore wind farms connected through VSC HVDC. *Wind Engineering*, *32*(2), 103–113. doi:10.1260/030952408784815781

Burton, T., Sharpe, D., Jenkins, N., & Bossanyi, E. (2001) Wind Energy Handbook, John Wiley, & Sons, Ltd, Chichester, ISBN 0 471 48997 2. doi:10.1002/0470846062

Conroy, J. F., & Watson, R. (2007). Low-voltage ride-through of a full converter wind turbine with permanent magnet generator. *IET RPG Proceedings*, *1*(3), 182–189.

Ekanayake, J. B., Holdsworth, L., & Jenkins, N. (2003). Control of doubly fed induction generator (DFIG) wind turbine. *IEE Power Engineering*, *17*(1), 28–32. doi:10.1049/pe:20030107

Eltra, (2004). *Wind Turbines Connected to Grids with Voltages Above 100 kV*, Technical Regulations TF 3.2.5, Doc. No. 214493 v3, Eltra, Skérbćk.

Erinmez, I. A., Bickers, D. O., Wood, G. F., & Hung, W. W. (1999). NGC experience with frequency control in England and Wales – provision of frequency response by generator, presented at the IEEE PES Winter Meeting. doi:10.1109/PESW.1999.747521

Erlich, I., Wilch, M., & Feltes, C. (2007). Reactive power generation by dfig based wind farms with AC grid connection, presented at EPE 2007 – 12th European Conference on Power Electronics and Applications, 2–5 September 2007, Aalborg, Denmark.

Glauert, H. (1935). Airplane propellers, Aerodynamic theory, (Ed. W.F. Durant), Vol. 4, Division L (Springer, Berlin)

Hansen, A. D., & Michalke, G. (2007). Fault ride-through capability of DFIG wind turbines. *Renewable Energy*, *32*(9), 1594–1610. doi:10.1016/j.renene.2006.10.008

Holdsworth, L., Charalambous, I., Ekanayake, J. B., & Jenkins, N. (2004). Power system fault ride through capabilities of induction generator based wind turbines. *Wind Engineering*, *28*(4), 399–409. doi:10.1260/0309524042886388

Holdsworth, L., Ekanayake, J. B., & Jenkins, N. (2004). Power system frequency response from fixed speed and doubly fed induction generator-based wind turbines. *Wind Energy (Chichester, England)*, *7*(1), 21–35. doi:10.1002/we.105

Howard, R. J. A., & Pereira, J. C. F. (2006). A study of wind turbine power generation and turbine tower interaction using large eddy simulation. *Wind and Structures*, *9*(2), 95–108. doi:10.12989/was.2006.9.2.095

Islam, M., Ting, D. S. K., & Fartaj, A. (2008). Aerodynamic Models for Darrieus-Type Straight-Bladed Vertical Axis Wind Turbines. *Renewable & Sustainable Energy Reviews*, *12*(4), 1087–1109. doi:10.1016/j.rser.2006.10.023

Ivanell, S. (2009). Numerical computation of wind turbine wakes. *PhD thesis, KTH Mechanics, Royal Institute of Technology*.

Le, H. N. D., & Islam, S. (2008). Substantial control strategies of DFIG wind power system during grid transient faults, IEEE/PES Transmission and Distribution Conference and Exposition, 2008, T&D, pp. 1–13.

Liu, Z., Anaya-Lara, O., Quinonez-Varela, G., & McDonald, J. R. (2008). Optimal DFIG crowbar resistor design under different controllers during grid faults, *Third International Conference on Electric Utility Deregulation and Restructuring and Power Technologies*, DRPT 2008. Wind Turbine Control for System Contingencies 239

Morren, J., & de Haan, S. W. H. (2005). Ride through of wind turbines with doubly-fed induction generator during a voltage dip. *IEEE Transactions on Energy Conversion*, *20*(2), 435–441. doi:10.1109/TEC.2005.845526

National Grid, (2008). *The Grid Code*, Issue 3, Revision 25.

Netz, E. O. N. (2006). *Grid Connection Regulations for High and Extra High Voltage, E.* Bayreuth: ON Netz GmbH.

Paraschivoiu, I. (2002). *Wind turbine design: with emphasis on darrieus concept*. Montréal, Québec, Canada: Polytechnic International Press.

Ramtharan, G. (2008). Control of variable speed wind turbine generators. PhD Thesis. University of Manchester.

Ramtharan, G., Ekanayake, J. B., & Jenkins, N. (2007). Frequency support from doubly fed induction generator wind turbines. *IET Renewable Power Generation*, *1*(1), 3–9. doi:10.1049/iet-rpg:20060019

Rogowski, K., & Maroński, R. (2015). CFD computation of the Savonius Rotor. *Journal of Theoretical and Applied Mechanics*, *53*(1), 37–45.

Rom, J. (1992). *High angle of attack aerodynamics*. New York: Springer. doi:10.1007/978-1-4612-2824-0

Winkler, J., & Moreau, S. (2008). LES of the trailing-edge flow and noise of a NACA6512-63 airfoil at zero angle of attack. *Center for Turbulence Research, Proceedings of the Summer Program*.

Xu, L., Yao, L., & Sasse, C. (2007). Grid integration of large DFIG-based wind farms using VSC transmission. *IEEE Transactions on Power Systems*, *22*(3), 976–984. doi:10.1109/TPWRS.2007.901306

KEY TERMS AND DEFINITIONS

CFD: Computational Fluid Dynamics, a branch of fluid mechanics that uses algorithms and numerical methods in order to solve problems dealing with fluid flows.

H-Type Vertical Axis Wind Turbines: One of three dominant types of vertical axis wind turbines, along with the Darrieus and Savonius turbines, sometimes referred to as a Gyro-mill turbine. Characteristic in relation to the other main types i.e. by a constant diameter along rotor length.

K-Omega SST Model: A variation of a two-equation eddy-viscosity model using the Shear Stress Transport (SST) formulation in order to alleviate a number of issues that may occur while using the classic k-Omega model for analyzing wind turbine cases e.g. an overly high sensitivity to inlet free stream turbulence.

Spalart-Allmaras Model: A one-equation turbulence model designed primarily for aerospace applications. It solves a transport equation for a modified form of the turbulent kinematic viscosity, avoiding calculating a length scale related to local shear-layer thickness.

Wing Speed to Wind Speed Ratio: An equivalent of the tip speed ratio used for Horizontal Axis Wind turbines, it is the difference between the speed of the wind and the speed of the moving wing or wing part, described as a factor.

XFOIL: Open-source software, created by Mark Drela, dedicated to analysis and design of subsonic isolated airfoils for both inviscid and viscous flows.

Chapter 3
Assessing the Profitability of Changing a Turbine for a Hydroelectric Power Plant Based on Long-Period Water Gauge Readings

Jan H. Wiśniewski
Warsaw University of Technology, Poland

Bartosz M. Olszański
Warsaw University of Technology, Poland

ABSTRACT

The chapter focuses on explaining the construction of author's own engineering-level model which calculates energy production based on historical water level and flow rate readings as well as economic factors such as net present value of the proposed investment on the example of a HPP on the Wkra River. The model methodology assumes the identification of location's hydrological features and translates them into a set of contingency scenarios. Various internal costs, such as maintenance or labor costs, related to normal HPP activity and taxation are discussed and incorporated into the economic part of the model. Test case results indicate that for a series of good years in terms of water flow and electricity production, full repayment of initial investment costs is possible after less than three years. Results for the chosen modernization parameters indicate that within 10 years of installing a new turbine, even the most pessimistic case would bring added value to the real estate valuation.

INTRODUCTION

Small, privately owned hydroelectric power plants (HPP) can be one of the most eco-friendly solutions for generating electric energy in light of the strong pressure exerted by the European Union officials in Renewable Energy Directive 2009/28/EC aiming to increase the share of renewable sources in the energy produced to 20% in 2020. Due to the fact that water density is roughly 800

DOI: 10.4018/978-1-4666-8222-1.ch003

times greater than air density, a turbine for a HPP is much smaller and consumes far less materials as well as energy for its manufacturing than its wind counterpart, which is particularly important regarding to the initial cost of such an installation. Another advantage of HPP is that it can usually boast a much longer longevity (typically a couple of decades) than wind turbines with assumed average lifetime in range of 20-25 years (Jungbluth, Bauer, Dones, & Frischknecht, 2005) and this is a helpful factor in assessing the economic viability of such an investment.

Moreover, well managed and maintained, hydroelectric power plants keep rivers and lakes from overflowing – which could cause effects such as spoiling of crops or taking in vast quantities of fertilizers from fields, an event which would lead to strong eutrophisation, loss of biodiversity, and overbreeding of algae, resulting in an oxygen deficiencies in water, dead fish, 5 types of algae toxins dangerous to the local population and so on. They also keep water levels from going too low in order to satisfy the needs of fish populations and industries located near the banks. Any person in an ownership of a HPP is conscious of these things, as well as the fact that every once in a while, the proprietor can get sued for keeping the water level too low (by the fishermen) or too high (farmers), and sometimes even for both these things at the same time. It is by far more demanding than owning a windmill, e.g. it cannot be left without care for the weekend or for holidays, but typically, if a small HPP is already installed somewhere, it is generally recognized as an economically feasible, socially acceptable, non-emitting source of electrical energy.

Far less known is the fact that small and privately owned hydroelectric power plants, which have been in use for more than a decade are usually equipped with Francis type or Pelton type turbines (at least in middle- and eastern-european countries), which for most installation conditions give a much lower annual energy production than a Kaplan turbine installed for the same water conditions would. Installing a new turbine is an important investment and not one to take lightly. Both costs and potential production increase have to be counted and weighted against one another. In some cases a change in turbine type will not be responsible for a large increase in energy production, while in others, like the case that became the reason to create a method for gauging production for a different turbine type, an investment can repay itself in less than 3 years, and generate high additional annual income for decades to come.

BACKGROUND

Increased need for "clean" electric energy in recent years is fundamental for raising a question whether a modernization of old and frequently dilapidated infrastructure will be more beneficial than building completely new installations in a different location, bearing in mind however, that production assessment without the necessary hydrological data production estimates may prove to be incorrect and usually overestimated. In cases in which the answer proves to be positive for the former option, the next step should be defining the technical capabilities and challenges (at the site) facing the expected investment. In that circumstances the cost of turbine that can be as high as couple of hundred thousand dollars for a 100 kW class machinery usually dominates overall investment expenditures for small HPP. It is therefore a necessity that some elements of economic analysis should be an integral part of an assessment model since a financing of investments requires realistic planning of income & costs, return rates as well as awareness of independent factors (natural disasters like drought or floods, regulation of the river upstream HPP activities) possibly acting on the profitability of modernization should be taken into account to some extent.

A helpful, preliminary tool in solving this issues can be for example a SWOT/TOWS analysis with strengths, weaknesses, opportunities and threats clearly defined and connected with each other by mutual dependencies. In the literature integrated technical & economic approach relating to the small HPP topic is rarely seen.

The method will be shown on the example of a hydroelectric power plant on the river Wkra, Poland. The history of this specific HPP traces back to 1864 when a water mill was built there. Its primary task at the time was grain threshing. After the fire of 1903 the mill was rebuilt and a new Francis type turbine was installed in place of old turbine. In 1995 another serious fire broke up and the mill was completely destroyed above water level. One of the few items not consumed by the fire was a nearly one hundred year old, 48 kW installed capacity rated Francis type turbine that became the core of a small HPP with a nominal drop of 2.5 m erected in place of the former, in those days loss-bringing mill.

BASIC TURBINE INFORMATION

Typically, water turbines used for electricity generation can be divided into two types depending on the whether the runner is being put into motion by having its elements barraged by jets of water(Impulse turbines), or whether it is immersed in a flowing stream, being turned by the current(Reaction turbines).

While impulse turbines are usually applicable for head values of 20 m or higher, they have a number of advantages for small installations which meet this condition. Because of the fact that the runner is not immersed in water, but only barraged by water jets discharged from nozzles, access to moving parts is much simpler, while at the same time sand particles dragged along by the water are less likely to hinder turbine operation and damage components. The turbine itself is much easier to fabricate, but the rotation speed of the runner requires including a drive system in order to increase the rotation speed at the generator.

Reaction turbines have wings shaped in order to impose lift forces, resulting from pressure differences between the upper and lower profile surfaces, which makes them harder to fabricate in some regions in developing countries, but is a practical necessity in order to reach high efficiency for low-head conditions. The Francis turbine, originally designed for use in mills is based on the mathematical theories of Leonard Euler and Johann Segner as well as Jean-Victor Poncenet's inward-flow turbine, was developed in 1849, with several improvements over its predecessor, the most important being that the flow of water in a Francis turbine is being guided before making contact with the runner, as to approach it at the best possible angle. The Kaplan turbine is a much newer design, and was developed by Professor Viktor Kaplan by the year 1912 (Quantz, 1924). It uses adjustable guide vanes to deliver the water onto a propeller-type runner, and was developed for use in power plants with low heads, providing efficiency higher than 90% for a wide range of operating conditions (Shuhong, Shangfeng, Michihiro, & Yulin, 2008).

ASSUMPTIONS

There is no single water turbine type that is unparalleled by others for every river, and therefore for every set of water conditions, however, for a given site, a water turbine type surpassing others in terms of energy production can be found.

As Figure 1 shows a Kaplan turbine is designed to operate with a high efficiency (in relation to nameplate capacity) in a much wider range of flow values than a Francis turbine so it should produce more energy in the long term. It should be mentioned that this assumption may not be true in all circumstances since hydrological conditions are strongly dependent on location and stable, predictable flow rates may favor a turbine with narrower, but slightly higher overall (including internal) efficiency.

Figure 1. Comparison of HPP turbine efficiency dependence on flow levels

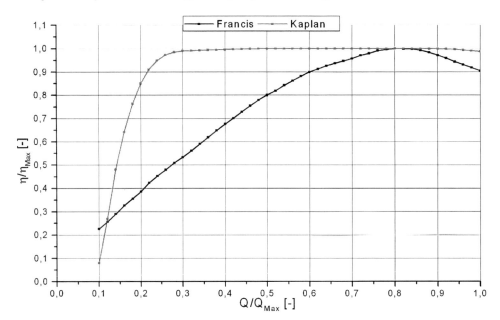

Accurately basing production after a technological change solely on the basis of the plant's production in previous years cannot be done reliably in most cases. Different turbines require different strategies for optimizing production, which leads e.g. to different levels of accumulation above the power plant. Also, as dry and wet years can vary greatly in relation to available water resources and their distribution, and above- or sub- optimal cycles can last for a series of years, a period of at least 10-20 years is believed to be needed for reliable approximation (although a river's characteristics may change permanently over a period of time). The easiest way to meet those expectations is to have a water gauge on the same river, not far above the power plant, with no forks or very large water users in between. This condition cannot be met for all hydroelectric power plants, but it happens for some installations. In cases where the waterway has not undergone major restructuration, old readings are also acceptable. This last fact is very important in countries like Poland, where investments along existing waterways are very rare, but the array of water gauges has been cut down from over 3300 to a little over 1000 over the course of the last three decades.

A comparison of preexisting flow conditions with production parameters of Kaplan reactive turbines shown in Figure 2 allows for making initial assumptions on the nominal capacity of a proposed Kaplan turbine. In the case of the analyzed turbine with an aforementioned drop of 2.5 m, an installed capacity of 120kW was chosen to be analyzed with a nominal volumetric flow of 6.5 m/s, a value much above the average flow rate through the power plant, but not unreasonable, due to the Kaplan turbine main advantage – retaining high efficiencies at subnominal flow rates.

MODEL

The model is composed of two main parts (modules). The first one is responsible for calculating the energy production based on daily gauge readings, the second one – recalculates the gross

Figure 2. Production parameters of Kaplan type reactive turbines

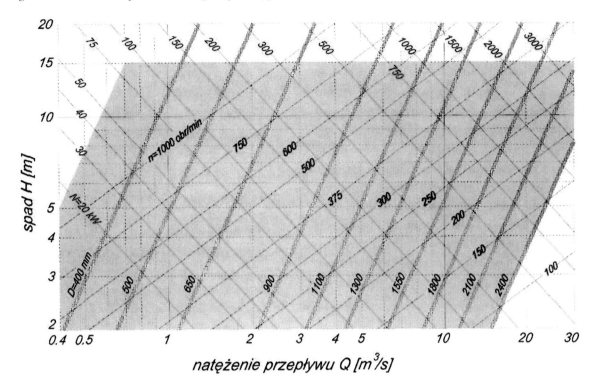

income minus taxes (social included), maintenance and other financial costs. Both of them are integrated into spreadsheets. Net profitability of Kaplan turbine equipped hydroelectric power plant and non-modernized version of the HPP are compared against one another. It is justified because the most popular method of investment funding (such as loans) and its associated costs (installments) effectively lessens net income in short- and mid-term so it may carry some risks with solvency and may force a temporary shut down or even a bankruptcy of HPP in very unfavorable circumstances. The model deals with this issue using a set of historical data derived scenarios for benchmarking the output data from assessment calculations.

Getting from the hydrological part, the actual gross head (H) of turbine is a count of baseline head value (H_0) and water level in river (h).

$$H = H_0 + h [m] \qquad (1)$$

Power output (P) is computed using the following formula.

$$P = \eta_0(Q) \cdot CF \cdot P_0 [kW], \qquad (2)$$

where nominal capacity (P_0) can be taken in a straight line from turbine specification and capacity factor (CF) is derived from equation:

$$CF = \frac{Q \cdot H}{Q_0 \cdot H_0} [-], \qquad (3)$$

where (Q_0) indicates nominal flow rate. One of the most important features of the model is that, the (CF) is calculated individually for each day, instead of monthly or annual statistical approximation or prediction (a fairly common approach nowadays) from the reasons outlined in previous paragraphs.

Dimensionless (relative to installed capacity) turbine efficiency dependent on volumetric flow rate ($\eta_0(Q)$) should be taken directly from specific turbine characteristics curve (similar to Figure 1) provided by manufacturer. When that data appears to be unavailable, an approximate curve, compliant with the general characteristics of the turbine type could be assumed. Note that the relative turbine efficiency is only a function of flow rate and its curve should not be misunderstood with overall (taking hydraulic, energy conversion, mechanical and turbo-machinery losses into account) efficiency ($\eta_0(Q)$) curve, that maximum efficiency of whole turbine usually peaks at ~60-80% of available power in water (P') for low head small HPP (Paish, 2002).

$$P' = Q \cdot \rho \cdot g \cdot H [kW], \qquad (4)$$

where ρ is water density and g is the gravitational constant. The model omits discussion of internal efficiency influence, as the (P_0) supplied by manufacturer should be the highest practically achievable value, rather than theoretically one from Formula (4). If there were still a strong need to it could be incorporated by replacing the Equation (2) with:

$$P = \eta(Q) \cdot P' [kW] \qquad (5)$$

In case of the analyzed HPP (equipped with determined earlier 120 kW Kaplan turbine) $P = P_0$, ($\eta(Q)$) equals ~75,3% which stands in good agreement with values reported in literature and it is likely to be realizable by decent quality turbine.

Finally, daily production (PR) is calculated from the simple equation:

$$PR = P \cdot 24 [kWh] \qquad (6)$$

Equation (2) is constrained from top and bottom. If the capacity factor drops below 15%, water exploitation index for this case is assumed to be 0 and the plant is not producing electricity ($PR = 0$). The main reason are fish passes, which consume a majority of flow for such poor conditions and very low value of ($\eta_0(Q)$), which is discouraging to running a turbine. Overproducing, id est. a hypothetical situation when the (CF) is over 1 because of high water level as well as flow rate simultaneously and the turbine works beyond its installed capacity, is forbidden so the model can be considered as a conservative one. In that situation the additional amount of water is bypassing the turbine. Efficiency of turbine above 1 is also not permitted, maintaining the model's conservative character, however it is not uncommon for manufacturers to downgrade the truly achievable installed capacity of turbine, so it may be possible to achieve some (couple of percent) surplus in practice.

Finally, generation of electricity for each day is then summed together to get the production estimate for weeks, months as well as years and provides a basis for further work.

The turbine is assumed to be robust and-well maintained so deterioration of installed capacity along time is not taken into consideration. Own consumption of electricity (lightning, control panels supply etc.) was preliminarily omitted in model also, in spite of the fact that it may contribute significantly (even couple of percent for a drought year) on the side of expenses for very small (in terms of produced energy) HPP units (not this one). Since own consumption is highly dependent on local conditions (state of power network, machinery requirements, level of automation and so on), it is difficult to standardize at ease, nonetheless if the appropriate data is provided, these costs would be relatively easy to incorporate into the maintenance box of model.

The economic part of the model recalculates the revenue and net income. A factor with some variability over a long period of time is the energy price for 1 MWh which in fact is composed of two elements – buyout price of produced energy and the certificate of origin – in case of renewable sources also known as "green certificate". Both of them are regulated in Poland by URE (Energy Regulator Office). In case of the preliminary study long-term deal with distributor and stable overall price were assumed.

The useful tool in assessment of economical effectiveness of investment is well-known NPV (net present value) method, which is described by equation:

$$NPV(i,N) = \sum_{t=1}^{N}\left(\frac{NCF_t}{(1+i)^t}\right) - \sum_{t=1}^{N}\left(\frac{I_t}{(1+i)^t}\right) - I_0 [PLN], \quad (7)$$

where i is the discount rate, t is time, N is number of considered time periods (years) and the value must be at least 1, (NCF_t) is net cash flow (revenue minus taxes, maintenance costs and other various financial commitments), (I_0) is initial expenditures (in "year zero") and (I_t) are investment costs incurred in subsequent years. This method is commonly utilized at the planning phase, well before the first shovel in the ground punching. The positive value indicates the profits for the company from the investment, the negative – losses. The general formula for calculation of (NCF_t) in a year cycle is:

$$NCF_t = \sum_{t=1}^{t=365(366)} (PR \cdot CpkWh) - e_t [PLN], \quad (8)$$

where ($CpkWh$) corresponds to the unit price of 1 kWh and (e_t) are all expenses in a year cycle that can be divided on two separate categories – steady (by means of value), "tight" commitments (f_t) described by Formula (9):

$$f_t = Employment\ costs + Maintenence\ \&\ Other [PLN] \quad (9)$$

and various taxes that are usually (but not always) directly linked with income or revenue. Note that the sum in Formula (8) takes leap years into account. Tax considerations (rules and rates) strictly depend on the legislation in the country concerned so the general formula for calculating (e_t) should be like:

$$e_t = f_t + Social\ tax + Health\ ins. + Taxation [PLN] \quad (10)$$

For small hydroelectric power plants in Poland the simplest as well as the most popular method of taxation is a registered lump sum at the rate of 5,5% ($tax\ rate$) reduced by the amount of health insurance (if applicable). ($Social\ tax$) is usually a predetermined fraction of revenue (for personal tax), however in case of running an economic activity, such as HPP, it is usually a fixed monthly amount, just as health insurance. For the Polish tax law aggregate formula to calculate ($Taxation$) is then:

$$Taxation = \left[\sum_{t=1}^{t=365(366)} (PR \cdot CpkWh) - Social\ tax\right]$$
$$\cdot tax\ rate - Health\ ins. [PLN] \quad (11)$$

If the calculated value is negative, the tax will not be paid. For simplicity annual settlement has been adopted, although the settlement of monthly and quarterly figures is the most commonly used in practice. Indexation of social security contributions ($Social\ tax$ and $Health\ ins.$) was neglected in the model as it is impossible to assess an average indexation rate, when its principles are often purely political decisions. Current rule is combined with the average growth of GDP and wages in the country's economy. Forecasts of growth of both of these indicators (even using the theory of economic cycles) might be highly biased so the idea of incorporating of indexation into the model was dropped.

Already possessing all the necessary data, the next step is arrangement of a set of different contingency scenarios. The basic time unit for building a scenario set was a year, because it covers the whole cycle of the seasons (6 meteorological seasons in case of Poland) with all their properties. From the data possessed (as it was previously mentioned, for at least a decade is preferred), two years were selected in the first step – the most productive and the poorest one in terms of produced energy. These outstanding years indicate severe droughts or big flood which may occur. Naturally, level of divergence is a feature of the location and it can vary significantly – from a few percent for the "L" year to several hundred percent for the "H" year in relation to the long-term average (codenamed "M"), which in turn includes all years (without exclusion of these outstanding ones). For the purpose of scenario sets, the productivity level of "L" year was established at precisely 105% of basic "L" value and "H" year at 95% of basic productivity of maximum year correspondently. Such procedure can be explained by the fact that years with a more close to long-term average production levels are slightly more likely to occur in two or even three consequent years than these of extreme ones. The analysis of historic data is essentially important in this situation and the adoption of appropriate correction factors for a model "L" and "H" year should arise directly from the past readings, e.g. the location with tendency to maintain above average levels, but not well-above medium, and incidental, severe one-year drought per two decades, has got higher possibility to obtain 1,1-1,2 factored "LLL" scenario rather than pure, un factored "LLL" and so on. The authors suggest cautiousness in selection of factors.

Three different model years are then arranged in 3-year basic cycles, for example "LLL", which means unusually bad situation with three years lasting drought or "HHH" for the most optimistic case. Amongst 27 possibilities of arrangement of 3-year scenarios, eleven conservative options (for example "LLM" was selected instead of "MLL" etc.) were considered and overall income as well as NPV was calculated. These stacks are helpful for initial assessment whether there is a risk of standing on the brink of bankruptcy in the first three critical years, and on the other hand, if full repayment of initial investment costs is achievable in such a short period.

Nevertheless, 3 years is generally too short timescale for final judgment of profitability of HPP, so the second step was the arrangement of 10-year scenarios utilizing, among others, previously prepared stacks. The decade lasting prediction requires a little more complex method than in previous stage. All historic years were assigned into 5 categories:

- Extremely low productivity ("L_y");
- Below average productivity ("U_y");
- Medium productivity ("M_y" value +/- 5%);
- Above average productivity ("O_y");
- Extremely high productivity ("H_y").

By definition, "L_y" and "H_y" categories consist of one year respectively. In this case, three different scenarios were produced using following rules:

- Pessimistic case with one model "H" year, the number of "L" years = $L_y + U_y$ The rest are "M" years. Arranged in the very unfavorable pattern (the majority of drought years occurring just after opening of modernized HPP).
- Optimistic case with one model "L" year, the number of "H" years = $H_{ay} + O_{yo}$ The rest are "M" years. Arranged in the very favorable pattern (the majority of flourish years occurring just after opening of modernized HPP).

- Realistic case with the number of "M" years = $M_y + 2/3\ U_y + 2/3\ O_y$, "H" years = $H_y + 1/3\ O_y$, "L" years = $L_y + 1/3\ U_y$. Arranged in balanced pattern with some conservativeness ("L" year at the beginning of operability, "H" at the end of 10-year cycle).

For these scenarios NPV and accumulated net income were calculated also.

CASE STUDY RESULTS

The input data for the method were daily gauge readings from years 1974-1983 however it should be born in mind that a hydrological year starts in the November of previous year Figure 3 and Figure 4 presents prepared data in monthly cycles. For each month red and green vertical bars indicate the lowest and the highest values in same order.

As the Figure 3 shows, the river Wkra tends to keep its level above 20 cm (which corresponds to 40% of water level necessary for achieving nominal head) even in the drought year 1981. On the other hand, the most productive year 1977 cannot boast the highest water levels for all of the history of readings.

Figure 4 points out, that the main driver of productivity is definitely flow rate with discrepancies far greater than in the case of water level, e.g. 4.1979 when severe spring thaw happened the maximum flow rate achieved nearly 700% of its nominal value, and dry summer of 1981 when (Q) dropped to only 6% of (Q_0). Despite months with mean flow rate greater than (Q_0) constitute approximately only 25% of all, the number of months without single day of sub-nominal flow is only 1 (12.1974). Thus, as can be seen even in case of the most productive month, sub-nominal days could be reckoned.

It is pretty obvious also, that there is a strong correlation between number of over-nominal months in a year and generated energy, nevertheless even the very robust analysis suggests, that sub-nominal flow rates is a dominating flow form for a location of the analyzed HPP. It is not uncommon that there is unambiguous relationship

Figure 3. Mean water level with minimum and maximum values (monthly)

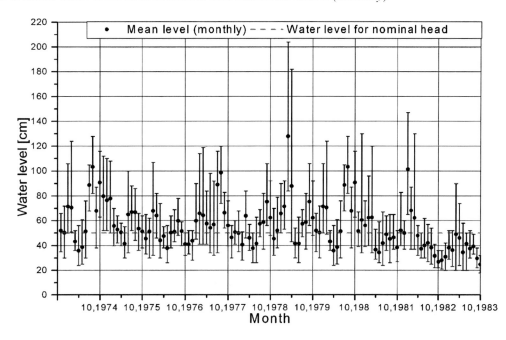

Figure 4. Mean flow rate with minimum and maximum values (monthly)

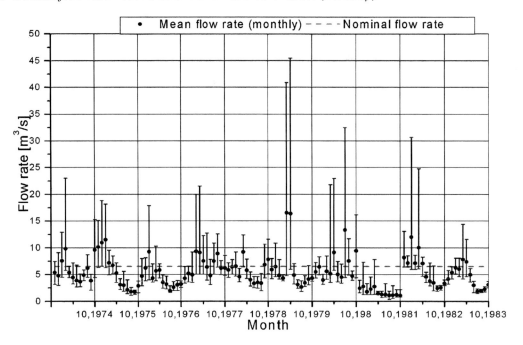

between the water level, flow rate as well as the seasons, but with some exceptions – as a proof of this thesis might provide the summer of 1981, when fairly high h (around H_0) accompanies very low Q (a good model of stagnant water).

Notice, that (h) and (Q) is subjected to considerable fluctuations in a month cycle, so the averaging of water level for each month as a base for further production calculations would introduce some error as a mean value is not lying in the middle of vertical bar for almost all cases. Appropriate comparison was performed to assess an impact of monthly as well as annual averaging of (h) and (Q) on the final results (see Table 1).

Figure 5 summarizes the production calculated from daily gauge readings. Average annual production (PR_a) over the considered decade is 726,255 MWh. It was observed that after the above average year, the following one or two years had generated less energy as compared to the directly preceding year (constant decrease). The end of that tendency is observed together with the advent of next (more profitable) year which opens a new cycle. Hence, the decade can be divided on 4 separate periods: 10.1973 – 10.1976, 10.1976 – 10.1979, 10.1979 – 10.1981 and 10.1981 – 10.1983.

There is nothing surprising in the fact that modernized HPP gives much more energy than non-modernized version with smaller, 48 kW rated turbine (Figure 6). Truly interesting information, however, brings Figure 7. It is a clear evidence for much more efficient work of Kaplan turbine in a duel with Francis turbine.

Modernized installation is advantageous particularly above 90% and below 60% of maximum achievable production level for each turbine type. Outside that range, the results are comparable, but it is still favoring Kaplan turbine. The gain of power for the same water conditions can be as high as 13,4%. It follows, of course, directly from the efficiency curves (Figure 9).

Various factors needed for the purpose of economic analysis in the model section of the chapter mentioned were assumed to be:

Table 1. Various methods for annual production calculation for Kaplan turbine equipped HPP

Year	Daily Readings [MWh]	Monthly Averaging [MWh]	Difference (in Relation to Daily Readings) [%]	Annual Averaging [MWh]	Difference (in Relation to Daily Readings) [%]
1974	832,212	847,152	1,80	986,176	18,50
1975	703,931	710,152	0,88	937,401	33,17
1976	670,876	688,413	2,61	731,620	9,05
1977	914,602	959,532	4,91	1 051,200	14,94
1978	821,327	844,320	2,80	918,800	11,87
1979	769,710	786,898	2,23	1 051,200	36,57
1980	888,864	917,626	3,24	1 054,080	18,59
1981	235,283	254,697	8,25	268,552	14,14
1982	775,607	780,648	0,65	966,336	24,59
1983	650,133	666,599	2,53	690,974	6,28

- I_0 = 700 000 PLN – estimated cost of turbine, new suction elbow pipe (1,3 m diameter) and all other associated works. Not applicable in case of abandoning a modernization plan (Francis turbine retained)
- I_t = 0 PLN – all small repairs are included in *Maintenence&Others*.
- $CpkWh$ = 0,36 PLN – long-term deal, rather conservative.
- i = 0,1 – average, conservative interest rate for such investments taking into account high as the conditions of European Union (0,05% in Eurozone now) benchmark interest rates in Poland (currently 2,5%).
- *tax rate* = 5,5% – registered lump sum rate for renewable energy producers in Poland as of 2014.
- *Maintenence&Others* = 2000 PLN per annum – conservation, mowing of algae, accident insurance, land taxes, upkeep etc.
- *Employment costs* = 1800 PLN per month – wage for a technician.

Current (as of August 2014) social security contributions for corporate tax payers are:

- *Social tax* = 772,06 PLN per month.
- *Health ins.* = 270,40 PLN per month.

Using the aforementioned values it is possible to calculate NPV for years 1974-1983. Figure 8 presents these results. Less than six years are sufficient for Kaplan turbine equipped HPP to surpass NPV of non-modernized hydroelectric power plan with Francis turbine, in spite of 700 000 accountant debt at the beginning. Constantly rising NPV reaches 0 early as the fourth year of operation. Initially, it only confirms the enormous potential for development inherently present in the analyzed location. Nevertheless the historical data cannot be transferred directly in 1:1 scale to future production levels, it provides a database and very useful guidelines and tips for further assessment.

Table 1 presents the results comparison for yearly and monthly averaging of water level and flow rate in relation to results obtained from daily accumulation. The general rule seems to be – the longer period of averaging for a single unit, the less accurate result is obtained. It is especially true for distinctive anomalies (1977, 1979, 1981) occurring throughout the year which translates into the biggest determined differences in conclusion.

Table 2 summarizes the results for a decade 1974-1983. Delivering too optimistic results (overestimation of production) might lead to the serious economic problems when the success of investment is directly dependent on long-term average of (h) and (Q).

Figure 5. Annual production over the years 1974-1983 for Kaplan and Francis turbines

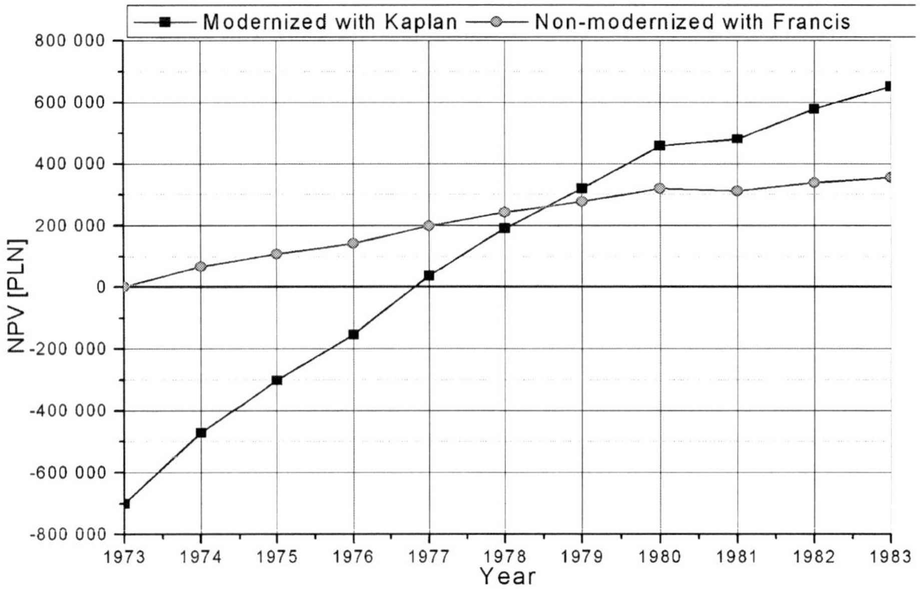

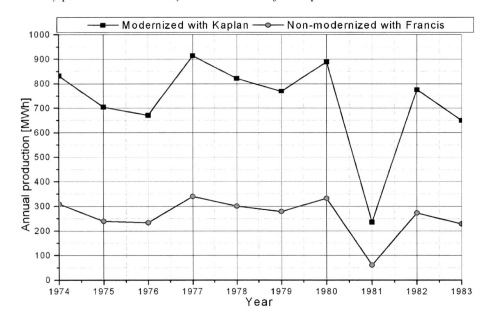

Figure 6. Monthly production over the years 1974-1983 for Kaplan and Francis turbines

Figure 7. Actual monthly production in relation to maximum achievable production over the years 1974-1983 for Kaplan and Francis turbines

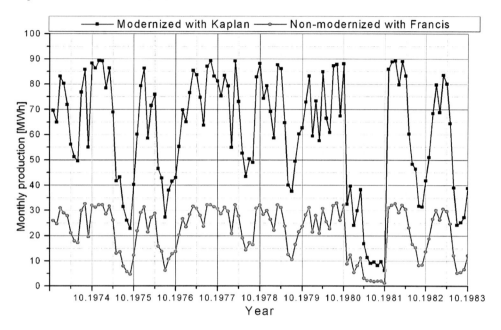

Figure 8. NPV for two versions of HPP on the river Wkra based on exact data from the years 1974-1983

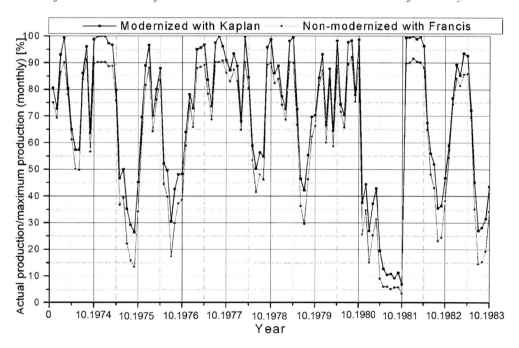

Figure 9. NPV over the years for different long period scenarios and turbine types

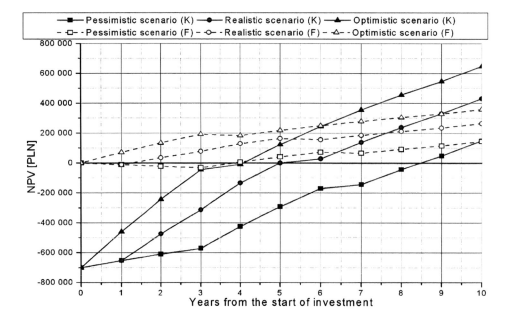

Using daily data gives much more reliable results than any form of deriving from monthly or yearly data and should be used when it is available. The monthly average is somewhat applicable, but still in case of location with very high discrepancies within a month, e.g. a change of seasons in the tropical climate zone, the exactness of production estimation would be apparently lost. Using averaged water level and flow rate for over a year causes radical, more than 20 percent error which translates into over a half-million difference in revenue within a decade! It is not recommended, however it is very often used nowadays as the simplest and the fastest way for obtaining a coarse production forecast.

Using data for gauge readings the hydrological year of 1977 was selected as the model "H" year and 1981 for the "L" year. (PR_a) amounts to 726,255 MWh for the "M" year. As it was outlined in previous section of chapter, conservative representative scenarios were chosen, for example scenarios "LLH" and "HLL" gave the same income, but from the NPV point of view, the "LLH" is worse, so that one was selected.

Calculated results of eleven 3-year scenarios mentioned in the section model are collected in the Table 3 and Table 4. The only one that generates overall losses is "LLL" scenario for Francis turbine, although interim losses were noted for every sce-

Table 2. Summary of various methods of calculation of accumulated production for Kaplan turbine equipped HPP

Method	10-Year Accumulated Production [MWh]	Difference (In Relation to Daily Readings) [%]
Daily readings	7 262,546	-
Monthly averaging of water level and flow rate	7 456,038	+2,66
Annual averaging of water level and flow rate	8 656,340	+19,19
Long-term average of water level and flow rate	8 851,012	+21,87

nario containing the "L" in its code for this turbine type. It indicates that modernization would bring a positive effect, enabling the reduction of the percentage of fixed costs (including social security contributions) on the expenditure side, which for the aforementioned scenario amount to even 100% (effective $Taxation = 0$). Worth noting is the substantial increase of income. Relatively pessimistic Kaplan turbine case "LLM" Kaplan turbine equipped HPP prove to be relatively safe even in case of unusually prolonged poor hydro conditions. Accumulated net income for the most optimistic scenario "HHH" allows the almost full repayment of initial investment cost. Internal rate of return (IRR) for this particular case is precisely 6,28%. It is few percent more than a good interest rate for bank deposits in Poland now.

The methodology of building a decade long scenarios was outlined in previous paragraph and it is similar to these shorter ones. Categorization of database years has brought following results:

- $L_y = 1$ (1981).
- $U_y = 3$ (1975, 1976, 1983).
- $M_y = 2$ (1979, 1982).
- $O_y = 3$ (1974, 1978, 1980).

Table 3. Accumulated net income and NPV for 3-year scenario sets of Francis turbine equipped HPP

Scenario	Accumulated Net Income [PLN]	NPV [PLN] (Without Investment)
LLL	-39 671,84	-26 679,68
LLM	30 852,44	23 392,03
LLH	53 617,34	39 554,94
LMM	101 106,02	78 470,92
LMH	122 618,84	94 633,82
MMM	167 751,47	139 057,69
HML	122 618,84	108 543,09
LHH	144 131,67	112 413,02
MMH	189 264,29	155 220,60
MHH	210 777,12	172 999,79
HHH	232 289,95	192 556,90

- $H_y = 1$ (1977).

Table 5 and Table 6 gather the results for three model scenarios – two extreme and one balanced should be sufficient to determine the main trends. The analysis of all possibilities (59 049) is very impractical and it is not necessary after the identification of river characteristics.

Table 4. Accumulated net income and NPV for 3-year scenario sets of Kaplan turbine equipped HPP

Scenario	Accumulated Net Income [PLN]	NPV [PLN] (Without Investment)	NPV [PLN] (Investment Included)
LLL	155 070,97	128 546,18	-571 453,82
LLM	318 097,29	251 030,27	-448 969,73
LLH	366 615,60	287 482,79	-412 517,21
LMM	4811 23,62	385 762,77	-314 237,23
LMH	529 641,93	422 215,29	-277 784,71
MMM	644 149,95	533 968,53	-166 031,47
HML	529 641,93	455 591,98	-244 408,02
LHH	578 160,23	462 313,07	-237 686,93
MMH	692 668,25	570 421,05	-129 578,95
MHH	741 186,56	610 518,82	-89 481,18
HHH	789 704,86	654 626,37	-45 373,63

Table 5. Accumulated net income and NPV for 10-year scenarios of Francis turbine equipped HPP

Scenario	Scenario Code	Accumulated Net Income [PLN]	NPV [PLN] (Without Investment)
Pessimistic	LLL MMM LMM H	304 119,97	144 456,83
Realistic	LMM HML MMM H	463 915,00	264 690,49
Optimistic	HHH LMM MMM H	576 071,86	358 155,75

Figure 9 can be treated as an envelope of NPV because the scenarios cover practically all expected possibilities. Maintaining a Francis turbine can be only justified for the first couple of years – from six for optimistic scenario comparison and up to ten for the same pessimistic conditions. When running a HPP much further time horizon is assumed however, the modernization will be much more beneficial than keeping the old turbine working according to the introduced in this chapter profitability assessment model. The income multiplication factor is 5,076 for the pessimistic case, 4,13 for the realistic case and 3,78 for the optimistic respectively. It is much more than the result of the comparison of the installed capacity of two turbines (2,5), thus it only confirms the scale of benefits associated with modernization. The Kaplan turbine has demonstrated resistance to unfavorable hydrological conditions, achieving $NPV > 0$ for the all considered scenarios. IRR for pessimistic, optimistic and realistic scenario is 13,7%, 21,52% and 29,91% respectively in case of modernized HPP.

CONCLUSION

The chapter presents a new, easy to implement scheme for assessment of the profitability of outdated hydroelectric infrastructure modernization. This method can be successfully used as a case study element of a business plan or in the long term fiscal forecast of a currently running installation. Utilization of huge amounts of data from gauge readings for identification of hydrological characteristics and using it as a base for a production long-term forecast afterward was one of the most innovative elements of the methodology shown above. However the detailed values of all factors could not be built because of natural hydrological diversity of various geographic locations and climates, the universal framework was provided, and a real test case of a HPP located on the river Wkra described in detail. It is the authors' belief that site specific data from everyday reading for the time period of no less than one decade, if available, can be used, by the method described, to create the most reliable modernization profitability analyses for small hydroelectric power plants.

Table 6. Accumulated net income and NPV for 10-year scenarios of Kaplan turbine equipped HPP

Scenario	Scenario Code	Accumulated Net Income [PLN]	NPV [PLN] (Investment Included)
Pessimistic	LLL MMM LMM H	1 543 579,49	148 966,14
Realistic	LMM HML MMM H	1 918 150,45	430 955,56
Optimistic	HHH LMM MMM H	2 178 213,38	647 355,43

REFERENCES

Jungbluth, N., Bauer, C., Dones, R., & Frischknecht, R. (2005). Life cycle assessment for emerging technologies: Case studies for photovoltaic and wind power. *International Journal of Life Cycle Assessment*, *10*(1), 24–34. doi:10.1065/lca2004.11.181.3

Paish, O. (2002). Small hydro power: Technology and current status. *Renewable & Sustainable Energy Reviews*, *6*(6), 537–556. doi:10.1016/S1364-0321(02)00006-0

Quantz, L. (1924). *Wasserkraftmaschinen - Eine Einführung in Wesen, Bau und Berechnung neuzeitlicher Wasserkraftmaschinen und Wasserkraftanlagen*. Berlin: Verlag von Julius Springer.

Shuhong, L., Shangfeng, W., Michihiro, N., & Yulin, W. (2008). *Flow simulation and Performance prediction of a Kaplan Turbine.* Paper presented at the 4th international Symposium on Fluid Machinery and Fluid Engineering, Beijing, China.

ADDITIONAL READING

Aggidis, G. A., Luchinskaya, E., Rothschild, R., & Howard, D. C. (2010). The costs of small-scale hydro power production: Impact on the development of existing potential. *Renewable Energy*, *35*(12), 2632–2638. doi:10.1016/j.renene.2010.04.008

Alexander, K. V., & Giddens, E. P. (2008). Microhydro: Cost-effective, modular systems for low head. *Renewable Energy*, *33*(6), 1379–1391. doi:10.1016/j.renene.2007.06.026

Alfieri, L., Perona, P., & Burlando, P. (2006). Optimal water allocation for an alpine hydropower system under changing scenarios. *Water Resources Management*, *20*(5), 761–778. doi:10.1007/s11269-005-9006-y

Anagnostopoulos, J. S., & Papantonis, D. E. (2007). Optimal sizing of a run-of-river small hydropower plant. *Energy Conversion and Management*, *48*(10), 2663–2670. doi:10.1016/j.enconman.2007.04.016

Cherry, J., Cullen, H., Visbeck, M., Small, A., & Uvo, C. (2005). Impacts of the north atlantic oscillation on Scandinavian hydropower production and energy markets. *Water Resources Management*, *19*(6), 673–691. doi:10.1007/s11269-005-3279-z

Da Deppo, L. (1984). Capacity and type of units for small run-of-river plants. *International Water Power and Dam Construction*, *36*(10), 33–38.

Fabritz, G. (1940). *Die Regelung der Kaplan-Turbine. Die Regelung der Kraftmaschinen.* Berlin: Springer. doi:10.1007/978-3-7091-9983-1

Fahlbush, F. (1983). Optimum capacity of a run-of-river plant. *International Water Power and Dam Construction*, *35*(3), 45–48.

Fulford, D. J., Mosley, P., & Gill, E. A. (2000). Recommendations on the use of micro-hydro power in rural development. *Journal of International Development*, *12*(7), 975–983. doi:10.1002/1099-1328(200010)12:7<975::AID-JID699>3.0.CO;2-K

Hosseini, S. M. H., Forouzbakhsh, F., & Rahimpoor, M. (2005). Determination of the optimal installation capacity of small hydro-power plants through the use of technical, economic and reliability indices. *Energy Policy*, *33*(15), 1948–1956. doi:10.1016/j.enpol.2004.03.007

Kaldellis, J. K. (2007). The contribution of small hydro power stations to the electricity generation in Greece: Technical and economic considerations. *Energy Policy*, *35*(4), 2187–2196. doi:10.1016/j.enpol.2006.06.021

Kaldellis, J. K., Vlachou, D. S., & Korbakis, G. (2005). Techno-economic evaluation of small hydro power plants in Greece: A complete sensitivity analysis. *Energy Policy, 33*(15), 1969–1985. doi:10.1016/j.enpol.2004.03.018

Karlis, A. D., & Papadopoulos, D. P. (2000). A systematic assessment of the technical feasibility and economic viability of small hydroelectric system installations. *Renewable Energy, 20*(2), 253–262. doi:10.1016/S0960-1481(99)00113-5

Kaunda, C. S., Kimambo, C. Z., & Nielsen, T. K. (2014). A technical discussion on microhydropower technology and its turbines. *Renewable & Sustainable Energy Reviews, 35*, 445–459. doi:10.1016/j.rser.2014.04.035

Leboutillier, D. W., & Waylen, P. R. (1993). A stochastic model of flow duration curves. *Water Resources Research, 29*(10), 3535–3541. doi:10.1029/93WR01409

Leigh, P., Aggidis, G., Howard, D., & Rothschild, B. (2007). Renewable energy resources impact on clean electrical power by developing the northwest England hydro resource model. *International Conference on Clean Electrical Power, ICCEP '07*, 315-322. doi:10.1109/ICCEP.2007.384230

Montanari, R. (2003). Criteria for the economic planning of a low power hydroelectric plant. *Renewable Energy, 28*(13), 2129–2145. doi:10.1016/S0960-1481(03)00063-6

Niadas, I. A., & Mentzelopoulos, P. G. (2008). Probabilistic Flow Duration Curves for Small Hydro Plant Design and Performance Evaluation. *Water Resources Management, 22*(4), 509–523. doi:10.1007/s11269-007-9175-y

Nouni, M. R., Mullick, S. C., & Kandpal, T. C. (2005). Techno-economies of micro-hydro power plants for remote villages in Uttaranchal in India. *International Journal of Global Energy Issues, 24*(1-2), 59–75. doi:10.1504/IJGEI.2005.007078

Ogayar, B., & Vidal, P. G. (2009). Cost determination of the electro-mechanical equipment of a small hydro-power plant. *Renewable Energy, 34*(1), 6–13. doi:10.1016/j.renene.2008.04.039

Paish, O. (2002). Micro-Hydro Power: Status And Prospects. *Proceedings of the Institution of Mechanical Engineers. Part A, Journal of Power and Energy, 216*(1), 31–40. doi:10.1243/095765002760024827

Tharme, R. E. (2003). A global perspective on environmental flow assessment: Emerging trends in the development and application of environmental flow methodologies for rivers. *River Research and Applications, 19*(5-6), 397–441. doi:10.1002/rra.736

Voros, N. G., Kiranoudis, C. T., & Maroulis, Z. B. (2000). Short-cut design of small hydroelectric plants. *Renewable Energy, 19*(4), 545–563. doi:10.1016/S0960-1481(99)00083-X

Willer, D. C. (1991). Powerhouse and small hydropower project cost estimated. In Gulliver, J. S. (Ed.), Hydropower engineering handbook (6.1–6.58), New York, NY: McGraw-Hill.

Williamson, S. J., Stark, B. H., & Booker, J. D. (2011). Low Head Pico Hydro Turbine Selection using a Multi-Criteria Analysis. Renewable Energy, 61(2014), 43–50.

KEY TERMS AND DEFINITIONS

Head: The height from which water drops onto a hydroelectric turbine runner.

HPP: Hydroelectric Power Plant.

Nominal Capacity: Capacity value reached for nominal conditions.

Nominal Conditions: Conditions in which a hydroelectric turbine reaches rated power production.

NPV: Net Present Value is the expected added value for an investment after taking into account all costs and profits over a certain time horizon, taking into account the change of value of funds by means of discounting each value over time.

Power Output: Alternating-current power measured in watts delivered by mans of an amplifier.

Runner: Main component of a water turbine or water mill, taking energy from water and using it for rotation.

Water Gauge: Basic instrument used for indicating water levels on rivers.

Chapter 4
An Overview to Thermal Solar Systems for Low Temperature:
Outlining the European Norm 12976

Vicente González-Prida
University of Seville, Spain

Anthony Raman
NTEC Tertiary Group, New Zealand

ABSTRACT

This chapter deals with those prefabricated systems with a steady state of operation (state in which the temporal variation of the thermodynamic properties is null), describing, in a brief manner, a methodology for testing the characterization of the thermal performance in accordance with the European normative. All of the previously mentioned form the justification for a foundation or base from which a testing installation is proposed in a later chapter that, at the same time, is compared to a real installation. Lastly, this chapter attempts to outline a simple mathematical methodology to analyze the future behavior of the reliability of a system (solar in this case), when it is still in an extremely early stage of its life cycle, such as the design phase.

INTRODUCTION

Along the last decade, a great number of researches, master degree projects and doctoral thesis regarding diverse aspects of the application of solar energy, have been developed. In particular, some projects set out installations and methodologies that represent an important support in the search for solutions to technical problems in the field of solar thermal energy of low temperatures. These form the base from which this research is developed.

Some of these projects and theses do not contemplate the European Standard EN-12976 (parts 1 and 2) for thermal solar energy systems (norm that has been implemented nationally in Spain since June 2001). This norm deals with thermal solar energy systems and their components. In particular, it deals with prefabricated systems for the production of sanitary hot water through solar energy. The first part (EN-12976-1, 2006) specifies the general durability, reliability and safety requirements for thermal solar energy systems of prefabricated heating as products. The installation of these systems is not considered in this regulation, but it does include all requirements in the documentation for the person performing the installation and for the final user. In other respects, the second part of this norm (EN-12976-2, 2006) expresses the distinct testing methods applied to domestic solar energy systems for the validation of the prior requirements.

The European Standard "Thermal solar systems and components" has been prepared by the European Committee for Normalization, with the aim to normalize and homogenize at a European level all diverse norms that have previously existed in each E.U country. Therefore, this normative is directly applicable to each member state and its interpretation (as well as possible omissions in the text) should have global and supranational aspirations. In short, the work developed in this book is intended to be applied as an example for emerging countries, following the next objectives:

- Analyzing the European normative regarding thermal solar energy systems prefabricated with a steady regime of performance.
- Adapting a real testing installation to this new normative.
- Describing an analytical procedure to forecast the reliability behavior of a thermal solar energy system.

BASIC CONCEPTS ON THE SOLAR ENERGY UTILIZATION

This section includes a series of concepts related to solar energy that will facilitate the understanding of the whole book. These concepts are extracted from the applicable European normative for that purpose, EN ISO 9488 "Solar energy. Vocabulary".

The main terminology regarding to radiation and outside conditions are as follows:

- **Radiation:** The emission or transfer of energy in the form of electromagnetic wave or particles.
- **Irradiance (G):** The radiant power incident on a unit area of a given surface. It is expressed in W/m^2.
- **Irradiation (H):** The energy incident on a unit area of a given surface, resulting from the integration of the irradiance during a given time interval, normally an hour or one day. It is expressed in MJ/m^2, for the specified time interval.
- **Direct Solar Radiation:** The solar radiation incident on a given surface, coming from a small solid angle centered in the solar disc. Direct radiation is measured generally under normal incidence.
- **Hemiospherical Solar Radiation:** The solar radiation incident on a given flat surface, received from a solid angle of 2π sr (of the hemisphere located above the surface). It is composed of direct and diffuse solar radiation.
- **Global Solar Radiation:** The hemispherical solar radiation received from a horizontal surface.
- **Diffuse Solar Radiation:** The hemispherical solar radiation minus the direct solar radiation. In solar energy technology, diffuse radiation includes the solar radiation dispersed in the atmosphere, as well as the solar radiation reflected by the ground, depending on the receiver surface inclination.
- **Solar Constant (I0):** The extraterrestrial solar irradiation[1], incident on a surface perpendicular to this radiation, when the Earth is located at a mean distance from the Sun (149,5 x 10 km).
- **Direct Solar Irradiance (Gb):** The quotient between the radiant flow[2] received from a given flat surface, which comes

from a small solid angle centered in the solar disc, and the area from that surface. It is expressed in W/m².

- **Hemispherical Solar Irradiance (G):** The quotient between the radiant flow received from a given flat surface, which comes from a solid angle of 2π sr, and the area of that surface.
- **Diffuse Solar Irradiance (Gd):** The radiation from the diffuse solar radiation 'over' a receiver flat surface.
- **Sky Temperature:** The equivalent temperature of a blackbody which globally emits the same long-wave radiation to the atmosphere on a horizontal surface.
- **Ambient Air:** Air (inside or outside) around a thermal energy accumulator, a solar collector or any other object that has been considered.

Regarding to components and related magnitudes:

- **Absorber:** A component of a solar collector which is used to absorb the radiant energy and transfer it to a fluid in the form of heat.
- **Cover:** Element(s), transparent or translucid, which cover the absorber to reduce thermal losses and protect it.
- **Aperture:** The surface through which the non-concentrated solar radiation is admitted by the collector.
- **Aperture Area (Aa):** The maximum projected area through which the non-concentrated solar radiation penetrates in the collector. This opening surface does not include temporary transparent parts that did not absorb solar radiation, when the direction of this surface is perpendicular to the projected surface that defines the aperture surface.
- **Collector Total Surface (AG):** Maximum area projected by the whole collector, excluding any supporting means and pipe connections.
- **Area of Absorber (A_A):** The maximum projection area of the absorber. This area does not include any other part of the absorber that does not get solar radiation, when its direction is perpendicular to the projection surface that defines the absorber area.

In Figure 1, we can observe:

1. **W1:** Maximum width excluding fixing supports and pipe connections.
2. **L2:** Maximum length excluding fixing supports and pipe connections.
3. **d:** Diameter of the tuve.

Other concepts are defined at the end of the chapter.

Figure 1. Opening surface of a tubular collector

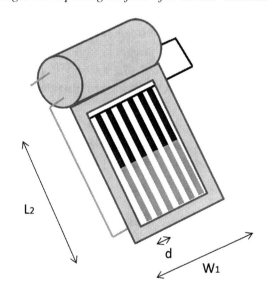

DESCRIPTION OF THERMAL SOLAR SYSTEMS FOR LOW TEMPERATURE

Following, there is a brief description of a thermal solar system at low temperature for hot water production.

Overview

Thermal collection of solar energy is understood as the procedure which transforms radiant solar energy to thermal energy. When the obtained thermal energy is used at temperatures lower than 80 °C, it is then called low temperature solar energy (Duffie et al. 1991). A thermal solar energy system is constituted by different subsystems (Figure 2). In their upmost general form, these subsystems are:

- Exchange.
- Collection.
- Accumulation.
- Hydraulic subsystem.
- Regulation and control.

Function and Composition of Subsystems

Practically, the subsystem components have been defined previously in this chapter. This section attempts to elaborate on some of the concepts defined beforehand.

Collector

The function of this subsystem is to transform the radiant energy coming from the Sun into thermal energy. Its main component is the collector, which in general, is constituted by:

- **Absorber:** The element in which the transformation of the radiant energy is produced. This energy comes first in the form of thermal energy that absorbs the heat transfer

Figure 2. System of solar thermal energy

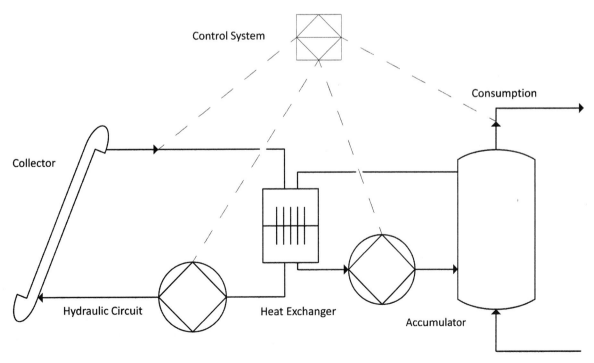

fluid. In general, it is constituted by a set of tubes superficially treated with special black paint or an absorber selective treatment (high absorption in low wavelengths and low emissivity in long wavelengths) that absorb on a large scale all the radiation that falls on them. The heat transfer fluid flows through the interior of the tubes, normally water, which carries the collected energy to the exchange system.
- **Cover:** The element which is transparent to the solar radiation and opaque to the longwave radiation emitted by the absorber, producing a greenhouse effect in the collector's interior, increasing, in this manner, its performance. The cover also serves to reduce losses by convection, as well as to protect the absorber against external and unwanted agents.
- **Casing:** The element used to constitute the collector. It contains and supports the rest of the collector elements.
- **Insulation:** The material used to reduce the collector thermal losses. Generally, it is set in the lateral zones and in the rear part of the collector.

Storage

It is the subsystem whose function is to store the generated thermal energy, increasing the water temperature for its subsequent utilization.

The main component is the accumulator, which has to be designed to endure the most difficult conditions of operation, as well as to forecast the thermal expansion of the fluid. Generally, it has to be insulated from the exterior to decrease its own thermal losses.

Exchange

It is the subsystem that makes energy exchange possible in the form of heat between the consumption water and the heat carrier fluid of the primary circuit.

The main component is the (heat) exchanger, and it is required when there is a need to separate the consumption water from the fluid that circulates through the sensors (fundamentally to prevent problems related to calcareous incrustations or in the case of utilizing antifreeze fluid).

Hydraulic System

The hydraulic subsystem is composed by the hydraulic circuit, which includes three parts clearly defined:

1. The set of pipes, valves, filters, etc. that let the fluid move from one part of the installation to another.
2. The working fluid is generally water due to its low cost and high thermal capacity.
3. The insulation, which is necessary to prevent thermal losses in the rest of the elements of the hydraulic circuit.

Regulation and Control

The regulation and control subsystem is composed of the control system, which makes it possible to carry out diverse functions. Among these are the ones that follow:

1. Antifreeze protection to sensors.
2. Maximum temperature limit in accumulators.
3. Operation of pumps.

CONCEPTS REGARDING PRODUCT QUALITY

As it has been previously mentioned, the normative EN 12976-1 deals with requirements of durability, reliability and security of prefabricated systems. Such concepts of durability, reliability and security are detailed in the EN ISO 9000 normative "Systems of quality management. Fundamentals and vocabulary". Therefore, this section refers to those terms of product quality facilitating a better comprehension of this document.

Terms Related to Quality

Quality is defined as the degree of how one product (that is its set of characteristics or permanently existing and distinguishing features) complies with all the applicable requirements. The requirements are defined as the necessities or expectations established for each specific product, generally in an implicit or mandatory manner; and, those requirements that can be generated by the different parties involved. One specified requirement is the one that is expressively stated, for example, in a document. Together with the term of "quality" we can also refer to:

- **Quality Control:** Refers to the compliance with the requirements,
- **Quality Assurance:** Refers to providing assurance in the compliance with such quality requirements.

Customer satisfaction will be the perception of the people who receive the product (that means the consumer or the final user) regarding the degree in which those requirements have been met.

Terms Related to the Product

The product is understood as the result of a process or a set of activities mutually related, or that interact with one another, which transform input elements into results. The design and development of a product can be defined by the set of processes that transform the requirements into specified characteristics or the specification of such product. The specification is a document in which all those requirements are outlined. The procedure is the specific manner in which an activity or a process can be carried out, this can be documented or not. The term of performance security is used to describe the performance of the availability and the factors that have an influence on it:

- Reliability performace,
- Maintenance capacity.

Traceability is described as the capacity to track the history, application and location of everything that is taken into consideration. When a product is considered, the traceability can be related to:

- The origin of material and parts;
- Process history (record);
- Distribution and location of the product.

Terms Related to Compliance and Review

Conformity refers to the compliance with a requirement, whereas defect refers to the non-compliance with a requirement associated to its specified expected use. An expected use just as the one expected from the consumer or user of the product can be affected by the nature of the information provided by the manufacturer, such as, instruction of operation or performance or maintenance for example. Inspection is understood as the conformity assessment by means of observation and judgment, associated, when applicable, to measurement, testing or pattern comparison. Objective evidence is considered as the set of data that support the existence or veracity of something. It can be obtained by means of observation, measurement, testing or by other means.

Through the attribution of objective evidence one can refer to:

- **Verification:** Confirmation of the compliance with all specified requirements,
- **Validation:** Confirmation that all requirements comply with an expected specified usage and application.

The confirmation can comprise of actions such as alternative calculations, comparisons, reviews, or development of tests and trials. The testing and trial processes are the determination of one or more characteristics according to a procedure.

Terms Related to Quality Assurance for Measurement Processes

A measurement system is an instrument, measurement pattern or reference material necessary to carry out a set of operations that will allow to determining the value of magnitude. Metrological confirmation consists in those fixed operations, necessary to assure that the system measurement complies with all the requirements for its expected function. This confirmation generally includes a calibration, adjustment, comparison, etc. The relative requirements for the expected utilization can include considerations such as range, resolution, maximum permitted errors etc. these requirements of measurement systems are normally different to those of the product, and are not specified in them.

FIRST LAW OF THERMODYNAMICS APPLIED TO A PREFABRICATED SOLAR SYSTEM

General Form of the First Principle

From a physics perspective, the prefabricated system is a transformer of energy coming from the sun in form of electromagnetic waves, in thermal energy that is absorbed by a heat transfer fluid (Fisch, 1995). Thermodynamically, the prefabricated system can be described as an open or closed system depending if there is or if there is not consumption. The process of energy transference is highly irreversible (just as almost all of the water heaters) since it is done at a very high temperature thermal source (approx. 6000 °C) to a system with a very low temperature (40-60 °C). Considering a general way of performance, a certain amount of fluid is taken into account, which is null when the consumption is null, and that comes in with a determined enthalpy and goes out with generally a higher one, while the system has been able to modify its internal energy and an energy exchange has been produced between the system and the surroundings basically by radiation, but also by conduction and convection. If we apply the first principle of thermodynamics, in its upmost general from, we get:

$$\frac{d\sum_i (m_i \cdot u_i)}{dt} = \sum_e \dot{m}_e \cdot h_e - \sum_e \dot{m}_s \cdot h_s + \sum I \cdot A - \sum P$$

where:

- $\dot{m}_e$, Mass speed of the inlet fluid (kg/s).
- h_e, Specific Enthalpy of the inlet fluid (kJ/kg).
- $\dot{m}_s$, Mass speed of the outlet fluid (kg/s).
- h_s, Specific enthalphy of the outlet fluid (kJ/kg).
- m_i, Mass of component i of the solar system (kg).
- u_i, Specific Internal energy of the component (kJ/kg).
- I, solar irradiance on the installation surface (W/m²).
- A, Aperture surface of the collector system (m²).
- P, Energy losses by a time unit (W)

If we consider that the fluids dealt with in this document are liquids, we can therefore relate the inlet and outlet enthalpies with the inlet and outlet temperatures respectively through a specific thermal capacity of the fluid under a constant pressure, c_p (Figure 3).

In particular, the first principle states that the amount of fluid that goes into the system is equal to the amount that goes out, thus the previous expression is now:

$$\frac{d\sum_i (m_i \cdot u_i)}{dt} = \dot{m} \cdot c_p \cdot (T_e - T_s) + \sum I \cdot A - \sum P$$

q_u is the useful energy that is extracted from the system, and it is precisely equal to:

$$\dot{m} \cdot c_p \cdot (T_e - T_s)$$

If the equation is then set in relation to useful energy, we will get:

$$q_u = \sum I \cdot A - \sum P - \frac{d\sum_i (m_i \cdot u_i)}{dt}$$

Simplification to a System Only Composed by the Collector

In order to make the analysis easier, we will consider firstly a thermal solar system composed of only one collection system. In other words, let's consider a flat solar collector and see how we can precisely define each term of the previous expression. Of all the solar energy that gets through the collection system by radiation, there is a part that is not able to go through the cover, as it is later reflected. Another aspect is that the absorber will only be able to absorb and transmit to the fluid part of the energy that gets into the cover or by a greenhouse effect. Both effects are taken into account through the coefficient $(\tau\alpha)$, which is the product of the transmittance of the cover by the absorbance of the absorber. This coefficient fundamentally depends on the physical proprieties of both cover and absorber, as well as their geometry. The coefficient $(\tau\alpha)$ will also depend on the angle in which the solar radiation falls on the collector surface. Generally, this dependency on the incidence angle is resolved by obtaining the coefficient $(\tau\alpha)_n$ for an incidence approximately normal to the collector surface and affecting it with a coefficient $K\tau\alpha(\theta)$ referred to as the incidence angle modifier (Gruber, 1984).

In another respect, the radiation that gets to the solar energy system can be divided into a direct component and a diffuse one that gets to the collector. At the same time, diffuse radiation will be the sum of the diffuse sky radiation and the radiation reflected by the surroundings of the collector. The coefficient $(\tau\alpha)$ will be a different

Figure 3. Heat flow in the solar thermal system

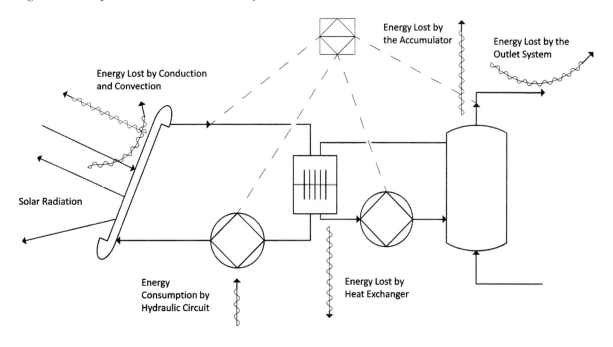

one for each type of radiation. More advanced energy models consider these effects introducing different incidence angle modifiers for direct and diffuse radiation. Taking into account the above mentioned, energy gains of the solar energy system can be expressed minus the reflection losses in the cover and in the energy transmission process in the absorber through the following expression:

$$K\tau\alpha_{direct}(\theta) \cdot (\tau\alpha)_n \cdot I_{direct} + K\tau\alpha_{diffuse}(\theta) \cdot (\tau\alpha)_n \cdot I_{diffuse}$$

where:

- **K$\tau\alpha$ (θ)**: Modifier of the incidence angle for direct radiation. Depends on the incidence angle, θ.
- **$(\tau\alpha)_n$**: Product resulting by multiplying transmittance and absorptance of the collector for normal incidence to the aperture surface.
- **I_{direct}**: Direct irradiance on the collector surface (W/m^2).
- **K$\tau\alpha$ $_{Diffuse}$**: Modifier of incidence angle for diffuse radiation.
- **$I_{diffuse}$**: Diffuse irradiance (W/m^2).

It is necessary therefore, to obtain data from the testing installation of both the direct radiation and the diffuse radiation. It can also be interesting to measure the transmittance of different materials to gain insight into enhancing new materials for covers and at the same time the absorbance for the absorber. The dependency of the modifier of incidence angle with the same incidence angle can be easily evaluated with the two testing banks, one identical to the collector in a fixed position and the other to a tracking structure so that the radiation normally falls on the collector surface at any instance during the day. In any case this evaluation is not easy, every time that it deals with term approximate to 1 and its variation is very small. The rest of thermal losses can be divided into conduction losses, convection losses and long wave radiation losses. Traditionally, it is proposed that the collector temperature can be considered approximately constant. Most importantly this is practically true, however, advanced models consider temperature distribution along the collector and divide the collector into nodes which each and every one of them present distinct coefficients of thermal losses.

In this first analysis we will consider, as it is considered in most of the models, that the collector has a constant temperature, T_p. Convection and conduction losses will be proportional to the difference of temperatures between the panel and the surrounding environment. Having said that, the coefficient of thermal losses, U_L, is proven that they depend mainly on that temperature difference between the panel and the surrounding environment and the air speed that circulates around it. In another respect, the solar collector, being at that temperature, T_p, different from the sky temperature will emit a long wave radiation that is translated into more losses. These losses will have the type of expression that follows:

$$P_{rad.O.L} = F\varepsilon\sigma\left(T_p^{\,4} - T_c^{\,4}\right)$$

being:

- **F**: Shape factor of the collector in respect to the surroundings or sky.
- **σ**: Constant of Stefan-Boltzmann, equal to 5.67 · 10^{-8} W/m^2 K^4.
- **T_p**: Solar collector temperature.
- **ε**: Equivalent Emissivity of the collector.
- **T_c**: Sky equivalent temperature.

In order to calculate these losses it is necessary to know the sky temperature which is very difficult to evaluate or measure. These losses could be evaluated experimentally by a specto-radiometer with an infrared range. A thermographic system can also be an interesting instrument for this type of study. Evaluating sky temperature and knowing in a more accurate way the losses due to long wave radiation can be an interesting line of research in future investigations. That is why it is left open as a possible line of research to follow.

The first models for solar collectors consider that all thermal losses can be included in the term $U_L(T_p - T_a)$, supposing a U_L constant. Other models consider that this coefficient depends, at the same time, on the temperature difference between the panel and the surrounding ambient, and that thermal losses can be expressed as: $U_0(T_p - T_a) + U_i(T_p - T_a)^2$. More advanced models consider also a lineal dependence on the wind speed and sky temperature, which is still a first approach. Dependency on wind speed can be studied considering in the testing installation the possibility to control this parameter through the vents. However, there is still a need to find a physical sense to the parameters. If we were able to know in a better way the distinct terms of thermal losses of a solar collector and if we relate them to the physical parameters and material properties that they are composed of, we could make advances in the design and the materials that are used in solar energy systems. Regarding the term that refers to the variation of the system internal energy, generally, it is considered through the product of an equivalent mass by a thermal capacity equivalent of the system giving an effective thermal capacity so that:

$$\frac{d\sum_i (m_i \cdot u_i)}{dt} = (M \cdot c)_e \cdot \frac{dT_p}{dt}$$

The effective thermal capacity and the time constant of a collector are fundamental parameters when there is a need to determine the behavior in a transitory regimen of the collector or when there is a need to study the solar energy system when it works as a closed system. A collector can be generally considered as a combination of thermal masses, each of them at a different temperature. When a collector is in operation, each component responds differently to any changes in the condition of operation, thus it is very practical to consider an effective thermal capacity to the entire collector. Unfortunately, the effective thermal capacity depends on the operation conditions and it does not have a unique value per collector. There are several methods to measure and calculate the effective thermal capacity of a solar collector, and it has been proven that similar results had been obtained from entirely different methods, being the methods based on experimental testing, the most contrasted ones. After the previous analysis we can write the equation of the first principle of thermodynamics for the collection system:

$$q_u = K\tau\alpha_{direct}(\theta)(\tau\alpha)_n \cdot I_{direct} + K\tau\alpha_{diffuse}(\theta)(\tau\alpha)_n \cdot I_{diffuse}$$
$$- A(v) \cdot (T_p - Ta) + B(v) \cdot (T_p - T_a)^2 - C \cdot (T_p^4 - T_c^4)$$
$$- (Mc)_e \cdot (dT_p / dt)$$

being:

- q_u: Useful energy ceded by the system per area unit
- $A(v) \cdot (T_p - Ta) + B(v) \cdot (T_p - T_a)^2$: Conduction and convenction losses. They depend on wind speed, v.
- $K\tau\alpha_{diffuse} \cdot (\tau\alpha)_n \cdot I_{diffuse}$: Diffuse radiation energy gain.
- $K\tau\alpha_{direct} \cdot (\tau\alpha)_n \cdot I_{direct}$: Direct use radiation energy gain.
- $C \cdot (T_p^4 - T_c^4)$: Long wave radiation losses.
- $(Mc)_e \cdot (dT_p/dt)$: Internal energy variation of the system.

Once the equation of the thermodynamics first principle is developed for the collection system, there will be a need to develop the same equations for the rest of the systems that compose the solar energy system. These equations are developed in most of the cases considering that the rest of the elements that integrate the solar installation are just tubes and hydraulic elements, the heat exchanger and the accumulation deposit.

Thermodynamic Description of the System Operation

If the system operates by thermosyphon, the amount of water that flows through it will be in respect to its density difference due to the temperature difference in distinct points of the system. Therefore, if the system operates with a forced flow, one can consider basically two modes of performance (if we do not take the transitory modes into account). These performance modes depend on the fact that if the pump is in operation or not. The control system operates the pump when the temperature differences between the water coming out from the collectors and the water in a specified location of the accumulator is more than a previously set value, and it will stop when such difference is less than another previously fixed value (normally several degrees Centigrade less than the other). When the pump is in operation, the internal energy variation of the collector can be considered as null, taking into account that it operates in a quasi-steady state. We therefore get that:

$$q_u = \dot{m} \cdot c_p \cdot (T_e - T_s) = \sum I \cdot A - \sum P$$

The energy supplied by the system will be therefore direct function of the flow and the thermal jump of the work fluid through the collector which, in another respect, will also depend on the fluid flow. If the pump does not stop, the term of the collector internal energy is not null, but there is no entrance or exit of fluid in the system, thus the term related to the water flow is null. Therefore we can get:

$$q_u = 0 = \sum I \cdot A - \sum P - (M \cdot c)_e \cdot \frac{dT_p}{dt}$$

Or, that is the same:

$$(M \cdot c)_e \cdot \frac{dT_p}{dt} = \sum I \cdot A - \sum P$$

The time in which the pump is not in operation will fundamentally depend on the consumption and environmental conditions, but also in a great scale on the effective thermal capacity of the solar collector system and the stratification in the accumulator. The time in which the pump is in operation will be a very relevant factor when it comes to dealing with optimizing supplied energy by the collector system.

PREFABRICATED SOLAR ENERGY SYSTEMS FOR WATER HEATING

In accordance with the normative UNE-EN 12976, prefabricated solar heating systems are defined as those solar energy systems that are used for hot water preparation, either as a complete pre-packaged system or a system made up from different individual parts (Gruber, 1984). The system consists of only one component or set of standardized components (Figure 4). It is manufactured under certain standard conditions and is offered on sale under one commercial product name.

That means that those systems are sold as complete systems ready for installation, with a registered commercial product name. The preparation is only of domestic hot water, even though they could also be used for heating, this option should not be considered during testing tasks. A single system can be tested as a whole in a laboratory, giving rise to results which are representative of systems that are sold under the same commercial product name, configuration, components and dimensions (Smith, 2001).

These products have a fixed configuration, therefore if some of the components are altered or changed in such a way, then they must be considered as a new product, and thus, as a new testing report[3]. Prefabricated systems for domestic water heating can be differentiated from custom built systems as following indicated:

Figure 4. Subsystems of a prefabricated system

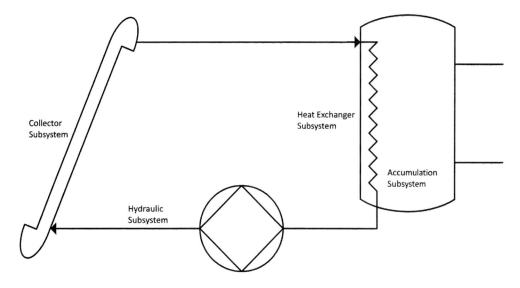

- **Prefabricated Solar Energy Systems:**
  - Systems with integrated collector-deposit for domestic hot water.
  - Thermosyphon systems for domestic hot water.
  - Forced circulation systems as a set of products with fixed configuration for domestic hot water.
- **Custom Built Solar System:**
  - Forced circulation system for hot water and/or heating, installed using components and configurations described in a component catalogue (mainly small systems).
  - Single systems of design and installation, used for water heating and/or heating (mainly in large systems).

In summary, the prefabricated solar heating systems are divided (as a set of fixed configuration products) into:

1. Systems with integrated collector-deposits.
2. Systems with thermosyphon circulation.
3. Forced circulation system.

Figure 5 shows the basic functional installation schemes that function by forced circulation and thermosyphon.

The auxiliary energy systems connected in series to the prefabricated solar energy system will not be considered as part of the prefabricated systems. Other parts that are not considered are: pipes of the cold water network, ducts from the system to the auxiliary energy one or two consumption points and any other integrated exchanger or heating pipes.

REQUIREMENTS APPLIED TO PREFABRICATED SYSTEMS

Most of the requirements here presented should be demanded from the manufacturer of the prefabricated system. This section, therefore, will be of special interest to the product manufacturer (UNE-EN 60335, 2001).

Thermodynamic Description of the System Operation

Frost Protection

The manufacturer should set the minimum permitted temperature for the prefabricated system. This way, the parts that are more exposed to the minimum temperatures will be able to withstand those set temperatures without causing permanent damage to the system. However, if some of the components of the prefabricated system are exposed to temperatures below 0 °C, they should be protected against frost conditions through a protective method previously described by the manufacturer. If an antifreeze fluid is used, the minimum temperature for the system will be the freezing point of such fluid; although in some cases that limit can be exceeded without causing damage to the system. Nevertheless, the manufacturer should take the necessary precautions to prevent possible deterioration to the condition of the antifreeze fluid as a consequence of high temperatures.

Overheating Protection

The system should be designed so that its prolonged exposure to high solar radiation without consuming hot water, does not result in a situation in which the user has to take any special actions to restore the system back to normal performance. If the system has the possibility of draining a certain amount of hot water as a protection against overheating, the drainage should be carried out in a way that makes sure that no damage is caused to the system or any other adjacent area of the residence on which the system is installed. Moreover, the installation of the system must be done in a manner that guarantees that no damage will occur to the habitants of the household that results from hot water or vapor drainage. In the case that the overheating protection depends on electrical and/or cold water supply, the manufacturer should expressively state it in the system instructions. Finally, the system must be designed so all materials never exceed the maximum permitted temperature.

Pressure Resistance

To begin with, the system should satisfactorily pass the pressure test given in EN 12897. Those system circuits that are not included in the previous normative, should withstand the minimum of the following pressure values:

Figure 5. Forced circulation (left) and thermosyphon circulation (right)

- 1,5 times the maximum work pressure specified by the manufacturer;
- The maximum work test pressure specified by the manufacturer.

When a pressure resistant test is carried out using the before mentioned pressures, there must be neither permanent damage nor leaks in the system components or interconnections. After the time taking for testing has passed, the hydraulic pressure should not drop more than a 10% of the value measured at the start of the testing period. When non-metallic materials are used in any circuit, these materials should be able to withstand the previously mentioned pressures for at least an hour. In these cases, the pressure resistant test should be carried out at high temperatures considering that the pressure of a non-metallic material can be adversely affected when its temperature is increased. The consumption circuit should withstand the maximum required pressure by the national or European norms for drinking water in open or closed consumption water installations. The system must be designed so that the properties of any of the materials exceed the maximum permitted limit of pressure.

Electrical Protection

The manufacturer should comply with the requirements set forth in the normative ENV 61024-1, which are expressed in Annexes E and F of the normative UNE-EN 12976-2, as these last ones merely informative.

Requirements for the Working Fluid

The heat transfer fluid (or working fluid) used in the prefabricated system during the tests will be that which is specified by the manufacturer. The manufacturer therefore, must define the composition of the heat transfer liquid permitted by the system, including additives. If the prefabricated system is designed to be used with an antifreeze fluid, the tests will be done using that fluid also following the manufacture's specifications. In this case, and as a consequence of all possible high temperature during the tests and during the normal use of the system, the manufacturer must take the necessary precautions in order to prevent possible deterioration[4] of the antifreeze fluid.

In another respect, the fluid directly obtained from the prefabricated system will be water fit for human consumption. Whilst other verifying European criteria can be implemented, the existing national regulations are effective and concern the usage and/or characteristics of prefabricated systems with respect to drinkable water quality. Regulations for this purpose are:

EN 1717: "Protection from drinkable water contamination in water installations and general requirements of reverse flow preventive devices."
EN 806-1: "Specifications for water conduction installations designated for human consumption inside buildings. Part 1: General."

Component Requirements

Collector

For those systems in which the collector cannot be tested separately from the whole system (just as the case studied in this document), then, the collector of the prefabricated system should comply with:

- High temperature resistance,
- External thermal impact,
- Rain penetration and,
- Mechanical load testing.

These tests are regulated through the normative UNE-EN 12975-2 ("Thermal solar systems and components. Collectors. Part 2: testing methods"). The compliance of the previous tests denotes that none of these sever failures occur:

- Leaks in the absorber or deformities that could lead to the permanent contact between the absorber and the cover.
- Cracks or permanent deformation of the cover and its fixtures.
- Cracks or permanent deformation of the fixtures of the collector casing.
- No-losses, so that the sub-atmospherical or vacuum collectors cannot be classified as such.
- Accumulation of damp in the collector.

Support Structure

In case the system has a support structure installed normally in the outer part, the manufacturer should then specify the maximum values for:

- s_k (snow load), and
- ν_m (average wind speed).

These values must be in accordance with the currently effective regulation:

Eurocode 1: "Project foundation and actions on structure".
Eurocode 1: "Project foundation and actions on structures."
ENV 1991-2-3: "Actions on structures. Snow Loads".
ENV 1991-2-4: "Actions on structures. Wind actions."

The system will be only be installed in locations where the previously mentioned values s_k (snow load), and ν_m (average wind speed) are less than the ones specified by the manufacturer. This must also be clearly stated in the installation documents. Some examples of these supports are shown in Figure 6 and Figure 7.

Pipes

The design of the entire piping system of the prefabricated system must prevent obstruction formation or accumulation of lime deposits in the piping circuits, considering that these conditions will drastically influence the system's correct performance.

Circulation Pump

In the particular case of a forced circulation system, and thus the use a circulation pump is necessary, such pump must comply with the effective normative for that purpose:

EN 809: "Pumps and motor pump groups for liquid. Common safety requirements".
EN 1151: "Pumps. Rotodynamic pumps. Circulation pump which energy consumption does not exceed 200 W, destined for central heating and domestic hot water distribution. Requirements, tests and marking."

Heat Exchanger

It is necessary for some prefabricated systems to use a heat exchanger. Such heat exchanger will be regulated through the currently effective normative for that purpose:

UNE-EN 307: "Heat exchangers. Guidelines to elaborate installation, operation and maintenance instructions, necessary to keep the performance of each of the types of heat exchangers."

When a system has been designed to be used in zones of highly harsh waters and temperatures higher than 60 °C, heat exchangers in contact with the consumption water must be able to prevent the formation of deposits or, for that matter a

Figure 6. Domestic solar energy system installed on a flat roof: 1. heating solar energy system; 2. support structure for flat roofs; 3. hot water pipes; 4. flat roof

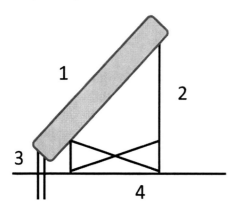

Figure 7. System installed on an inclined roof: 1. heating solar system; 2. inclined roof; 3. hot water pipes; 4. internal temperature

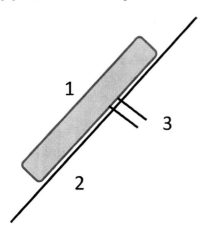

cleaning system instead. As a recommendation for the manufacturer, there is the possibility to avoid deposit formation by increasing the exchanger surface, since normally such deposit formation is caused by high temperatures between the metallic surface of the exchanger and the surrounding consumption of water.

Control System

When there is a temperature sensor in the accumulator, it should be able to withstand 100 °C with no alterations of more than 1 K. Similarly, if there is a temperature sensor in the collector, it must withstand the blockage conditions with no alterations of more than 1 K. The blockage temperature is the collector's temperature during non-extraction periods of useful heat from the collector with high solar radiation and high surrounding ambient temperature. To calculate this temperature any of the two procedures described in Annex C ("Stagnation temperature of liquid heating collectors") of the norm UNE-EN 12975-2 ("Solar thermal systems and components. Solar collectors. Part 2: Testing methods"). In any case, temperature sensors must be isolated from the environment.

Requirements in the Safety System

Safety Lines and Expansion Lines

When the prefabricated system is designed with a safety line and an expansion line, these lines must be dimensioned in such a way that, for the case of the greater possible flow of hot water or vapor, in any point of the collector's loop exceeds the maximum permitted pressure due to a pressure decrease of such lines. Both lines must be connected and located in points which prevent dirt accumulation, deposits or any other unwanted material. If the system is designed with only one safety line, this line will not be able to be closed.

Purge Lines

These must be installed (if the system is enabled for it) in location points where water cannot accumulate nor freeze inside the lines. The orifices will be placed so that the outgoing liquid or steam from the safety valves cannot represent a risk of danger to people, materials and the environment.

Safety Valves

The system must be supplied with at least one safety valve:

- Any integrated system, in which the valve can be integrated in combination with an inlet.
- Each section of the collectors field that can be isolated, if there is a collector accumulator.

The safety valve must withstand:

- Temperature conditions to which it is exposed (specially higher temperatures that can occur),
- The medium of the heat exchanger transfer.

Moreover, it must be dimensioned so that it can discharge the greatest flow of hot water or vapor possible.

SUBSYSTEMS TEST PROCEDURES FOR VALIDATION OF PREFABRICATED SYSTEMS

Frost Protection

The objective of this section is to assure the correct operation of Frost protection devices. The verification is carried out as it will be later explained and in accordance with the manufacturer recommendations. As previous considerations have made clear, those components of the system installed in the outside and inside and that may be exposed to temperatures below 0° C, should be fully protected against frost.

The manufacturer must:

- Set a minimum permitted temperature for the system, so that those external components exposed to such conditions can withstand the set temperature without causing any damage.
- Describe the applicable Frost protection methodology for the system.

Test authorities must:

- Identify which methodology has been used (described by the manufacturer).
- Assure the correct performance of all the system parts for Frost protection.

Configuration of Frost protection devices will be verified based on the manufacturer recommendations and in accordance with the proper section of the following list.

Systems with Antifreeze Fluid

The components of the system that are exposed to low temperatures are filled with antifreeze fluid, normally is a mixture of glycol/water, with a sufficiently low freezing point. No test procedures are needed for these types of systems. However, there could be an insufficient amount of data related to the freezing point of the antifreeze fluid. In this case, we must proceed as follows:

1. Measure the FREEZING POINT of the antifreeze fluid, unless this information is already available. The measurement is carried out according to the glycol concentration and using a portable refractometer.
2. Make sure that:
   a. FREEZING POINT = SYSTEM MINIMUM TEMPERATURE the minimum temperature of the system is provided by the manufacturer.
3. Make sure that:
   a. FREEZING POINT < LOCAL AMBIENT MINIMUM TEMPERATURE the minimum ambient temperature refers to the air temperature.

If the freezing point has been provided by the manufacturer, steps 2 and 3 must be directly applied. If the concentration of glycol exceeds a certain limit, it is possible that the system can reach that freezing point without causing any damage. In this case it is permitted that:

FREEZING POINT > SYSTEM MINIMUM TEMPERATURE

The composition of the antifreeze fluid and additives must be coherent with that which is specified by the manufacturer.

Drainage Recovery System

The fluid from the components exposed to low ambient temperatures, is drained into an accumulative deposit for subsequent use in the event of dangerous frost conditions.

1. Make sure that the INCLINATION OF THE HORIZONTAL PIPES corresponds to the manufacturer's recommendation, at least by 20 mm/m. The check is done with an air bubble level.
2. Start the PUMP.
3. Make sure that the FILLING is done in accordance with the operations manual. This verification can be observed through:
   a. A manometer,
   b. A water level indicator,
4. Any other method provided by the manufacturer.
5. Turn the PUMP off.
6. Make sure that the DRAINAGE is done in accordance with the operations manual. This verification can be observed through a value decrease in the reading of:
   a. The manometer,
   b. The water level indicator,
   c. Or the method provided by the manufacturer.

Step 5 can be REPEATED but at a high accumulation temperature (90°C) to ensure drainage in any eventuality. However, that verification would correspond to an Overheating protection test.

Systems for External Drainage

In this type of systems, the fluid of the components exposed to low ambient temperatures are drained and removed outwards to the outside when there is a danger of frost conditions. The trial procedure is carried out as follows:

1. Perform a check of the correct opening and closing of the DISCHARGE VALVE.
2. In the case of:
   a. SOLENOID VALVE independent from the control unit:
      i. Simulate the opening temperature.
   b. Non-electrical POWER VALVE:
      i. Spray with a freezing-spray material.
      ii. Make sure that the thermometers are in the correct location.

iii. Make sure that the temperature measured in the open valve is equal or higher than the nominal temperature specified by the manufacturer.
3. Make sure that the INCLINATION of the horizontal pipes corresponds to that of specified by the manufacturer, at least by 20 mm/m. This check is done with an air bubble level.
4. In the DRAINAGE VALVE:
    a. Open (if it is electrical, the power is interrupted).
    b. Measure the amount of drained fluid using a deposit and a chronometer.

Systems of Combined Freeze Resistant Devices

This refers to those cases in which the solar energy system is designed to combine protective devices with control functions against frost. With this test the control unit is verified.

1. Adjust the SIMULATED TEMPERATURE OF THE PROTECTION SENSOR with the PROTECTION DEACTIVATION TEMPERATURE.
2. Slowly decrease the SIMULATED TEMPERATURE.
3. Measure the TFP (frost-protection) of the actuator.
4. Compare the measured value to the MANUFACTURER NOMINAL VALUE.

Systems with an Exterior Deposit

These are the systems with the deposit also exposed to external low temperatures, in which we must also take into account during the test:

- Connection pipes,
- Antifreeze protection devices (active or not).

The objective of this testing, in this particular case, is to evaluate the frost protection devices of integrated systems with an external deposit. The European Norm EN 12976-2 proposes a procedure in Annex C (section C.1), which is merely informative and it attempts to verify if determined conditions of certain European climate regions will cause damage to the system or negatively affect its correct performance. This procedure is based on the use of a CLIMATE CAMERA, which will generate the necessary specified conditions of temperature, wind speed and irradiance.

Other Systems

For the rest of the systems, the following items must be checked:

- Pump control system,
- Outside drainage valve,
- Any other item or frost protection device.

These tests are carried out:

- In accordance with the manufacturer specifications,
- At the minimum permitted temperature (set by the manufacturer).

Overheating Protection

The objective is now to determine if a system is secured against damage, and if the user is protected against the possibly reheated water. Moreover, it will be relevant to find out if the system is able to withstand extreme overheating conditions. As previous considerations have made clear, the system must be correctly designed so that in the event of high and prolonged solar radiation with no hot water consumption, the user does not have to take any special action to put the system back to its normal status of operation. Its design must contemplate, as well, that the materials constituting the system do not exceed the maximum permitted temperature.

If the system can drain a small amount of hot water as a protective measure against overheating, this drainage must be done in a way that it does not cause damage to the system or the residence. That means that the installation must be done in a way that the drainage of hot or vapor water does not represent a danger for the inhabitants. During the test, there cannot be any vapor coming out from any of the consumption points. In the case that the water in such consumption points exceeds 60° C, the documents for the person in charge of the installation must specify the assembly of a mixture automatic system or another that limits the extraction temperature to 60 °C. However, the system must be able to withstand the maximum possible temperature of system extraction.

If non-metallic materials are used in a certain circuit, the maximum permitted temperature in the circuit should be measured during the overheating protection test, so it can also be used in the pressure resistance test. The solar energy system should be tested in accordance with section 5.1 of the regulation ISO/DIS 11924 ("Solar heating. Domestic water heating systems. Testing methods for reliability and safety assessment"). However, the total values of minimum solar radiation in the solar collector surface as well as the minimum solar intensity in the solar collector to be applied in such test will be expressed as follows (Table 1):

If the test is carried out with radiation conditions which do not correspond to the maximum permitted, it should be expressively stated in the user's manual. The European Norm EN 12976-2 proposes a procedure in Annex C (section C.2), which is merely informative and it attempts to verify if the system is able to withstand determined conditions of extreme overheating without causing damage. This procedure is based on the use of a SOLAR SIMULATOR that generates the necessary specified boundary conditions of temperature, wind speed and irradiance, which really makes no sense in a climate region such as south of Spain or Mediterranean Sea.

Pressure Resistance

The objective of this test is to assure that no permanent damage is produced and that leaks in components of the system and their interconnections do not occur. The system will be installed previous to the test, just as specified in the installation manual. Each closed circuit of the system is supplied with a safety valve that must be able to withstand the maximum temperature value that can be reached in its placement.

First to be tested will be the safety of the pressure system, in other words, to make sure that all safety valves and other overheating protection devices are present and correctly located, if there are no valves between components, discharge valves, etc... the prefabricated system must be tested in accordance with section 6.1 of the regulation ISO/DIS 11924 ("Solar heating. Domestic water heating systems. Testing methods for reliability and safety assessment"). If non-metallic materials are used in a circuit, the maximum temperature in the circuit must be measured during the overheating protection test, to be subsequently used in the pressure resistance test. In this case, the pressure test will be applied for an hour at the highest temperature measured during the overheating protection test + 10 °C.

Electrical Protection

In case that the solar energy system has an electrical device, this must be tested in accordance with the regulations currently in effect:

Table 1. Radiation and intensity by climate zone

Climate Zone	Solar Radiation [MJ/(m^2d)]	Solar Intensity [W/m^2]
Northern European	20	700
Central European	20	850
Mountain	25	1050
Mediterranean	25	1050

UNE-EN 60335-1: "Safety of electrical appliances and analogues. Part 1: General requirements".

UNE-EN 60335-2-21: "Safety of electrical appliances and analogue devices. Part 2: Particular requirements for thermal electrical devices".

Regarding protection against electrical discharge, the system must comply with the regulation ENV 61024-1. Alternatively, Annex E of the Euro-Norm EN 12976-2 provides a testing methodology for domestic water heating systems based on ENV 61024-1. Said annex is merely informative.

Mechanical Stress of the Support Structure

If the system has a supporting frame structure installed in the external part, the manufacturer must specify the maximum values for:

- s_k (snow load), and
- v_m (average wind speed).

Such values must be verified during the design stage, calculating support structure stress in accordance with the regulations specified in the requirements section. However, Annexes G and H of the Euro-Norm EN 12976-2 describe an optional methodology for mechanical stress evaluation of the support structure.

Protection Against Reverse Flow

A visual inspection must be done in order to assure that there is a retention valve or another type of similar protection. This protective device is necessary to prevent an increase in heat losses due to a reverse flow in the circuit.

Water Pollution

There is a need to confirm the applicable regulation: UNE-EN 1717. "Protection from drinkable water contamination in water installations and general requirements of devices that prevent contamination by reverse flow".

Safety System

Safety Lines and Expansion Lines

The SYSTEM DOCUMENTATION must demonstrate (in order to verify) that safety and expansion lines:

- Cannot be closed,
- Are connected and located to prevent any dust accumulation, deposits or other similar unwanted materials.

It must be checked, in case it exists, the INTERIOR DIAMETER of the expansion line in respect to the previously mentioned requirements.

Purge Lines

The HYDRAULIC SCHEME and the SYSTEM DOCUMENTATION must be checked (in order to verify) that the purge lines, if they exist, comply with the previously mentioned requirements.

Safety Valves

The regulation currently in effect must be checked, on which compliance with will therefore logically affect the pressure resistance of the system. The regulation for that purpose is:

UNE-EN 1489: "Building valves. Safety valves, testing and requirements".

UNE-EN 1490: "Building valves. Combined temperature and pressure release valves. Testing and requirements".

The SYSTEM DOCUMENTATION must demonstrate that each solar collector circuit is fed by at least one safety valve. The SAFETY VALVES SPECIFICATION, in the terms of the materials they are made of, must be checked to verify they comply with the previously mentioned requirements. There must be a check if the FLUID TEMPERATURE, at a discharge pressure of the safety valve, exceeds the maximum permitted temperature of the heat transfer device. To check the applicability of the MAINTENANCE FREQUENCY specified in the thermostatic valve, a THERMOSTATIC VALVE AGING TEST must be carried out, which is described in Annex D in the Euro-norm EN 12976-2, Annex D being merely an informative annex.

CATEGORIZATION OF THERMAL PERFORMANCE

Previous Considerations

At the same time that complete products for domestic hot water production have been launched into the market, the need to categorize the whole set as a unique entity arises. In the particular case of thermal categorization, a mathematical model is set out which considers all the prefabricated systems as a "black box", from which inputs generate outputs. The current testing model for the thermodynamic categorization of a steady state operation prefabricated system is based on this mathematical model. Following there is a test of thermal performance of a prefabricated system in steady state operation, according to the Euro-Norm EN 12976, which also makes reference to the norm:

ISO 9459-2: "Solar heating. Domestic water heating system. Part 2: Testing method in the exterior for the categorization of solar-only systems and annual performance prediction".

Before beginning the thermal performance categorization, the other specified tests in the same norm EN 12976-2 should be completed. If there is a malfunction or defect found in such tests, this should be eliminated by the manufacturer before proceeding to the thermal categorization of that solar power system. If this is not possible, the malfunction must be registered in the performance test log.

Some systems allow variations in the installation of their components; this however, could affect system performance. In such cases where the reference conditions are not clearly defined, the most unfavorable conditions must be selected in order to perform the test. This way, there is more flexibility for the manufacturer to limit energy input. An example could be the case of forced circulation (Figure 8), which should be tested at a lower position from the related deposit of the collector and the maximum pipe length between the solar collector and the deposit specified by the manufacturer.

In the same way, thermosyphon circulation system tests are performed with the deposit at its lowest position on the collector and with the minimum pipe length between the solar collector and the deposit, which is specified by the manufacturer (Figure 9). The testing method used does not consider which type of system is being tested, that means, it considers the system as a black box and thus, the method is applicable to any type of factory-made system for water production with no auxiliary energy input.

Installation Conditions

The test will be carried out with all the components of the system located in their corresponding places. For this reason the system will be mounted in the testing lab according to the specified mounting instructions of the manufacturer (González-Prida & Crespo, 2012). If there are no specific mounting instructions set out by the manufacturer, the prefabricated system will be installed as follows[5]. The system will be installed

Figure 8. Forced circulation system

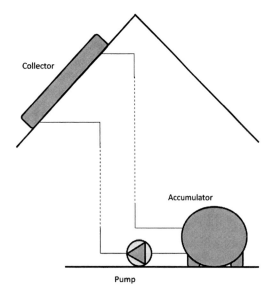

Figure 9. Circulation by thermosyphon system

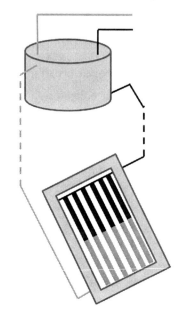

so that the safety of the personnel executing the test is assured. Some safety precautions must be taken related to sharp sticking-out edges in the outer part of the system. Also, some special attention must be paid to possible hot liquid leaks and to scenarios in which the collector cover can break or fall.

If the system is equipped for hot water drainage, the mounting must be done in such a secure way that the drainage will not represent a dangerous risk. The assembly of the system must guarantee that there will not be steam coming out of any of the points during the over-temperature test. Hot air currents are not permitted (for example those that rise through building walls) to flow through the solar energy system. That is why systems that are tested up on a roof of a building should be placed at a minimum distance of 2 m from the edge. The following is a list of the particular requirements for prefabricated system components.

Accumulator Installation

Unlike those systems in which the accumulator is joined, in any way, to the collectors, the accumulator will be installed at the lowest possible point permitted by the manufacturer instructions. For those cases in which the accumulation deposit is separated from the collectors (parted system) and it's a forced circulation system, the hydraulic circuit pipes that connect them must have a total length of 15 m. the diameter and the isolation of those pipes must adjust to the instructions supplied by the manufacturer and to the specifications established by the norm currently in effect for that purpose.

Collector Installation

The collector of prefabricated systems must be located where no shade can affect it during the thermal performance test; it must be installed in a fixed position orientated to the south[6], with a tolerance of ± 10°. The collector will have a determined inclination that must be included in the results document. This inclination must remain constant throughout the thermal performance test. Moreover, it must be placed so that it does not reach in a significant way the reflected radiation from the surroundings and there will not be significant obstructions in its range of view. In some collectors, the performance is influenced by wind speed, of a range that goes from 0 a 3 m/s. To simulate this situation, collectors sensitive to wind speed must be installed so that air can flow through the aperture, the rear zone and the collector laterals, at an average speed between 3 and 5 m/s. Collectors designed to be integrated into a roof must be protected from the wind in their rear parts, in any case this information should be included in the test results.

Installation of the Structure Supporting Frame

The prefabricated system will be installed in the same assembly structure provided by the manufacturer. If the manufacturer does not provide such a structure, and also if it is not clearly specified on which structure the system should be mounted (for example if the system is an integrated part of a roof), an open and lightweight structure is then used so that it influences, in the least possible way, the heat exchanges of the surrounding system. The mounting must also be done in such a way that the structure is able to withstand any wind draft or snow, in other words, it must be able to withstand the shocks indicated in the regulation currently in effect for that purpose. The structure where the system is installed must not block the opening area of the collector and must not affect the thermal isolation of the solar energy system in its rear part, lateral parts nor the accumulator.

Installation of Temperature Sensors

There is a possibility that the solar energy system includes, among its components, temperature sensors for the accumulator and collector. In this case, location of installation of these temperature sensors must ensure a satisfactory thermal contact with the temperature measuring part.

Material Conditions

The materials[7] which are part of the system must ensure that there is no possibility of blockage or lime deposits forming in the circuits of the system, since those conditions can negatively affect system performance (González-Prida et al. 2012). Every component installed in the external part of the system must be able to endure ultraviolet radiation as well as other atmospheric conditions during the maintenance interval planned for that system. In other words, they must be designed to withstand climate conditions. If there is a need for repairs or spare parts of the system to guarantee its correct operation for a longer period than 10 years, this must be clearly specified in the user's manual.

Test Conditions

The following are considered as test conditions:

- **Storage Capacity:** The test procedure is applicable only to systems with a storage capacity of 0.6 m^3 or less.
- **Irradiance:** Test results must be obtained for at least 4 days with daily radiation values distributed uniformly between 8 MJ/m^2 and 25 MJ/m^2.
- **Temperature Difference:** During these 4 days, the variation of ($T_{a(day)} - T_{inlet\ water}$) should be approximately the same. The range of values must be between -5° C y +20° C. Similarly, results of 2 more days can be obtained but with a temperature difference of 9° C higher or lower than the previous related value.
- **Surrounding Air Speed:** During the system's exposure to solar radiation, the air speed upon the collector must not exceed a range between 6 m/s ± 2 m/s. It is recommended to use a fan to keep the speed at 4 m/s ± 1 m/s.

- **Waterflow Extraction from the Accumulator:** The volumetric flow of water extracted must be 0.6 m³/h. That is to say, that the volume kept in the accumulator can be fully extracted in one hour.

Test Period

Regarding test periods, the following operations must be carried out:

1. Preconditioning
   a. The fluid must be CIRCULATED until the system interior temperature, T_s, is uniform. For that reason the following must be maintained:
      i. The fluid temperature, T_c, constant.
      ii. The collector aperture protected from solar radiation (through a reflective solar cover).
2. Exposure
   a. The cover must be REMOVED.
   b. The solar radiation system must be EXPOSED from 6:00 until 18:00 (solar hour), leaving it OPERATING normally.
   c. During this period, the solar radiation data H must be OBTAINED.
3. Extraction
   a. 3 times the volume of water must be EXTRACTED from the deposit from 18:00 onwards. As previously mentioned, the extraction water volume must be 10 l/ min.
   b. During the extraction, the deposit must be REFILLED with cold water at a temperature, T_c, (the same at the starting of the test) and the water flow must be MONITORED coming out in relation to the extracted volume, $T_{out}(V)$.
4. Night Losses
   a. The losses coefficient[8] of system (U_s) must be MEASURED during the night.

Calculations and Test Results

Regarding calculations and test results (Chaves Repiso, 1999):

Extraction Profile Temperature

The temperature of output water in relation to the extracted volume, $T_{out}(V)$. This can be represented in an adimensional form normalized with this function:

$$f(V) = \frac{T_{out}(V) - T_c}{T_{out}(V=0) - T_c}$$

In regards to extreme determined cases, its graphic representation is shown in Figure 10.

Available Energy in the Accumulator

This will be the area below the curve defined by the extraction temperature profile.
Extracted energy in a differential volumen dV element:

$$dQ = Pc_P \left[T_{out}(V) - T_c \right] dV$$

Available energy relative to tc, in the case that the deposit has a homogeneous temperature before the extraction is produced:

$$Q_0 = V_{sp} c_P \left[T_{out}(V=0) - T_C \right]$$

Energy Fraction

The energy fraction that remains in the accumulator in relation to the extracted volume is:

$$\gamma(V/V_s) = 1 - \int_0^V \frac{dQ}{Q_0}$$

An Overview to Thermal Solar Systems for Low Temperature

where the integral is defined between 0 and V. These are the 2 extreme situations in which the real systems will be:

Perfect mix:

$$\gamma(V/V_S) = \exp(-V/V_S)$$

Perfect stratification:

$$\gamma(V/V_s) = \begin{cases} 1 - \int_0^V \frac{dQ}{Q_0} \\ 0 \end{cases}$$

$$0 < V < V_S$$
$$0 > V > V_S$$

Energy fraction varies considerably in a real installation for distinct days; therefore, a systematic behavior has not been found which expresses such variations (Ullmer, 1995). However, the function y (V/Vs) influences the installation performance prediction very little.

Energy Supplied by the System

The operation of the solar energy system can be represented by the following equation:

$$Q = \alpha_0 + \alpha_H H + \alpha_T \left(T_{a(day)} - T_{inlet\ water}\right)$$

where:

- H is the daily irradiance.
- $(T_{a(day)} - T_{inlet\ water})$ is the diffference between ambient temperature and the inlet water temperature,
- $\alpha_0 + \alpha_H H + \alpha_T$ are the system coefficient resulting from the minimum square method.

The results obtained from the multivariable lineal correlation can be graphically represented by the values of (Figure 11):

$$\left(T_{a(day)} - T_{inlet\ water}\right) = -10K, 0K, 10K \text{ and } 20K$$

Maximum Increase of Water Temperature

Similarly, this expression is used:

$$T_{d(max)} - T_{inlet\ water} = b_1 H + b_2 \left(T_{a(day)} - T_{inlet\ water}\right) + b_3$$

where:

- b_1, b_2, b_3 are obtained from the previous system coefficients,
- $T_{d(max)}$ is the maximum water temperature before the extraction.

The obtained expressions for the system's supplied energy and the maximum water temperature are 2 functions of daily irradiance and the difference between temperatures of the ambient and the water input, that are used to estimate the system's behavior with the surrounding conditions. In the same manner as before, we can graphically represent the results from the multivariable lineal correlation obtained for the same values of $(T_{a(day)} - T_{inlet\ water})$.

1. $T_{a(day)} - T_{inlet\ water}$ = 20 K.
2. $T_{a(day)} - T_{inlet\ water}$ = 10 K.
3. $T_{a(day)} - T_{inlet\ water}$ = 0 K.
4. $T_{a(day)} - T_{inlet\ water}$ = -10 K Results from the tests.

Other Calculations

In respect to the extraction temperature profile, at least two normalized profiles must be obtained for two different conditions of daily

Figure 10. Representation of extreme cases: 1. ideal system, perfect stratification in accumulation; 2. system with a small amount of mixture in the accumulator; 3. system with the perfect mixture; 4. extraction system through a heat exchanger

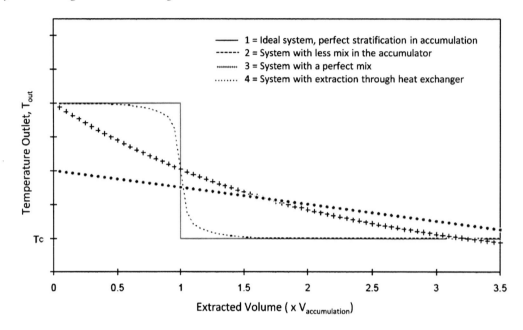

Figure 11. Results from the multivariable lineal correlation

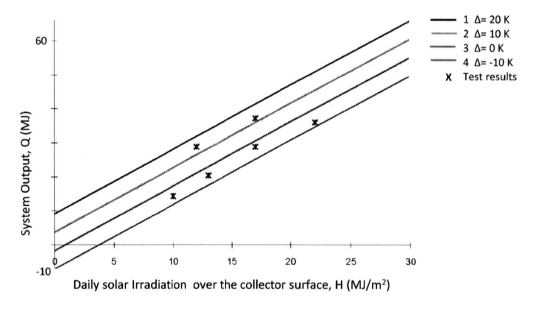

irradiation, the first between 8 MJ/m² and 16 MJ/m², and the second between 16 MJ/m² and 25 MJ/m². Additionally, it is recommended to obtain two more profiles of extraction temperature in case that a midday solar extraction is produced, both for an irradiation between 16MJ/m² and 25 MJ/m². The first corresponds to the extraction of half the accumulation volume at midday and the second for an extraction volume equal to 1.5 times the accumulation volume.

The calculated extraction temperature profile for any irradiation conditions can be obtained from the normalized extraction temperature profile. Mean ambient temperature during the day and input water temperature. The test must also provide the extraction temperature profile with normalized mixture, g(V); such a test consists only of filling the accumulation deposit, at a constant temperature higher than 60 °C and proceeding to a volume extraction equal to three times the deposit volume. Finally, a night test will be carried out in order to obtain the coefficient of thermal losses of the system during the night. It could also be significant to obtain the coefficient of thermal losses in the accumulator independently; this will provide a qualitative measurement of the reverse flow that can be produced in the system and the losses caused by long-wave radiation that are produced in the collector (DIN 4757).

Prediction of Yearly Performance Indicators

Uniform reference conditions for performance prediction (included at the end of this section) can be used to obtain the following performance indicators based on the results from solar-only systems performance:

Heat Produced by the Heating System QL

It corresponds to the value Q, previously calculated from the test, and it is represented through the specified equation:

$$Q = \alpha_0 + \alpha_H H + \alpha_T \left(T_{a(day)} - T_{inlet\ water} \right)$$

Solar Fraction Fsol

The calculation involves an energy balance of the system using data and reference conditions (which will be provided further on in the text) on an annual basis that also includes energy quantities:

- Q_D, heat demand,
- Q_L, heat supplied by the solar heating system (load),
- Q_{par}, parasitic energy (electricity) by pump and controls.

Such an energy balance refers to the following scheme (Figure 12):

1. Collector.
2. Pump.
3. Accumulator.
4. Control unit.
5. Cold water.

Reference conditions for the calculation of the load Q_L are the power point from the deposit and the power points of load side in the heat exchanger, if it exists. The reference temperature to calculate loads is the cold water temperature. Thermal losses in the circulation line are not included in the loads. The solar fraction is defined as the supplied energy by the solar part of the energy system divided by the system total load (that is to say, heat demand):

$$f_{sol} = \frac{Q_L}{Q_D}$$

Parasitic Energy Qpar

The norm EN 12976-2 states that in case of parasitic energy, pumps annual electricity consumption, control system and system electrical valves will be obtained from the same conditions than the ones specified for thermal performance, assuming a pump operation time of the circulation pump of 2000 h, that means:

Parasitic energy = Pump power or 2000 h

In another respect, the reference conditions for performance prediction, related to the system will then be:

- **Orientation:** The collector must be orientated to the south[9].
- **Inclination Angle:** For the collector test, the collector must be inclined (45 ± 5)°, or according to the value specified by the manufacturer.
- **Total Length of the Circuit:** In case that all pipes are not supplied by the system or specified by the manufacturer, the total length of collector circuit must be 20 m.
- **Location of the Pipes in the Circuit:** They must be located as far as possible from the testing bank:
  ◦ If the system has the accumulator in the interior, the pipes must also be in the interior.
  ◦ If the system has the accumulator is in the external part of the system, pipes must also be on the exterior.
- **Ambient Temperature of the Accumulation:** The temperature will be 15 °C. In case that the accumulator is located on the exterior, weather-related data should be used.
- **Hydraulic Circuit Material:** The material supplied by the manufacturer for the prefabricated system must be copper, unless otherwise specified in the installation instruction manual. The reason that copper is used is because it is the most commonly used material in hydraulic domestic installations; therefore, if different materials are mixed with the hydraulic circuit, galvanic corrosion can be produced, including the subsequent damage and deterioration from said corrosion. That is why, it is also recommendable to use copper in the hydraulic circuit during the testing installation.
- **Pipes Diameter and Thickness:** If the manufacturer supplies, together with the rest of the system, the pipes and the isolator or that their diameter and thickness values are clearly specified, we must then use this material or the values that the manufacturer supplies. Otherwise, the values to be used are the ones that follow (Table 2, Table 3).

Regarding weather conditions, the Reference[10] locations will be: Stockholm, Würzburg, Davos and Athens, even though a different location can be selected in the report. In terms of climate data:

- **For Stockholm:** CEC and Test Reference Year.
- **For Würzburg, Davos, and Athens:** Test Reference Year.

In relation to the conditions for heat loads, the daily water profile must be 100% approximately 6 h after the solar midday. For the test, daily loads must be specified in the test procedure. The desired temperature of the mixing valve will be 45 °C.

If daily or annual loads are calculated in terms of energy, this energy must be calculated using the temperature of the cold water supply and the desired temperature. In another respect, the supply temperature of cold water for the test must be specified in the procedure, the cold water temperature must be calculated in relation to:

An Overview to Thermal Solar Systems for Low Temperature

Figure 12. Energy balance for solar-only energy systems

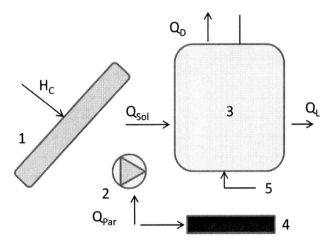

$$T_{cw} = T_{average} + T_{amplitude} \cdot sen[2\pi(Day - D_S)/365]$$

where:

- T_{cw} is the cold water temperature to be used for performance representation.
- $T_{average}$ is the average annual temperature in the reference point.
- $T_{amplitude}$ is the average range of seasonal variations in the reference point.
- Day is the day number of the year.
- D_S is the adjustment term.

Values of $T_{average}$, $T_{amplitude}$ and D_S that must be used are detailed in Table 4.

The daily load volume must be selected between one of the following series:

50, 80, 110, 140, 170, 200, 250, 300, 400, 600... Liters/day.

This series can be extended multiplying by $\sqrt{2}$, and rounding the multiple closer to 10. The manufacturer must provide a system design load, so that the closest value to the previous series is used, as well as the next highest and lowest values (values within the range of 0.5 times or 1.5 times the design load are recommended).

In the test those load volume values must be specified in the test procedures. Finally, in relation to the extraction water flow of the accumulator must be 10 liters/minute. If the maximum water flow extraction is less than 10 liters/minute, the water flow extraction must be used.

Other Method of Performance Categorization

The following is a test method that is briefly described to categorize the thermal performance (Chaves Repiso, 1999), (González-Prida, 2002). The European Norm makes no reference to this method. However, it is a significant comment due to the fact that it has been applied on some occasions (above all occasions previous to the European Norm currently in effect) in the test for prefabricated systems. In any case, this method is practically the same as that in the previously developed European norm EN 12976-2. In this section therefore, those existing differences will be only analyzed between this method and the European Norm. The difference according to the field of application consist in that, whereas the European Norm is considered to be only applicable to prefabricated systems that have an accumulation volume equal or less than 600

Table 2. Diameter and thickness of forced circulation systems

Waterflow/Level in the Hydraulic Circuit (l/h)	Exterior Diameter of Pipes (mm)	Thickness of Pipes (mm)	Thickness of the Insulator (mm)
Less than 90	10 ± 1		20 ± 1
Between 90 and 140	12 ± 1		20 ± 1
Between 140 and 235	15 ± 1		20 ± 1
Between 235 and 405	18 ± 1		20 ± 1
Between 405 and 565	22 ± 1		20 ± 1
Between 565 and 880	28 ± 1	1,5	30 ± 1
Between 880 and 1445	35 ± 1	1,5	30 ± 1
Between 1445 and 1500	42 ± 1	1,5	39 ± 1
More than 1500	So that fluid speed is 0,5 m/s.	1,5	As the interior diameter of the pipe.

Table 3. Diameter and thickness of thermosyphon circulation systems

Total Opening Area of the Collector (m^2)	Exterior Diameter of Pipes (mm)	Thickness of Pipes (mm)	Thickness of the Insulator (mm)
Between 1 and 2	15 ± 1	1	20 ± 2
Between 2 and 6	18 ± 1	1,5	30 ± 2
Between 6 and 10	22 ± 1	1,5	39 ± 2

I, the pre-normative method was considered applicable to domestic solar energy systems that complied with:

- The useful surface of the collector is less than 15 square meters.
- The solar accumulation volume is less than 1500 liters.

Depending on the accumulation volume of the prefabricated system, the pre-normative establishes several extraction water flow values, in accordance with Table 5.

The European Norm establishes that a water flow extraction of 600 l/h for all solar energy systems can be tested by this norm that must have therefore an accumulation volume less than 600 l. Thus, the pre-normative method widens the application range to larger systems. In relation to the differences in testing conditions, it is noted that in contrast to the six days of valid tests required by the European Norm, the pre-normative method considered a larger number of valid tests although under the same general conditions. In short, this demanded the minimum test conditions of twelve valid tests that would comply with the condition of $(T_{a(day)} - T_{inlet\ water})\ 0 \pm 2$ C. In such twelve tests,

Table 4. Average values according to reference locations

Reference Location	$T_{average}$ (°C)	$T_{amplitude}$ (°C)	(Days)
Stockholm (Sweden)	8,5	6,4	137
Würzburg (Germ.)	12,0	3,0	137
Davos (Switzerland)	5,4	0,8	137
Athens (Greece)	17,8	7,4	137

Table 5. Water-flow extraction in relation to the accumulation volume

Accumulation Volume (Litres)	Water Flow Extraction (Litres/Hour)
0 – 500	600
500 – 800	900
800 – 1500	1200

valid results must be obtained for a determined number of days with the following global radiation values (Table 6).

An additional minimum of twelve days of test must be carried out that must comply with the formula for each valid test day: $(T_{a(day)} - T_{inlet\,water})$ between -15 C and +25 C. Moreover, from these additional tests valid results must be obtained for each of the conditions of $(T_{a(day)} - T_{inlet\,water})$ with the global radiation values shown in Table 7.

In short, there is a considerable increase of the minimum number of valid tests in the pre-normative method in respect to the European Norm, therefore, the results will be even more representative of the solar energy system's behavior. However, this also represents a considerable increase in both the test period of a solar energy system and the related test costs. The testing procedure is exactly the same in both cases except for the consideration taken from the water-flows during the extraction period. There is another slight difference in the preconditioning and in the thermal losses of the accumulator test. Whereas in the European Norm the system considered has reached steady conditions, where the difference between the outlet water temperature of the solar energy system and the inlet water is less than 1K during a period of 15 minutes, in the pre-normative test, a period of only 5 minutes is required. In other respects, for separable deposit systems, the European Norm demands that the test is carried out using 15 m of pipe length between the deposit and the collector, in comparison to the 10 m length required by the pre-normative method.

There is a reduced pre-normative method which is basically the same as the one previously mentioned, except for that fact that it is done under specified conditions. Such conditions would be:

- Daily irradiance > 20 MJ/m^2.
- Final extraction of the day will be done for values[11] of:
  - $T_{a(day)} - T_{inlet\,water}$ 10 C°
  - Water-flow = 9 l/min.

With the valid day values and using the minimum square method for an independent variable, we can then obtain a straight line that corresponds to the heat produced by the system:

Table 6. Global radiation values during test days

Number of Days	6	3	3
Daily Global Radiation (MJ/m2)	between 15 and 20	More than 20	Less than 15

Table 7. Water temperature in relation to global radiation values

$(T_{a(day)} - T_{inlet\,water})$ (°C)	+ 20 ± 5	+ 10 ± 5	- 10 ± 5
Daily Global Radiation (MJ/m2)	Less than 15	between 15 and 20	More than 20

$$Q_u = aH + b$$

Regarding the system extraction profile obtained under the previous conditions, and resulting in a normalized curve f(V) so that:

$$\int_0^\infty f(V)dV = 1$$

It is not necessary to know the analytical function expression f(V), since it is only necessary to know the integrated values on which the measures are based in the extraction procedure. This can be determined in the following manner:

- Energy Q_i contained in a small volume of extracted water ΔV_i, being tdi the mean temperature of such water volume ΔV_i:

$$Q_i = P_a C_p (T_{di} - T_f) \Delta V_i$$

- Qi is calculated for each volume ΔV_i and the total energy will be given by the sum of all the Qi:

$$Q = \Sigma Q_i$$

- Values of f associated to each ΔV_i are calculated therefore by:

$$f_i = \frac{Q_i}{Q}$$

In the same way, the mean solar performance can be obtained as the mean of daily performances, defined by the expression:

$$\eta = \frac{Q_u}{H}$$

The system night losses coefficient, U_s, are calculated by a separate test. In this test, the deposit is filled at a constant temperature (50 and 70 °C in two different tests), and it is left in progress for a determined period of time. At the end of this period of time, that should correspond to the night time, the heat contained in the deposit Q_f is measured, obtaining the coefficient of thermal losses with the following expression:

$$U_s = \frac{Q_0 - Q_f}{T_{in} - T_{an}} \cdot \frac{1}{\Delta T}$$

where:

- Q_0 is the energy contained in the deposit at the start of the test.
- Q_f is the energy contained in the deposit at the end of the test.
- T_{in} is the water temperature at the start of the test.
- T_{an} is the average ambient temperature during the test period.
- ΔT is the duration of the test time period.

CONCLUSION

This chapter analyzes the requirements that a prefabricated system must comply with, in regards to components, work flow and the security system, as well as the requirements demanded to specified protection devices (1st part, UNE-EN 12976). With that goal, throughout this document we have defined the basic concepts related to solar energy utilization and regarding to product quality. In addition to this, the different components that constitute thermal solar systems for low temperature have been distinguished. After a short development referred to the First Law of Thermodynamics, it has been implemented to prefabricated solar energy systems for water heating.

In another respect, distinct testing projects are presented to validate the previous mentioned requirements and the thermodynamic characterization of the prefabricated system, in accordance with the currently effective normative (2nd part, UNE-EN 12976). Test procedures are depicted for the validation of prefabricated systems, resulting finally a categorization of the thermal performance. At the end, this research presents a test method which has been developed to provide a reliable answer, but at the same time, fast as possible and at a minimum cost.

REFERENCES

Chaves Repiso, V. M. (1999). Instalaciones de energía solar térmica con acumulación distribuida. Proyecto Fin de Carrera, University of Seville.

Duffie, J. A., & Beckman, W. A. (1980). Solar engineering of thermal processes (Vol. 3). New York: Wiley.

Fisch, N. (1995). *Manuskript zur Vorlesung Solartechnik I*. University of Stuttgart.

González-Prida, V. (2002). Influencia de la normativa europea en el procedimiento de ensayos de sistemas solares térmicos prefabricados: Propuesta de adaptación de la instalación de ensayos del Instituto Andaluz de Energías Renovables. Proyecto Fin de Carrera, University of Seville.

González-Prida, V., & Crespo, A. (2012). A reference framework for the warranty management in industrial assets. *Computers in Industry, 63*, 960–971.

González-Prida, V., Crespo, A., Pérès, F., De Minicis, M., & Tronci, M. (2012). *Logistic support for the improvement of the warranty management*. Advances in Safety, Reliability and Risk Management. Taylor & Francis Group.

Gruber, E. (1984). *Energieeinsparung und Solarenergienutzung in Eigenheimen: Forschungberich T84-287 des Bundesministeriums für Forschung und Technologie*. German Ministry of Research and Technology.

Smith, D. (2001). *Reliability, maintainability, and risk: practical methods for engineers*. Newnes.

Ullmer, E. (1995). *Theoretische Validierung einer Prüfmethode für solare Brauchwasserwärmungsanlage*. University of Stuttgart.

UNE-EN 1151. (1999). *Bombas. Bombas rotodinámicas. Bombas de circulación cuyo consumo de energía no excede de 200 W, destinadas a la calefacción central y a la distribución de agua cliente sanitaria doméstica. Requisitos, ensayos, marcado*. International Standard.

UNE-EN 12976-1. (2006). *Sistemas solares térmicos y sus componentes. Sistemas prefabricados. Parte 1: Requisitos generales*. International Standard.

UNE-EN 12976-2. (2006). *Sistemas solares térmicos y sus componentes. Sistemas prefabricados. Parte 2: Métodos de ensayo*. International Standard.

UNE-EN 1489. (2001). *Válvulas para la edificación. Válvulas de seguridad. Ensayos y requisitos*. International Standard.

UNE-EN 1490. (2001). *Válvulas para la edificación. Válvulas de alivio de presión y temperatura. Ensayos y requisitos*. International Standard.

UNE-EN 1717. (2001). *Protección contra la contaminación de agua potable en las instalaciones de agua y requisitos generales de los dispositivos para evitar la contaminación por reflujo*. International Standard.

UNE-EN 60335-1/A15. (2001). *Seguridad de los aparatos electrodomésticos y análogos. Parte 1: Requisitos generales*. International Standard.

UNE-EN 60335-2-21/A1. (2001). *Seguridad de los aparatos electrodomésticos y análogos. Parte 2-21: Requisitos particulares para los termos eléctricos*. International Standard.

UNE-EN 806-1. (2001). *Especificaciones para instalaciones de conducción de agua destinada al consumo humano en el interior de edificios. Parte 1: Generalidades*. International Standard.

UNE-EN 809. (1999). *Bombas y grupos motobombas para líquidos. Requisitos comunes de seguridad*. International Standard.

UNE-EN ISO 9000. (2005). *Sistemas de gestión de la calidad. Fundamentos y vocabulario*. International Standard.

UNE-EN ISO 9488. (2001). *Energía solar. Vocabulario*. International Standard.

UNE-ENV 1991-2-3. (1998). *Eurocódigo 1: Bases de proyecto y acciones en estructuras. Parte 2-3: Acciones en estructuras. Cargas de nieve*. International Standard.

UNE-ENV 1991-2-4. (1998). *Eurocódigo 1: Bases de proyecto y acciones en estructuras. Parte 2-3: Acciones en estructuras. Acciones del viento*. International Standard.

ADDITIONAL READING

American Institute of Chemical Engineers Center for Chemical Process Safety. (1989). *Guidelines for process equipment reliability data, with data tables*.

Barberá, L., González-Prida, V., Moreu, P., & Crespo, A. (2010). "Revisión de herramientas software para el análisis de la fiabilidad, disponibilidad, mantenibilidad y seguridad (RAMS) de equipos industriales". Revista Ingeniería y Gestión de Mantenimiento. Vol. Abril/Mayo/Junio 2010, Spain.

Crespo, A., & Iung, B. (2007). A estructured approach for the assessment of system availability and reliability using Montecarlo simulation. *Journal of Quality in Maintenance Engineering*, *13*(2), 125–136. doi:10.1108/13552510710753032

DIN 4757, Teil 4. (1995). "Sonnenkollektoren Bestimmung von Wirkungsgrad, Wärmekapazitat und Druckabfall". German Standards.

González-Prida, V., Barberá, L., & Crespo, A. (2010). "Practical application of a RAMS analysis for the improvement of the warranty management". *1st IFAC workshop on Advanced Maintenance Engineering Services and Technology*, Lisbon, Portugal.

González-Prida, V., & Crespo, A. (2010). Book review: Reliability engineering. Warranty management and product manufacture (By D.N.P. Murthy and W.R. Blischke). Production Planning & Control. *The Management of Operations*, *21*(7), 720–721.

Gonzalez-Prida, V., Parra, C., Gómez, J., Crespo, A., & Moreu, P. (2009). Availability and reliability assessment of industrial complex systems: a practical view applied on a bioethanol plant simulation. In *Safety, Reliability and Risk Analysis: Theory, Methods and Applications* (pp. 687–695). Taylor & Francis.

Henley, E., & Kumamoto, H. (1992). *Probabilistic risk assessment, reliability engineering, design and analysis*. IEEE Press.

Høyland, A., & Rausand, M. (1994). "System reliability theory, models and statistical methods". J. Wiley. University of Michigan. ISBN 0-471-59397-4.

ISO 9459-2. (2013). "Solar heating. Domestic water heating systems. Part 2: Outdoor tests methods for system performance characterization and yearly performance prediction of solar-only systems". International Standard.

ISO/DIS 11924 (1995). "Solar heating. Domestic water heating systems. Testing methods for reliability and safety assessment". International Standard. Alonso Aguilar, P. (1997). "Instalación integral de ensayos para la evaluación energética de sisitemas solares térmicos de baja temperatura". Proyecto Fin de Carrera, University of Seville.

ISO/DIS 14224. (2004). "Petroleum, petrochemical and natural gas industries - Collection and exchange of reliability and maintenance data for equipment". International Standard.

Moreu, P., González-Prida, V., Barberá, L., & Crespo, A. (2012). "A practical method for the maintainability assessment using maintenance indicators and specific attributes". Reliability Engineering and System Safety. 100; 84-92. Elsevier. ISSN: 0951-8320.

Parra, C., Crespo, A., Cortés, P., & Fygueroa, S. (2006). On the consideration of reliability in the life cycle cost analysis (LCCA). In *A Review of Basic Models. Safety and Reliability for Managing Risk* (pp. 2203–2214). Taylor & Francis.

SINTEF Technology and Society. (2002). *Offshore Reliability Data Handbook* (4th ed.). OREDA.

UNE-EN 307. (1999). "Intercambiadores de calor. Directrices para elaborar las instrucciones de instalación. Funcionamiento y mantenimiento, necesarias para mantener el rendimiento de cada uno de los tipos de intercambiadores de calor". International Standard.

Vintr, M. (2007). Reliability Assessment for Components of Complex Mechanisms and Machines. 12th IFToMM World Congress. Besançon, France.

KEY TERMS AND DEFINITIONS

Absorbance (α): Also called absorption factor. Is the relation between the energy flow, absorbed by an element of a surface, and the incidence radiation. Absorbance can be applied to a single wavelength or a range of wavelengths.

Incidence Angle (δ): The angle that is formed by the line that joins the core of the solar disc with the surface point that is exposed to the Sun and the norm to the related surface at that point.

Pyranometer: A radiometer designed to measure the solar irradiance on a receiver flat surface.

Pyrheliometer: The radiometer which uses a collimator to measure the direct solar irradiance under normal incidence.

Radiometer: The instrument used to measure radiation. Depending on its design, its output can be given in units of irradiance or radiation.

Reflectance (θ): Also called, reflection factor. It is defined as the relation between the energy flow reflected by a surface and the incident radiation.

Solar Azimuth (Υs): The angle that is formed with the South[12] and the horizontal projection of the straight line that connects the position of the sun to the observation point, measured clockwise using the projections on the horizontal surface of the observation point.

Surrounding Air Speed: Air speed measured at a specified point located in the vicinity of a collector or a solar installation.

Transmittance (τ): Also called transmission factor. It is defined as the relation between the energetic flow that goes through a body and the incident radiation.

World Radiometric Reference (WRR): Standard measurement[13] that defines the total radation unit, in the International System, with an incertanty of $\pm 0,3\%$.

ENDNOTES

[1] That is the received within the extraterrestrial atmospheric limits.
[2] Emitted, transferred or received power in the form of radiation.
[3] Unless the configuration is only modified to a less favorable situation from an energy point of view. In that case it would not be necessary to do a new test.
[4] Mainly antifreeze property losses.
[5] Considerations taken from the norm ISO 9459-2.
[6] In the northern Hemisphere.
[7] Considerations taken from the norm ISO 9459-2.
[8] This is determined in a distinct test, measuring temperature decrease in the accumulator during the night when the deposit is filled with hot water.
[9] In the northern Hemisphere.
[10] Take into account that this chapter is based on an European Norm, so the reference location are European cities which try to represent Nordic, continental, mountain and Mediterranean climates.
[11] $(T_{a(day)} - T_{inlet\ water})$ is the difference between the ambient temperature and the water going into the system.
[12] From the point of view of the northern Hemisphere.
[13] Adopted by the World Metereological Organization (WMO), and effective since July 1st of 1980.

Chapter 5
A Reliability Test Installation for Water Heating Solar Systems:
Requirements and Design According to the European Norm 12976

Vicente González-Prida
University of Seville, Spain

Anthony Raman
NTEC Tertiary Group, New Zealand

ABSTRACT

This chapter analyzes the requirements that a test installation design must comply with in order to carry out the test procedures of prefabricated systems mentioned in a previous chapter and based on the norm EN 12976. In other words, the authors consider for the test installation design, firstly, the requirements that the hydraulic circuit of such an installation must meet, followed by the specifications required by the custom-built systems; all of this has the aim of certifying that the characteristics of the prefabricated system are the applicable ones. Subsequently, the chapter then directly proceeds to design a test installation, which is to be later compared to a real installation.

INTRODUCTION

This chapter continues with the statements provided in the previous chapter "An overview to thermal solar systems for low temperature: Outlining the European Norm 12976". In order to propose a reliability test installation for solar systems, it is necessary to know the requirements for such test installations (González-Prida, 2002). These requirements come from the international norm ISO 9459-2, which makes reference to the European Norm EN 12976.

DOI: 10.4018/978-1-4666-8222-1.ch005

Hydraulic Circuit

Pipes

Pipes must be constituted of a material which allows the use of water as the working fluid, as well as to be able to withstand maximum temperatures limited to 95 °C. Considering that most of the solar energy installations utilize pipes made of copper to prevent galvanic corrosion, all the test installation circuits must be made out of this material. In order to minimize the possible effects of the environment on the temperature of the water that goes into the prefabricated system, the longitude of the related pipes will be the minimum permitted; trying that the distance between the outlet temperature regulator[1] (upstream to the prefabricated system) and the inlet of such system is the allowable minimum possible[2]. A drainage pipe must be installed in a fashion that is the bifurcation of the inlet pipes into the prefabricated system, just before such entrance.

Insulation

The circuit must be insulated with an outdoor reflective coating, in this manner, thermal losses will be less than 0,2 W/K. Similarly, the pipes that connect the temperature sensors, in and out of the system, must also be protected with a reflective coating so that the loss or gain of temperature in each section does not exceed the 0,01 K of the test conditions.

Working Fluid

Water is used as the working fluid for the test installation. The quality of the water extracted from the prefabricated system must not have any adverse effects and must be fit for human consumption. However, the water used in the tests must be able, under any circumstances, to be used for human consumption.

Mixing Devices

The mixing devices (for example orifices, elbow joint or mixers) have these purposes:

- That the meassure in the temperature sensors corresponds to that of a homogeneous fluid,
- That the temperature of the cold water coming in can be controlled in order to carry out tests under different conditions,
- That the temperature of the hot water going out can be limited.

In reference to the first case, it will be necessary to install mixing devices immediately upstream of the temperature sensors (Fisch 1995). For that, is recommended that the location of the temperature sensors and mixing devices (joints in this case) as they are shown in the following illustration (Figure 1).

In the second case, and as an alternative to a temperature regulator, the degree of fluid mix coming from a hot water tank and from a cold one, are both kept at a constant temperature and can be controlled, thus obtaining the desired temperature of the fluid as it enters the system.

Finally, the third case refers to systems in which the hot water temperature in the consumption points can exceed the 60 °C maximum and therefore it must be limited. To do this, the assembly instruction should mention the installation of an automatic mixing system or another system that limits the extraction temperature to 60 °C, either in the components of the solar energy system or in the hot sanitary water. However, this device must be able to withstand the maximum extraction temperature possible.

Figure 1. Location of the joints in relation to the temperature sensors

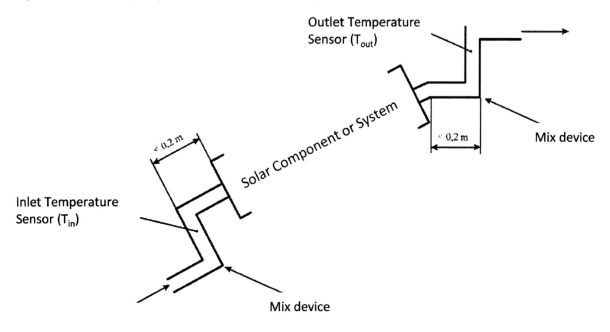

Requirements of Measurement Systems

Temperature

Ambient Temperature

The ambient temperature will be measured with a sensor equipped with a device that screens solar radiation and at the same time which lets the airflow circulate freely. Thus it is important to highlight that water entrance and damp in such sensors must be prevented. This related device must be installed far from the collector, at least at a distance of 1.5 m as well as from the rest of the system components, but never more than 10 m from the solar energy system. It must be located approximately at 1 m from the ground. The temperature of the surface around the solar power system must be as similar as possible to the ambient air temperature. That is the reason why they must not be installed near chimneys, refrigeration towers or hot air outlets. To avoid this last one, it is recommended to use cases for the ambient temperature sensors with forced ventilation during the collection of measurement values.

Temperature of Inlet Water

The temperature of the water supply into the prefabricated system that is being tested, inlet water, must be regulated in a way that it makes it possible to carry out tests procedures at different temperature conditions. To achieve this, it would be then be necessary to locate a temperature regulator element in the testing installation. Such a regulating device must be located in the testing installation circuit at the entrance of the accumulation deposit of the prefabricated system. It would be also necessary for the device to be of high potency due to the elevated water volume required for the thermal performance test. As an alternative, the temperature regulation can be obtained by controlling the fluid mix coming from a hot water tank and another from a cold water tank, both of them kept at a constant temperature. The temperature regulator (or mix valve in the case of using an alternative solution) must be able to control temperature entering to the solar energy system within a range of 0.2 K during the extraction period in the thermal performance test, when the water-flow is 600 l/h. There could be oscillations between ± 0.25 K in the inlet fluid temperature caused by the temperature regulator hysteresis. Such oscillations are permitted if the hysteresis is the only cause and also additional deviations are not produced.

Accuracy, Precision, and Response Time

Accuracy and precision of the temperature measurement instruments (including Reading devices), must be limited by the values shown in Table 1(Gruber, 1984). In all cases, the response time is less than 5 seconds.

Fluid Water-Flow (Volume)

Measurements of the fluid water-flow are required to have an accuracy more than, or equal to, ± 1% of the measured value of mass units by a time unit. To control and measure the water-flow of the used water in the testing installation, a water-flow regulator will be installed as well as a flowmeter at the cold water inlet to the prefabricated system. This way, measurements will not be influenced by temperature variations. During the thermal performance test in forced circulation systems, the flowmeter must be installed in the hydraulic system so that water-flow can be measured with a precision of ± 5%. In these forced circulation systems, the fluid water-flow will be the one recommended by the manufacturer.

Table 1. Accuracy and precision of temperature sensors

	Accuracy	Precision
Ambient air temperature (°C)	± 0.5	± 0.2
Inlet cold water temperature (°C)	± 0.1	± 0.1
Difference between inlet and outlet temperatures of the solar energy system (K)	± 0.1	± 0.1

Delay Time

It is important to highlight that the delay time must be measured with a precision of ± 0.20%.

Solar Radiation

Instruments used for irradiance, both global and diffuse, measurement are the pyranometers. These instruments must be of a first class degree (of the highest quality) according to the specification of the norm ISO 9060. This way, recommendations specified in the norm ISO/TR 9901 must be taken into account and carried out. Pyranometers will have to be calibrated by a standard pyrheliometer in accordance with the norm ISO 9846 or with another reference pyranometer according to the norm ISO 9847. In the event of important damage to the pyranometer, it must be then calibrated again in order to verify the stability calibration factor as well as the time constant. If a response variation occurs of more than ± 1% during a one-year period, the measuring equipment must be calibrated as frequently as possible; if such instability persists, the equipment must be replaced.

Wind Speed

Wind speed surrounding the solar energy system is to be measured using instruments and reading devices which can specify the mean speed of the air in each thermal performance test, with a precision of ± 0.5 m/s.

Data Collecting System

Data collecting devices (analog as well as digital) must have a precision of ± 0.5% or more than the total reading scale and at a time constant equal or less than 1 second. The signal peak must be located between 50% and 100% of the total scale. In case of using digital techniques or electronic integrators, they must have a precision of 1% or higher than the measured values (Barberá et al., 2010). The entrance impedance of each data collecting system must be 1000 times higher than the impedance of the sensor, or of 10 MΩ, depending on the highest one. The minor division scale of the instrument or measuring system must never be two times higher than the specified precision. That means, if the precision value is higher than ± 0.1 °C, the minor division scale must be more than ± 0.2 °C.

Other Requirements

In parallel to European regulations for prefabricated testing systems, we have mentioned a previous methodology that has been on occasions applied, whose thermal performance test method is based on the same guidelines set by the European normative (Gonzalez-Prida et al., 2009). This section therefore, presents an analysis of those existing differences between the pre-normative method and the European one in respect to testing installation and their requirements:

Maximum Water-Flow Capacity

The maximum water-flow capacity in the pre-normative testing method is of 1200 l/h in contrast to the 600 l/h of the European norm. However, in most of the cases, the water-flow capacity will be the same for both procedures. Therefore, the testing installation can be designed to fit a water-flow of 600 l/h, which is the capacity most generally used, even though the possibility of performing tests with a water-flow capacity of up to 1200 l/h must also be seriously considered.

Thermal Losses

The demands in relation to thermal losses between sensors and temperature regulators and systems to be tested are more demanding in the European Norm (for example the maximum variation between the inlet and outlet temperatures of the system and temperature sensors permitted by the European norm is of 0.01 K in contrast to 0.04 of the pre-normative method), thus this will be taken into account in the testing installation.

Precision and Tolerance

In respect to the precision and tolerance specifications of the measuring systems and additional considerations for testing installations, both methods require basically the same demands.

DESIGN PROPOSAL FOR THE TEST INSTALLATION

Preliminary Considerations

Testing installations for prefabricated systems in steady state operation are proposed based on the norm (within the field of low temperature thermal solar energy) EN 12976. In general, it would be of the upmost interest that the installation could carry out the following tests:

- Frost and overheating protection tests of a prefabricated system.
- Load loss tests of the hydraulic elements (pressure resistance).
- Water contamination study.
- Safety protection test from electrical discharges.
- Mechanical stress test of the supportive structure.
- Safety system and protection from reverse flow assessment.
- Thermal loss test in the prefabricated system components.
- Thermodynamic categorization test of a prefabricated system.

Design Conditions

Testing installation designs (Henley & Kumamoto 1992) are conditioned by the general requirements for the previously expressed types of test installations (American Institute of Chemical Engineers Center for Chemical Process Safety. (1989). These requirements would then have to be considered together with the detailed conditions shown in Figure 2.

Figure 2. Test installation scheme of the valve aging process

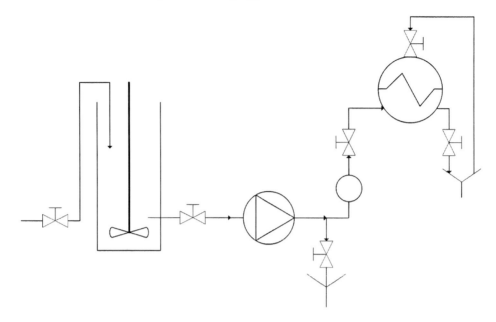

Necessary Test Elements

Frost Resistant

- Filling pump.
- Portable refractometer to calculate glycol concentration.
- A manometer to check the filling.
- A manometer to check the drainage.
- Freezing spray to check the drainage valve. Flowmeter to measure the drain.
- Air bubble level to calculate the inclination of the pipes.

Aging of the Valves

- Mixer deposit.
- Filling valve.
- Drainage valve.
- Flowmeter.
- Electrical heating deposit.
- Water stirring device. Circulation pump.

Electrical Safety

- Frame where the complete prefabricated system is installed.
- Instruments to measure distances.

Mechanical Stress

- Frame where the complete prefabricated system is installed.
- Rigid over-structure to apply the testing loads in the support structure.
- One or two devices enabled to apply loads to the over-structure.
- One or two dynamometers.
- A recording device of the dynamometer.
- A weighting system.
- Force line.
- Tools to install and disassemble the test parts.
- Safety panels to separate the testing area from the control area.

Figure 3. Testing installation scheme for thermal performance

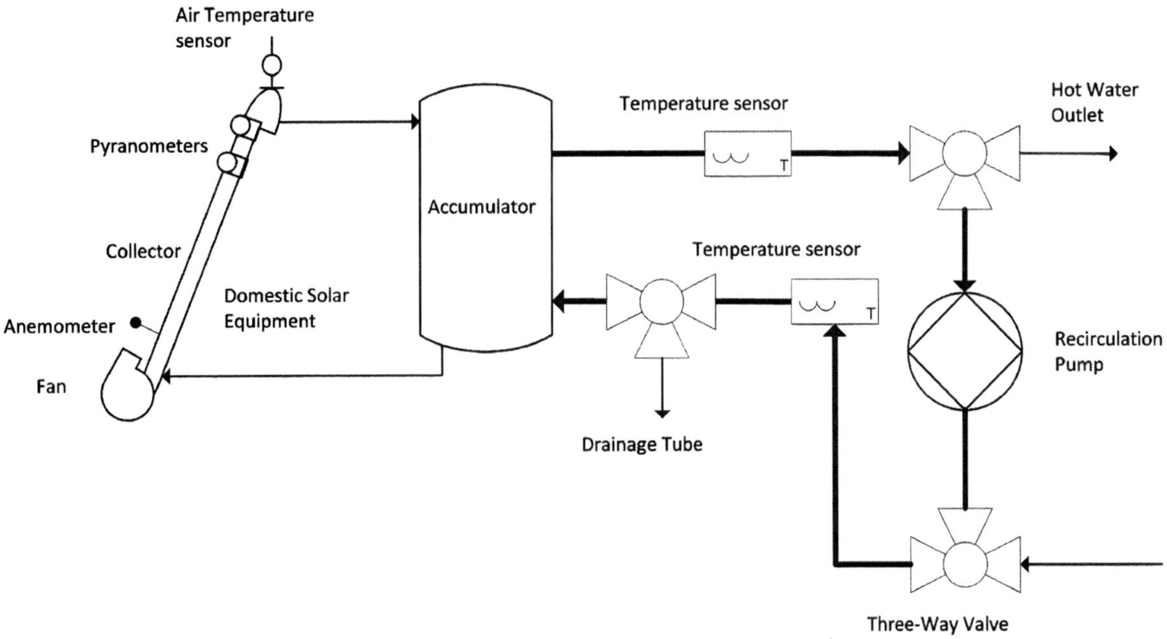

- Supply pipes and drainage installation.
- Instruments to measure angle, distances, etc.

A couple of schemes for two possible installations for mechanical stress tests can be directly taken from the EN 12976.

Thermal Performance
- Temperature sensor of system inlet and outlet fluid.
- Ambient temperature sensor.
- Temperature regulator.
- Three-way valves.
- Pyranometers.
- Recirculation pump.
- Waste pipe.
- Flowmeter of inlet fluid.
- Water-flow regulator.
- Anemometer.
- Vents.

The Figure 3 shows a testing installation scheme for thermal performance.

Given the importance of the thermodynamic categorization test, we can consider as a starting point for a test installation design, the basic scheme of the installation for such a test (González-Prida et al. 2010). Therefore, we have considered then the type of installation (together with their specifications) proposed by the European norm and the pre-normative method for the thermal performance test of domestic solar power systems, prefabricated in steady state operation (that means, prefabricated systems).

Measure Limitations

The installation must be able to obtain the desired characteristics of pressure, water-flow and inlet system fluid temperature within the most rigorous range limits specified by the norm. In another respect, the precise inlet and outlet fluid measurements of temperature and pressure are necessary as well as the ambient temperature, global and diffuse radiation on the desired surface and wind speed and direction. The limits and tolerance points of the measuring and controlling systems have been designed within the ranges of the previously mentioned norms, trying to achieve the most precise accuracy possible (Parra et al. 2006).

Maximum Water-Flow Capacity

The maximum water-flow[3] capacity required by the testing systems of which accumulation volume is between 800 and 1500 liters is 1200 l/h. However, most of prefabricated systems have an accumulation volume less than 500 l and therefore the maximum required water-flow for the test procedure would be of 600 l/h. This is the reason why, a water-flow of 600 l/h capacity is used, in accordance with the specification of the European norm.

Maximum and Minimum Temperature

- The maximum temperature that the hydraulic system can bear is also determined by the specifications of the testing normative, which is of 95°C.
- The minimum temperature is determined by the climate conditions of the location of the installation; in this case, we consider the city of Seville (Spain).
- The minimum ambient temperature for Seville is -3.2 °C; now, considering that the installation is set on an urban area and at the top of a roof of a building, we can then consider that the minimum ambient temperature for the installation will never be below the +3 °C. Consequently, risks of frost are discarded from the test procedure.

Maximum Pressure

The maximum pressure design of the installation will be 8 bar, that will be used in tests to analyze the solar energy system resistance in extreme temperature and pressure conditions. Nevertheless, safety systems will be tared at 6 bar to prevent over-pressure.

Maximum Heating Capacity

The testing installation is designed for a maximum heating capacity in the main circuit accumulation volume, using water as the working fluid, from 10 °C to 80 °C, during a 10 hour period. The possibility of installing a tank to reduce the heating time of the same volume to an hour is forecast.

Cooling Circuit

In the same way, a cooling system has been designed through a heat exchanger for the water-flow design, 600 l/h, from a water temperature of 85 °C hasta 15 °C.

Other Conditions

Valve System

The valve system of the hydraulic circuit of the testing installation is designed for:

- The manual activation of those valves that are to be operated only at the beginning of a test or to change any other element of the circuit,
- The automatic activation (with a possible manual activation) in such valves that are is necessary to be regulated during the test.

Deposit for Fluid Losses

The installation of a deposit has been presupposed in order to fill all those fluid losses that can be produced during the test procedure. This deposit also allows certain independence of operation in case of a failure in the water supply network. This way it permits the atmospheric pressure operating circuit test to function without the need of any other additional element, as well as the working fluid variation (if necessary) in a simple way. Even though other fluid could be used, the testing installation is prepared to basically function with water as the working fluid, since this is the most common among domestic solar energy systems.

Measuring and Control Systems

Measurement and regulation systems are chosen so that they comply with the strictest requirements of accuracy and tolerance (Vintr, 2007).

These requirements have been expressed previously in the chapter, in which also reference to the way of operation and location were made (UNE-EN 806-1, 2001).

Measurement Systems

Ambient Temperature Sensor

Their measurement is carried out by using thermopairs, resistance thermometers or thermisors. The ambient temperature sensor must be equipped with a screening device for solar radiation that permits the air flow freely. In particular, the ambient temperature sensor must be a tube of Pt 100 of tolerance class A or better, of four strings. The tube will have a 10 m cable length individually isolated and in a set of tephlon, interiorly shielded and twisted and stainless spring at the cable outlet. This tube will be installed in the interior of a standard meteorological case to protect it from solar radiation. The case must let the air flow in its interior and it consists of preferably a forced circulation device.

Fluid Temperature Sensor

The fluid temperature sensors must be installed at no more than 200 mm from the solar energy system inlet and outlet. If it is necessary to locate the temperature sensor to a distance of more than 0.2 m from the system or component to be measured, then it must be verified that the temperature measurement of the fluid is not affected. In this case, the temperature sensors are tubes Pt 100 of tolerance class 1/10 DIN B, four threads, with a case made of stainless steel of 3 mm of diameter by 100 mm of length without a nut. Each tube will have a 10 m length of isolated cable and in a set of Teflon, interiorly screened and twisted, with a connection racor at a process of 1/8'' of "male" gas and stainless springs at the cable outlets. In the same way, the pipes that connect temperature sensors in the inlet and outlet to the system must be isolated and protected using a reflective outdoor cover so that the gains or losses of temperature in each of the sections do not exceed 0.01 K in the testing conditions. Mixing devices are necessary, for example elbow joints, orifices or mixers, immediately upstream of the temperature sensors. The maximum calibration process must be of a year, this being done following a pattern that keeps traceability with other international patterns.

Flowmeters

The water-flow will be measured through any system that is within the demanded tolerances. The tolerance in this case must be of $\pm$ 6 l/h since the design water-flow is 600 l/h and the required accuracy in the measurements must be more than, or equal to, $\pm$ 1% of the water-flow measured value. The flowmeter must be installed in the cold water inlet in the case that the measurements or the system's own conservation could be affected by the high outlet temperatures or by possible sudden temperature variations. The fluid temperature must be known when it circulates through the flowmeter, with a tolerance of $\pm$ 2 °C. The calibration of the measuring device will be by comparison with another flowmeter internationally traceable or through a calibrated piston bank. There are different types of devices to measure liquid water-flow. The most common are:

- Turbines,
- Mass water-flow measuring device,
- Gravimetric procedures,
- Vortex water-flow measuring device,
- Positive displacement flow measuring device,
- Ultrasonic water-flow measuring device,
- Magnetic flow measuring device,

Generally, it is better to discard all those devices whose operation consists of a mechanical movement since they require shorter calibration periods. Also, we aim to reach the desired precision with a device that has an outlet signal that can be read or sensed by the data gathering system automatically, so we will have to resort to the last four alternatives. Table 2 expresses the general characteristics of these types of flowmeters.

Ultrasonic water-flow measurement devices as the Vortex present inferior technical characteristics than mass or magnetic ones; they do not meet all the requirements for the testing installation since their minimum water-flow capacities are generally larger. For fluids with a minimum electrical conductivity, such as water or any other heat-transfer fluid that is used in solar energy systems, magnetic flowmeters adjust best to the necessities of the testing installation. However, the mass water-flow devices, although technically they present better characteristics in contrast to the magnetic ones, their supply exceeds the necessities regarding testing of domestic solar energy systems. In the hydraulic circuit of the testing installation there will be a water-flow of 600 l/h, with an exterior pipe diameter of 22 mm, and 20 mm in the interior diameter. Under these conditions the best electromagnetic flowmeter will be the one with the nominal diameter of 15 mm, which is DN 15. Furthermore, the water measuring device has a precision of 0.25 to 0.5% of the real water-flow, which is within the required testing ranges.

Table 2. General flowmeter characteristics

	Magnetic	Mass	Ultrasonic	Vortex
Measure principle	Electromagnetic	Coriolis	Ultrasonic, transit time	Vortex
Measure of	Water-flow volume	Mass water-flow Total mass Density Temperature Water-flow volume Total volume Fraction percentage	Volume water-flow Mass water-flow Sound speed	Volume water-flow Mass water-flow Normal volume Water volume Water and vapor volume
Installation	Liquid filled tubes	Liquid filled tubes	Liquid filled tubes	Vapor, liquid, gas filled tubes
Means	Conductive means	Means able to be pumped	Sound transmitting means	Low viscosity, gas and vapor liquid
Main sectors	Chemical products, food, water treatment, electrical/heating	Food and drinks, chemical products, petrochemical electrical/heating	Chemicals, water treatments, electrical/heating	Food and drinks, chemical products, petrochemical electrical/heating
Applications	Control, regulation, registry, dossification, mix	Control, regulation, registry, dossification, weighting, supervision	Regulation, registry, supervision, Counting	Regulation, registry, supervision
Dimension	DN 6-2000	DN 3-50	DN 10-4000	DN 15-3000
Measure range	0,025-113.000 m^3/h	1-80.000 kg/h	0,2-500.000 m^3/h	Gas/vapor: 1,5-21.300 m^3/h Liquid: 0,25.2.300 m^3/h
Liquid temperature	-20°C a +200°C	-50°C to180°C	-200°C to +250°C	-40°C to +400°C
Maximum pressure	350 bar	500 bar	160 bar	100 bar

Pyranometers

Pyranometers must be of class 1 and must be mounted as it is later indicated. Instruments must be protected from the effects of condensation coming from the ambient humidity. It is important to clean the exterior dome and it must be verified that the there is no dust or humidity particles in the interior of the domes.

The measurement of hemispheric solar radiation must be carried out using a fisrt class pyranometer. The surface on which the sensor is located must coincide with the surface of the collector system with a tolerance of $\pm 1°$, the direction East-west is the most critical one. The sensor must be mounted on a metallic (or any other material) panel, protected from the outdoors, which cannot change its shape or texture due to humidity or temperature. It is recommendable to let the air flow through the base and the dome of the pyranometer.

The measurement must be carried out, preferably, by difference of the given indications from the pyranometer and the pyrheliometer. As an alternative solution, a first class pyranometer is used and it is equipped with a direct radiation collecting device. In this case, the shade band will be adjusted at least once a week, depending on the sensor geometry and of the shade band itself. The surface on which the sensor is located must coincide with the surface of the collector with a tolerance of $\pm 2°$ (Figure 4).

Anemometers

Anemometers are used to measure wind speed. These can be cupped anemometers or with wind-facing blades and always with a precision better than 0.5 m/s. It is recommended that the same element is also enabled to measure and record wind direction through a weather-vane installed on the same anemometer's axis.

The sensor must be located on a fixed surface (of which minimum dimension must be of 1m x 1m) on the same collecting surface. The blades or cups of the anemometer must be at a distance of 15 cm from the surface. The anemometer must be as close as possible to the collectors, at a distance of no more than 1 m. Its location can be either on the upper profile of the test system solar panels or approximately at the center of the collectors, out of the range of a possible turbulence effect caused by an adjacent obstacle. The maximum calibration period must be a year.

Pressure Measuring Devices

Pressure measuring devices must allow us to measure the circuit pressure within its own performance range, up to 8 kg/cm2. The performance temperature range must be the same as the one employed in the rest of the circuit; this is from 3 °C to 95 °C. In another respect, they must generate a digital output that can be read by the data collecting system. Therefore, for the pressure input measurement, a standard pressure transmitter will be used with a range of 0-10 bar which complies temperature and precision specifications, with a digital output of 4-20 mA.

Data Collecting System

Table 4 shows the minimum permitted sampling times used by the data collecting system in relation to the variables to be measured and the measurement period.

Table 3. Class 1 pyranometer technical specifications per WMO

	Tolerance	Units
Response (Resolution)	± 1.0	W/m^2
Stability	± 1.0	%
Thermal compensation	± 1.0	%
Espectral sensibility (0.3--3um)	± 20	%
Lineability	± 1.0	%
Response time 95%	30	s
Coseno response (0°-70°)	± 1.0	%

Figure 4. Measuring equipment on the prefabricated system

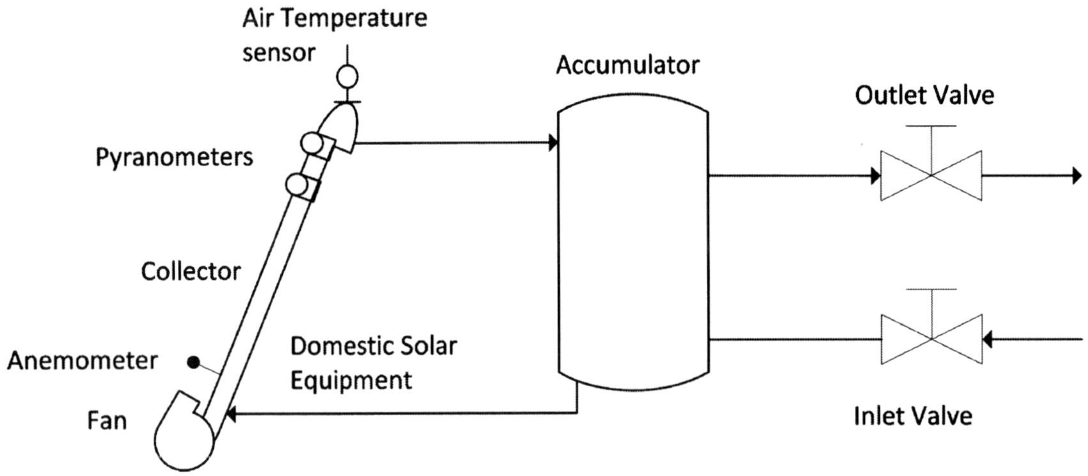

Regulation Systems

Regulation of the Inlet Fluid Temperature

In specific thermal loss tests, it is necessary to have a system which is able to heat the inlet water into the prefabricated system. The inlet temperature control mechanism going into the testing system will need then a certain capacity and accuracy due to the elevated water-flow used during the testing procedure of prefabricated systems. In order to regulate the inlet temperature, the mix level of the fluid can be controlled; this fluid comes from a hot water tank and from another that is cold, both are kept at a constant temperature, and therefore it is possible to obtain the desired temperature of the inlet fluid. The distance between the outlet temperature regulator, which is located at the water supply inlet to the circuit, and the solar energy system inlet, must be minimized in order to reduce environmental effects on the entrance water temperature. This part of the circuit must be isolated to ensure minor thermal losses of less than 0.2 W/K and, moreover, must be protected with an outdoor reflective cover. The procedures proposed for fluid heating are the ones that follow:

Electric Heater

Using electrical energy through a resistor for water heating is not recommended, this because, in a way it represents a degradation of high quality energy when there are other processes that are cheaper and equally effective. However, the electric heater

Table 4. Sample times

	Symbol	Maximum Testing Intervals	
		In Extractions	Without Extractions
Hemispheric irradiance (W/m^2)	H_t	5s	5s
Diffuse irradiance (W/m^2)	H_{dif}	5s	5s
Inlet temperature value (°C)	T_E	2s	15s
Outlet temperature value (°C)	T_S	2s	15s
Volumetric water-flow (m^3/s)	v'_i	2s	2s
Temperature	Ta	15s	15s

would be constituted by one or several electrical resistors located in the interior of a deposit at the entrance of the hydraulic system (UNE-EN 307, 1999).

These resistances must be controlled by an immersion thermostat located at the height of the water deposit outlet in the direction of the testing system. Supposing the deposit had a volume of 1500 liters of water, that is to be heated from 10 °C to 80 °C and with a heating period of 10 hours (that is to say the entire night), then, the necessary heating power would be:

$$W' = m\, C_p\, (T_{ac} - T_{af})\, /\, \text{Time} =$$

$$= [1500 \text{ kg} \cdot 4.19 \text{ (kj/kg K)} \cdot (80 - 10) \text{ K}]$$

$$/\, [10 \text{ h} \cdot 3600 \text{ s / h}] = 12 \text{ kW}$$

We can employ several resistors at the same time in order to reach this satisfactory electrical power. This will enable the system to gradually start the operation of each resistance according to the needs. The system is then supplied with four resistances of 3 kW each, and with independent control.

Gas Boiler

As an alternative, the use of a gas boiler is proposed. In this case, the boiler would be attached to the system inlet circuit; the nominal power of the gas boiler will depend on the heating time and the temperature ranges that are considered appropriate. If we consider one hour as the maximum heating time, for the water heating fluid, and the same range of temperature previously considered, 10 °C cold water temperature and 80 °C the hot water temperature, the power of the boiler must be more than 122.2 kW.

Additional Temperature Regulator

An additional temperature regulator is recommended in those testing installations that could have thermal losses in the hydraulic circuit that goes from the electric heater or gas boiler (mentioned before), to the testing system. The additional temperature increase will be carried out in this case through an electric heater connected in series with the mentioned element. The heater must be able to increase the fluid temperature up to 2 K for a water-flow of 600 l/h. This involves a nominal heating power of at least:

$$W' = Q \cdot p \cdot C_{pwater} \cdot T =$$

$$= 600\, (l_{water} / h) \cdot (1 \text{ h} / 3600 \text{ s}) \cdot (1 \text{ kg} / 1\, l_{water})$$

$$\cdot 4.18 \text{ (kJ/ kg K)} \cdot 2 \text{ K} = 1.4 \text{ kW}$$

To get within a security margin, a 2 kW heater can be chosen. If the distance in the hydraulic circuit which goes from the electric heater or the gas boiler is minimized, then there is no need for an additional temperature regulator. Nevertheless, this element would provide a more precise adjustment of the inlet fluid temperature.

Regulation of the Oulet Fluid Temperature

The manufacturer must limit the temperature for systems in which hot water temperature, at the consumption points could exceed 60 °C. Nevertheless, it is possible, depending on the testing installation, to restore this output temperature to its initial value in order to work in a closed circuit (Figure 5).

To serve this purpose, water from the outlet hydraulic circuit must be cooled through a heat exchanger. This will be an exterior plate exchanger, using outlet water of the prefabricated system as the fluid for the primary circuit whereas the cooling fluid is used for the secondary circuit. The amount of heat extracted from the test fluid will depend on the water-flow and temperature of the cooling fluid. The secondary circuit has been designed to perform heat evacuations using water from the network or the filled water deposit. There is a possibility of regulating the water-flow of the

Figure 5. Regulation of the oulet fluid temperature

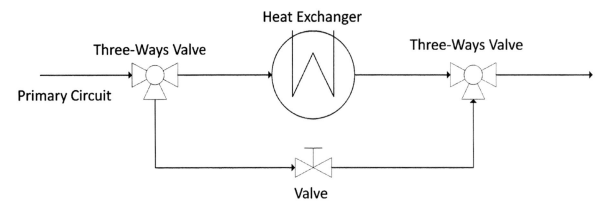

secondary circuit using a regulation automatic valve and consequently controlling the amount of evacuated heat from the primary circuit. This valve would be controlled starting from the temperature measurements of the freezing fluid, and of the water-flow as well as the temperatures of the test fluid at the heat exchanger's inlet and outlet (assigned parameter). The exchanged heating power must be less than 50 kW in the most unfavorable condition which is during the first extracted volume in the test of a prefabricated system.

Water-Flow Regulation

Water-flow regulation mechanisms will be installed in the cold water inlet pipe so that outputs resulting from it are not affected by temperature changes. The water-flow design will be of 600 l/h with a maximum vatiation of ± 50 l/h. for which a water-flow control mechanism is necessary to ascertain a stable performance. For this purpose, an automatic valve is proposed to regulate the water-flow circulating through the testing element. The water-flow that is deflected through this valve must be able to serve as a mixer element in the deposit and then homogenize the fluid temperature in such deposit to as great an extent as possible (Figure 6).

The valve must be able to be activated from the control room, although, it should also be able to be activated manually in case a manual water-flow regulation is preferred (UNE-EN 1490, 2001).

Figure 6. Scheme for the water-flow regulation

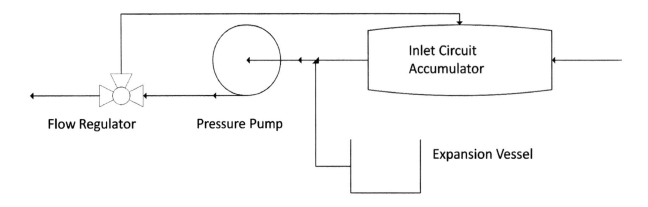

Table 5. Exchanger heating parameters

	Primary Circuit	Secundary Circuit
Inlet temperature (°C)	85	10
Outlet temperatue (°C)	15	30
Water-flow(litres/hour)	600	2100
Work fluid	Water	Water

Air Speed Regulation

The norms presented recommend using vents to keep air speed on the solar collector cover between 3 and 5 m/s. In order to guarantee this air speed on the collector's surface, a fan will be installed in the lower zone of the solar collectors, conveniently separated from them to avoid shade over them. Due to the utilization of wind generating fans, it must be verified that the anemometer is inside the air stream and that the air speed circulating through the collectors is the same. The fan must produce a homogeneous air stream over the collector surface, so that at any other point of the surface there is a deflection more than 25% over the air speed measured by the anemometer. It is very likely that it will be necessary to channel the air from the fan outlet for a better stream distribution. Air temperature in the fan outlet in any case must never exceed the ambient temperature value by more than 1°C.

Pressure Regulation

The testing installation has been designed to work at pressures higher than the atmospheric pressure since it can be noteworthy in certain tests (for example the pressure resistant test). That is why, there will be a pressure elevating pump located at the circuit inlet of the solar energy system; this pump will be activated in relation to the manometer pressure (or any other chosen pressure measuring device) at the system entrance. This pump will make pressure elevation possible of the primary circuit from the atmospheric pressure up to a maximum pressure of 8 bar.

Hydraulic Circuit

Just as it can be deduced from the above mentioned, the installation must be constituted of:

- An inlet circuit to the system that must be tested, where the water-flow and the fluid temperature are regulated;
- An outlet circuit from the prefabricated system, where the fluid can be cooled once it has been heated by such a system, and this way, can work in a closed circuit;
- A filling circuit that allows the installation to perform in an open circuit and independently from the water supply.

Pipes and Isolation

Most of the domestic solar energy systems use copper pipes. Therefore, all the circuits from the testing installation must be made out of this material and thus prevent galvanic corrosion. Furthermore, being the water-flow design of 600 l/h and the maximum water-flow for the testing installation of 1200 l/h, we can utilize pipes with an exterior diameter of 22 mm. Just as the pipes like the rest of the elements in the hydraulic circuit must be isolated to minimize thermal losses. The insulating material must be:

- Of a 20 mm thickness with outdoor and fire protective;
- With a thermal conductivity coefficient less than 0.045 W/mK at a temperature of 40°C.

With this isolation, water-flow design and the most unfavorable difference between the working fluid and the ambient temperatures, less than 15.2 W losses by lineal meter of pipe can be guaranteed. Therefore, to make sure that there is a temperature loss in the working fluid less than 0.01 K in a section of the pipe, this section's length must be less than 0.45 m. This value establishes the maximum

Figure 7. Scheme of the prefabricates system inlet circuit

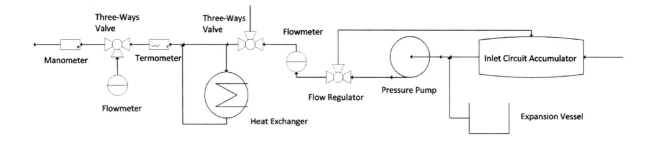

tolerable clearance between the two temperature sensors of the fluid and the inlet and outlet from the prefabricated testing system.

Inlet Circuit to the Prefabricated System

This circuit must be installed as close as possible to the location of the systems that are to be tested. This way, thermal losses in connecting pipes sections are minimized. Moreover, the elements of this circuit must be accessible for their handling and control by the operators, making it easier to read the measuring devices as well as the supervision and detection of possible leaks or installation irregularities. As it is shown in the illustration on the following page, a manometer has been situated at the prefabricated system inlet with the aim of calculating the inlet pressure of the fluid. Also, there is a flowmeter in the drainage pipe that allows the calculation of the water-flow as well as the drainage time of the prefabricated system. The drainage pipe outlet must be at a lower height than the plain on which the prefabricated system is supported. These elements are, therefore, necessary to carry out the antifreezing and over- temperature protection tests (Figure 7).

Inlet Circuit Deposit

The function of this deposit is to help obtain the necessary conditions at the entrance of the testing solar energy system, that is to say, to adjust the inlet water temperature during the tests. Because most of the domestic solar energy systems have an accumulation volume less than 500 liters and, in test procedures, an extraction of 3 times the accumulation volume is required at a constant inlet temperature, a circuit inlet deposit of 1500 liters has been designed. This deposit must be supplied with a heating system which will be through an electric heater (an electric resistance), or else a gas boiler to be able to provide a more instantaneous power. All these systems are controlled by a thermostat located in the interior of the deposit and at the same height as the test fluid outlet. In other respects, it will be laid out horizontally to reduce stratification effects in its interior, and this way will help keep the fluid contained in its interior at a same temperature. Moreover, it will be conveniently isolated with an isolation coefficient of 0.018 kcal/h m^2 °C. The deposit will be made of carbon steel interiorly protected against corrosion. The maximum pressure that it will have to withstand will be the maximum pressure of the circuit design, which is 8 bar.

Inlet Circuit Pump

This pump will have a water-flow design of 600 l/h. Moreover, it must bear the maximum and minimum design temperature values, that is to say, from 3 °C to 95 °C. Thus, one must resort to:

- Accelerator pumps of hot water in heating or centralized installations,
- Cold water pumps for climate control, or
- Industrial acceleration systems.

The maximum work pressure must be at least of 8 bar. Given these design parameters, we can utilize pumps which are the wet rotor circulating type for heating installations or sanitary hot water, with a power contained between 25 W and 60 W (UNE-EN 1151, 1999).

Expansion Tank

The hydraulic circuit design must take water dilation into account when it is produced to increase temperature. This requires an expansion system that absorbs the water volume increase in open circuits, and the pressure increase in closed circuits. In the inlet circuit of the system being tested, serving that purpose there is an expansion tank membrane type which is pressurized by air. This tank will be located in the pump aspiration zones, and this way it will respond to the starting and stopping requests of such pump. This will allow the pump to function at a basically constant pressure. In another respect, it must be directly connected with the circuit deposit of the inlet without an interfering blockage valve. This is due to the fact that the deposit is the largest water volume element in the testing installation. The expansion tank capacity must be in relation to the water volume in the circuit and to the dilation capacity of the water within the temperature range in which such circuit can correctly operate.

Outlet Circuit of the Prefabricated System and Recirculation Pump

Just as previously mentioned, fluid cooling is possible in the inlet circuit of the prefabricated system (once it has been heated by such system), and this way, it can operate as a closed circuit. The elements that make working fluid cooling possible have been also previously mentioned. However, the inlet circuit must reach the filling deposit[4] as well, where we must consider the fact that the inlet return pipe of the outlet hydraulic circuit is located below its water level. It can be observed that in the testing installation scheme provided by the norm for thermal performance categorization[5], it is necessary to locate a connection between the outlet and inlet circuit where a recirculation pump must be installed. The connection will be made possible through a three-way valve located upstream to the heat exchanger. This pump must be a water-flow design of 600 l/h, and must be able to withstand temperatures between 3 °C and 95 °C, and a maximum work pressure of at least 8 bar.

Therefore, a pump with similar characteristics to the pump installed in the inlet circuit can be used (UNE-EN 809, 1999). That means a circulation pump with a wet rotor for heating installations of sanitary hot water that uses a power between 25 W and 60W. The utility of such a recirculation pump must be to homogenize fluid temperature inside the prefabricated system, in other words, to be able to break the possible unwanted stratification formed inside the system. Apart from the recirculation pump, the manometer location can be also observed in the scheme shown on the next page; this location assures the measurement of the outgoing fluid pressure as well as air entrance in order to be able to fully empty the prefabricated system and this way, the flowmeter located in the drainage pipe[6] can calculate the system volume and drainage time. These elements therefore, are necessary for antifrost protection tests and over-temperature protection tests. The location of an expansion tank has been considered at this air inlet which is sufficiently high to prevent fluid overflow. Nevertheless, to prevent leaks of fluid out of the inlet air conduct, a retention valve can be also used to impede fluid in that direction (Figure 8).

Filling Circuit

Each filling circuit will also be composed by filling deposit pipes, a filter, a floating cut valve and several cut valves (Figure 9).

This deposit must have several functions:

Figure 8. Scheme of the prefabricated system outlet circuit

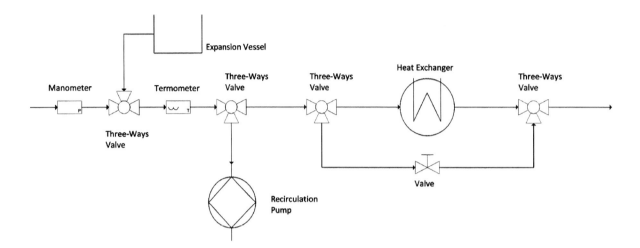

- To serve as filling for the inlet hydraulic circuit of the testing system, allowing at the same time the open circuit performance of the installation;
- It makes possible the supply of reserves of the fluid that will be used, in case this was outside the water network, it permits the testing installation be supplied with a certain Independence.

The filling deposit must be of glass fiber with an outdoor endurance and must have a sufficient capacity to be supplied with fluid reserves during the prefabricated system test in a steady state of operation. This capacity therefore is of the testing authorities' free choice (as they consider convenient). This deposit must include a coating or lid to prevent unwanted objects or susbtances going into it. This coating must allow its total openning for deposit cleaning procedures.

Water Fluid

Exactly at the filling outlet deposit there is a filter of 200 m to prevent unwanted particles going into the hydraulic circuit which can cause deterioration or mal-functioning of any of the circuit elements.

Valve for Network Water Supply

To fill the filling deposit and to keep it at a desired water level, a closing valve will be located at the supply connection of the network; this valve will be controlled by a floating device

Figure 9. Scheme of the filling circuit of the installation; filling deposit

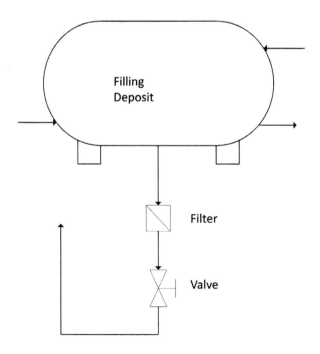

Figure 10. Scheme of the complete installation design

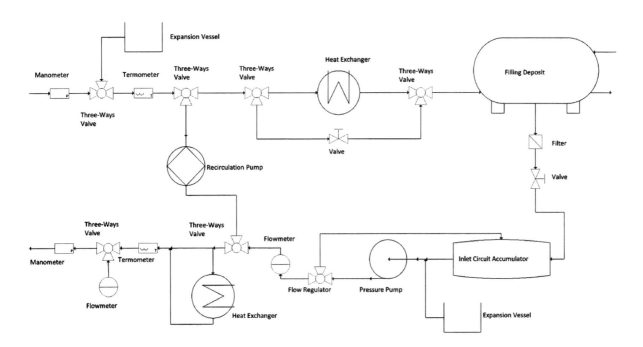

(buoy). The water level in the filling deposit must be considered, since the entrance to the return valve of the return outlet hydraulic circuit from such deposit, must be located below its water level.

Complete Scheme of the Hydraulic Circuit Testing Installation

The three previously explained circuits are jointly shown in Figure 10.

As it can be observed, the conditioning of the fluid used in the secondary circuit of the outlet heat exchanger circuit (for cooling), as well as the additional temperature regulator in the system inlet circuit (for heating), can be chosen freely by the testing authorities.

ADAPTATION OF A REAL TESTING FACILITY

Description of the Real Installation

In the following section, we will now describe a testing installation of a thermal solar energy installation that would enable us to investigate new different configurations, test different systems, etc., all this with the aim of keeping moving forward towards this technical field (Alonso Aguilar, 1997). This installation is basically constituted by:

- Four testing modules of components and thermal solar energy systems designed for its totally independent operation,
- A centralized system for the working fluid conditioning, and
- The data collecting system and control of the installation.

Figure 11. Vertical and horizontal representation of the test module

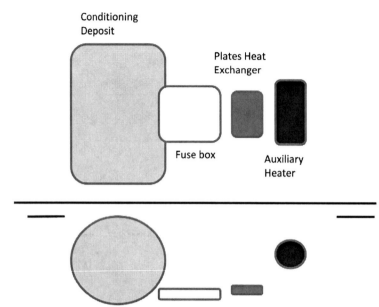

The design has been made in accordance with the prefabricated testing systems, currently active at the moment of such installation's design and previous to the Euronorm (UNE-EN 12976, 2006). The installation has a very versatile design, which lets it adapt to a greater amount of different configurations and to carry out tests of many types. This section will cover in a general way a description of the installation and all the permitted possibilities, but without specifying any configuration in particular.

Testing Modules

There are 4 testing modules in the installation. Each one of them admits diverse variants and configurations in accordance to the test which is carried out (Figure 11). Each module includes:

- 1 fuse box.
- 1 plates heat exchanger.
- 1 conditioning deposit.
- 1 auxiliar heater.
- Measure sensors.

It also includes:

- 2 pipes open to the exterior that let each module connect with the testing system.
- A set of pipes that connect together the diverse parts of the modules.
- A set of regulation, cut-off, and three-way valves, that allows the system to confiġúrate the module in several ways.

Conditioning Deposit

It is a deposit that keeps the accumulated water at a homogenious temperature to supply the systems that are tested in the modules. The reason why in tested systems there is no direct reception of water from the network is because it is supplied with water that complies with the established limits set by the testing norm in reference to temperature (Figure 12).

Figure 12. Fitting deposit representation

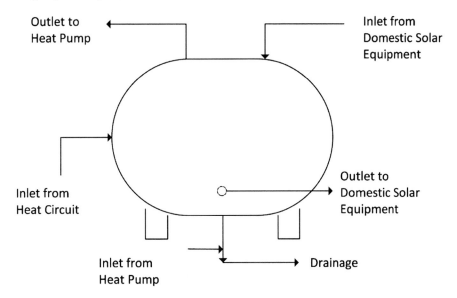

Also, it is necessary to have a previous buffer container that absorbs pressure and temperature shocks in the water that comes into the systems and that could negatively affect the tests. This depostit has a storage capacity of 1000 l. It has an inlet pipe in the lower part that lets it fill with water from the network and also the recirculation of water through other conditioning deposits of other modules. In the top part, it has an outlet pipe that makes such circulation possible. Also in the lower part, there is a deposit drainage pipe. It has an outlet pipe in its front part to supply the prefabricated system with water and in the top part an inlet pipe that comes from the prefabricated system.

Figure 13. Representation of exchanger plates

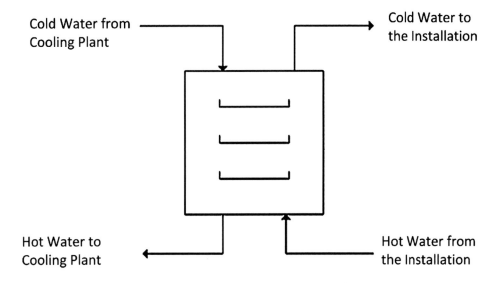

Plate Exchanger

This is the element that lets the installation adapt the temperature of the water to those demanded by, and necessary for, the tests. It is a plate exchanger that acts counter-currently, in which, a conducter circulates water coming from a cooling plant and through the other conducter circulates the water which is to be cooled. The cooling plant is really a heat pump that produces cold water or hot water according to the particular test (Figure 13).

Auxiliar Heater

It is the element that does not appropriately belong to the test installation, but that permits (if necessary) the heating of water. There are two distinct models:

- Modules 1 and 2 have an electrical heater: the electrical heater permits water accumulation, allowing a progressive heating of such water. One of them has a capacity of 75 l and the other of 100 l. Power resistance of the thermos is of 1500 W, when its performance controlled by a thermostat between 30 and 70°C (Figure 14).
- Modules 3 and 4 have a gas heater: the gas heater only lets the fluid pass through with no accumulation; thus it only produces an "instant" fluid heating.

Connection Pipes to the Exterior

The two system connection pipes that each module has let the installation gather diverse solar elements for its testing. Since there are connecting pipes for each module, the tests of each of them can be done independently, so that different conditions can be established in each testing module. All that has been mentioned up to this point about the testing module can be summarized in the scheme shown in Figure 15.

Hydraulic Circuit

Each of the modules is connected with pipes to the circuits that pass below all modules, these are:

- Filling, mixing and distribution circuits.
- Heat circuit.
- Cold circuit.
- Cooling circuit.

As it has been previously mentioned, each module is able to communicate with other circuits through diverse valves. These circuits pass through under the 4 circuits (thus are all common) and in the East-West direction in a 2 m wide pipe tray. A number is assigned to each pipe, which 1 corresponds to the pipe located at the upmost Northern location, and from then on in a sucessive fashion. This set of pipes, common to the 4 modules is shown in Figure 16.

This set of pipes is intended for the filling and mixing of the water in the systems and deposit in the modules, as well as the distribution of water between the modules if necessary (in yellow). According to the test that is carried out, the modules

Figure 14. Representation of an elecrical or gas heater

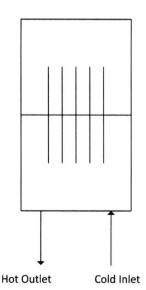

A Reliability Test Installation for Water Heating Solar Systems

Figure 15. Testing module summary scheme

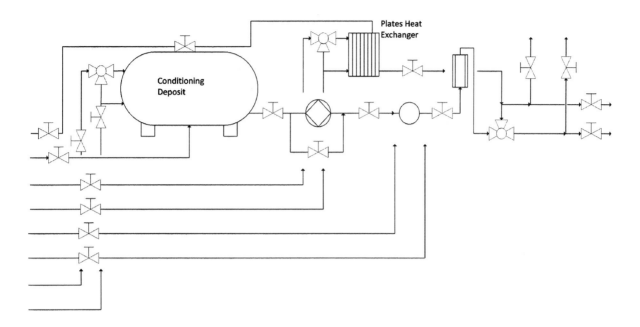

considered convenient will be connected to some of these circuits. To serve this purpose there is a set of cut-off valves, regulation valves and three-way valves that make all sorts of variants possible. Non-numbered pipes make possible the connection of prefabricated systems to the testing modules. The connection of each module must be set so that such tests can be carried out in an independent manner and under different conditions.

Circuit of Filling, Mix, and Distribution

This circuit corresponds to pipes 7 and 8 (in yellow) in the scheme of the real installation. The installation for tests is connected to the water supply network. There is a main pipe that brings the water in from the exterior and connects to pipes 7 and 8 (of the common pipes) to the 4 modules. Pipe 8 is connected at the same time to the lower inlets that hold the conditioning deposits. When in the conditioning deposits or in its circuits, a leak or a water depression is produced, or even if some of the deposits are empty, the network water comes in until the system is fully filled. The work pressure of the network is 2 bar. Since everything is interconnected and at the same time connected to the network, is the network that keeps the constant work pressure. The drainage of deposits is carried out by opening the cut-off valve located in the lower part of the deposit.

Cold Circuit

This circuit corresponds to pipes 5 and 6 (in blue) in the installation scheme, and it is connected to a cooling plant. A pipe branch comes out from pipe 5 to each module, containing cold water to the plate exchanger. The cold water goes through the plate exchanger and comes out incorporating itself to pipe 6, which sends back hot water to the cooling plant (Figure 17).

The cooling plant, as it has been mentioned before, is a heat pump which during the thermal performance test, receives hot water from the installation through the pipe 6 of the modules lower tray; this same water comes from the plate exchanger, and has been heated by the cooling down of the water of the installation. The water reaches the cooling plant and it is cooled; then, it comes out pumped

Figure 16. Summary of a complete installation

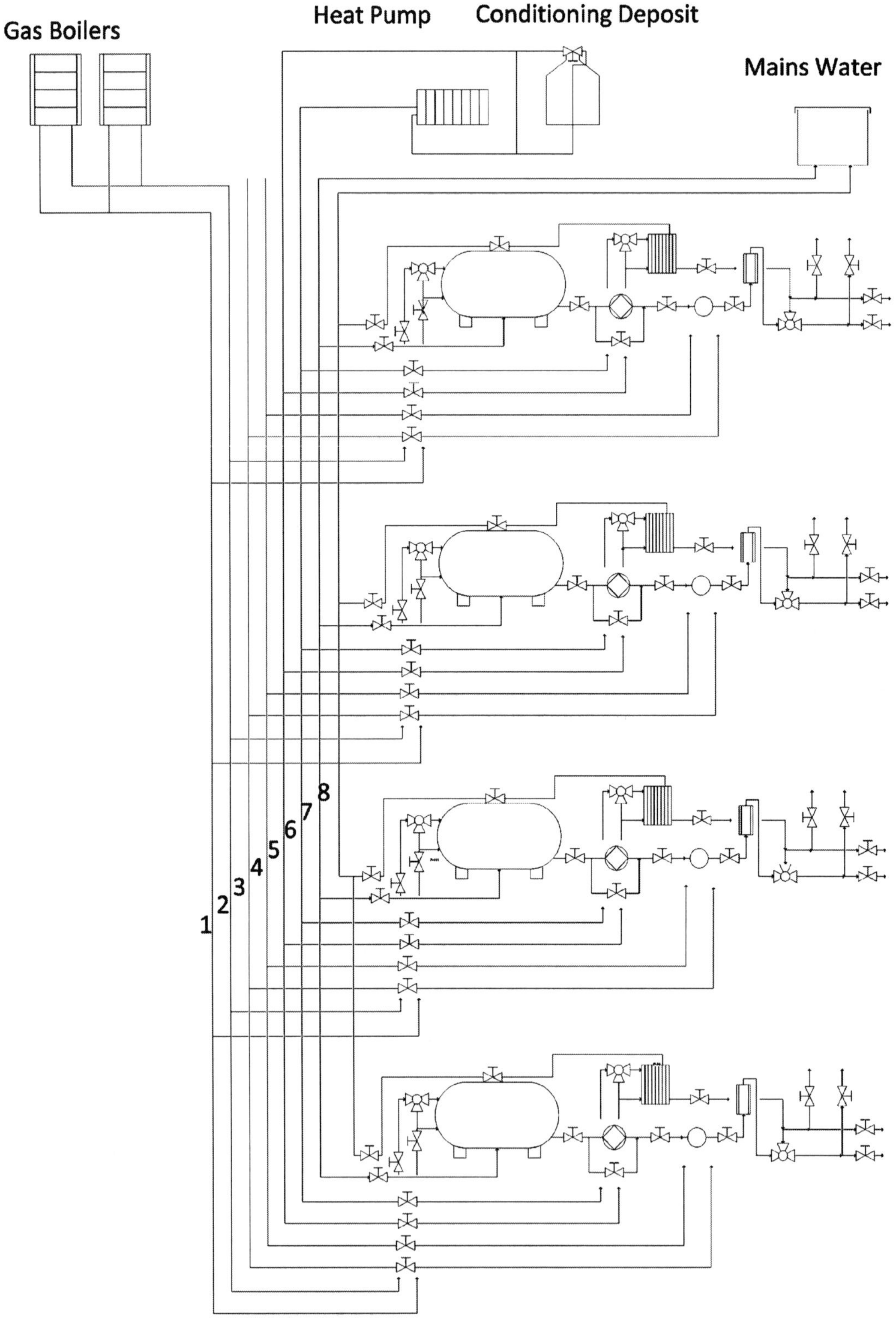

Figure 17. Cooling circuit summary

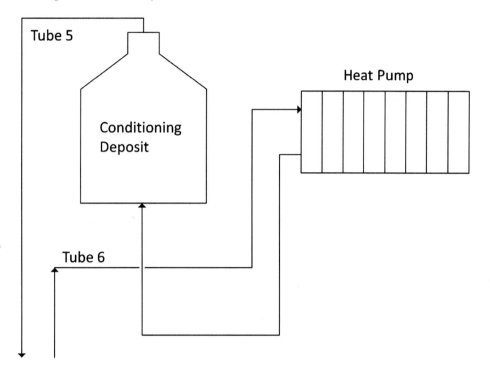

by the plant towards the inertia deposit, which has a volume of 1000 liters. From the deposit a pipe comes out that connects pipe 5 of the tray and takes the cold water back to the plate exchanger.

This water cools the installation and comes out heated through pipe 6 and this way successively. The cooling plant therefore permits water cooling of the installation. The fact of having water stored represents a disadvantage of energy loss towards convection; on the other hand, it makes it possible for the plant to have a much lower power since it does not have to work exclusively in the moment when cooling is necessary, but it can continuously keep on working at a lower power and storing such cold water in the deposit for its subsequent use when needed. The cooling circuit is connected to the network through the heating pipes set (extension of pipe 6 towards the cooling plant). This connection is always open, which means that the cooling plant has a network working pressure of approximately 2 bar.

Cooling Circuit

The cooling circuit is constituted only by the pipes 3 and 4 (in green) of the installation scheme, which are thought to be used for a possible connection to another cooling system (tubes, convectors, cooling towers, etc.).

Table 6. Functions of the set of pipes

Pipes 1 and 2	Pipes 3 and 4	Pipes 5 and 6	Pipes 7 and 8
They connect the modules of a water heating system in boilers 20 (in red).	They connect the modules of a water cooling system 21 (in green).	They connect the modules to a cooling plant 22 (in blue).	For water filling, mixing and distribution in the system (in yellow).

Heating Circuit

The 2 heating pipes 1 and 2 (in red) in the installation scheme are connected to each one of the conditioning deposits of the modules. These pipes are thought for a possible subsequent connection to a centralized heating system formed by one or two natural gas boilers, so that the heating supply is regulated through a three-way valve, of an "all-or-nothing" type action, controlled by a thermostat located in the conditioning deposit. However, hot water is currently produced with a heating pump that operates inversely to the cooling production circuit[7].

Data Collecting System and Installation Control

It is possible to use the installation for tests for a multitude of different configurations. Through the data collecting system and installation control we can obtain:

- The measurement of variables and actions taken on different elements of the installation,
- The storage of the information collected and
- Its subsequent use.

It is likewise possible to know the water temperature at different points of each module. There are a great number of thermal pipes[8] disposed to serve that purpose in almost all the joints of the circuit. Also we can measure the direct global radiation that falls on the solar collectors; the water-flow that goes through the consumption circuit, as well as that can act on the electrovalve that allows extractions to be carried out.

Equivalent Elements between the Designed and the Real Installation

Equivalence is understood as the quality of those components which are equally applicable to perform a particular function. It is important to stress the fact that the different device configurations and the pipes in a real installation do not necessarily have to coincide with the functional scheme of the proposed installation. More clearly, the objective here is to adapt the existing components in the depicted installation to use them in a similar manner (although logically not identical) to the configuration of the previously proposed installation and thus carrying out the already specified tests. This way we could take advantage of instruments, systems, hydraulic circuits, etc. which are currently available for the testing of the prefabricated systems in steady state operation according to the effective related norm.

Filling Deposit, Inlet Filling Deposit, and Conditioning Deposit

Table 7 is presented to compare the filling circuit together with the inlet filling circuit (corresponding to the proposed installation) in relation to the conditioning deposit for each module of the real installation.

As it can be seen, with the conditioning deposit we get the same results expected from the inlet circuit deposit together with the filling deposit of our proposed installation. Furthermore, a capacity problem is detected in the components of the real installation, since its real storage capacity is of 1000 liters, which is not sufficient in comparison to the 1500 liters required for the proposed installation. However, the stored volume can circulate from one module to another, thus in total we would obtain an accumulation volume of:

Table 7. Functions of the set of deposits

	Filling Deposit	**Inlet Circuit Deposit**	**Conditioning Deposit**
Functions	Supply water to the system; Accumulate fluid reserves;	Fluid accumulation and obtainment of its initial conditions.	Accumulate a homogeneous water temperature. Cushions pressure and temperature peaks.
Capacity	Capacity must be according to the testing authorities considered applicable.	Minimum capacity of: *3 extractions x [maximum water volume in a prefabricated system] = 1500 l*	Has a storage capacity of 1000 liters.
Other Characteristics	This deposit must have a cover lid.	- Has a heating system. - Maximum pressure of 8 bar.	Can recirculate water through the other conditioning deposits.

4 modules x 1000 liters / module = 4000 liters

A volume of 1500 liters is required in each test of a prefabricated system; we can observe that with the previous accumulation volume of 4000 liters, we can only test 2 prefabricated systems simultaneously:

2 systems x 1500 liters / system = 3000 liters < 4000 liters

That means that, with the real test module capacity, we can carry out:

- The test of 1 prefabricated system of 500 liters with 2 modules, or

- The test of 2 systems of 500 liters each with 3 modules, leaving the 4th testing module unused.

Nevertheless, these calculations are carried out considering that there will be an extraction of 3 times the volume of prefabricated systems of 500 liters capacity. In the instance of running tests to systems whose volume is lower than:

1000 liters / 3 extractions = 333.33 liters

We can test 4 prefabricated systems simultaneously in the 4 modules and with no necessity of recirculate the fluid from a module to another.

Table 8. Function of the outlet regulator and the plate exchanger

	Outlet Fluid Temperature Regulator	**Plate Exchanger**
Functions	Cools the fluid back to its initial temperature through a heat exchanger.	Element that let the water of the installation cool.
Other Characteristics	The heat exchanger is an exterior plate exchanger.	It is a plate exchanger that acts against the current.

Table 9. Functions of the regulator and auxiliary heater

	Additional Inlet Regulator of Temperature	**Auxiliary Heater**
Functions	Heats the water when there are thermal losses in the inlet circuit of the system.	Element that lets the water heat.
Other Characteristics	It is an electrical heater of approximately 2 kW, able to increase to 2 K a water-flow of 600 liters/hour in series with the inlet deposit.	Modules 1 and 2 have an electrical heater of 1.5 kW. Modules 3 and 4 have a gas heater.

Table 10. Functions of the filling, mixing, and distribution circuits

	Filling Circuit	**Filling, Mixing, and Distribution Circuit**
Components	• Filling deposit, • Filter, • Cut-off valve, • Pipes.	Main pipe that carries the water and connects to pipes 7 and 8 (refer to section 5.4.1.1. of this document).
Functions	• Supply water to the system; • Accumulate fluid reserves;	Water comes in from the net until there is a pressure balance in the modules.

Fluid Temperature Regulation

Outlet Regulator and the Plate Exchanger

Table 8 is presented containing the characteristics of the temperature regulator of the fluid at the outlet of the proposed installation, in comparison to the existing real plate exchanger.

Additional Inlet Regulator and the Auxiliary Heater

Similarly other characteristics are also presented in Table 9 of the additional inlet regulator of fluid temperature in the proposed installation jointly with the auxiliary heater that is part of the real installation.

The auxiliary heater complies with the presented requirements in the design of the test installation even in respect with the nominal power of the electrical heater, since in the proposed installation such power would be:

$$W' = Q \cdot p \cdot c_{pwater} \cdot T =$$

$$= 600 \, (l_{water} / h) \cdot (1 \, h / 3600 \, s) \cdot (1 \, kg / 1 \, l_{water}) \cdot 4.18 \, (kJ/ \, kg \, K) \cdot 2 \, K = 1.4 \, kW$$

Even though in the end a decision was made to use a heater of 2 kW in the proposed installation in order to have a sufficient safety margin as well as to use a standardized heater; we can observe that a power of 1.5 kW (as in the case of the real installation) is sufficient for this water-flow and this temperature increase. Currently, module auxiliary heaters are not used. Nevertheless, if there was a necessity of using them, we would need a pipe circuit adaptation.

Hydraulic Circuit Equivalence

The Filling, Mixing, and Distribution Circuit

Table 10 is presented to compare the filling circuit of the proposed installation and the real installation's filling, mixing and distribution circuits. Both circuits have an equivalent utility.

The Outlet Circuit and the Cooling Circuit

In the synoptic chart (Table 11), it can be jointly observed the components and their functions (in a simplified way) of the outlet circuits (of the proposed installation) and the cooling circuit (of the real installation).

Table 11. Functions of outlet and cooling circuits

	Outlet Circuit	**Cooling Circuit**
Components	Plate exchanger, recirculation pump, three-way valve.	Pipes 5 and 6 (refer to section 5.4.1.1. of this document). Plate exchanger.
Functions	Cool the fluid so that it can work in a closed circuit. Recirculation of the fluid back to the system.	It cools down the water of the installation.

Table 12. Functions of inlet, filling, and heating circuits

	Inlet Circuit	Filling, Mixing, and Distribution Circuits	Heating Circuit
Components	Inlet circuit deposit. High pressure pump. Water-flow regulator. Additional temperature regulator. Expansion basin/tank.	Pipes 7 and 8 and (See section 5.4.1.1. of this document).	Pipes 5 and 6, operating the heat pump to produce heat.
Functions	The same as the fitting deposit, impulses the fluid into the system, Absorbs water dilation.	Fitting of the fluid into the system.	Fitting of the fluid into the system.

It is observed that the circuits are basically identical except for the recirculation pump. In the proposed installation, the authorities of the test are free to choose the obtainment and conditioning of the fluid used in the secondary circuit of the heat exchanger. In the real installation, we can observe that such fluid comes from the inertia deposit, once it is cooled down (in this case) by the heating pump.

The Inlet Circuit and the Filling, Mixing, and Heating Distribution Circuits

Finally, a comparative chart is presented between the inlet circuit of the installation design and the filling, mixing, and distribution circuits together with the heating circuit of the real installation (Table 12).

CONCLUSION

The European Standard EN-12976 (parts 1 and 2) for thermal solar energy systems deals with thermal solar energy systems and their components. In particular, it deals with prefabricated systems for the production of sanitary hot water through solar energy. Along this chapter, we have analyzed mainly the second part of this norm (EN-12976-2, 2006) which expresses the distinct testing methods applied to domestic solar energy systems for the validation of the prior requirements. With that purpose, the chapter starts introducing the requirements needed for the design of a test installation. Based on a real testing facility and according to the proposal design, the installation is adapted in order to be performed the corresponding reliability tests.

For instance, the objective wished for the inlet circuit in the proposed installation is accomplished by the real installation of components, serving that configuration for the same purposes. In fact, in the proposed installation, hot water is produced through electrical resistances submerged in the inlet circuit deposit and/or through a heat transfer with hot fluid in the secondary circuit of a heat exchanger. It is also observed in the real installation, where we obtain the same result by using a hot fluid coming from an inertial deposit, once is has been heated (in this case) by the heating pump and used as a fluid of the secondary circuit in the plate exchanger.

In any case, the adaptation of real facilities to accomplish new standards is currently the order of the day, being in the presented case not complicated to achieve. The interest in thermo-solar energy is growing worldwide and it is usual the setting of new norms in order to commercialize related products in different markets. This European Norm and its adaptation to an operating installation may be a good example and aid for those emerging countries who are nowadays applying new standards in this field along with the growth of the global solar market.

REFERENCES

Alonso Aguilar, P. (1997). *Instalación integral de ensayos para la evaluación energética de sistemas solares térmicos de baja temperatura*. Proyecto Fin de Carrera, Universidad de Sevilla.

American Institute of Chemical Engineers Center for Chemical Process Safety. (1989). Guidelines for process equipment reliability data, with data tables. Author.

Barberá, L., González-Prida, V., Moreu, P., & Crespo, A. (2010). *Revisión de herramientas software para el análisis de la fiabilidad, disponibilidad, mantenibilidad y seguridad (RAMS) de equipos industrials: Revista Ingeniería y Gestión de Mantenimiento. Vol. Abril/Mayo/Junio 2010*. España.

Fisch, N. (1995). *Manuskript zur Vorlesung Solartechnik I*. Academic Press.

González-Prida, V. (2002). *Influencia de la normativa europea en el procedimiento de ensayos de sistemas solares térmicos prefabricados: Propuesta de adaptación de la instalación de ensayos del Instituto Andaluz de Energías Renovables*. Proyecto Fin de Carrera, Universidad de Sevilla.

González-Prida, V., Barberá, L., & Crespo, A. (2010). *Practical application of a RAMS analysis for the improvement of the warranty management*. Paper presented at the 1st IFAC workshop on Advanced Maintenance Engineering Services and Technology, Lisbon, Portugal.

Gonzalez-Prida, V., Parra, C., Gómez, J., Crespo, A., & Moreu, P. (2009). Availability and reliability assessment of industrial complex systems: a practical view applied on a bioethanol plant simulation. In *Safety, reliability and risk analysis: Theory, methods and applications* (pp. 687–695). Taylor & Francis.

Gruber, E. (1984). *Energieeinsparung und Solarenergienutzung in Eigenheimen*. Forschungberich T84-287 des Bundesministeriums für Forschung und Technologie.

Henley, E., & Kumamoto, H. (1992). *Probabilistic risk assessment, reliability engineering, design and analysis*. IEEE Press.

ISO 9459-2. (2013). *Solar heating. Domestic water heating systems. Part 2: Outdoor tests methods for system performance characterization and yearly performance prediction of solar-only systems*. International Standard.

Parra, C., Crespo, A., Cortés, P., & Fygueroa, S. (2006). On the consideration of reliability in the life cycle cost analysis (LCCA). In *A review of basic models: Safety and reliability for managing risk* (pp. 2203–2214). Taylor & Francis.

UNE-EN 1151. (1999). *Bombas. Bombas rotodinámicas. Bombas de circulación cuyo consumo de energía no excede de 200 W, destinadas a la calefacción central y a la distribución de agua caliente sanitaria doméstica. Requisitos, ensayos, marcado*. International Standard.

UNE-EN 12976-1. (2006). *Sistemas solares térmicos y sus componentes. Sistemas prefabricados. Parte 1: Requisitos generales*. International Standard.

UNE-EN 12976-2. (2006). *Sistemas solares térmicos y sus componentes. Sistemas prefabricados. Parte 2: Métodos de ensayo*. International Standard.

UNE-EN 1490. (2001). *Válvulas para la edificación. Válvulas de alivio de presión y temperatura. Ensayos y requisitos*. International Standard.

UNE-EN 307. (1999). *Intercambiadores de calor. Directrices para elaborar las instrucciones de instalación. Funcionamiento y mantenimiento, necesarias para mantener el rendimiento de cada uno de los tipos de intercambiadores de calor.* International Standard.

UNE-EN 806-1. (2001). *Especificaciones para instalaciones de conducción de agua destinada al consumo humano en el interior de edificios. Parte 1: Generalidades.* International Standard.

UNE-EN 809. (1999). *Bombas y grupos motobombas para líquidos. Requisitos comunes de seguridad.* International Standard.

Vintr, M. (2007). *Reliability assessment for components of complex mechanisms and machines.* Paper presented at 12th IFToMM World Congress, Besançon, France.

ADDITIONAL READING

Chaves Repiso, V. M. (1999). *Instalaciones de energía solar térmica con acumulación distribuida.* Proyecto Fin de Carrera, Universidad de Sevilla.

Crespo, A., & Iung, B. (2007). A estructured approach for the assessment of system availability and reliability using Montecarlo simulation. *Journal of Quality in Maintenance Engineering, 13*(2), 125–136. doi:10.1108/13552510710753032

DIN 4757, Teil 4. (1995). "Sonnenkollektoren Bestimmung von Wirkungsgrad, Wärmekapazitat und Druckabfall

Duffie, John A., Beckmann, William A. (1991). "Solar energy thermal processes".

González-Prida, V., & Crespo, A. (2010). Book review: Reliability engineering. warranty management and product manufacture (By D.N.P. Murthy and W.R. Blischke). Production Planning & Control. *The Management of Operations, 21*(7), 720–721.

González-Prida, V., & Crespo, A. (2012). "A reference framework for the warranty management in industrial assets". Computers in Industry 63: 960–971. ISSN: 0166-3615.

González-Prida, V., Crespo, A., Pérès, F., De Minicis, M., & Tronci, M. (2012). *Logistic support for the improvement of the warranty management.* Advances in Safety, Reliability and Risk Management. Taylor & Francis Group.

Høyland, A., & Rausand, M. (1994). "System reliability theory, models and statistical methods". J. Wiley. University of Michigan. ISBN 0-471-59397-4.

ISO/DIS 11924 (1995). "Solar heating. Domestic water heating systems. Testing methods for reliability and safety assessment". International Standard.

ISO/DIS 14224. (2004). "Petroleum, petrochemical and natural gas industries - Collection and exchange of reliability and maintenance data for equipment". International Standard.

Moreu, P., González-Prida, V., Barberá, L., & Crespo, A. (2012). "A practical method for the maintainability assessment using maintenance indicators and specific attributes". Reliability Engineering and System Safety. 100; 84-92. Elsevier. ISSN: 0951-8320.

SINTEF Technology and Society. (2002). *Offshore Reliability Data Handbook* (4th ed.). OREDA.

Smith, D. (2001). *Reliability, maintainability, and risk: practical methods for engineers.* Newnes.

E. Ullmer. (1995). "Theoretische Validierung einer Prüfmethode für solare Brauchwasserwärmungsanlage".

UNE-EN 1489. (2001). "Válvulas para la edificación. Válvulas de seguridad. Ensayos y requisitos". International Standard.

UNE-EN 1717. (2001). "Protección contra la contaminación de agua potable en las instalaciones de agua y requisistos generales de los dispositivos para evitar la contaminación por reflujo". International Standard.

UNE-EN 60335-1/A15. (2001). "Seguridad de los aparatos electrodomésticos y análogos. Parte 1: Requisitos generales". International Standard.

UNE-EN 60335-2-21/A1. (2001). "Seguridad de los aparatos electrodomésticos y análogos. Parte 2-21: Requisitos particulares para los termos eléctricos". International Standard.

UNE-EN ISO 9000. (2005). "Sistemas de gestión de la calidad. Fundamentos y vocabulario". International Standard.

UNE-EN ISO 9488. (2001). "Energía solar. Vocabulario". International Standard.

UNE-ENV 1991-2-3. (1998). "Eurocódigo 1: Bases de proyecto y acciones en estructuras. Parte 2-3: Acciones en estructuras. Cargas de nieve". International Standard.

UNE-ENV 1991-2-4. (1998). "Eurocódigo 1: Bases de proyecto y acciones en estructuras. Parte 2-3: Acciones en estructuras. Acciones del viento". International Standard.

KEY TERMS AND DEFINITIONS

Collector Circuit: It is the circuit that includes collectors, pumps or vents, tubes or pipes, and heat exchanger (if it exists), to transfer the heat extracted from the heat accumulator collectors.

Collector Performance (n): The quotient between the thermal energy extracted by the energy-carrier fluid during a determined time interval, and the product of the total area (of the absorber or aperture) of the collector by the solar irradiation that impacts on the collector at the same time interval, in a permanent regimen condition[9].

Energy-Carrier Fluid: The fluid used to transfer thermal energy between the components of an installation.

Flat Collector: A solar collector without concentration in which the absorbent surface is basically flat.

Parasitic Energy (QPAR): The energy consumed by the pumps, vents and controls in a thermal solar installation.

Permanent Regime: The state of a collector when the sum of the extracted heat and the losses of heat are equal to the solar power contribution.

Solar Collector: A device designed to absorb the solar radiation and then transmit this produced thermal energy to an energy carrier fluid that circulates in the interior of the system.

Solar Contribution: The energy supplied by the solar part of the installation. For its better definition, it is necessary to delimit precisely the solar part of the installation, as well as all losses associated to it.

Solar Fraction (f): The relation between the energy supplied by the solar part of the installation and the total supplied energy by said installation. It is necessary to accurately delimit the solar part of the installation, as well as all losses associated to it, in order to better define the solar fraction.

Thermal Load: The heat supplied by the user, for example in form of hot water. Due to heat losses in the distribution system, the heat extraction point must be exactly specified in order to to define accurately the heat supplied.

ENDNOTES

[1] Test installation element that conditions thermally the water that comes into the tested system.

[2] The norm does not provide a particular maximum value.

[3] In case of following the pre-normative test specifications.

4   Refer to the following section: "Filling circuit".
5   Refer to the chapter "An overview to thermal solar systems for low temperature: Outlining the European Norm 12976".
6   Refer to the functional circuit scheme of the prefabricated inlet circuit.
7   Refer to "Cooling circuit" in this same section.
8   Elements in which temperature sensors are introduced pt-100.
9   The collector performance can be also defined by the conditions of transitory regime.

Chapter 6
Electricity Production from Small-Scale Photovoltaics in Urban Areas

Constantinos S. Psomopoulos
Piraeus University of Applied Sciences (TEI of Piraeus), Greece

George Ch. Ioannidis
Piraeus University of Applied Sciences (TEI of Piraeus), Greece

Stavros D. Kaminaris
Piraeus University of Applied Sciences (TEI of Piraeus), Greece

ABSTRACT

The interest in solar photovoltaic energy is growing worldwide. Today, more than 40GW of photovoltaics have been installed all over the world. Since the 1970s, the PV system price is continuously dropping. This price drop and the adaptation of feed-in tariffs at governmental or utility scale have encouraged worldwide application of small-scale photovoltaic systems. The objective of this chapter is to present the potential for electricity production focusing mainly on the benefits of small-scale installations in urban areas, along with the growth of the global photovoltaics market. The types of installation alternatives are described but the focus is on the rooftop installations due to their simplicity and relatively low cost for urban areas. Electricity production data are presented together with their technical characteristics. Furthermore, analysis of the cost reduction is attempted and the benefits gained from the implementation of small-scale systems are also presented, demonstrating the sustainability role they will play.

INTRODUCTION

The photovoltaic systems consist of cells that convert sunlight into electricity. Each of these cells is manufactured by layers of a semi-conducting material, specially produced in order to develop electric field across the layers when the sun light is falling on the cell. This electric field causes the electricity flow and thus power is produced. The electric field is generated by the following mechanism (EPIA-Greenpeace, 2011; Luque & Hegedus, 2003; Sen, 2008): When the cell is

DOI: 10.4018/978-1-4666-8222-1.ch006

exposed to light, photocarriers are generated and this carrier separation within the cell layers produces a photovoltage and charge motion produces a photocurrent, which runs in reverse through the diode junction as described in Goetzberger and Hoffmann (2005). The sun light intensity determines the amount of electrical power each cell generates. A photovoltaic system does not need bright sunlight in order to operate. It can also generate electricity on cloudy and rainy days from reflected sunlight (Rockett, 2010).

Photovoltaics present high interest in community due to fact that the sun is freely providing light in huge quantities and it is expected to be shining for the next hundred million years. There is more than enough solar irradiation available to satisfy the world's energy demands. According to several studies (EPIA-Greenpeace, 2011; Luque & Hegedus, 2003; Sen, 2008; Šúri, Huld, Dunlop, & Ossenbrink, 2007), the Earth is exposed to enough sunlight to generate electricity with average value equal to 1,700 kWh/m^2 of land per year, using currently available technology. A large amount of statistical data on solar energy reaching earth's surface is collected globally for many years and for many areas. For example, in the US National Solar Radiation, a database with 30 years of solar irradiation and meteorological data from 237 sites in the USA is available. The European Joint Research Centre (JRC) also collects and publishes European solar irradiation data from 566 sites (EPIA, 2011; Greenpeace, 2008). Figure 1 presents the solar irradiation and photovoltaic power generation potential in EU in kWh/kWp of installed power, as it is given by Šúri et al. (2007) and EPIA (2011). Higher solar irradiation corresponds to higher power generation potential. Europe receives around 1,200 kWh/m^2 of average annual energy while Middle East 1,800 to 2,300 kWh/m^2. Even though only a rather small part of solar irradiation can be used for electricity production, this "efficiency loss" does not actually waste a finite power resource, as the use of fossil fuels. However, the reduced efficiency has an impact on the cost of the PV systems (EPIA-Greenpeace, 2011; Sen, 2008; Šúri et al., 2007; EPIA, 2011).

In this Chapter, special issues related to the PV systems are presented. In particular, types of PV systems are described and a review of the global installed capacity is presented. Furthermore, following these sections, monthly values of electricity produced by typical small scale building roof top installations from different areas of Europe, operating in actual conditions are presented. Evaluations related with the power production of kWh/kWp of installed power are also presented. Additionally, a brief economic analysis and cost evaluation based on international and local market values are presented, and finally in the last section the benefits of PV power generation focusing on small scale installations are described. This chapter is addressed to policy makers in the fields of energy, energy and buildings engineers and technicians and of course potential small-scale PV owners.

BACKGROUND

International Energy Agency (IEA) calculations revealed that if 4% of the world's very dry desert areas were used for PV installations; the global total primary energy demand could be met. Following this report and considering the fact that there are huge free areas without any use, the untapped potential is already very high. These vast areas, such as: roofs, building surfaces, fallow land and desert, could be used to support solar power generation. For example, if all suitable roofs and facades were covered with solar panels, 40% of the European Union's total foreseen electricity demand in 2020, could be covered by from PVs (EPIA-Greenpeace, 2011; Šúri et al., 2007; EPIA, 2011; Greenpeace, 2008; IEA PVPS, 2003; IEA, 2010; Krauter, 2006).

Measuring the cumulative nameplate capacity of all PV systems installed worldwide present difficulties and significant uncertainties, due to the

Figure 1. Solar irradiation and photovoltaic power generation potential in EU (Source: Šúri et al., 2007).

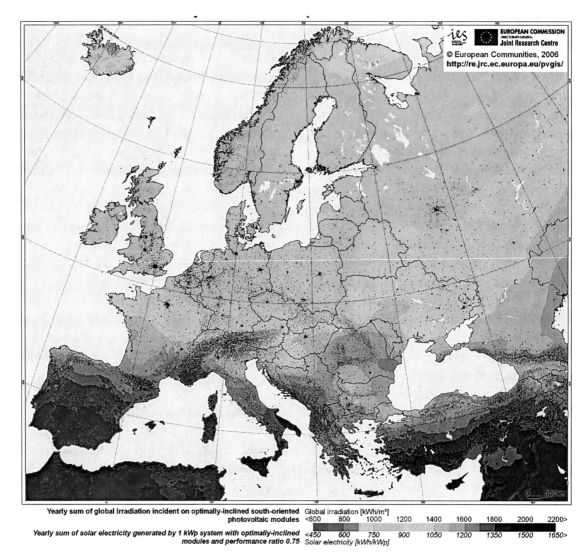

lack of adequate data in many nations globally, and especially for the less developed countries. The consideration of a broad range of parameters to determine the amount of PV installed can also explain these difficulties and uncertainties. For example, not all the purchased and shipped PV modules are installed until the end of each year, since the installations projects face a lot of difficulties, especially the large scale installations. Thus, this assumption seems to lead to overestimations on the cumulative installation capacity. On the contrary, only utilizing data from reported installations tends to underestimate the total amount of installed PV, on account of the difficulty of tracking off-grid installations, installations by companies that no longer exist, and other capacity not captured by the measure. Despite these difficulties the European Photovoltaic Industry Association (EPIA) estimated that global cumulative installed PV capacity reached 39.5GW by the end of 2010, as shown in Figure 2. In 2010 approximately 16.6 GW of additional PV capacity installed worldwide,

Figure 2. Global cumulative installed PV capacity through 2010, with market share (%); ROW: rest of world, ROEU: rest of EU
(IEA, 2010; EUROBSERV'ER, 2012).

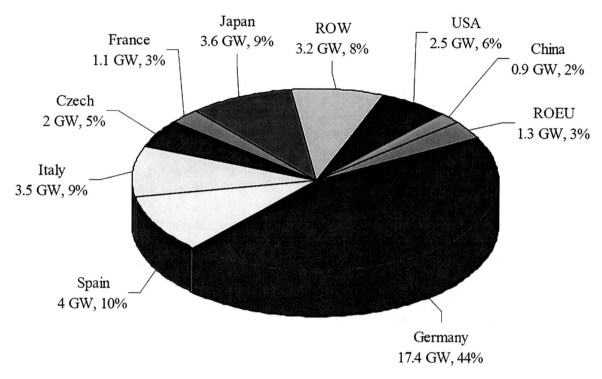

constituted a 131% increase over the 7.2 GW installed in 2009. These 16.6GW corresponded to a 71% increase in global cumulative installed PV capacity. European Union was the leader in installations for 2010 with 13 GW of the total 16.6 GW installed last year. Germany led new installed capacity with 7.4 GW installed, followed by Italy 2.3GW, Czech Republic with 1.5GW, Japan with 0.99GW, United States with 0.88GW, France with 0.72GW, China with 0.52GW, and Spain with 0.37GW. Estimates of global solar capacity vary widely across data sources, while accurate data are considered the ones published by (EPIA, 2011; Krauter, 2006; SMA, 2012; Zahedi, 2006; EUROBSERV'ER, 2012; EPIA, 2010).

Examining the evolution of this cumulative installed capacity of PV systems considering the type of installations, since 2005 the grid-connected systems have steadily gained market share relative to off-grid systems. One reason for the shift is that government subsidies worldwide tend to promote grid-connected PV. EU Member States with Germany and Spain were the first to promote these subsidies, while rest of EU and the world followed making adequate adaptations to the local conditions (SMA, 2012; EUROBSERV'ER, 2012).

While grid-connected PV systems present significant increase the last years dominating the global PV market, there are still a number of countries where smaller, off-grid systems comprise the majority of their local PV market. The inequalities in the different countries' market distribution between grid-connected and off-grid installations reflect the various types of subsidies, the stages of market maturity, demand for particular applications, geographical differences and other economic or not factors. The majority of the countries listed in Figure 2 had installed mainly grid-connected PV in 2010, including Germany, France, Italy,

Spain, Japan, and the United States. In contrast to the world leaders in installed PV, the PV market in countries like Sweden, Turkey, Mexico, and Norway dominated by off-grid systems, mainly domestic and small scale installations. Generally, domestic off-grid applications tend to be more common than off-grid industrial or agricultural systems (EPIA-Greenpeace, 2011; EPIA, 2011; Greenpeace, 2008; IEA, 2010).

The huge increase of installed PV systems in EU resulted in the increment of the share of photovoltaics in the EU's electricity mix. Now PV systems produce power which is almost 2% of the demand in the EU and roughly 4% of peak demand, on average. In selected regions and countries this amount is notably higher. Italy is a country example where PV covers 5% of the electricity demand, and more than 10% of peak demand, in the electricity mix. Looking at Regions, Bavaria, a federal state in southern Germany, the PV installed capacity amounts to 600 W per habitant, corresponding roughly to three panels per capita – an astounding figure, especially if it is considered that Germany do not have high PV generation potential (Figure 2) (EPIA-Greenpeace, 2011; EPIA, 2011; Greenpeace, 2008; IEA PVPS, 2003; IEA, 2010).

SMALL-SCALE SOLAR PV MARKET TRENDS

The solar PV industry is forecast to reach 49 GW in 2014, representing strong growth compared to 36 GW in 2013. However, the 49 GW total addressable market is really the combination of two distinct served addressable markets: small and large-scale PV installations.

Aside from Japan and Europe, only the US and Australia provided any appreciable levels of demand to the small-scale segment during 2013. Excluding Japan, Europe accounted for 72% of the small-scale global PV market in 2013.

Between 2010 and 2012, the growth in small-scale solar PV installations was driven by Feed-in-Tariff (FiT) policies within Europe, with approximately 75% of all small-scale solar PV deployed in Europe.

This is shown clearly in Figure 3, with the decline in the global small-scale served addressable market in 2013 caused by the reduced rooftop PV installed across major European markets (Germany and Italy). In 2014, the contributions from Europe are forecasted to be at just above 40% of the global small-scales served addressable market. Until 2012, Europe's small-scale solar market contribution was below 50%. This is a trend that will continue also in 2014 (Strong, 2010).

Figure 4 provides further details of the small-scale solar PV segment during 2013. Japan was the clear leader in the small-scale segment with almost 40% of global demand. Collectively, Japan and Germany accounted for over half of the global capacity installed.

During 2013, seven of the top-10 countries for small-scale solar PV were European, illustrating that there are still relatively few countries globally that have managed to create a sustainable residential and small commercial solar PV market.

TRENDS IN 2014

In order to create sustainable long-term growth, every country should develop the small-scale and large-scale photovoltaic markets. However, this has been a challenge to administrators for some time now.

More specifically, a vibrant small-scale market offers the scope for local installers, focused on relationship-driven sales activity with repeat business often secured through customer referrals. Annual growth levels in the megawatt level for small installers are often enough to keep this segment of the market, with steady business which is the key issue.

Figure 3. Small-scale solar PV from 2010-2014, RoW: rest of world (Strong, 2010).

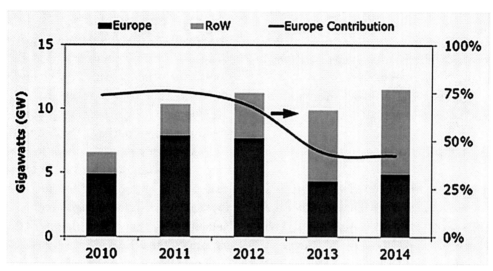

Globally, the small-scale segment has challenges like: competing sales models related to upfront capital and rooftop ownership variants, threats to net-metering and distributed generation arrangements, and policies that can change quickly based upon local elections or fiscal budget alterations.

From all this follows that suppliers of primary components (modules, inverters and mounting) require greater market visibility when selling into the small-scale PV market, compared to the large-scale segment. The decision concerning which countries, or states/territories, should be prioritized in 2014 will once again be essential for success.

Figure 4. Small-scale solar PV in 2013 by country (Strong, 2010).

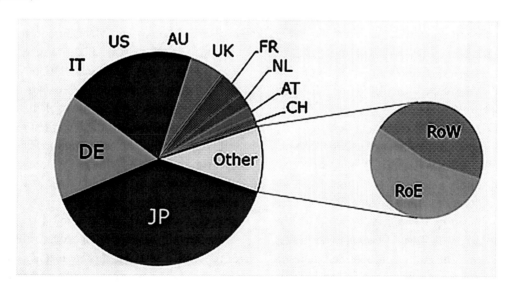

TYPES OF INSTALLATION AND BASIC EQUIPMENT OF PV SYSTEMS

In the following sections the different available types of installation are outlined and discussed. Moreover, the basic equipment which is necessary for the proper and efficient operation of photovoltaic systems is presented.

Types of Installations

The PV systems can be divided in two main types based on their installation considering if they are grid connected or "off-grid" installations. The "off-grid" or else "stand alone" installations can be divided further considering if they are used for domestic consumptions or non domestic consumptions. The grid-connected systems can be divided in distributed PV systems and centralized PV systems (EPIA-Greenpeace, 2011; SMA, 2012).

As it was mentioned above in the cases where a PV system is not connected to the electricity grid, then the system is called as "stand-alone" or "off-grid". These photovoltaic systems provide electricity to households and villages located in rural areas that are not connected to the national electricity grid. They are usually designed to provide electricity for lighting, refrigeration and other low power domestic loads. These types of PV systems have been installed in thousands places worldwide and they are often the most appropriate technology to meet the electricity demands of off-grid communities. Off-grid domestic systems are typically around 2-3 kW per household, in size and generally offer an economic alternative to extending the electricity distribution grid (EPIA-Greenpeace, 2011; SMA, 2012; EUROBSERV'ER, 2012). In order such a system to supply electricity consumption, additional systems are required for storage the excess energy produced by PV system, which it will be used when the production is lower or even there is no production. The equipment required includes batteries and charge controllers. Lead acid batteries are typically used in this kind of installations due to lower cost and relative high operational life, while new high-quality batteries, designed specifically for solar applications and with a life of up to 15 years, are now available. Batteries are connected to the PV array via a charge controller. The charge controller protects the batteries from overcharging or discharging and monitors the condition of the batteries. The controller can also provide information about the state of the system or the electricity consumption. The controller has a significant role on the system because manages the batteries charging and discharging conditions, and thus affects the actual lifetime of the batteries (Krauter, 2006; SMA, 2012; Zahedi, 2006).

The off-grid non-domestic PV systems are common in many applications requiring small amount of electricity and the distribution network is quite far and the PV commercially cost is competitive with other small generating sources. These systems were the first commercial application for terrestrial PV systems. They provide power at a low maintenance for a wide range of applications, such as telecommunication, water pumping, vaccine refrigeration and navigational aids. In recent years this type of installation is used to supply illumination installations such as street lights, plazas' lights, etc inside urban areas, in an effort for reducing the use of electricity from fossil fuels (SMA, 2012; Zahedi, 2006).

The grid-connected distributed photovoltaic systems are typically installations up to few hundreds of kW in size. These installations are usually residential or commercial ones and are connected to the electricity grid. In most of the cases the electricity produced by the PV is fed back to the utility grid, while the consumption is covered by the grid. In some cases the PV system covers part of the whole of the consumption, and if there is additional need for electricity this is covered by the grid, or electricity is fed back into the utility grid when the on-site generation exceeds the demand of the load. These systems are usually integrated into the built environment and supply electricity to residential houses, commercial and industrial buildings. The costs of these systems are lower compared to an off-grid installation because there is no need for battery storage (EPIA-Greenpeace, 2011; SMA, 2012; EPIA, 2010).

The grid-connected centralized photovoltaic systems or large industrial and utility-scale power plants, as the called also, are large PV installations requiring huge free spaces. These types of PV electricity generation plants present installed power from many hundreds of kilowatts (kW) to several megawatts (MW). The solar panels for industrial systems are usually mounted on frames on the ground or are installed on large industrial buildings such as warehouses, airport terminals or railway stations. In this way, these systems can make double-use of an urban space and provide electricity into the grid where energy-intensive consumers are located, strengthening the utility distribution network locally. These systems can produce enormous quantities of electricity at a single point with respect for the environment. They are also installed as an alternative to conventional centralized power generation, or next to central power stations and fossil fuel open minds for rehabilitation reasons and for reducing the emissions by the stations (EPIA-Greenpeace, 2011; SMA, 2012; EPIA, 2010).

Basic Equipment

The basic and key parts of a solar energy system are the photovoltaic modules, to collect and transform the sunlight and the inverter, to transform the produced direct current (DC) to alternate current (AC). The support structures to orient the PV modules towards the sun and the system components required for the connection of the PV to the grid are also required. In case that the PV system is a stand-alone one a set of batteries with a regulator – charger will be additionally required. The system components, excluding the PV modules, are referred to as the balance of system (BOS) components according to (EPIA-Greenpeace, 2011; Luque & Hegedus, 2003; Greenpeace, 2008).

The solar cell is the basic unit of a PV system. As it is well known, the PV cells are generally made either from crystalline silicon, sliced from ingots or castings, from grown ribbons or from alternative semiconductor materials deposited in thin layers on a low-cost backing (Thin Film). Cells are connected together to form larger units called modules. Thin sheets of Ethyl Vinyl Acetate (EVA) or Polyvinyl Butyral (PVB) are used to bind cells together and to provide weather protection. The modules are normally enclosed between a transparent cover (usually glass) and a weather-proof backing sheet (typically made from a thin polymer). Modules can be framed for extra mechanical strength and durability. Thin Film modules are usually encapsulated between two sheets of glass (EPIA-Greenpeace, 2011; Luque & Hegedus, 2003; Greenpeace, 2008; SMA, 2012).

Typical connection type between modules is the array, where they are connected to each other in series in order to increase the output voltage produced by the system. The current produced by the system is increased when the arrays are connected in parallel. The connections present similarities with the batteries connection types to increase the output voltage and current of a battery system as demonstrated (EPIA-Greenpeace, 2011; SMA, 2012). The power generated by PV modules depends on the module size, the power density measured in Wp/m^2 and the technology used and varies from a few watts (typically 20 to 60 Wp) up to 300 to 350 Wp per module. Low wattage modules are typically used for stand-alone applications where power demand is generally low. Standard crystalline silicon modules contain about 60 to 72 solar cells and have a nominal power ranging from 120 to 320 Wp depending on size, efficiency and power density. Standard Thin Film modules have lower nominal power (60 to 150 Wp) and their size is generally smaller. Modules can be sized according to the site where they will be placed and installed quickly, but usually typical standard dimensions, almost common to any manufacturer, are used when a system is designed. They are robust, reliable and weatherproof. Module producers usually guarantee a power output of at least 90% and 80% of the Wp, even after 10 and 20 to 25 years of use, respectively. Module lifetime is typically considered of 25 years, although it can easily reach over 30 years (EPIA-Greenpeace, 2011; Luque & Hegedus, 2003; Greenpeace, 2008; SMA, 2012).

The inverters are required in every PV installation. They convert the DC power generated by a PV module to AC power in order to make the system compatible with the electricity distribution network and most common electrical appliances. The inverter is essential for grid-connected PV systems. Inverters are offered in a wide range of power classes ranging from a few hundred watts (normally for stand-alone systems), to several kW (the most frequently used range) and even up to 2MW central inverters for large-scale systems. Usually inverters of up to 8kW are one-phase meaning that the output AC voltage is one-phase, while over 8kW the inverters usually provide three-phase symmetrical output AC voltage. In case of large inverters, over 250kW, these central inverters have symmetrical three-phase medium voltage output, ready for connection to the local medium voltage distribution network (EPIA-Greenpeace, 2011; Luque & Hegedus, 2003; Greenpeace, 2008; SMA, 2012; Rigby, et al., 2011).

Inverters are produced by many manufacturers around the world, and they present typical values of efficiency around 95-98% depending on the manufacturer and input voltage and current of the PV system. Differences of 2% are critical for the total power produced by the PV system, especially for the grid-connected ones. If the inverters include transformers then their total efficiency is decreased by a small percentage of around 3-5% depending on the transformer power losses, and thus the total efficiency of such inverter is around 91-93% (EPIA, 2011; Greenpeace, 2008; SMA, 2012; Zahedi, 2006).

The output of the PV module increases with the solar radiation, and decreases with the temperature increase, while other parameters like: ambient temperature, wind velocity, atmospheric conditions, etc. affect also the operation. The maximum output power of the module is different and depends on the specific conditions. The inverters should have an important function: be able to convert the maximum achievable energy under different condition. In order to be able to perform this operation an intelligent control system is required to set the optimum operating point for every inverter under different conditions, as seen literature (EPIA-Greenpeace, 2011; Sen, 2008; Rigby et al., 2011).

Modern inverters are able to always achieve the maximum performance from the solar modules despite different radiation levels. A procedure that measures the PV generator's current potential at fixed intervals, thus allowing the generator to be operated as continuously as possible at its Maximum Power Point (MPP) is included in the inverters. During MPP tracking, the inverter's internal resistance undergoes very small changes at specific time intervals, which simultaneously change both the voltage and the current values of the PV generator. The change in both of these parameters directly affects the generator's output power. If this output power increases, the inverter retains the new voltage and current values. If the PV generator loses power, the inverter continues operating with the original values until the next measuring interval. MPP tracking operates extremely reliably and all well-known inverter manufacturers use it. In the last two years some high power inverters (over 8kW) have divide their power input connector in two sections and incorporated two MPP systems (EPIA-Greenpeace, 2011; Krauter, 2006; Rigby et al., 2011; Fthenakis & Bulawka, 2004).

SMALL-SCALE INSTALLATIONS IN URBAN AREAS AND ELECTRICITY GENERATION

Power Range and Type of PV Installations in Residential Sector

Orientation and elevation of modules is the key element for the correct installation of the PV station, independently from the installed power. Every solar module will deliver its maximum power when it is oriented perpendicular to the incoming radiation. This can only be achieved by tracking systems that follows the sun, but in most applications the modules are installed in a fixed position. This is usually predominant in the case of small installations in urban areas, such the ones under consideration in this chapter. Then the best annual averaged irradiation has to be chosen as a compromise, and at the same time the correct inclination should be chosen. In general, in all locations in the northern hemisphere, a solar generator should be oriented as much as possible toward the south and in the southern hemisphere toward the north. For PV generators located in Europe the recommended orientation is towards the south, with a permissible deviation of 10 to 15 degrees that does not appreciably reduce output. The inclination should be toward the horizontal 25–45 degrees for an optimized output over the full year, depending on the geographic latitude. Other geographic locations require different optimization rules. The general rule for the optimized annually power production for south-facing PV systems, the optimal inclination is approximately the geographic latitude with a tolerance of ± 10 degrees. In order to avoid shading in these installations an analytical calculation should be executed based on the latitude, the module width, the obstructions relative height (height from the basement of the module) and of course the day and time for which the shading should be avoided, which is usually the winter solstice, for the northern hemisphere. Also some empirical approaches exist depending on the latitude that can be used directly. For example to avoid shading at the winter solstice between 9:00 and 15:00 solar time, the separation distance between PV arrays and obstructions should be at least 2 times the obstruction height at latitudes around 30°, 2.5 times the height at latitudes around 35°, 3 times at 40° latitude and 4 times at 45° latitude (EPIA-Greenpeace, 2011; EPIA, 2011; Greenpeace, 2008).

The same conditions apply for the calculations to all installations, independently form the installed power and connection type (off-grid or grid connected). In built or urban areas, PV systems can be mounted on top of roofs (Building Adapted PV systems –BAPV) or can be integrated into the roof or building facade (Building Integrated PV systems – or BIPV). The majority of these systems in urban areas are small scale PV installations. These systems are always systems used for the decentralized supply of electricity and in principle, the electrical power produced can be used in the building directly. But in many countries government or utility regulations exist for feed in tariffs which are higher than the cost of electrical power supplied by the grid, and thus the entire PV electricity is fed into the grid. Especially in urban areas, where networks are extended and strong enough, this is the common practice. Modern PV systems are not restricted to square and flat panel arrays but technology development and modern manufacturing methods and materials have allowed them to be curved, flexible and shaped to the building's design. Innovative approaches in the field are constantly developing new ways to integrate PV into the infrastructures, creating buildings that are dynamic, beautiful and reducing their carbon footprint (EPIA, 2011; Greenpeace, 2008).

Usually the PV systems installed in buildings have a power range of around 5-20 kWp. This is equivalent to a solar generator area of nearly 50-170 m^2, depending on the technology and power density of the modules. Mostly, the PV systems are installed on the roof of single family

Figure 5. Rooftop installation over stone rooftiles in Kifisia, Attica; installed power 4,94kWp, year of installation 2011
(Courtesy of ArvisSolar Ltd).

houses, even though the cases where multihouse buildings with installed PV systems are increasing through the last years. PV systems which are integrated into building facades often have higher power because the usable areas are larger, but the numbers of such installations are quite smaller than the roof-top installations as demonstrated in literature (Luque & Hegedus, 2003;Greenpeace, 2008; SMA, 2012).

Considering the integration possibilities of a solar generator on the building surface, the installation can be on a slope roof (see Figures 5 and 6), on a flat roof (Figure 7) or on a facade. If the type of installation is considered, then the PV arrays could be mounted on racks (usually on flat roofs), mounted at the building surface or even integrated into the roof or façade where the generator replaces part of the building envelope. Additional possibilities including but not limited to sun shades are added quite often as the technology development is continuous (EPIA, 2011; Greenpeace, 2008).

Focusing on the type of the roof, the installation is affected by the slop existence. The installation on a sloped roof is fast, very strong and allows cooling of the modules. Using specific standardized supports for each type of roof cover like: stone tiles as in Figure 5, or ceramic tiles as in Figure 6, the modules are installed at a distance of around 5cm from the existing roof tiles providing cooling through natural convection. In such installations there is no need for adequate distance between the rows of modules in order to avoid unwanted shading. The inverter in such installations cannot be installed bellow the supports and it is usually installed in a protected position and then the electricity is supplied to the grid. In the case of installation on a flat roof as Figure 7 presents, the modules are mounted on a support structure on the surface of the flat roof similar to the one used in large scale or landscape installations. Today, mounting frames, available on the market, present high stability and do not

Figure 6. Rooftop installation over ceramic rooftiles in Kalivia, Attica; installed power 9,46kWp, year of installation 2011
(Courtesy of ArvisSolar Ltd).

Figure 7. Rooftop installation over flat cement roof in Xanthi, North Greece; installed power 7,60kWp, year of installation 2011
(Courtesy of ArvisSolar Ltd).

affect the integrity of the roof of the structure. Furthermore, in most cases the inverter is located directly underneath the solar modules thus it is not necessary to puncture the roof for the cable connection between the solar generator and the inverters. The inverters should be adequately protected against human and weather conditions, depending on their specifications. The installation of the inverter directly on the flat roof has the additional advantage that the DC cable is very short, has no connection with the house and safety issues related to dc current are reduced. In these installations, it is possible to mount the modules in the optimum positions regarding orientation and inclination, but at the same time the minimum space between the rows to prevent shading must be retained (Greenpeace, 2008; SMA, 2012).

When the roof-integrated PV systems are installed into sloped roofs, the modules replace part or all of the normal roofing material, having a double function. They perform the function of a roof and, in addition, produce electric electricity. In recent years, new developments in manufacturing also provide the possibility to use special solar roof tiles that are shaped like a regular roof tile, and presenting a few Watts peak power. The main advantage of the solar roof tiles is that integration in a conventional roof is possible without problems, but the implementation cost is very high due to the connection of a large number of solar tiles to one solar generator. Solar roof are very interesting solutions for the integration of solar generators into buildings, even in historic monuments, especially in areas where dark colored, usually brown or black, roof tiles are used (Greenpeace, 2008).

The facade-integrated PV systems when they are used, not only produce electricity but also play additional functions including but not limited to protection from weather (e.g., rain, wind, humidity), sun protection (overheating, glare), use of day lighting, protection from noise, thermal insulation, etc. Existing experience form the last 10 years has shown that solar modules are able to fulfill all the above listed functions, and they are also able to meet the properties and the technical characteristics of known building materials. Facade-integrated solar generators do not have the optimal orientation and inclination, as they are practically always vertical and thus present reduced production compared with a BAPV system with the same installed power. The loss in output power depends on geographic latitude and the output increases the further north (or south on the southern hemisphere) the building is located, if it is assumed a southern orientation of the facade. Nevertheless, building facades are very interesting examples of solar and modern architecture. The construction of an optically demanding building facade can be combined with an innovative environmentally beneficial electricity generation. Facade-integrated PV systems have been installed in the most cases globally, in banks, administrations, and environmental organizations. Such facades can even be economical considering facade material replaced by PV, depending on the quality, even without considering the electricity production (Greenpeace, 2008; SMA, 2012).

Residential PV Systems in EU

In the following section, the current situation about the residential photovoltaic systems in the European Union is presented revealing the barriers and legislation each member state (National Renewable Energy Lab, 2011).

Portugal

Since February 2013 small-scale PV installations are subject to the same legislation, the DL 25/2013 (the local grid operator is obliged to buy the energy produced, regardless the supplier of the energy for consumption). The only segment is micro-generation, with a maximum capacity of

3.68 kW (nominal inverter). Each consumer can install a system with a capacity of 50% of the capacity contracted for consumption. The production capacity is limited to 25% of the medium voltage transformer capacity.

Spain

In Spain, the legislation gives an exact definition of what is a small PV rooftop (named Type I.1). The RD 1699/2011 says that small PV rooftop must be located in covered structures or facades of fixed, closed buildings, made from resistant materials, dedicated to residential, service, commercial and industrial use, including those of a agricultural nature. It is not possible to install legally a small PV rooftop which does not comply with all the possibilities of this definition.

Although RD 1699/2011 helped to achieve a more agile administrative procedures for small PV rooftops, when the RDL 1/2012 came into force set a moratoria that avoided any new PV installations from perceiving the Feed-in Tariff (FiT). In 2013, a new regulatory framework was proposed. It only allows the most restricted form of self-consumption by: limiting to individual projects, limiting the building typology where to install the project and not taking into consideration any form of net-metering. It´s also proposed a "Backup Toll", a tax on self-consumed electricity.

France

Most of the French PV market in the residential segment is constituted of PV installations under 3 kWp set up on residential buildings. PV Installations on rooftops of residential individual houses are supported by the Feed-in Tariff (FiT) on the produced electricity. The most developed segment is the one which concerns PV systems up to 3 kWp. Since March 2011, building integrated PV (BIPV) installations on residential houses up to 9 kWp are eligible to the highest FiT level. The main barriers are related to obtaining the administrative permissions and to the grid connection costs. Technical barriers are less hindering.

Italy

In Italy PV residential market, 3kW rooftop mounted PV systems are the typical installation. The Feed in Tariff mechanism in Italy finished in July 2013. Authorization procedures are simple enough, even though there can be differences at territorial level.

Bulgaria

The construction of small rooftop PV systems on residential buildings in Bulgaria is hampered by numerous complicated procedures. A recent amendment to the RES Act implements more formal simplified regime for building of small PV on rooftops and surrounding areas by reducing the number of procedures, relevant to large and middle scale ground mounted PV plants. According to this, the procedure should not last more than 60 days, but in practice it takes over 4 months because of several controversial provisions. A transitional provision of the RES Act implements rescheduling of the grid connection of larger scale PV plants after 2016, which the Electricity Distribution Companies are trying to apply also for the small residential PV segment.

Slovenia

In Slovenia, rooftop PV systems on residential buildings fall under the general legislation for RES Systems up to 1 MWp. Since 2007, small rooftop PV systems on residential buildings in Slovenia have been developing significantly. The installed power has almost doubled in each consecutive year; however, changes in the support mechanism have brought with it an almost complete stop in terms of investments into the PV industry. As the main obstacle for poor development of this area are not so much the legal-administrative barriers, but rather the fact that small PV systems are too expensive in comparison to larger ones per kW yield and therefore their return is not as attractive enough for investors. Also, there are no regional differences in Slovenia.

Austria

About one third of the installed PV capacity in Austria is small-scale PV systems smaller than 5 kWp. This percentage has already been higher, but because of changes in the budget for the Feed-in Tariff for bigger plants (since 2012 more money is available) also a lot of bigger plants have been built. Main barriers are uncertainties about costs for the grid connection, uncertainties caused by limited promotion budgets as well as long and costly approval procedures. Rooftop PV systems of 3 kWp (up to 5 kW) are the typical installations for the residential market in Austria.

Poland

According to the Energy Law (11 Sept. 2013), many significant changes for the installations up to 40 kW (called micro-installations) were introduced. Systems up to 40 kW do not require a building permit and installations up to 3 meters high do not require building notification. Connection of micro-installations to the grid is done on a notification and there is no charge for connecting. Individuals can sell the produced energy at a price of 80% of the average selling price of electricity on the competitive market. If the owner of such system is a legal entity, the concession is required.

Slovakia

Residential (roof-top) PV systems represent the smallest share of PV installed in Slovakia. The whole administrative process of installation takes up to 10 months. Since January 1st 2014, the only support scheme is a Feed in Tarrif (89.94, - Eur/MW) which is the biggest barrier because of the too long term of a return on investment. Furthermore, important barriers are also the unstable and constantly changing legal framework, the unpredictable and untransparent regulation and finally a lot of unnecessary bureaucracy. Most of these kind PV installations are usually up to 20 kWp.

Germany

In Germany (2012), small rooftop systems on residential buildings up to 10 kWp in size covered 13% of the total PV market. When applying for grid connection, the experiences with the individual grid operators differ. The application is often easy and quick. In an increasing amount of cases, however, PV system developer report that the process takes too long and involves high connection fees.

Czech Republic

Residential rooftop PV systems represent the only segment that is still in development. The installed capacity has reached 80 MWp by the end of 2012 (PV systems under 20 kWp). At present, there are over 12,000 systems over the country. The overall majority of residential systems have capacity under 10 kWp. There are slightly different administrative requirements for PV systems with capacity over 20 kWp (producers have to prove their professional competence to be allowed to operate a 20 kWp or larger system). From January 2013, the support is administered by the market operator, not the grid operators. There are also several technical difficulties (such as high-order harmonics emission, or ripple-control signal distortion), which tend to delay, or even block higher grid penetration.

Netherlands

The realization of small residential rooftop PV systems (average size of 3 kWp) has seen strong growth in the Netherlands over the last year, thanks to its economical attractiveness in comparison with prices for electricity from the grid and the many collective purchasing initiatives. The standard application for residential rooftop PV systems consists of building applied PV panels mounted on the surface of new and existing non-listed buildings.

Presently, the realizing a PV system on a non-listed building is almost without barriers. Barriers still exist for realizing a PV system on the rooftop of a listed building or in an area of monumental protection, however improvements to minimize these barriers for this particular market segment are on its way.

Belgium

Residential PV installations (maximum 10 kVA low voltage connected and yearly counting) had a pretty well development in Wallonia (residential PV systems represents 97% of PV systems in Wallonia)- Characterized by an average power of 5 kVA, it is mostly rooftop and sometimes on the ground (garden). The total installed residential PV is estimated to be up to 640 MWp (over 125,000 systems) end 2013. Currently the market is nearly totally stopped because of the uncertainties about the new support scheme (Qualiwatt) many specifications still have to be defined. This political context is the current main barrier.

United Kingdom

The latest estimations (November 2013) of the amount of installed PV power in the UK (by the Department of Energy and Climate Change, DECC) indicated that there are about 2.6 GW of installed PV capacity in operation, mainly concentrated in the residential and commercial market segment. Larger installations are also present but to a lesser extent. Data from DECC (November 2013) indicate that 466,918 installations had a capacity lower than 4 kW.

Residential systems can count on a quite swift procedure with minor and relatively few barriers in the installation process, as it takes between 2 and 4 weeks to complete the whole process.

Greece

Greece ranks 5th worldwide with regard to per capita installed PV capacity. 4.5 billion € were invested in PV in Greece during the last 5 years. In 2014, PV is expected to cover 7% of electricity demand in Greece.

For a small country in an unprecedented financial crisis, this performance is indeed amazing. There is a very good reason for this: a brave support mechanism. Greece has been offering high feed-in-tariffs (FiTs) for PV since 2006. This has skyrocketed the market, especially during the period 2011-2013, and reached a cumulative installed PV capacity of 2.5 GWp in 2013.

However, this very mechanism has also overheated the market. High feed-in-tariffs and dramatic decreases of PV costs since 2011 have led to a boom that the Electricity Market cannot sustain anymore. The Electricity Market Operator cannot raise the necessary funds for compensating the 2.5 GWp of PV installed so far (hence there are delays in payments of the feed-in-tariffs). As a result of cash flow problems of the Market Operator, the Greek authorities have taken drastic measures against existing and future PV installations, since August 2012:

- **A Temporary Tax:** Ranging between 25% and 42%, has been imposed to all operating PV plants (residential systems excluded).
- **Licensing Process:** For new PV projects has been put on halt (residential systems excluded).

Figure 8. Greek PV market development

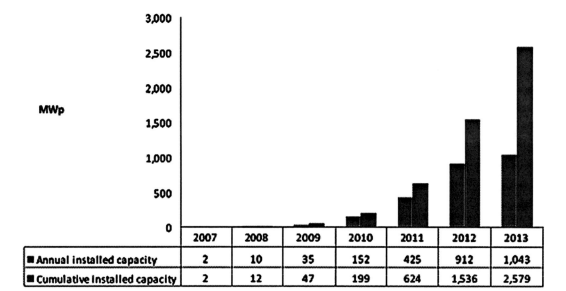

- **Feed-in-Tariffs (FiTs):** For new PV plants have been reduced in 2013 to 125 €/MWh for residential systems (<10 kWp), 120 €/MWh for small commercial systems (<100 kWp) and 95 €/MWh for large systems (>100 kWp), with a further drastic digression planned for the years to come.

The average wholesale cost of electricity in Greece is 0.75 €/MWh, with the cost of gas plants exceeding 110 €/MWh. Obviously, a solar kWh compensated with the new FiTs currently costs less than a KWh generated by a gas-fuelled power plant, reducing, simultaneously, country's CO_2 emissions.

In October 2013, the Greek Parliament approved a new support scheme based on net-metering (in parallel with the existing feed-in-tariff scheme). Projects making use of the net-metering scheme will not be affected by the freeze of the licensing process.

According to statistical date from HELAPCO (Hellenic Association of Photovoltaic Companies) the PV market development is shown in Figure 8.

Furthermore, in Table 1 the grid-connected PV systems are presented by size for the case of Greece (source: HELAPCO, 2013)

Power Production from Typical Residential PV Systems in EU

The power production in such installations presents differences even for nearby areas in the same city. These differences, in such case are explained mainly due to the differences in inclination and orientation that are obvious, especially in the case of sloped roof. Additionally to these reasons the existence of shading due to nearby buildings cannot be neglected when these data are collected analyze. Tables 2 and 3 present 15 in total different small scale PV installations – BAPV which were observed for over one year, along with their basic installed components. The seven (7) small scale urban PV installations of Table 2 are located in different areas of Greece, extending from some of the northern regions (e.g. Chalkidiki) to the southern ones as the island of Crete. Figures 5, 6, and 7 present three of these installations. The other eight (8) small scale installations presented

Table 1. Grid connected PV systems by size in Greece

Grid-Connected PV Systems by Size	Rooftop <10kWp	<20kWp	20-150kWp	150kWp-2MWp	>2MWp
Cumulative installed capacity (MWp)	372.7	65	917	843.2	380.9

(Source: HELAPCO).

in Table 3 are located in UK, Germany, France and Spain, extending from areas of EU with lower solar irradiance such us UK to countries with high one such as Spain. Tables 4 and 5 present the monthly produced electricity for one last year as this was measured by the energy meter of each system. As it can be seen the production seems adequate considering monthly values. Tables 6 and 7 present the normalized values of electricity production per installed kWp (kWh/kWp). As it can be seen the normalize values of the monthly electricity production shows differences for installations located in the same area (Attica). These differences even though cannot be neglected it can be easily attributed to the already mentioned before differences in each installation. Additionally since these measurements have been executed in each systems output in AC, include the inverters efficiency. Even though the inverters in all these installations are manufactured by the same company, the difference on their type corresponds to differences in their efficiency. 1% can be easily attributed in such efficiency difference. Figures 9, 10, and 11 present typical measurements of the daily production for two typical months, the monthly production for one year and the hourly production for two typical days respectively, taken by the inverter and stored to the memory. These are typical data that can be collected by each system. These data can provide useful information to the owner of the PV system showing for example in figure that on May 13, 2012, the solar irradiation was reduced for some reason, while the solar irradiation during May 6th 2012 had the expected shape. These assumptions can be determined by the fact that the production each time interval measured in a PV generation system, corresponds to a solar irradiation transform in to electricity by the PV generator.

A detailed comparison of the actually measured data will show that the experimental estimated electricity production per installed kWp is higher but in close agreement with the

Table 2. Small scale PV installations in different locations in Greece

PV Location	Installed Power (kWp)	Inverter Type	PV Panels' Type
Acharnes – Attica	7.76	3 Sunny Boy 2500	33pcs Sharp NU E235W
Karpenishi	9.83	Sunny Tripower 10000TL-10	41pcs Solyndra SL
Chania – Crete	9.66	Sunny Tripower 10000TL-10	42pcs Aleo S_18 230W
Anchialos – Volos	9.89	Sunny Tripower 10000TL-10	43pcs Schüco MPE 230
Chalkidiki	9.90	Sunny Tripower 10000TL-10	44pcs SILCIO SE 225Wp
Larissa	9.88	Sunny Tripower 10000TL-10	52pcs Suntech STP190
Ag.Varvara – Attica	7.99	Sunny Tripower 10000TL-10	34pcs Solon Blue 230/7 235W

(Courtesy of ArvisSolar Ltd).

Table 3. Small scale PV installations in different locations in EU

PV Location	Installed Power (kWp)	Inverter Type	PV Panels' Type
Werne, Germany	9.84	Sunny Tripower 10000TL-10	41pcs CanadianSolar CS6P-240P
Bergisch Gladbach, Germany	9.80	Sunny Tripower 10000TL-10	40 pcs aleo S_19 245W
Guildford County, UK	17.00	Sunny Tripower 17000TL-10	76pcs Sharp 235W
Gloucester, UK	16.17	Sunny Tripower 15000TL-10	66pcs Sharp 245 Watt
Villamartin, Spain	5.17	Sunny Tripower 5000TL-20	22pcs ATERSA A-235P
Madrid, Spain	2.64	Sunny Boy 2500	12pcs SolarWorld SW 220 poly +
Chatellerault, France	9.60	5pcs Sunny Boy 2000HF-30	40 pcs Sillia SeT240G
Joue les Tours, France	35.49	1 pcs Sunny Tripower 10000TL-10 2pcs Sunny Tripower 12000TL-10	169pcs Kyocera KD210GH-2PU

(Source: www.sunny.portal.eu).

theoretical ones mentioned by JRC maps (Figure 1). This agreement supports that the expected efficiencies of modules and systems are continuously increased as the technology improvements becoming commercially available and will lead to higher electricity production using the less area for installations.

ECONOMIC ANALYSIS AND COST EVALUATION

The share of legal-administrative costs over total project development costs (excluding PV equipment and other materials, Figure 12) can provide an idea of the economic burden that

Table 4. Monthly electricity production for selected small scale PV installations in different locations in Greece

Date	Acharnes (kWh)	Karpenisi (kWh)	Chania (kWh)	Anchialos (kWh)	Chalkidiki (kWh)	Larissa (kWh)	AgiaVarvara (kWh)
May 2011	1234.05	1703.69	1533.04	1547.45	1487.12	1254.05	1214.51
Jun 2011	891.69	1770.98	1570.72	1762.12	1554.38	1646.26	1227.88
Jul 2011	1337.75	1834.23	1599.70	1952.48	1704.9	1761.11	1364.92
Aug 2011	1312.51	1093.09	1572.17	1906.97	1834.47	1868.31	1300.54
Sept 2011	1130.92	1331.88	1178.08	1560.8	1389.79	1384.54	1093.77
Oct 2011	907.41	903.18	1096.59	998.58	854.71	983.97	805.77
Nov 2011	690.75	700.17	720.83	633.32	547.84	778.67	470.2
Dec 2011	599.82	505.99	870.11	606.57	688.35	686.84	450.2
Jan 2012	572.05	608.27	678.96	604.43	952.72	686.84	455.2
Feb 2012	552.55	608.27	900.84	690.43	762.67	1057.87	490.19
Mar 2012	1029.62	994.49	1431.37	1355.01	1451.61	687.10	945.3
Apr 2012	1218.39	1404.94	1491.78	1295.73	1304.71	1321.47	1186.86
Annually	11477.51	13459.18	14644.19	14913.89	14533.27	14117.03	11005.34

(Courtesy of ArvisSolar Ltd).

Table 5. Monthly electricity production for selected small scale PV installations in different locations in EU

Date	Werne (kWh)	Bergisch Gladbach (kWh)	Guildford County (kWh)	Gloucester (kWh)	Villamartin (kWh)	Madrid (kWh)	Chatelle-Rault (kWh)	Joue les Tours (kWh)
May 2013	1153	980	2157	2001	923	424	1008	4216
Jun 2013	1241	1067	2114	2127	963	458	1127	4861
Jul 2013	1354	1305	2667	2540	979	467	1512	5830
Aug 2013	1341	1035	2231	1831	849	443	1467	5438
Sept 2013	908	645	1459	1285	691	372	982	3438
Oct 2013	744	397	987	831	554	279	610	2120
Nov 2013	300	163	862	535	426	217	353	1214
Dec 2013	339	129	624	388	337	186	419	1384
Jan 2014	443	196	375	450	310	154	301	1093
Feb 2014	604	316	1083	745	351	192	560	942
Mar 2014	1199	724	1887	1376	668	331	1145	4998
Apr 2014	1111	885	1917	1590	749	394	1131	4317
Annually	10737	7842	18363	15699	7800	3917	10615	39851

(Source: www.sunny.portal.eu).

Table 6. Monthly electricity production in kWh/kWp for selected small scale PV installations given in Table 4

Date	Acharnes	Karpenisi	Chania	Anchialos	Chalkidiki	Larissa	Agia Varvara
May 2011	159.0	173.3	158.7	156.5	150.2	126.9	152.0
Jun 2011	114.9	180.2	162.6	178.2	157.0	166.6	153.7
Jul 2011	172.4	186.6	165.6	197.4	172.2	178.3	170.8
Aug 2011	169.1	111.2	162.8	192.8	185.3	189.1	162.8
Sept 2011	145.7	135.5	122.0	157.8	140.4	140.1	136.9
Oct 2011	116.9	91.9	113.5	101.0	86.3	99.6	100.8
Nov 2011	89.0	71.2	74.6	64.0	55.3	78.8	58.8
Dec 2011	77.3	51.5	90.1	61.3	69.5	69.5	56.3
Jan 2012	73.7	61.9	70.3	61.1	96.2	69.5	57.0
Feb 2012	71.2	61.9	93.3	69.8	77.0	107.1	61.4
Mar 2012	132.7	101.2	148.2	137.0	146.6	69.5	118.3
Apr 2012	157.0	142.9	154.4	131.0	131.8	133.8	148.5
Annually	1478.9	1369.3	1516.1	1507.9	1467.8	1428.8	1377.3

(Courtesy of ArvisSolar Ltd).

Table 7. Monthly electricity production in kWh/kWp for selected small scale PV installations given in Table 5

Date	Werne	Bergisch Gladbach	Guildford County	Gloucester	Villamartin	Madrid	Chatelle-Rault	Joue les Tours
May 2013	117.2	100.0	126.9	123.7	178.5	160.6	105.0	119.1
Jun 2013	126.1	108.9	124.4	131.5	186.3	173.5	117.4	137.3
Jul 2013	137.6	133.2	156.9	157.1	189.4	176.9	157.5	164.7
Aug 2013	136.3	105.6	131.2	113.2	164.2	167.8	152.8	153.6
Sept 2013	92.3	65.8	85.8	79.5	133.7	140.9	102.3	97.1
Oct 2013	75.6	40.5	58.1	51.4	107.2	105.7	63.5	59.9
Nov 2013	30.5	16.6	50.7	33.1	82.4	82.2	36.8	34.3
Dec 2013	34.5	13.2	36.7	24.0	65.2	70.5	43.6	39.1
Jan 2014	45.0	20.0	22.1	27.8	60.0	58.3	31.4	30.9
Feb 2014	61.4	32.2	63.7	46.1	67.9	72.7	58.3	26.6
Mar 2014	121.8	73.9	111.0	85.1	129.2	125.4	119.3	141.2
Apr 2014	112.9	90.3	112.8	98.3	144.9	149.2	117.8	121.9
Annually	1508,9	1483,7	1105,7	1125,7	1508,9	1483,7	1105,7	1125,7

(Source: www.sunny.portal.eu).

project developers have to bear in order to secure the authorizations needed to build and connect a PV system. This burden is normally reflected in national PV system prices.

The total labour required for accomplishing the permitting and grid connection procedures (Figure 13) can instead serve as a measure of the complexity and lack of transparency hidden within these administrative procedures.

The total duration of the development process for a PV project (Figure 14) is another measure of the economic risk faced by investors: the more it takes to build and connect a PV system, the longer investors are financially exposed without earning revenues. Additionally, the waiting time spent uselessly by a developer waiting for an answer from an authority or a grid operator can be a measure of the inefficiency shown by such parties in dealing with their tasks.

The technology developed by PV industry was among the most expensive ones from the first stages of development and use. Over the past 30 years of operation the PV industry has achieved impressive price decreases in the modules costs, which has lead the consumers to consider the PV installations as a possibility for electrification. According to EPIA – Greenpeace report of 2011 (EPIA-Greenpeace, 2011), the price of PV modules is reduced by 22% each time the cumulative installed capacity (in MW) is doubled. This reduction in manufacturing costs and retail prices of PV modules and systems (including electronics and safety devices, cabling, mounting structures, and installation) is a result of the economies of scale and experience gained by industry and installers. This has been brought about by extensive innovation, research, development and ongoing political support for the development of the PV market, especially from the EU. This effort of EU has as a result that EU Member States have cumulative installed power representing almost the 75% of the global volume (see Figure 2) (EPIA-Greenpeace, 2011; EPIA, 2010; Rigby et al., 2011; Fthenakis & Bulawka, 2004). However, the module is only part of the total cost picture. Additional costs are introduced with the inverter; the Balance of Sys-

Electricity Production from Small-Scale Photovoltaics in Urban Areas

Figure 9. Daily production of a roof-top installation for two typical months: a. Agia Varvara, Greece, installed power 7.99kWp; b. Werne, Germany, installed power 9.84kWp; c. Villamartin, Spain, installed power 5.17kWp

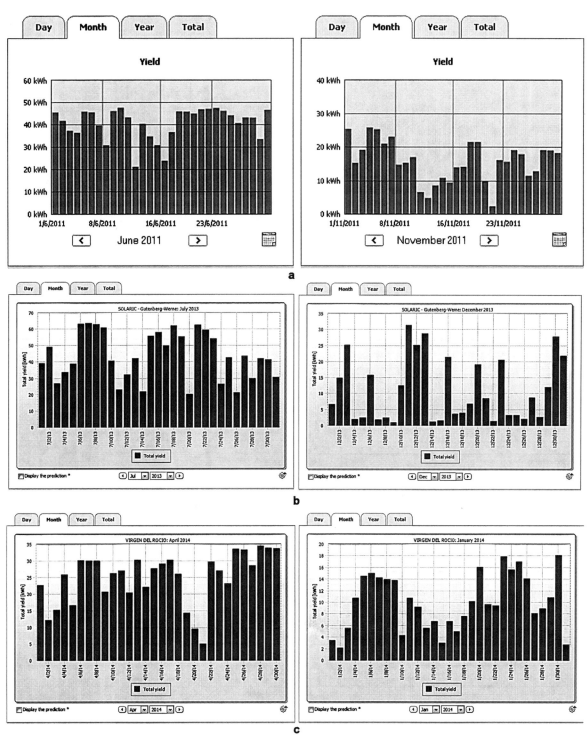

Figure 10. Total monthly production of a roof-top installation: a. Agia Varvara, Greece. Installed power 7.99kWp; b. Werne, Germany, installed power 9.84kWp; c. Villamartin, Spain, installed power 5.17kWp

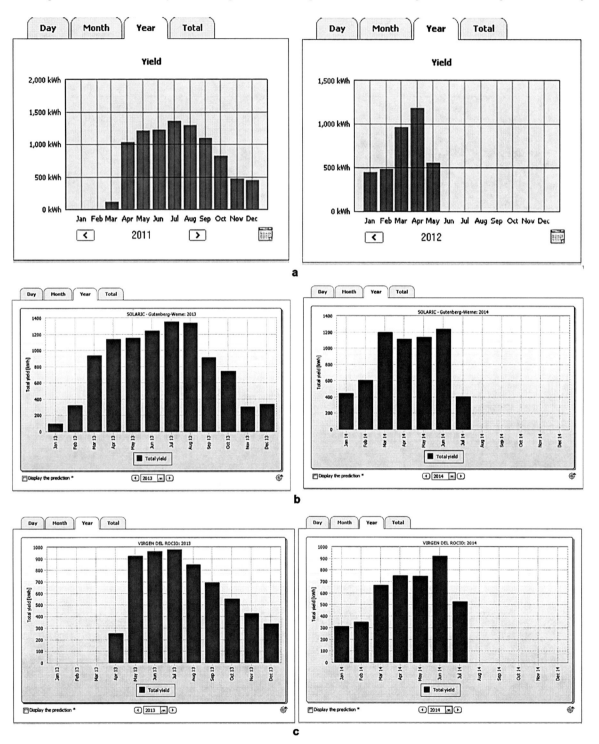

Electricity Production from Small-Scale Photovoltaics in Urban Areas

Figure 11. Typical hourly production of a roof-top installation: a. Agia Varvara, Greece, installed power 7.99kWp; b. Werne, Germany, installed power 9.84kWp; c. Villamartin, Spain, installed power 5.17kWp

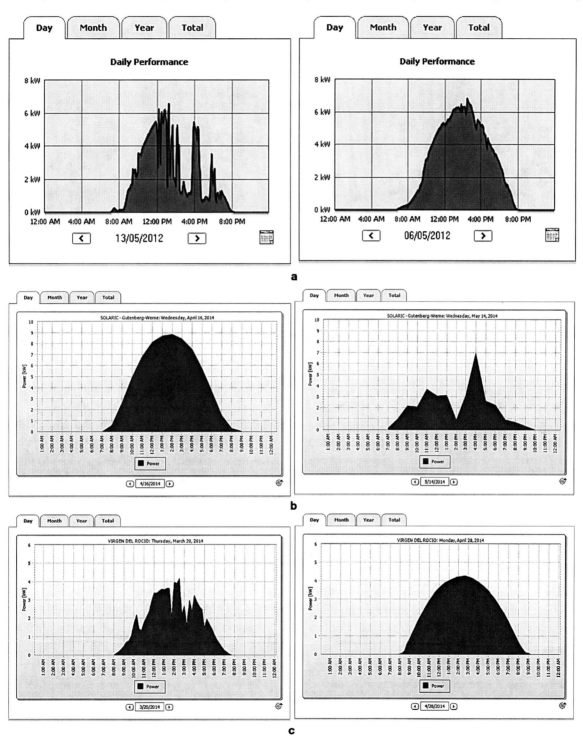

Figure 12. PV project development: legal-administrative cost share
(Source: HELAPCO, 2013).

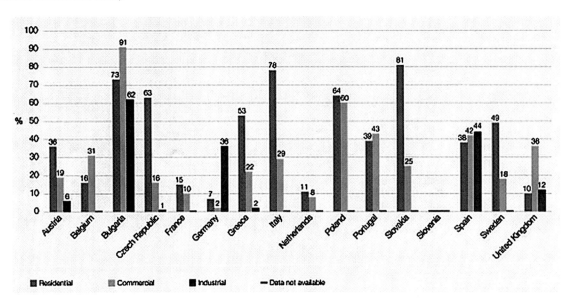

tem costs BOS which includes the other system components and technical costs, installation and engineering / permitting. The module represents typically 50-60% of total installed systems costs as it can be seen from Figure 16 (EPIA-Greenpeace, 2011; Rockett, 2010).

The PV system price was decreased over the past 30 years, even though this decrease was not as impressive as the PV modules one. The price of inverters has followed a similar price learning curve to that of PV modules, for the same reasons. Innovation, development and economies of scale resulted a significant reduction in the inverters cost (EPIA-Greenpeace, 2011; EPIA, 2010). Prices for some balance of systems (BOS) elements have not decreased with the same pace, due to the fact that the majority of the presented already economies of scale, while non-technical costs do not present such decreasing. In particular the price of the raw materials used in these elements (typically copper, steel and stainless steel) has been more volatile, as they are brokerage products. Installation costs have decreased at different rates depending on the maturity of the market and type of application. For example, some mounting structures designed for specific types of installations (such as typical bases for horizontal cement roofs in rooftop installations, as shown in Figure 7) can be installed in half the time it takes to install a more complex version (as bases for stone tales covered roofs with inclination in rooftop installations as shown in Figure 5). Reductions in prices for materials (such as mounting structures), cables, land use and installation account for much of the decrease in BOS costs. This of course lowers the total installation costs. Another contributor to the decrease of BOS and installation-related costs is the increase in efficiency at module level, which is directly related with the power density that the module presents as demonstrated (Luque & Hegedus, 2003). More efficient modules imply lower costs for balance of system equipment, installation-related costs and land use (EPIA-Greenpeace, 2011; EPIA, 2010).

Figure 15 shows as an example the price structure of PV systems for small rooftop (3 kWp) installations in mature markets. In only 5 years' time, the share of the PV modules in the total system price has fallen from about 60-75% to as low as 40-60%, depending on the technology, according to the EPIA-Greenpeace report of 2011

Figure 13. PV project development: legal-administrative labour requirements
(Source: HELAPCO, 2013).

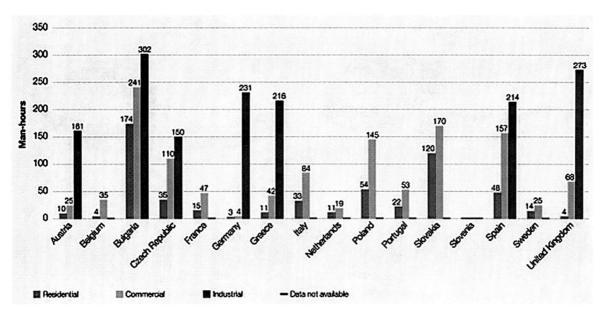

Figure 14. PV project development: duration and waiting time
(Source: HELAPCO, 2013).

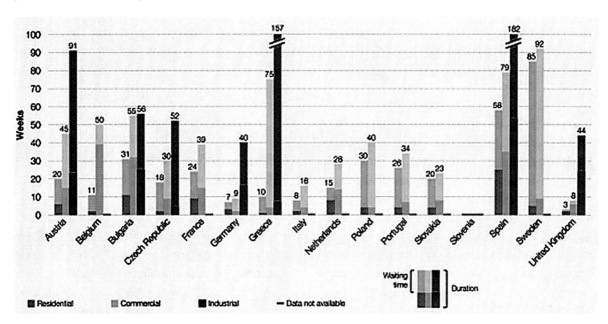

Figure 15. Price structure of a small PV rooftop installation 3kWp using c-Si or thin film (TF)

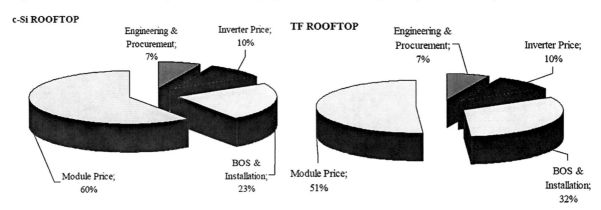

(EPIA-Greenpeace, 2011). The inverter accounts for roughly 10% of the total system price, while the cost of engineering and procurement makes up about 7% of the total system price (EPIA-Greenpeace, 2011; EPIA, 2010). The remaining costs represent the other balance of system components and the cost of installation. In 2010, the range represents prices for large systems in Germany. The rate at which PV system prices will decrease depends on the installed PV capacity. By 2030 prices could drop to between €0.70/Wp to €0.93/Wp and by 2050, the price could be even as low as €0.56/Wp, according to EPIA-Greenpeace, (2011); Greenpeace, (2008) and EPIA, (2010).

Considering the conditions in developing market such as the Hellenic one this price structure is not always apply. For example, in the case of VAT 23% in the total cost is applied in all small-scale rooftop installation that belongs to persons – citizens and not to legal entities. This administrative cost is extremely high. As an example, it is presented the cost per Wp from the lowest financial offer given by an installation company for two small scale rooftop installations. The first was for a 6,86kWp PV system and the second for a 7,36kWp PV system, both installed in a horizontal cement roof Athens. For the 6,86kWp installation the cost per installed Wp including VAT was between 3,08€/Wp - 3,32€/Wp depending on the module producer (both c-Si, 245Wp with same dimensions), while the cost without the VAT was vary between 2,51€/Wp and 2,70 Wp. For the 7,35kWp installation the costs are between 2,98€/Wp and 3,2€/Wp including VAT, while net prices between 2,42€/Wp and 2,63 Wp (VAT not included). Figure 16 shows the cost breakdown for these rooftop installations were the PV installation company provides 2 different types of modules the Bosh 245M and CNPV245P in each installation.

Rapidly falling prices have made solar more affordable than ever. The average price of a completed PV system has dropped by 33 percent since the beginning of 2011.

BENEFITS FROM PV INSTALLATIONS

The benefits from the PV systems electricity production are quite important. These systems produce energy from sunlight which is freely available. The use of solar photovoltaic reduces the greenhouse emissions obviously by replacing fossil fuels. Life cycle studies have proven that the energy required to manufacture a solar power system can be recouped by the energy costs saved over one to two years, while some new generation technologies can even recover the cost of the energy used in less than a year, depending on their

Figure 16. Cost of PV elements for two installation calculated using two different modules (Courtesy of ArvisSolar Ltd).

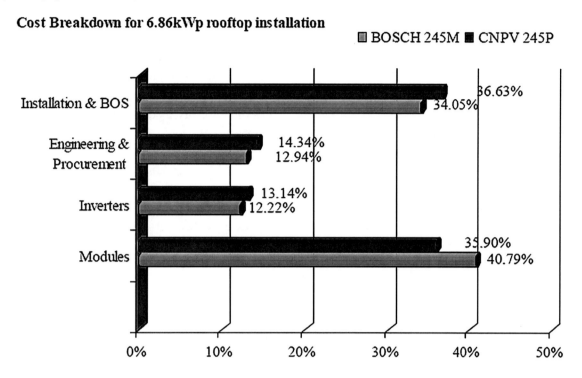

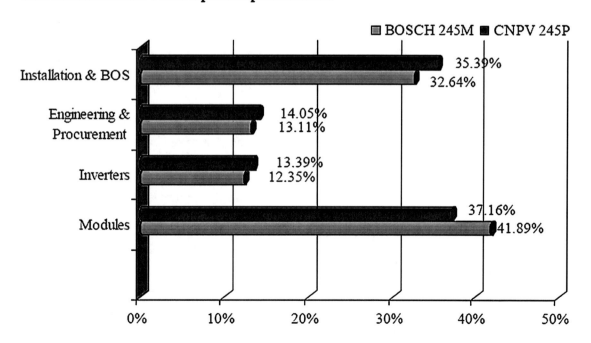

location. As it is already mentioned above the typical operational life time of the PV systems is at least 25 years; thus it is ensured that each panel generates many times more energy than it costs to be manufactured (EPIA-Greenpeace, 2011; Zahedi, 2006; EPIA, 2010; Colville, 2014). Further to this operational life, the fact that there are no substantial limits to the massive deployment of PV could enhance this effort. Material and industrial capability are plentiful, mainly due to fact that there alternative materials to crystalline silicon. The PV industry has demonstrated its ability to increase production quickly enough to meet growing demands. This ability has been demonstrated in Germany, Italy and Japan which have implemented proactive PV policies. This ability will become a significant advantage for the effort to increase renewable energy share in the power mix in order to support sustainability in the power production (EPIA-Greenpeace, 2011; Šúri et al., 2007; EPIA, 2011; Zahedi, 2006; EPIA, 2010).

Small-scale PV systems can be installed in every urban area as long as there is no shading. This possibility is extremely helpful where decentralized electricity generation is considered. Decentralized electricity production systems are among the key points of the smart grids, and can support heavily the power demand which is increasing continuously. Since small scale photovoltaic systems can be connected directly to low voltage distribution network, they can be easily spread out the network providing support to specific network positions and reducing the need for local capacity increase, due to increased demand in many cases according to literature (EPIA, 2010; EPIA-Greenpeace, 2011; EPIA, 2011; EUROBSERV'ER, 2012). This deployment of small-scale PV systems will support the effort of making cities greener. EU, where the majority of studies have been executed, has a capacity of over several GWp of PV systems to be installed at the 40% of all building roofs and in the 15% of all facades that are suited for. If the over 22,000km2 total ground floor area in the EU-27 cities are added to the above mentioned areas, the potential capacity of PV systems that could be installed in EU-27 will reach or exceed the 1.5TWp. If this scenario will be applied then the resulted electricity generation could potentially exceed the 1,400TWh, an amount corresponding on the 40% of the electricity demand predicted for 2020. Domestic and tertiary consumptions including but not limited to lighting, air-conditioning, office equipment, refrigerators, food processing equipment, etc, are responsible for a significant amount of energy consumed globally; thus, resulting to significant amounts of greenhouse gases emissions considering fossil fuels. If the abovementioned scenario for PV installation will be applied, as a total or even partially, then these emissions will be decreased notably. Additionally, the small-scale photovoltaic systems installation in buildings is one of the key factors in the effort of developing sustainable, low or even zero emission buildings (Zahedi, 2006; Šúri et al., 2007; EPIA, 2010; EPIA-Greenpeace, 2011; EPIA, 2011; Colville, 2014).

FUTURE RESEARCH DIRECTIONS

As it has been already mentioned, the photovoltaics are one of the most promising renewable energy technologies. Photovoltaics (PV) are a truly elegant means of producing electricity on site, directly from the sun, without concern for energy supply or environmental harm as it was mentioned in literature (Jelle & Breivikb, 2012).

Research laboratories and companies are in a "race" to manufacture more and more efficient PV cells. This completion was presented very clearly in graphic produced by NREL (Figure 17). There as it can be seen the increment in the cells efficiency is increasing 0.1-0.7% in many technologies while the most promising seems to be the multi-junction cells and single-junction GaAs cells. In addition the last years emerging PV cells have been developed such as dye-sensitized, organic and organic tandem cells, inorganic, even quantum dot cells. These technologies are mainly developed in Universities' research labs and seem to be yet at their earliest stages with scientists globally starting to pay attention to these technologies. It seems that the next years the research on improving further the cells' efficiency seems to be focused in these types of cells. The research paths for potential new PV technologies can be found in miscellaneous fields, such as but not limited, ultralow cost and low-medium efficiency organic based modules, ultra-high efficiency modules, solar concentrator and/or solar trapping systems embedded in the solar cell surface and material beneath, and flexible lightweight inorganic thin film solar cells, and of course the PV technologies yet to be discovered (Scognamiglio & Røstvik, 2013; EU, 2010).

One recent research strategy utilized to achieve high solar cell efficiencies is to make so-called sandwich or stack or multilayer solar cells, which use several different material layers and cells with different spectral absorbencies to harvest as much as possible of the solar radiation in a wide wavelength range. Hence, a much larger portion of the solar radiation is utilized. This technique is very common both in EU and USA research laboratories with notable results as can be seen in Figure 17 (Scognamiglio & Røstvik, 2013; EU, 2010).

The installation of PV in buildings will also be enhanced by the recent recast of the European Directive 2010/31/EU, which establishes that starting from the end of 2020, all new buildings will have to be Nearly Zero Energy Buildings (Nearly ZEBs). According to this directive, 'Nearly ZEB' means a building that has a very low energy yearly energy consumption, which can be achieved by both the highest energy efficiency and by energy from renewable sources, which shall be 'on-site' or 'nearby' (Voss & Musal, 2012). A relevant

Figure 17. A timeline for reported best research-cell efficiencies, depicting all verified records for various PV conversion technologies
(Source: National Renewable Energy Lab, 2011).

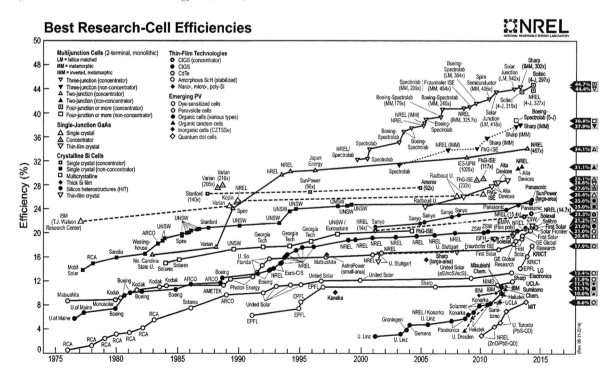

international effort on the subject can be found in the literature 'Towards Net Zero Energy Solar Buildings' (Argan, 1993; Rossi, 2011).

Both from the theoretical and practical points of view, this new approach for the buildings might be a revolution for architecture and for Photovoltaics (PV), too (IEA SHC, 2008) The engineering only research taking into account mainly the energy aspects seems to be not sufficient to ensure the diffusion of ZEB models: in achieving the ZEB target a major role will be played by architects and designers, who are amongst the main actors of this revolutionary change. The way architects will take up the challenge of designing ZEBs and include PVs is crucial, as architects are highly responsible of the form of the city and of its buildings (Ventur, Scott Brown, & Izenour, 1977, Koolhas & Mau, 1997; Scognamiglio, Ossenbrink, & Annunziato, 2011; Voss, Kiefer, Reise, & Meyer, 2003).

In a near future, buildings will be designed to need very little energy (passive design strategies for energy efficiency) and to integrate active surfaces for generating energy and the PVs have already developed to meet this challenge and new design concepts are developed almost "daily". This approach requires a new thinking, especially from the architects, able to use the energy needed as an input for design. The energy should be seen as a variable able to be related to the form of the buildings (or clusters of buildings or even cities and landscapes), instead of being seen as a variable that design cannot deal with. In the future, design has to consider not only the space used directly by the people directly but also the space required to provide for electrical and thermal energies from renewable sources: the surface necessary for placing the energy generation devices (Scognamiglio, Bosisio, & Di Dio, 2009). Because the renewable energy generation systems, in contrast to conventional energy sources, are visible, for the first time in the tradition of architecture, energy can take a 'form' a shape, colors and features of a PV generator, and architects are responsible for designing these forms (Jelle & Breivikb, 2012; EU, 2010; IEA SHC, 2008).

Photovoltaics have many potentialities in a ZEB scenario, thanks to its features and huge decrease in cost. Because of the high energy consumption of the European countries, PV can contribute significantly to the reduction of the primary, conventional energy supply, as well as to the reduction of the CO_2 emissions. PV is a mature technology and seems to be technically the easiest way to obtain the zero energy balance, as the recent, sharp, drop in prices makes it competitive even with active solar thermal collectors and building materials in general (IEA SHC, 2008).

Photovoltaics can be used exactly where the energy is consumed ('on-site' energy generation). It can be easily integrated anywhere into the building envelope, allowing for a number of functions: that is, on/in rooftops, opaque and semitransparent envelope surfaces, having a structural function as well as sun-shading and cladding function and so forth, and enabling also a construction costs reduction. This condition advantages PV over other renewable sources because only few of them are suited for being used very close to the building and only even fewer can be integrated into the building envelope (PV and solar thermal). PV can be used also in combination with such technologies, allowing for an optimal energy design of the building (Jelle & Breivikb, 2012; EU, 2010; IEA SHC, 2008).

Furthermore, thanks to the dramatic decrease in costs, today PV can be considered a 'standard' material for buildings with the advantage of generating energy: a PV module costs in the order of $100 EUR/m^2$ and generates about 80/150kWh/m2/year at an efficiency of 120 W/m^2 which can be standard low cost power density for different types of PVs (Scognamiglio & Røstvik, 2013; EU, 2010; IEA SHC, 2008).

These considerations on the potentialities of PV in achieving the ZE balance suggest a very simple architectural implication: PV is going to become an indispensable material for buildings, with the consequence of being in a near future a very visible part of the building composition. Because of the mandatory request for buildings to achieve the Net ZE balance, if today PV plays a minor role in the composition of the most of the existing buildings envelopes—small surfaces—in a near future, it will have a main role, as PV surfaces in buildings will likely become bigger and bigger. The consequent influence of the use of PV on the architectural image of the building, and on the way the city itself can look like, is very considerable, opening up a new wide perspective for the relationship between PV and architecture (Jelle & Breivikb, 2012; EU, 2010; IEA SHC, 2008).

Until now, in fact, much research has been carried out on how to use PV in buildings focusing on technical, aesthetical (and economical) aspects. Nevertheless, the relationship between PV and the energy balance of the building was not the main concern (IEA SHC, 2008). The need to meet the ZEB balance opens new perspectives for the use of PV: PV might, in fact, be used into the building envelope, and also, it might be used close to the building ('on-site' or 'nearby' energy generation) accounting for the building energy balance. Furthermore, the boundary of the building's balance could be also extended to a cluster of buildings, with the consequence that the use of PV should be described and accounted not for a single building but for a cluster of buildings. In this condition, it seems unpostponable to open a design investigation on the use of PV in Net ZEBs, where ZEB balance is the main target of design (Jelle & Breivikb, 2012; EU, 2010; IEA SHC, 2008).

Considering this bright future of the implementation of PVs in buildings, from the engineering and from the architectural point of view the need to create a bridge over scientific knowledge and architecture's practice seems the "biggest challenge". This challenge will be the new frontier both architecture and PV industry, along with the new technologies under development in the research labs.

CONCLUSION

The interest in solar photovoltaic energy is growing worldwide resulting in more than 40GW of photovoltaic systems cumulative capacity all over the world. Environmental concerns and price decreasing along with Feed-in Tariffs have support the implementation of photovoltaic systems. The major advantage of solar photovoltaic energy systems is that they generate electricity pollution-free and can be easily installed on residential and commercial buildings as grid-connected PV applications. Globally, the potential for electricity production from photovoltaic systems is enormous, but the small-scale installations in urban areas are the most interesting case as they provide additional benefits and can help highly in the overcome of demand coverage issues among others. The types of the roof-top installations, due to their simplicity and relatively low cost they present, seem to be the key factors for a more sustainable future, especially for urban areas. Electricity production data, measured in installations operating for at least one year in different installations in Hellenic cities verify the theoretical values and demonstrate this significant support to production that these systems can provide. Experimental results from field measures are in close agreement with theoretical ones mentioned in literature and demonstrate the increment of photovoltaic systems efficiency as new and improved materials becoming commercially available. The cost analysis showed the reduction considering the global market as well the local market of Greece. The benefits of the implementation of small-scale photovoltaic generators are notable: they provide support to the low voltage distribution network, contribute in the green of cities and urban areas, they can contribute significantly in the effort of reducing the greenhouse gases emissions. All these data presented and discussed, support the opinion that the photovoltaic systems can help to achieve sustainability and the role of small- scale low cost installation can be an additional advantage, since they can be easily implemented in urban areas.

ACKNOWLEDGMENT

The authors would like to thank ArvisSolar Ltd for the data provided regarding urban installations in different areas of Greece.

REFERENCES

Argan, G. (1993). *Storia dell'arte come storia della città (IT)*. Roma, Italy: Riuniti.

Colville, F. (2014). *Small-scale solar PV market trends in 2013*. Retrieved January 2, 2014, from http://www.solarbuzz.com/resources/articles-and-presentations/small-scale-solar-pv-market-trends-in-2013

EPIA. (2010). *Global market outlook for PV until 2014*. Brussels, Belgium: European Photovoltaic Industry Association. Retrieved January 2, 2014, from http://www.epia.org/fileadmin/EPIA_docs/public/Global_Market_Outlook_for_Photovoltaics_until_2014.pdf

EPIA. (2011). *Global market outlook for PV until 2015*. Brussels, Belgium: European Photovoltaic Industry Association. Retrieved March 10, 2014, from http://www.epia.org/fileadmin/EPIA_docs/public/Global_Market_Outlook_for_Photovoltaics_until_2015.pdf

EPIA-Greenpeace. (2011). *Solar generation: Solar photovoltaic electricity empowering the world*. Brussels, Belgium: European Photovoltaic Industry Association. Retrieved March 10, 2014, from web site http://www.epia.org/EPIA_docs/documents/SG6/Solar_Generation_6_2011_Full_report_Final.pdf

EU (2010). Directive 2010/31/EU on the energy performance of buildings. *Official Journal of the European Union*, L153/14 -L153/34.

EUROBSERV'ER. (2012). *Photovoltaic barometer*. Retrieved March 10, 2014, from http://www.eurobserv-er.org/pdf/photovoltaic_2012.pdf

Farkas, K., Frontin, I. F., Maturi, L., Roecker, C., & Scognamiglio, A. (2013,). *Designing architectural integration of solar systems: criteria & guidelines for product developers*. Retrieved August 18, 2014, from web site: http://infoscience.epfl.ch/record/197098/files/task41A3-2-Designing-Photovoltaic-Systems-for-Architectural-Integration.pdf?version=1

Fthenakis, V., & Bulawka, A. (2004). Environmental impact of photovoltaics. Encyclopedia of Energy, 5, 61-69.

Goetzberger, A., & Hoffmann, V. U. (2005). *Photovoltaic solar energy generation*. Heidelberg, Germany: Springer.

Greenpeace. (2008). *Implementing the energy [R] evolution*. Retrieved March 10, 2014, from http://www.greenpeace.org/international/PageFiles/25109/energyrevolutionreport.pdf

HELAPCO. (2013). *PV grid: Initial project report*. Retrieved March 10, 2014, from http://www.helapco.gr/pv-grid/pv-grid-initial-project-report/

IEA. (2010). *JRC, PV status report*. Luxembourg: Office for Official Publications of the European Union. Retrieved March 10, 2014, from http://re.jrc.ec.europa.eu/refsys/pdf/PV%20reports/PV%20Report%202010.pdf

IEA PVPS. (2003). Energy from the desert, feasibility of very large scale photovoltaic power generation (VLS-PV) systems. London: James & James (Science Publishers).

IEA SHC. (2008). *Task 40-ECBCS annex 52, towards net zero energy solar buildings*. Paris, France: International Energy Agency. Retrieved May 22, 2012, from http://www.iea-shc.org/task40/

Jelle, B., & Breivikb, C. (2012). The path to the building integrated photovoltaics of tomorrow. *Energy Procedia, 20*, 78–87. doi:10.1016/j.egypro.2012.03.010

Krauter, S. (2006). *Solar electric power generation - Photovoltaic energy systems*. Heidelberg, Germany: Springer.

Luque, A., & Hegedus, S. (2003). *Handbook of photovoltaic science and engineering* (1st ed.). West Sussex, UK: John Wiley & Sons. doi:10.1002/0470014008

National Renewable Energy Lab. (2011). *Best research-cell efficiencies, rev. 12-2011*. Retrieved May 15, 2014, from http://www.nrel.gov/ncpv/images/efficiency_chart.jpg

Papadopoulou, E. (2012). *Energy management in buildings using photovoltaics*. London: Springer. doi:10.1007/978-1-4471-2383-5

Rigby, P., Fillon, B., Gombert, A., Ruenda, J., Kiel, E., Mellikov, E., . . . Warren, P. (2011). *Strategic energy technology plan: Photovoltaic technology, scientific assessment in support of the materials roadmap enabling low carbon energy technologies*. Retrieved March 10, 2014, from http://setis.ec.europa.eu/system/files/Scientific_Assessment_PV.pdf

Rockett, A. (2010). The future of energyn-photovoltaics. *Current Opinion in Solid State and Materials Science, 14*(6), 117–122. doi:10.1016/j.cossms.2010.09.003

Rossi, A. (2011). *L'architettura della città (IT)*. Macerata, Italy: Quodlibet.

Scognamiglio, A., Bosisio, P., & Di Dio, V. (2009). *Fotovoltaico Negli Edifici (IT)*. Milano, Italy: Edizioni Ambiente.

Scognamiglio, A., Ossenbrink, H., & Annunziato, M. (2011). Forms of cities for energy self-sufficiency., In *Proceedings UIA 2011, The XXIV World Congress of Architecture*. Tokyo: Academic Press.

Scognamiglio, A. R. H., & Røstvik, H. N. (2013). Photovoltaics and zero energy buildings: A new opportunity and challenge for design. *Progress in Photovoltaics: Research and Applications, 21*(6), 1319–1336. doi:10.1002/pip.2286

Sen, Z. (2008). *Solar energy fundamentals and modeling techniques* (1st ed.). London: Springer.

SMA. (2012). *SMA technology brochure 7, OptiTrac*. Retrieved March 10, 2014, from http://files.sma.de/dl/3491/TECHOPTITRAC-AEN082412.pdf

SMA. (2014). *Sunny portal*. Retrieved July 10, 2014, from http://www.sunnyportal.com/Templates/PublicPagesPlantList.aspx

Strong, S. (2010). *Building integrated photovoltaics (BIPV), whole building design guide*. Retrieved November 11, 2014, from http://www.wbdg.org/resources/bipv.php

Šúri, M., Huld, T., Dunlop, E., & Ossenbrink, H. (2007). Potential of solar electricity generation in the European Union member states and candidate countries. *Solar Energy, 81*(10), 1295–1305. doi:10.1016/j.solener.2006.12.007

Ventur, R., Scott Brown, D., & Izenour, S. (1977). *Learning from Las Vegas: The forgotten symbolism of architectural form*. Cambridge, MA: MIT Press.

Voss, K., Kiefer, K., Reise, C., & Meyer, T. (2003). Building energy concepts with photovoltaics—concept and examples from Germany. *Advances in Solar Energy, 15*, 235–259.

Voss, K., & Musal, E. (2012). Net zero energy buildings - International projects on carbon neutrality in buildings (2nd ed.). Institut für Internationale Architektur-Dokumentation.

Zahedi, A. (2006). Solar photovoltaic (PV) energy latest developments in the building integrated and hybrid PV systems. *Renewable Energy, 31*(5), 711–718. doi:10.1016/j.renene.2005.08.007

ADDITIONAL READING

Alafita, T., & Pearce, J. (2014). Securitization of residential solar photovoltaic assets: Costs, risks and uncertainty. *Energy Policy, 67,* 488–498. doi:10.1016/j.enpol.2013.12.045

Aubrecht, G. (2012). *Dr. Gordon Aubrecht talks renewables at TEDxColumbus*. Retrieved from http://osumarion.osu.edu/news/dr-gordon-aubrecht-talks-renewables-tedxcolumbus

Branker, K., Pathak, M., & Pearce, J. (2011). A Review of solar photovoltaic levelized cost of electricity. *Renewable & Sustainable Energy Reviews, 15*(9), 4470–4482. doi:10.1016/j.rser.2011.07.104

Cristóbal López, A., Vega, A., & López, A. (2012). *Next Generation of Photovoltaics, New Concepts*. Heidelberg, Germany: Springer. doi:10.1007/978-3-642-23369-2

EPIA. (2011). *Solar Photovoltaic Electricity Empowering the World*. EPIA. Retrieved March 10, 2014, from http://www.greenpeace.org/international/Global/international/publications/climate/2011/ Final%20SolarGeneration%20VI%20full%20report%20lr.pdf

EPIA. (2012). *Global Market Outlook for Photovoltaics until 2016*. EPIA. Retrieved March 10, 2014, from http://www.epia.org/fileadmin/user_upload/Publications/Global-Market-Outlook-2016.pdf

EPIA. (2013). *Global Market Outlook for Photovoltaics 2013-2017*. EPIA. Retrieved March 10, 2014, from http://www.epia.org/fileadmin/user_upload/Publications/GMO_2013_-_Final_PDF.pdf

Fraunhofer, I. S. E. (2009). *41.1% efficiency multijunction solar cells*. (2009). Retrieved March 10, 2014, from http://www.renewableenergyfocus.com

Fraunhofer, I. S. E. (2013). *Photovoltaics Report*. Freiburg: Fraunhofer ISE. Retrieved March 10, 2014, from http://www.ise.fraunhofer.de/en/downloads-englisch/pdf-files-englisch/photovoltaics-report-slides.pdf

Harris, A. (2011). *A Silver Lining in Declining Solar Prices*. Retrieved March 10, 2014, from http://www.renewableenergyworld.com/rea/news/article/2011/08/a-silver-lining-in-declining-solar-prices

Hoen, B., Wiser, R., & Cappers, P. a. (2011). *An Analysis of the Effects of Residential Photovoltaic Energy Systems on Home Sales Prices in California*. Berkeley National Laboratory. Retrieved March 10, 2014, from http://emp.lbl.gov/sites/all/files/lbnl-4476e.pdf

IEA. (2014). *Snapshot of Global PV 1992-2013 - Photovoltaic Power Systems Programme*. IEA. Retrieved March 10, 2014, from http://www.iea-pvps.org/fileadmin/dam/public/report/statistics/PVPS_report_-_A_Snapshot_of_Global_PV_-_1992-2013_-_final_3.pdf

IEA/OECD. (2011). *Deploying Renewables 2011: Best and Future Policy PracticeBest and Future Policy Practice*. Paris: International Energy Agency.

IEA/OECD. (2014). *The Power of Transformation - Wind, Sun and the Economics of Flexible Power Systems*. Paris: International Energy Agency.

Jacobson, M. Z. (2009). Review of solutions to global warming, air pollution, and energy security. *Energy & Environmental Science, 2*(2), 148–173. doi:10.1039/B809990C

Jeff, S. J. (2012). *Solar Electronics, Panel Integration and the Bankability Challenge*. Retrieved March 10, 2014, from http://www.greentechmedia.com/articles/read/solar-electronics-panel-integration-and-the-bankability-challenge

McDonald, N., & Pearce, J. (2010). Producer responsibility and recycling solar photovoltaic modules. *Energy Policy, 38*(11), 7041–7047. doi:10.1016/j.enpol.2010.07.023

Mojiri, A., Taylor, R., Thomsen, E., & Rosengarten, G. (2013). Spectral beam splitting for efficient conversion of solar energy—A review. *Renewable & Sustainable Energy Reviews, 28*, 654–663. doi:10.1016/j.rser.2013.08.026

Oliwenstein, L. (2010). *Caltech Researchers Create Highly Absorbing, Flexible Solar Cells with Silicon Wire Arrays*. Retrieved March 10, 2014, from http://www.caltech.edu/content/caltech-researchers-create-highly-absorbing-flexible-solar-cells-silicon-wire-arrays

Pathak, M., Sanders, P., & Pearce, J. M. (2014). Optimizing limited solar roof access by exergy analysis of solar thermal, photovoltaic, and hybrid photovoltaic thermal systems. *Applied Energy, 120*, 115–124. doi:10.1016/j.apenergy.2014.01.041

Pearce, J. (2002). Photovoltaics – a path to sustainable futures. *Futures, 34*(7), 663–674. doi:10.1016/S0016-3287(02)00008-3

Petrova-Koch, V., Hezel, R., & Goetzberger, A. (2009). *High-Efficient Low-Cost Photovoltaics, Recent Developments*. Heidelberg, Germany: Springer. doi:10.1007/978-3-540-79359-5

Psomopoulos, C. (2013). Solar Energy: Harvesting the Sun's Energy for Sustainable Future. In I. J. Kauffman & K.-M. Lee (Eds.), *Handbook of Sustainable Engineering*. Dordrecht, Holland: Springer Science & Business Media. doi:10.1007/978-1-4020-8939-8_117

Quiggin, J. (2012, January 3). *The End of the Nuclear Renaissance*. Retrieved March 10, 2014, from http://nationalinterest.org/commentary/the-end-the-nuclear-renaissance-6325

REN21. (2013). *Renewable Energy Policy Network for the 21st century*. Paris: Renewables 2013 Global Status Report. Retrieved March 10, 2014, from http://www.ren21.net/Portals/0/documents/Resources/GSR/2013/GSR2013_lowres.pdf

Renewable Energy World Network Editors. (2011). *Renewables Investment Breaks Records*. Retrieved March 10, 2014, from http://www.renewableenergyworld.com/rea/news/article/2011/08/renewables-investment-breaks-records

Roberts, S., & Guariento, N. (2009). *Building Integrated Photovoltaics, A Handbook*. Basel, Switzerland: Birkhäuser. doi:10.1007/978-3-0346-0486-4

Schultz, O., Mette, A., Preu, R., & Glunz, S. (2007). Silicon Solar Cells with Screen-Printed Front Side Metallization Exceeding 19% Efficiency. *22nd European Photovoltaic Solar Energy Conference and Exhibition*, (pp. 980-983). Milan, Italy. Retrieved March 10, 2014, from http://publica.fraunhofer.de/eprints/urn:nbn:de:0011-n-735217.pdf

Swanson, R. M. (2009). Photovoltaics power up. *Science, 324*(5929), 891–892. doi:10.1126/science.1169616 PMID:19443773

Vick, B., & Clark, R. (2005). Effect of panel temperature on a Solar-PV AC water pumping system. *Proceedings of the International Solar Energy Society (ISES)*, (pp. 159–164). Orlando, Florida.

KEY TERMS AND DEFINITIONS

Balance of System (BOS): Represents all components and costs other than the PV modules. It includes design costs, land, site preparation, system installation, support structures, power conditioning, operation and maintenance costs, batteries, indirect storage, and related costs.

BAPV, Building Adapted Photovoltaics and BIPV, Building Integrated Photovoltaics: The BAPV term is used for PV panels anchored down onto the roofs of buildings, while the term BIPV is used for the design and integration of PV into the building envelope, typically replacing conventional building materials. This integration may be in vertical facades, replacing view glass or other facade material; into roofing systems, over windows; or other building envelope systems.

Green Tariff: Special rates at which purchase electricity from facilities that use RES (solar, wind, geothermal, wave and tidal energy, hydro power (installed capacity less than 10 MW), biomass, biogas, natural gas, etc.).

Irradiation: The solar radiation incident on an area over time. Equivalent to energy and expressed in J/m^2. In technical practise is the solar radiation incident on a surface. Usually expressed in kW/m^2. Commonly the term "insolation" is used which is corresponds to the sunlight, direct or diffuse; from "incident solar radiation". It is consist of two parts: a) diffuse insolation which is the sunlight received indirectly as a result of scattering due to clouds, fog, haze, dust, or other obstructions in the atmosphere and b) direct insolation which is sunlight falling directly upon a collector.

Operating Point: The current and voltage that a module or array produces when connected to a load. The operating point is dependent on the load or the batteries connected to the output terminals of the array. The interested is focused on the Maximum Power Point (MPP) which is the point on the current-voltage (I-V) curve of a module under illumination, where the product of current and voltage is maximum. For a typical silicon cell panel, this is about 17 volts for a 36 cell configuration. The devices that have been developed to automatically operate the modules or arrays in the MPP under all conditions are called Maximum Power Point Trackers (MPPT). An MPPT will typically increase power delivered to the system by 10% to 40%, depending on climate conditions and battery state of charge. Most MPPT controllers are down converters - from a higher voltage to a lower one.

Peak Sun Hours: The equivalent number of hours per day when solar irradiance averages 1,000 W/m^2 - one peak sun hours means that the energy received during total daylight hours equals the energy that would have been received had the irradiance for one hour been 1,000 W/m^2.

Photovoltaic (PV) Cell: The smallest semiconductor element within a PV module to perform the immediate conversion of light into electrical energy (dc voltage and current). It can be monocrystalline: material that is solidified at such as rate that single or few crystals (crystallites) form. The atoms within a single crystallite are symmetrically arranged, whereas crystallites are jumbled together. It can be polycrystalline, (sometimes referred to as multicrystalline or semicrystalline), when the material that is solidified at such as rate that many small crystals (crystallites) form. The atoms within a single crystallite are symmetrically arranged, whereas crystallites are jumbled together. These numerous grain boundaries reduce the device efficiency. A material composed of variously oriented, small individual crystals. A more recent type is the thin-film PV module (commonly called amorphous) where the PV module constructed with layers of thin film semiconductor materials. A layer of semiconductor material, such as copper indium diselenide, cadmium telluride, gallium arsenide, or amorphous silicon, a few microns or less in thickness, used to make photovoltaic cells.

Photovoltaic (PV) System: A complete set of components for converting sunlight into electricity by the photovoltaic process, including the array and balance of system components. Each system includes a number of modules or panels interconnected electrically in series, named string. Photovoltaic (PV) module or module is the smallest environmentally protected, essentially planar assembly of solar cells and ancillary parts, such as interconnections, terminals, and protective devices intended to generate DC power under unconcentrated sunlight. The structural (load carrying) member of a module can either be the top layer (superstrate) or the back layer (substrate). Photovoltaic (PV) panel refers to a physically connected collection of modules (i.e., a laminate string of modules used to achieve a required voltage and current). Array corresponds to any number of photovoltaic modules connected together to provide a single electrical output. Arrays are often designed to produce significant amounts of electricity. For the evaluation of the maximum possible output of the photovoltaic system the photovoltaic (PV) peak watt is used, which corresponds to the maximum "rated" output of a cell, module, or system. Typical rating conditions are 1000 watts per square meter of sunlight, 20 degrees C ambient air temperature and 1 m/s wind speed.

Small Scale PV Installations: Small scale installations of PV arrays that do not exceeds 100kW and usually are less than 15kW in case of residential installations.

Tracking Array: PV array that follows the path of the sun to maximize the solar radiation incident on the PV surface. The two most common orientations are (1) one axis where the array tracks the sun east to west and (2) two-axis tracking where the array points directly at the sun at all times. Tracking arrays use both the direct and diffuse sunlight. Two-axis tracking is the system capable of rotating independently about two axes (e.g., vertical and horizontal) and following the sun for maximum efficiency of the solar array. Two-axis tracking arrays capture the maximum possible daily energy. Typically, a single axis tracker will give you 15% to 25% more power per day, and dual axis tracking will add about 5% to that. Depends of course, somewhat on latitude and season.

Chapter 7
The Topicality and the Peculiarities of the Renewable Energy Sources Integration into the Ukrainian Power Grids and the Heating System

Vira Shendryk
Sumy State University, Ukraine

Olha Shulyma
Sumy State University, Ukraine

Yuliia Parfenenko
Sumy State University, Ukraine

ABSTRACT

The chapter proposes to approach the problem in the following ways: assess the problems in the existing power system, analyze the current state of the RES using, and determine the existing ways of computer modeling of the grid. The chapter also discusses the topicality of renewable energy use in the construction of distribution grids and the ability to model their work. It explores issues including current research attempts to identify the existing methods, which can be applied in the Decision Support System (DSS) for calculation and evaluation RES. The problem of making decisions for energy saving in district heating requires such measures as energy audit and planning. These activities require monitoring the current energy consumption in real time.

INTRODUCTION

Nowadays the energy of gas, oil or coal is the integral component of the technology to create comfortable living conditions in modern cities, thus, their saving is a priority for every modern human. Today the fact that the world's energy resources are limited is indisputable. Experts estimate, that modern planetary energy, a lot of which is concentrated in the cities, absorb fuel resources million times faster than nature produces them.

The problem of limited energy resources and their wasteful usage becomes more important from year to year. The primary energy resources are essentially used for production of two types of energy: heat and electricity. In Ukraine, as in the whole world, more than 40% of the primary energy is consumed by non-industrial buildings (houses, schools, hospitals, etc.). This is the largest sector of the energy consumption in the national economy. Today the long and unsuccessful search of the ways of reforming housing and communal services is largely linked to the inefficiency of energy using in buildings in Ukraine.

No surprise, that one of the most obvious ways of saving resources is based on using the renewable energy sources (RES). Today there are several ways to obtain renewable energy, which are different due to the type of source. Big Energy Complexes of Ukraine are focused on the use of the traditional energy sources. Ukraine has a significant potential for the RES, but they are generally used in private households. Therefore, the introduction of the RES as an energy source in a unified energy system of Ukraine is currently the actual task and it requires thorough investigation.

The high cost of the new energy grids construction, the locality of their placement and the differences between regions based on the RES potential require the preliminary assessment of the possibility and the necessity of their introduction into the grid. This needs efficient tools for energy management including the using of the information technologies, such as Decision Support Systems (DSS) concerning the RES integration into the existing power grid.

This chapter is dedicated to the research of the RES implementation into the unified energy system of Ukraine. The issues of the RES potential will be considered, the analysis of barriers for their use will be carried out, the current state of the use of the RES in terms of heat and electricity production will be described, as well as the issues of the information modeling of the energy systems will be explored.

BACKGROUND

According to the national power company Ukrenergo (Safe Mode the work of DEC Ukraine, 2014) the Unified Energy System of Ukraine is represented by thermal, nuclear, hydro, wind and solar power units which are interconnected in the parallel operation. It produces and distributes heat and electric power. Units for power production include (Ministry of Energy and Coal Industry of Ukraine, 2013):

- 5 electricity generators that produce electricity at thermal power plants;
- 4 nuclear power plants;
- 12 hydroelectric power stations;
- 27 local, mostly private, energy distribution companies;
- 104 thermal powers in larger cities;
- 4 solar power plants;
- 5 wind farms.

At the end of 2010 the power production in Ukraine reached 191.5 billion kWh, 47.27% of which came from nuclear power plants, 36.89% came from thermal generation, and approximately 7% came from Thermal Power Plants (TPP) and Hydro Power Stations (HPS), a small part was produced from the RES (Consulting Company NIKO, 2014).

Units for heat supply are:

- 250 cogeneration plants;
- More than 35000 centralized boiler plants;
- Industrial boiler plants of large individual enterprises;
- Individual heating systems of the apartments.

The units for heat supply incorporated in DH system provide total heating capacity of 127 thousand Gcal/h. As at 2013 the total amount of produced heat energy was 96.5 million Gcal. These include 91 million Gcal which were produced for the needs of cities (the State Statistics Committee of Ukraine, 2014).

Figure 1. The distribution of the power production among different resources

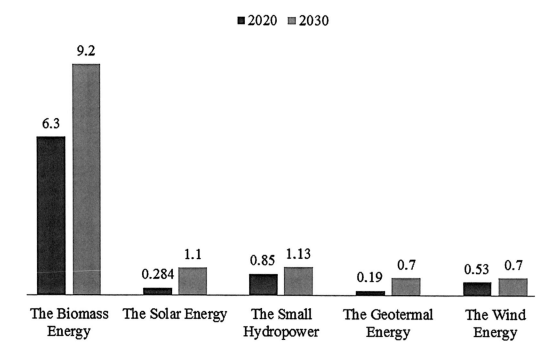

Ukraine consumes about 210 million tons of the oil equivalent (tce) of energy resources and refers to the energy deficient countries as it covers its own needs for energy consumption at approximately 53% and imports 75% of the required amount of natural gas and 85% of crude oil and oil products (Law of Ukraine on Electricity, 2014). The structure of the primary energy sources in power generation is presented in Figure 1.

The main fuels for cogeneration plants are: natural gas, 80%, fuel oil, 15% and coal, 5% (Shevtsov, Barannik, Zemlyaniy, & Ryauzova, 2010). The structure of the primary energy sources for heat generation in centralized boiler plants is presented in Figure 2.

The most problematic fact in the Ukrainian energy sector is a significant power equipment deterioration, which is based on the blocks of thermal and nuclear power plants. The vast majority of the power grid's units was built in the 50's and 70's of the last century. Thus, the most of the equipment that is required for the grid is coming to the end of the operation period. The losses during the transmission and the distribution of the electrical energy include transmission losses between power sources and the points of the distribution, and the distribution to consumers. In general, losses are estimated

Figure 2. The structure of energy resources in heat production

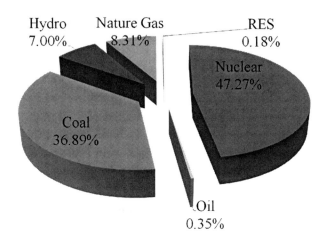

nearly at 11.8% of the total produced energy in 2010 (SE4ALL, 2013). This makes it unprofitable to transport electricity to the consumers. In this regard, the one of the promising areas of the energy industry is the creation of the distributed generation and introduction of the RES into the grid.

Nowadays the usage of the renewable energy sources in heat production is relevant for Ukraine, because it allows reducing its dependence on imported energy and improving energy security. Almost all of the heat energy produced from the RES is extracted from biomass. This is caused by its high energy potential and renewability.

The measures to provide energy saving lie in the increasing consumption of energy produced from alternative sources. The RES could play a crucial role in solving the problems connected with depletion of the stocks of the primary energy resources. The part of the RES in the balance of The Unified Energy System of Ukraine does not exceed 1.5%. The share of biomass in total energy consumption in EU by 2020 should be at least 24%. The process of the RES development in Ukraine is slow. But the potential is rather high.

In this regard the study of the possibilities of using the renewable energy is an urgent problem. The potential of energy savings from the renewable energy sources is about 40% of the total primary energy consumption in housing system. (Babaev, Malyarenko, & Orlova, 2012).

Moreover, the implementation of the RES can solve many economic and environmental issues as the grade of energy efficiency is measured by the social impact (the improvement of the quality of services), the environmental impact (the reduction of the CO_2 pollution), and the economic effect (the reducing of the costs and the power consumption).

THE FEATURES OF EXISTING ENERGY SYSTEMS

The Features of Existing Heat Supply Systems

District heating system is a variety of the large power systems designed for the production, distribution and consumption of heat energy (Zerkalov, 2012). The heat production for the needs of the economic activity sector and population of Ukraine is carried out by the district heating systems (DH systems), covering about 60% of consumers and individual heating systems (IH systems). The latter include own generators of heat energy – gas and the electric boilers, convectors, water heaters, heat pumps. The DH system in Ukraine is represented by the branched heat networks. Its length (in 2010) was 37.3 thousand km of double-pipe calculation (Collegium of the Ministry Energy and Coal Industry of Ukraine, 2012). There are more than 11 million apartments, mainly in cities and towns connected to the DH systems. Thus, in most modern cities of Ukraine residential and public buildings receive heat energy from the DH systems.

The DH systems have their own specific features that distinguish them among the large energy systems. The structural features include a greater length and branching of the heating networks with using of multiple sources of heat energy. The functional features associated with the necessity of the adjusting of the coolant parameters (pressure and temperature), which differ in physical nature and are characterized by the influence of the environmental conditions on them. The heating process is characterized by irregularity, so its management should be focused on the end user who is the most remote from the heating source. The quality and the efficiency of the DH systems in Ukraine are far from being perfect. The DH systems were created more than 50 years ago. Now in general they need modernization. Heat losses are very large; sometimes they exceed the useful heat energy consumption. Every year Ukraine spends a third part more fuel to produce one Gcal of heat energy than other European countries.

The main causes of the low energy efficiency of the district heating are listed as follows (Pavliuk, 2011; Ministry of Energy and Coal Industry of Ukraine, 2013):

- The boiler equipment practically has exhausted all allowable service life (service life over 57% of boilers is over 20 years);
- The poor insulation of pipes causes large heat losses (about 30% of the generated thermal energy);
- Significant economic losses due to the significant amounts of repairs of the heat pipes;
- The lack of proper metering devices (in 2012 more than 40% of the DH enterprises and 70% of consumers were not equipped with the heat counters);
- Large heat losses through the building envelope;
- The state funds are allocated for the modernization of the heat energy generation, rather than for reducing heat consumption;

Thus, the process of the heat production is characterized by the energy intensity. The primary energy deficit and its high cost require at least a partial transition to the alternative fuels.

The Features of the Existing Power Supply Lines and Technical Requirements to the Accession of the RES

The creation of a distributed power grid through the implementation of energy derived from the RES in Ukraine has its own characteristics. Initially, we should determine the types, the features and the limitations of the existing power grid.

Power grids are distinguished by voltage: up to 1 kV and above 1 kV. Power grid voltage meter above 1 kV can be divided into the grids of medium MV, high HV and extra high EHV voltage (Hase, 2008).

Low-voltage distribution grids operate in a range up to 35 kV. The distribution voltage level is typically used to supply electricity to local consumers, including people and businesses.

The distribution grids of high-voltage work mainly with a voltage of 110 kV, but cover the range of about 110 kV to 220 kV. These grids are the intermediary between the district distribution grids and the backbone grids of supply. Sometimes this class is associated with the transportation of electricity over long distances, but it's specifically separated in order to be able to connect the renewable energy facilities to the grid.

The EHV level lies in the range of 330 kV and above. The EHV level provides the transportation over long distances with fewer losses in the system and connects the supply chain of the large generating capacity and the consumption centers.

Most of the generating capacities of Ukraine, that use heat to produce electricity, such as thermal and nuclear power plants, are connected to the high voltage grid. However, the remote location, or a smaller number of the projects on the renewable energy may require them to be connected to the low voltage distribution network. Objects that are connected with the local distribution networks to provide local electricity needs should not have more power than the level of the minimum load. Only in this case it is possible to avoid leakage of the excess energy in the high-voltage distribution networks and supply chains.

According to the report of the experts of the EBRD the ratio of the consumption and the transmission in the regions of Ukraine is the following (USELF, 2013).

The transmission line of 330 kV passes through Odessa region and connects Ukraine with Moldova. Also from Moldova electricity can be delivered to the northern and the north-eastern regions of the country. In the region the peak load is nearly 700 MW, and Usatovska and Adzhalikska substations have the maximum current capacity of 4.000 MW.

Now Kherson region produces about 200 MW more than it is required in the region. The area has power lines in good condition of 750 kV. The good supply of wind and solar resources for the further development of these industries requires the connection to the Kakhovska substation, which is able to maintain generation capacity at the level of 4150 MW.

Zaporozhie region has more bandwidth lines; the supply grid has the voltage of 750 kV. The energy generated from the RES can be exported to the neighboring regions. Only two substations Molochansk and Melitopol have the total capacity of 1.500 MW. Wind energy potential is estimated at about 3,000 MW and solar energy at 1,000 MW.

In Donbass region, only the substation Peremoga is able to maintain the power of 1900 MW. Substations Zorya and Myrna are able to process 1.500 MW of wind energy received from the sources.

There is the shortage of the electricity of about 200 MW in the Zakarpattya region, which can be supplied from the production in the neighboring Lviv and Ivano-Frankivsk regions. There is energy excess of about 2100 MW in Ivano-Frankivsk area, which can be exported to the neighboring regions and Poland. The Carpathians region has the best conditions for the development of wind power; under condition of the correct coordination of the wind generating stations' work with the small HPSs, we can get over 1,400 MW of the excess that can be exported to the neighboring countries.

Tripolskaya TPP is the main source of electricity in the power supply system of central Ukraine; its capacity is about 4,500 MW. Despite the presence of Kiev region, as a powerful energy consumer, there is an excess of nearly 2,500 MW of electricity.

The data about the transfer capabilities of the existing transmission lines and the existing load constrain the opportunity for the using renewable energy, which can be developed in different parts of the country. Thus, the potential of the RES needs to be considered and it is described below.

THE ASSESSMENT OF THE RENEWABLE ENERGY POTENTIAL IN UKRAINE

Ukraine has a considerable potential of the renewable energy that can be technically achieved, which is more than 43.0 million tons of coal equivalent per year (for comparison, the total energy consumption for electricity generation in the country in 2009 was 21.5 million tons of coal equivalent) (State Agency on Energy Efficiency and Energy Saving of Ukraine, 2014).

According to the Energy Strategy of Ukraine till 2030 it is planned to achieve the RES parameters that are presented in Figure 3 (the data are presented in the billion tons of oil equivalents).

The forecast of the National Agency for Efficient Use of Energy Resources (Forecast of the Renewable Energy Agency) shows the trajectory of the renewable energy sources development, based on the technical potential of 15 TW/h, set a target - 150 TW/h in 2030 and 250 TW/h in 2050 (Konechenkov, 2013).

The possibility of use and the potential for the renewable energy in Ukraine are different and depend on the specific type of the renewable energy. The characteristics of the renewable energy are below.

- **Biomass Energy:** In 2010 from using different kinds of biomass were produced 10.6 TW/h which corresponds to approximately 0.65% of the national primary energy demand (Chaly, 2013). Bioenergy is currently produced mainly from the burning wood, straw, peat and used in the decentralized mode of the heat production and/or hot water.

Figure 3. Energy strategy of Ukraine until 2030 for RES

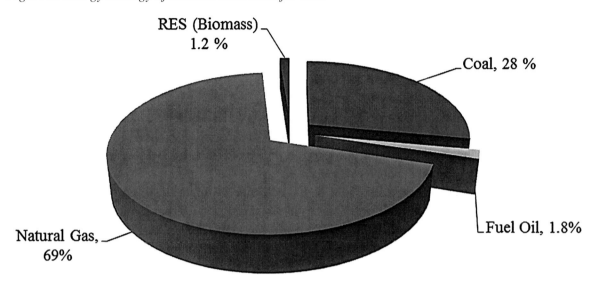

To use this raw material the modernization of the existing facilities is often carried out (e.g., coal-fired boilers) or the installation of the small and micro setting (e.g., for the production of biogas in agriculture or for burning straw). However, the problem is the low efficiency and the high emissions of re-equipped facilities.

The technical potential of biomass in accordance with the relevant data is from 126 to 162 TW/h, the highest proportion falls to the straw (45 TW / h) and energy crops (41 TW / h).

With the current price for fossil fuels (primarily natural gas), heat and biomass, the introduction of biomass boilers for heat production is economically feasible and can be recommended for thermal power facilities in the industrial and public sector. The implementation of such projects in the housing economy is on the verge of profitability. Payback period for the implementation of boilers for wood and straw is 2-3 years for industrial and public sector and more than 7-10 years - for the housing and communal services (Geletukha, Zheliezna, & Oliinyk, 2013).

The quantity of the energy potential of biomass varies from year to year and depends mainly on the yield of major crops (wheat, corn, sunflower, etc.). Over the past 10 years, fluctuations in the economic potential of biomass ranged from 25 to 38 million tons of standard fuel/year (s.f./year). The major components of the potential are agricultural waste and biomass of the energy crops. Depending on the yield of major crops the economically feasible potential of biomass varies between 5-35 million tons of standard fuel / year, representing 13-18% of primary fuels in Ukraine. The energy potential of the main types of biomass in Ukraine in 2013 is shown in the Table 1. (Geletukha, Zheliezna, Kucheruk, & Oliinyk 2014).

The total area of the unused agricultural land in Ukraine in 2012 is 3.5 million hectares. This land could be used for growing crops that can be used as the RES.

At present time, approximately 300 hectares of land is used to grow oilseed rape with an annual growth rate from 50 to 80% over the past three years. If oilseed rape will grow on 3 million hectares of land with an annual yield of about 1.5-3.0 t / ha, 75% of the crop would be sufficient to produce 2.7 million tons of biofuel. This type of biofuel is widely used in Zhytomyr, Sumy, Vinnytsia, Khmelnytsk regions.

Table 1. The energy potential of biomass in Ukraine, 2013

Type of Biomass	The Theoretical Potential, Million Tons	Share Available for Energy,%	Economic Potential, Million TCE
Production waste of corn and sunflower	68.1	40	7.28
Straw of cereals and oilseed rape	34.8	40	5.38
Woody biomass (waste wood) and peat	4.2	90	1.77
Biogas	6.5 bln m³	35	5.18
Energy crops (willow, poplar, miscanthus)	11.5	90	6.28
Total:	-	-	25.89

According to the set of indicators of peat and forest, Rivne, Volyn, Lviv, Kiev regions and the northern part of the Chernihiv region are suitable for the construction of HPP using peat and biomass of the fuel wood of the local origin. Figure 4 represents the data about the wood biomass potential.

Thus, woody biomass is unevenly distributed between regions.

Figure 4. The potential of the wood biomass, thousands of tons of the standard fuel/year

- **Biogas:** In Ukraine there are currently only a few plants for the biogas production. The structure has allowed developing animal husbandry in 2008 from 1.39 to 2.78 TW/h of electricity from the animal waste (Kalenska, Rahmetov, Kalenskiy, Yinik, & Kachura, 2012).

The biogas potential of the energy crops, such as corn, is estimated from about 6.28 to 12.57 TW/h, and this figure corresponds to the volume of grain exports in 2008. Thus, the technical potential of biogas production is now about 4-8% from the current electricity production in Ukraine.

In 2009, only 3% of households were able to self-assemble organic feedstock for biogas plants with the capacity of 500 kW. Consequently, the development of the agricultural sector will also determine for which enterprises such a facility will be effective.

- **Wind Energy:** According to the reports, the current installed capacity is about 360×10^3 billion kW×h per year (Kuznetsova & Kutsenko, 2010), is operated by the mostly small installations and having the capacity of 107.5 kW. The potential of wind energy that can be used in 2030 is estimated at 16 GW, and can annually produce from 25 to 30 TW/h of the electric energy.

The use of the wind turbines for the electricity production in Ukraine is the most effective in those regions where the average wind speed exceeds 5 m/sec. The Black and Azov seas, the mountains of the Crimea (especially the North-Eastern coast) and the Carpathians, Odessa, Kherson, Zaporizzya, Donetsk, Luhansk and Mykolaiv regions, places along the Dnieper River in central Ukraine are the most suitable for the construction of the wind power plants. But mountainous areas with steep slopes above 20° are considered to be technically difficult for implementation of such projects.

- **Geothermal Energy:** The technical potential of the geothermal reserves is estimated at about 14×10^3 billion kW×h per year. According to Konechenkov's work (2013) with the same technical capacity up to 2030 about 55 TW/h will be available annually, and in 2050 – about 75 TW/h of thermal energy through the using of the geothermal energy. Today about 13 MW of the installed capacity is in use.

The energy potential of the soil heat and the underground hot water heat in thousands MW/h per year is presented in Figure 5 (Renewable Energy Institute NAS of Ukraine, 2014).

In Figures 5-7 the all future data are the illustrated data among regions without Crimea, through the present political situation in Ukraine.

- **Solar Energy:** The Institute of the Renewable Energy of the National Academy of Sciences of Ukraine created the atlas of the energy potential of the RES, according to which the total potential energy of solar radiation in Ukraine is equivalent to 89.4×10^9 tons of oil equivalent (toe) per year, while technically achievable capacity is 42.6×10^7 toe per year, the rational economic potential is equivalent to 5.99×10^5 toe per year (Kudrya, Ryeztsov, & Surzhyk, 2010). This potential of solar energy is sufficient for the widespread introduction of heat and power as well as photovoltaic equipment virtually in all areas. The term of efficient operation of solar power equipment in Ukraine ranges from 5 months (May - September) in the North to 7 months (April - October) in the South. Photovoltaic equipment can operate throughout the year.

The most promising areas for the development of solar energy are in the South of Ukraine. The Central and the Eastern areas are characterized by

Figure 5. The energy potential of soil heat and the underground hot water heat

the moderately good level of resources available for the development of solar energy, but in each such indicators as economic feasibility, the presence of wavy/mountainous areas and the choice of the optimal technology should be calculated and studied carefully.

In Figure 6 the data about the total solar potential (common potential) are presented. This is the amount of energy that theoretically can be used and the economically justified at the present technical level of the amount of solar energy that can be used (Kudrya et. al., 2010). As it's presented sufficiently high level of insolation and, as results, the building of solar power can actually be in Ukraine and it's a serious argument for European Companies, which are interested in new markets.

- **Hydropower:** Currently hydropower is the only renewable energy source that is used in larger quantities. It is estimated that its technical capacity is about 6.4 billion kW×h per year. In fact, in 2010, it was produced 13.152 TW/h (SE4ALL, 2013). There are several river basins that are suitable for the future construction or reconstruction of the small hydropower. They are:

  ○ The Dniester River.
  ○ The Tisza River.
  ○ The tributaries of the Dnieper River in the central part of Ukraine.

Figure 6. The total annual solar energy potential of Ukraine, MW ×h/year

The information about The Hydropower Potential of The Small Rivers in millions kW/h is presented in Figure 7 (Kudrya et. al., 2010).

Unfortunately the information about the exact locations of the potential construction in these pools of the small hydropower plants is not available.

According to this analysis, the energy potential of the renewable resources in Ukraine in fact is (Konechenkov, 2013):

- Solar energy 130×10^3 billion kW×h per year.
- Wind energy 360×10^3 billion kW×h per year.
- Small Hydro energy 6.4 billion kW×h per year.
- Geothermal energy 14×10^3 billion kW×h per year.
- Biomass energy 6.1 billion kW×h per year.

Thus, Ukraine has enough potential of the RES, but the process of their implementation is slow. We are going to consider the obstacles that slow down the implementation of the RES into the unified energy system of Ukraine.

THE OBSTACLES FOR THE USING OF THE RENEWABLE ENERGIES

Despite the rather large energy potential indicators of the RES, a few points actively hinder the implementation and the development of the renewable energy in Ukraine.

Figure 7. The hydropower potential of small rivers

The obstacles for the use of the RES and the ways to overcome them should be considered at all stages of the use of the RES: the implementation of the RES into the existing power grid and the production of electricity and heat energy, the energy transfer to the end consumer and the saving of the energy received by the end consumer. Also it should be noted, that each of the renewable energy sources has some limitations such as existing capacity, weather conditions and spatial arrangement.

Some of the RES are better suited for the production of electricity and the others – for the district heating. For example, solar panels and wind turbines are easy to install and can be installed during the short period of time, while hydro and geothermal systems require more installation time, especially in the large projects. Biomass energy is used mainly in the production of heat energy.

Solar energy production depends on the availability of sunlight during the day and its accumulation at night; wind energy depends on the wind speed and also depends on the accumulation; hydropower depends on the drought; biomass energy depends on the supply of the fuel and can contribute to the development of the greenhouse effect. Thus, all the renewable energy sources are limited.

One of the main problems is the outdated energy distribution grids. As for the heating systems, this problem is especially acute. The obsolete pipes without insulation cause significant heat losses during its transmission.

The Carpathians are in the west of the country – the region that is passable with difficulties, that's why it is necessary to solve the problem of the delivery of the energy produced in other regions for their consumption or to create separate infrastructure projects under the generated energy.

Within the construction of the united distributed systems we should take into account the fact that the local load should not exceed the minimum possible network load. Thus, in Table 2, the estimated ability to transfer between the regions is presented with the data on the annual energy needs, peak load and generating capacity in the region in 2010 (Chaly, 2013). As it was shown in Table 6, the vast majority of regions in Ukraine have tremendous transfer capability and the excess of the generation capacity to export to neighboring regions; nevertheless, the neighboring regions with their own limited load and excess generation will not be able to consume all of the exported renewable energy from their neighbors. Following that, two different approaches can be applied for evaluation:

- The theoretical maximum for estimation of the export capacity of the renewable energy potential from a region is the transfer capability, but it's not realistic to apply except for a few cases.
- A more conservative approach says about the limits of the renewable resources, that is the availability of the resources in the region and/or about the load in the same region: the energy consumers. It is selected depending on which level is lower.

Also one of the serious problems of the renewable energy use is the lack of stability in the energy production. The volumes of production are highly dependent on the weather conditions. The common limitations of the wind and solar power are correspondence of the low-power generators with their expensive cost. Also in both cases it is necessary to accumulate power at night (for solar power) and windless time (for wind energy) and it requires additional devices.

The disadvantage of the geothermal power plants is the geographical limitations of their use: it is cost-effective to build the geothermal power plants only in the areas of tectonic activity, that is, where the most natural sources of heat are available.

One of the main obstacles for the use of the RES in power system is that the heat supply enterprises have limited funds to invest in boiler refurbishment for the use of biomass. The existing mechanism of the tariff discourages manufacturers of the thermal energy to use local biofuels, as the subventions are currently distributed only for gas. Another obstacle is the lack of biomass as a fuel. The heating value of biomass is about twice lower than that of the coal. The infrastructure of the storage and the supply of biomass are underdeveloped. Also, the environmental requirements for boilers fired with biomass currently are overpriced (Geletukha & Zheliezna, 2013).

One way to overcome the barriers that slow down the use of the RES is relevant legislative policy in this area. Let us consider the laws of Ukraine regarding the implementation of the RES in more detail.

THE LEGISLATION IN THE FIELD OF THE RENEWABLE ENERGY

At present, the legislative support is developing and improving in this field and the investment climate for the renewable energy projects is also improving. These actions are considered as the energy-saving policy, i.e. the administrative, legal, financial and economic regulation of the extraction, processing, transportation, storage, production, distribution and use of the energy resources for their management and economic consumption. Today in the field of energy saving there are about 100 of the legal acts, system of standards and significant number of the regulatory guidance documents.

Table 2. The transfer capability estimation between regions

Regional Electric Power Systems	Region	Load Cap		Total Generation Capacity (MW)	Net Export/ (Import) (MW)	Transfer Capability (MW)
		Annual Demand (GWh)	Weak Load (MW)			
Central	Cherkasy	3.593	51	1.373	1.022	1.900
	Chernihiv	2.141	16	200	-16	3.000
	Kyiv	9.465	63	3.944	2.981	21.000
	Zhytomyr	2.721	70	-	-270	2.250
Dnipro	Dnipropetrovsk	30.090	.670	5.334	2.663	11.250
	Kirovohrad	3.238	27	7	-320	4.150
	Zaporizhia	10.514	83	10.973	9.990	21.000
Donbass	Donetsk	25.674	.389	10.685	8.296	19.900
	Luhansk	12.019	.137	1.928	791	12.650
Northern	Kharkiv	8.115	04	2.985	2.181	12.000
	Poltava	5.845	43	274	-269	2.250
	Sumy	2.523	38	125	-113	13.500
Southern	Kherson	2.629	52	455	203	10.900
	Mykolaiv	3.372	29	3.495	3.166	15.800
	Odessa	7.018	00	-	-700	18.400
Southwestern	Chernivtsi	1.562	68	1.053	885	3.750
	Khmelnytskyi	2.578	52	2.000	1.748	14.250
	Ternopil	1.457	49	-	-149	3.000
	Vinnytsia	3.443	25	1.818	1.493	16.500
Western	Ivano-Frankivsk	2.675	29	2.401	2.172	6.050
	L'viv	5.047	08	729	221	18.200
	Rivne	2.765	61	2.880	2.619	9.800
	Volyn	1.692	68	9	-159	3.450
	Zakarpattia	2.226	42	32	-210	7.600

The main laws, which regulate renewable energy in Ukraine, are:

- The Law of Ukraine "On Alternative Energy Sources", which the Verkhovna Rada of Ukraine adopted in February 2003 (20.02.03, N° 555IV). It was drafted by the State Committee of Ukraine for the Energy Conservation performance objectives of the Renewable Energy Programme (1997). The law regulates the management and the regulation of the RES, the organizational support, standardization and some common features of the alternative energy sources.
- The Law of Ukraine "On alternative fuels". This law defines the legal, social, economic, environmental and organizational principles of the production and using of the renewable fuels and incentives to increase the using up to 20 percent of the total fuel consumption in Ukraine to 2020.
- The Law of Ukraine "On Electric Power Industry". The new version of the law

strengthened the position of the major players in the alternative energy industry. The law also supports the individuals who install the RES at their farms.
- The Law of Ukraine "On Energy". This law is the first to define the legal, economic, social and environmental pillars of energy efficiency for all the companies, associations, organizations and citizens of Ukraine. The new version of the law is "On energy efficiency" It provides that the part of the enterprises, which focuses on energy efficiency are exempt from taxation. Also, the bill provides for the issuance of loans for energy-efficient technologies at a zero rate.
- The Law of Ukraine "On the cogeneration of heat and power and the use of the potential energy faults". For the first time the real incentives of the introduction of the cogeneration technologies are provided, in particular, to 2015. The target bonus is not included in the tariffs for electric and thermal energy produced by cogeneration special units. Thus, the price of energy supplied must be 10-15% lower, which means that it will be more attractive for the consumer. This creates conditions for the development of the cogeneration technologies in Ukraine.

The introduction of the "green" tariff for electricity produced from the renewable energy sources was one of the most effective measures for the renewable energy in Ukraine. It's used in order to encourage the consumer to use the renewable energy. The size of the "green" tariff, according to the legislative acts shall take the account of the special rate, which depends on the capacity and the type of the renewable energy (in the range 0.8-4.6). For the plants using wind energy the factor ranges from 1.2 to 2.1 depending on the power, for the generators of energy, biomass is 2.3, for the stations, that produce the current by the solar energy is 4.4-4.6, for the small HPP the lowest coefficient was set as 0.8 (Ryasnoy, 2010).

In April 2009 the Ukrainian parliament has adopted The Law "On Amendments to The Some Laws of Ukraine" (to encourage the use of the RES), which proposes the number of amendments to The Law "On Electric Power", about the green tariff and its calculation.

There is a supplement to the Law "On Electric Power Industry", a new article - Article 17-1, which provides the procedure for fixing the "green" tariffs and changes the order of the calculation.

Table 3 shows the "green" tariff scheme and tariff indexes, fixed minimum charge and its associated equivalent in Euros (The Law of Ukraine On Electricity, 2014).

The legal procedure associated with the preparation and the launch of the renewable energy projects includes several stages. First, the producer of the energy from the alternative sources, if he wants to work on the "green" tariff, must necessarily apply to the National Electricity Regulatory Commission of Ukraine to obtain the license for the production of electricity, and then ask for the introduction of the "green" tariff. Also for the sale of energy produced by the RES into the grid it is required to become a member of the Wholesale Electricity Market. To establish the "green" tariff it is necessary to comply of the basic practical conditions:

- The availability of the ownership or the using of the power generating equipment, that runs on the alternative energy sources (license);
- The availability of the built power generating facility connected to the grid (in the case of the construction of a new facility);
- Certificate of compliance of the constructed power facility that it generates electricity using the alternative energy sources, has the project documentation and meets the requirements of state standards, building codes and regulations.

Table 3. Green tariff scheme

The Source of Electricity	Tariff Rate in 2009	Tariff Level in 2009, Excluding VAT UAH kop. / kWh)	Pegged to the Euro Minimum Fare in 2009 without VAT (Euro Cents/ kWh)	Tariff Rate in 2015 (-10%)	Tariff Rate in 2020 (-20%)	Tariff Rate in 2025 (-30%)
Wind (plant capacity of 600 kW)	.2	70.15	6.46	1.08	0.96	0.84
Wind (600 kW - 2000 kW)	.4	81.84	7.54	1.26	1.12	0.98
Wind (over 2000 kW)	.1	122.77	11.31	1.89	1.68	1.47
Biomass	.3	134.46	12.39	2.07	1.84	1.61
Solar energy (ground equipment)	.8	505.09	46.53	4.32	3.84	3.36
Solar energy (equipment installed on roofs of 100 kW)	4.6	484.05	44.59	4.14	3.68	3.22
Solar energy (equipment installed on the roof, with a capacity below 100 kW)	.4	463.00	42.65	3.96	3.52	3.08
Small Hydropower Plants	.8	84.18	7.75	0.72	0.64	0.56

Since the 1st of January 2012 for those wishing to get the "green" tariff another limitation is introduced: the so-called rule of "local content". Only those power plants are allowed to work at reduced rates which were constructed with the help of the domestic raw materials and equipment and whose work was performed by the local contractors. The share of Ukraine in the value of fixed assets should be at least 15%, and from the 1st of January 2013 is 30%, and since the 1st of January 2014 is 50% (Law of Ukraine on Electricity, 2014).

This rule should support the domestic manufacturers of the equipment for the alternative energy, and the construction organizations. But investors, which may plan to build power station from renewable sources, don't like this rule, since it's much easier to import ready-made equipment. The world's major manufacturers also support investors, who are interested in the deliveries of the finished equipment rather than placing the new production facilities in Ukraine.

However, our domestic manufacturers have proven that they can organize the necessary production, both independently and with the support of foreign companies. For example, "Turboatom" in Kharkov now produces the equipment for small hydropower plants, and "Pilar" in Kiev, "Prologue Semikor", "Quasa" and "Silicone" in Svetlovodsk produce the components and equipment for the photovoltaic plants.

Also the bill is registered and will be adopted the soonest «Draft Law on Amendments to Some Laws of Ukraine to stimulate the production of electricity from the alternative energy sources." According to this bill the reduction of the "green" tariff index is provided for the electricity generated from the alternative energy sources, particularly from solar energy.

However, the legislation supporting the implementation of the RES in Ukraine is developing slowly. Despite the outlined barriers and limitations, several projects have already been implemented on the territory of Ukraine.

THE CURRENT STATE IN THE SCOPE OF THE RENEWABLE ENERGY USING

In 2011 - 2012 years in Ukraine the renewable energy sector was developing more intensive than traditional. As of the 1st of April 2013, 86 companies work at 139 power plants using the renewable energy sources. The installed capacity of the domestic alternative power in 2012 increased approximately from 57.5% up to 645.1 MW×h, and the power generation in 2.2 times to 815.9 million kW×h. However, despite more than the twofold increase of the production, their share in the balance of power remained very low, and was about 0.45% in 2012 (0.22% in 2011) (Chaly, 2013).

In present time in Ukraine there is a variety of installations that use the RES, and each year it expands, also the investment climate in the field improves.

The vast majority of the hydroelectric power plants in Ukraine are large facilities, which have their own water reservoir, located mainly in The Dnieper and The Dniester rivers. In addition to the big station, there is a concept of the small HPP - with power below 10 MW. In Ukraine there are about 70 small hydro power plants with the total capacity of 72 MW, which produce from 275 to 400 million kW×h of electricity per year. As of 2012 in Ukraine there are about 30 private companies that invest in the renewable energy. The largest of them are: Foreign Economic Association "Novosvit", "Energoinvest" and others. The investments were mainly directed to Vinnytsia, Khmelnytsky, Cherkasy, Zhytomyr and Ternopil regions. Today 64% of the total stations numbers are concentrated there, while in those technical areas the small hydro potential of the rivers is only 14% of the total. Lviv is as promising as Zakarpattia region, where about 70% of the hydropower potential of small rivers are concentrated (National Report on the man-made and natural security, 2012).

The wind power capacity existing in Ukraine exceeds nearly 51 MW, and since the launch of the first domestic wind power generator more than 80 million kW×h of electricity is generated. According to the experts, the total potential capacity of Ukrainian wind power is 5000 MW (Kudrya, Ryeztsov, & Surzhyk, 2010). According to UkrEnergo's data, as the operator of the national electricity network in Ukraine, the proposals for the construction of the wind power plants with total capacity of 14 MW were received. Also the energy facilities with the capacity of 1,150 MW have been built in accordance with UkrEnergo's requirements.

The Viennese company "Activ Solar" built in December 2011 the solar power plant in the Crimean village Perovo; it took the area of 200 hectares. According to the executive director of the company "Active Solar", the solar park was built in seven months and supplies electricity in the amount of 132.500 MW. Perovo solar park is the one of the largest plants of its kind in Europe. In February, in the village Rodnikove the first major solar power plant in Ukraine was put into operation, its area is 15 hectares. In October 2011, also in the Crimea the second solar park was built in the village Ohotnikovo, its capacity is 80 MW, photovoltaic panels for parks are partly manufactured in Ukraine (Rosenberger, 2012).

Currently biomass in Ukraine is mainly used to produce heat. In the rural areas, firewood is used traditionally for space heating. Also over 1000 boilers are transferred on woody biomasses that are operated at the forestry enterprises. About 2,000 of boilers use wood waste (wood chips, pellets). More than 70 boilers serving the oil extraction plants and fat-plants use sunflower husk as fuel (Geletukha, Zheliezna, Kucheruk, & Oliinyk 2014).

The potential of the biofuels usage for the large thermal power stations in Ukraine remain undiscovered. Only one CHP in Cherkassy (Cherkassy region) in 2013 was completely transferred to use biomass as fuel. Two industrial CHP are built in LLC "Cargill Plant" and JSC "Kirovogradoliya" that use their manufacturing waste as fuel.

The most of the biomass boilers are installed in businesses and produce energy for both the production process and for the heating of the enterprises themselves, using manufacturing waste as biofuel. An important direction is spreading the use of the energy potential of biomass across the country. This requires setting up the production of fuel pellets such as the briquettes are.

The most powerful enterprise in the construction and maintenance of the plants for the production of solid biofuels in Ukraine is the enterprise "Bioresource Ukraine." As of January 2014 60 plants in Ukraine, Russia, Lithuania, Moldova are built (The Official Website Company Bioresource, 2014). One of the largest manufacturing plants, wood pellet plant is "Eco Prime" in the town Hluhiv in Sumy region (with a production capacity of 30 000 tons per year). Also the plants were built in Dnipropetrovsk, Zhytomyr, Kyiv, Chernivtsi regions close to the primary database. Thus, the market for biomass fuel production is developing projects for the construction of CHP on biofuel is developing. They are in need of the state support and appropriate investment.

There is the program of the European Bank for Reconstruction and Development (EBRD) to finance renewable energy in Ukraine (USELF). Within USELF the loans and the assistance in developing projects that meet the commercial, technical and environmental criteria of the program are provided. Program funding is € 50 million to promote the RES projects in Ukraine.

According to the Association of the market participants of the alternative fuel and energy of Ukraine (ATEU), the volume of investments in the solar industry in the first half of the year exceeded 360 million euro. Also according to the quarterly reports of Ernst & Young, in May 2014 Ukraine ranked the 39 place in the world ranking position of the market attractiveness for the renewable energy sources (Ernst & Young, 2014).

The compilers of the rating note, that Ukraine has a significant potential for the development of the renewable energy, in particular in the field of wind energy. At the same time the company's specialists have noted the complexity of the licensing procedures for potential investors, as well as the problems associated with the connection of the renewable energy power plants to the overall energy system. In addition, the investors have difficulty arising due to the uncertainty in the economy and the unstable political situation.

Despite the outlined problems, the well-thought-out combination of the RES facilities in conjunction with batteries and the fuel cells are capable to almost completely satisfy the needs of the consumers of the grid, maintaining low costs. Usually the methods of energy management are used for these tasks.

ENERGY MANAGEMENT APPROACHES

The classical energy networks were not created to work with the complex tasks of management of energy flows, which appear when using the RES. The construction of a new network or upgrading the existing one to use the RES requires the detailed due investigation. This is to ensure proper distribution of energy derived from different sources in order to meet the needs of the end users of heat and electricity. For these tasks, you can use the Decision Support System for Energy Management System. This will allow making the recommendations regarding the transfer of energy and its consumption, and do it in the real time with maximum efficiency and the use of new technologies. Therefore, the task of improving efficiency of energy management systems requires special attention.

DSS for power grid based on data from operational monitoring of meteorological and geographical conditions, modes of consumption structures, should solve problems of analysis of the RES implementation efficiency and issue recommendations regarding their operation modes. The process of making the management decision in DH systems needs an efficient energy audit in this area to evaluate the efficiency of the RES implementation.

The improvement of the energy management technology for the power systems with the RES for heat and power generation through the use of information technologies will be discussed below.

THE DSS IS BASED ON THE GEOGRAPHIC INFORMATION SYSTEM FOR EVALUATION OF THE RENEWABLE ENERGY IN THE DISTRIBUTED ELECTRICITY GENERATION

According to Ramachandra (2009) DSS may be described as comprising three subsystems: The Database Management Subsystem, The Model Management Subsystem, and The Dialogue Management Subsystem. The regional energy planning program mainly consists of the next blocks: Energy Scenarios, Computation Models, Aggregation and The Energy Database.

The existing DSS differ depending on the features: a variety of approaches based on the multi-criteria analysis, calculation methods of the technical characteristics, as well as on the method of data presentation.

The input parameters for the calculation can be interdisciplinary, relating to the meteorology, geography, architecture, urban utility systems, etc. Three directions are able to distinguish the development of DSS in the energy supply of buildings on the number of the renewable energy sources. One can use one source of the alternative energy, the other - several, others combine the sources, both heat and electricity.

Information is obtained at the "almost real" time, can be imported from the geographic information system (GIS) to conduct various types of spatial analysis. The information about the location of the equipment, the users and the potential of the renewable sources can be transmitted to the system to provide better understanding of the location.

The potential of the renewable energy description, the system must contain some information on the existing renewable energy facilities and the energy systems in the region. To construct the GIS there are several tools that are presented in Table 4, their analysis has been carried out depending on their functionality.

Some projects have already been built relating to DSS construction based on the geographic information system for the evaluation of the renewable energy in the distributed electricity generation. The following are some of the existing.

Tiba (2010) proposed a methodology to determine the best locations for the use of the new energy systems, but without the study of the network. The DSS for regional planners was

Table 4. Comparison of information systems working with geodata

	The Prevalence	Tools for Creating an Electronic Map Material	Support for Geo-Data Storage	The Conversion of Data Formats	Visualizing Data in Different Formats	The Tools of Geo Analysis
GRASS (Bivand, & Neteler, 2000).	+	+	+	+	+	+
MapGuide Open Source (MapGuide Project Home, 2014).	+	+	+	+	+	+
gvSIG (gvSIG, 2014).		+	+	+	+	+
Q GIS (Quantum GIS, 2014).		+	+	+	+	
Geo server (GeoServer, 2014).				+	+	
OpenMap (OpenMap, 2014).	+				+	
Post GIS *(Post GIS, 2014)*.	+		+	+		+
KOSMO (KOSMO, 2014).		+	+		+	+

developed for economic analysis of the set of wind power projects for private investors in Belgium (InterPSS Communities, 2014). In Yi, Lee and Shim's work (2010) GIS is used for the evaluation of hydropower plants. In Ramirez-Rosad and others' work (2008) DSS is designed for wind power. This decision support system using GIS can determine the right place for the construction of the renewable energy sources, search for the potential sites for the renewable energy installations, and then analyze the behavior of the networks and distributed generation systems, but the system does not use the idea of the Smart Grid (Lazarou, Oikonomou, and Ekonomou, 2012).

SMART GRID SIMULATION

In the power grid systems the electricity supply from the distributed sources is difficult, because the RES have the irregular schedule of power generation. Stable work in this case is provided by the integration of the monitoring and control. Substation automation allows controlling distribution time energy and, if it's necessary, planning the connection of the reserve sources of energy. Thus, it's suitable to use Smart Grid technology that allows the power system to better adapt to a dynamic distributed generation sources.

The Smart Grid is an electricity network that can intelligently integrate the actions of all the users connected to it – generators, consumers and those that do both - in order to efficiently deliver sustainable, economic and secure electricity supplies (SmartGrids Strategic Deployment Document finalized, 2014). Roncero (2008) distinguishes the key features of Smart Grid:

- Self-healing.
- Incorporates and empower the user.
- Tolerates security attack.
- Offers power quality enhancement.
- Accommodates various generation sources.
- Fully supports energy market.
- Optimizes asset utilization and reduces the expenses for system operations and maintenance.

The list of the possible logical systems reflecting the capabilities of the power system based on the concept of Smart Grid is presented below (Massel, & Bakhvalov, 2012):

- Distributed Monitoring and Control System.
- Distributed System of Monitoring Substation Automatic System Shutdown.
- Distributed Monitoring System for The Generation.
- Automatic Measurement of the Running Processes.
- Measurement Control System.
- Distributed Forecasting System.
- Operating Management System Smart.

The Smart Grid construction requires the prior optimal planning of work. There can be used three main approaches; their features are clearly illustrated by the way of describing the different tools for operation:

- **The Agent-Based Modeling (ABM):** The basic point is that the system consists of some sub-systems, which are called agents (Nissen, 2001). Agents are characterized by the ability to cooperate, autonomously act and the ability to learn. The combinations of these characteristics depend on the agent's functions. For ABM such software as NetLogo (Tisue & Wilensky, 2004), JADE (Java Agent Development Framework, 2014), Cormas (Cormas: Natural resources and agent-based simulations, 2014), AnyLogic (AnyLogic, 2014) and etc. can be used:
- **Simulators for Dynamic Model:** Simulation models are used to describe and analyze the behavior, ask "what if" ques-

tions about the real system, and aid in the design of the real systems. A simulation model can be static, if it represents the system life cycle in a single time period; and dynamic, if the current period depends on the evolution of the previous periods (Tisue & Wilensky, 2004). Modeling method enables to build the formal computer simulations of complex systems and use them to design more effective policies and organizations. A simulation model is used for representing cause-and-effect relationships with the use of the causal-loop diagrams, flow diagrams with levels and rates, and equations. It's the perspective and the set of conceptual tools that enable us to understand the structure and the dynamics of the complex systems. The set of simulators in the field of engineering system are such as NEPLAN (NEPLAN Desktop Overview, 2014), MATLAB (MatLAB, 2014), PSS NETOMAC & SINCAL (PSS Product Suite, 2014), VTB (Monti, Ponci, Smith, & Liu, 2009), etc.

- **Network Simulators:** "Network" is meant in a broader sense and includes wired and wireless communication networks, on-chip networks, and queuing networks and so on. Domain-specific functionality such as support for sensor networks, wireless ad-hoc networks, Internet protocols, performance modeling, photonic networks, etc., is provided by model frameworks, developed as independent projects.

Tools providing professional network planning tools, simulators for communication systems are NS-3 (The NS-3 Consortium, 2014), OPNET (OPNET Modeler Suite, 2014), GloMoSim (Takai, Martin, & Bagrodia, 2001), QualNET (The QualNet communications simulation platform. 2014) etc. smart grid simulator must be added to some of these tools; it may be OMNeT++ (Varga & Hornig, 2008).

Despite these methods, it's important to achieve the suitable level of interactivity of the future planning system, thus, the structure of system becomes more complex, and it's possible to build Decision Support System (DSS) with graphical shell.

THE ENERGY OPTIMIZATION OF THE CALCULATION AND EVALUATION OF THE RES

The construction of the electrical grid based on the RES requires prior optimal planning of work. At present there is a large amount of different approaches to plan grid's work. Thus, Ramachandra (2009) presented a tool for resources planning and management, which uses a Multi-Criteria Decision Analysis (MCDA), and the method NAIADE to find a compromise solution in the fuzzy decision-making environment. NAIADE approach involves the construction of the matrix of evaluation that includes, on the one hand, the limited set of alternatives, and on the other hand, the limited set of different criteria by which such alternatives must be evaluated on comparison of the pairs of alternatives.

The approach was used to determine the optimal use of wind turbines in the island of Salina (Aeolian Islands, Italy). Ramachandra (2009) proposed the use of five alternatives, wind turbines to initially determine the power and the quantity:

- Plan "A" (wind 150 kW).
- Plan "B" (wind five 15 kW).
- Plan "C" (wind two 150 kW).
- Plan "D" (PV + wind five 15 kW).

The following criteria are used:

- Investment cost.
- Operating and maintenance costs.
- Energy production capacity.
- Fuel savings.
- Technological maturity.

- Realization times.
- CO_2 emissions avoided.
- Visual impact.
- Acoustic noise.
- Impact on ecosystems.
- Social acceptability.

Research described the approach to Decision Support System construction that uses software HOMER, as the way of calculating and determining the best grid design (Muslih & Abdellatif, 2011). With the help of an integrated system it is proposed to calculate solar and wind energy that is used every hour. As emergency funds for the isolated areas of the electricity supply a diesel generator and a battery are offered. The basis of the selection of the final set is the function to minimize costs, and it takes into account the whole life cycle cost of the installation.

In other research provides the comparative analysis of such decision support system for the grid construction such as HOMER, Hybrid2 and RETScreen (Georgilakis, 2006).

HOMER and RETScreen are selected to choose the alternative source of energy in the economic terms, and Hybrid2 is a technical package. HOMER and Hybrid2 are designed for hybrid systems, while RETScreen is used for a single source. HOMER also includes the methods of optimization. These software products do not support multiple criteria analysis, and do not allow the calculation of biogas and geothermal energy to generate heat.

The system of the distributed energy management in buildings in real time Capo Vado (Liguria Region), which allows determining the optimal flow of energy in the building and is characterized by the combination of the renewable resources was constructed (Dagdougui et al., 2010). Solar collector, photovoltaic modules, one device for converting biomass, wind turbine and a battery are used for the calculations. DSS allows determining the time of the greatest generation of energy from different sources.

To calculate the energy produced by the wind turbine the technical data of the wind turbine power and the changes of the power of the wind data region that come every hour are used. The energy from the photovoltaic modules and the solar collector depend on the characteristics of the plants and the hourly data of solar radiation. Biomass energy depends on the quality (the percent of the humidity of biomass) and quantitative characteristics. The optimization of the objective function calculation is based on the using of the weighted sum of deviations from the requirements of different sources; to optimize the tool Lingo was used.

According to the analysis of the existing methods that can be applied in DSS and tasks of the information simulation of the distributed power grid, it can be concluded that there are many disparate ways of solving the problems.

THE TOOLS OF ENERGY MANAGEMENT IN HEATING SYSTEMS

The measures to improve the management of heat energy consumption are related to the use of the information technologies for providing real-time monitoring and regulation of the heat consumption modes of the particular building. The objects of the DH system are geographically distributed. Therefore to implement the DSS it is necessary to use web-technologies for sending data using the resources of the Internet for quick data acquisition. The development of such information system for the realization of the management of heat supply is intended to support the solution of the problem of rational combination of the RES with primary energy resources.

As an efficient tool of energy management the "HeatCAM" information system (the Heating Control, Analysis and Management system) may be used. The proposed information system is directly linked to the monitoring and prediction of heat consumption. Using this system for monitoring heat supply systems using the RES allows controlling the quality of services with the heat supply. It can also help in processing the specific information for the decision-making on heat consumption regulation (Parfenenko, Shendryk, Nenja, & Okopnyi, 2013).

The information system "HeatCAM" is organized in the combination of three-tier architecture with software agent's technology. The system consists of several subsystems. Their functionality is described briefly in Table 5.

The Data Collection subsystem is equipped with the hardware components. A heat meter with digital output is mounted on the incoming heating pipeline of the every building. The heating sections with forward and reverse pipes are equipped with pressure and thermal sensors. Some sensors are designed to control the temperature in the heated indoors and others to monitor ambient temperature. The heat meters, thermal and pressure sensors are connected to the terminals. Every terminal transmits collected data to a web-server through GPRS (Parfenenko, Okopnyi, Nenja, 2012). The acquisition agent collects data from monitoring devices (sensors, heat meters) and places them in a temporary data store. Every acquisition agent performs its main task autonomously during certain time intervals (as a rule, an hour). The monitoring agent provides data verification and data transmission to database. It also performs records of all actions into the log file.

The Data Presentation subsystem provides an ability to view monitored data for ambient weather conditions, together with the current and calculated parameters of the heating in definite building for the selected period. The representation of the monitoring data on the web-page in the form of tables and diagrams is supplemented with the interactive interface.

The categories of users are listed as follows:

- The system administrator has full rights for data access, giving the user rights to all categories of users and creating database backups;
- The engineer or energy auditor performs analysis of forecast data of the heating consumption and on the basis of this information makes decisions about heat-supply regulation.
- The accountant has an access to the accounts for heat supply services which are formed on the basis of current heat energy consumption;

Table 5. The functionality of HeatCAM subsystems

The Subsystem Name	The Functionality
The Data Collection subsystem	The data acquisition from terminals and transmission their into the temporary data store
The Monitoring subsystem	The data extraction from the temporary data store, data checking and a transfer into unified database of information necessary to assess the current level of short-term limits and calculation of heat consumption
The Data Storage subsystem	The storage of current monitoring data, prediction data, data on the structure of heat supply. Archiving the data.
The Data Presentation subsystem	The representation of the monitoring data on the web-page in the form of tables and diagrams supplemented with the interactive interface, which displays the current data on the user's web page
The Forecasting subsystem	The short-term heat consumption forecasting

Table 6. The inputs and outputs of neural network

The Element of Neural Network	The Data Caption	Meaning
X (t)	t	Air temperature
X (t)	v	Wind speed
X (t)	P	Air pressure
X (t)	h	Air humidity
Y(t)	G	Amount of heat demand

- The operator has an access to the monitoring data and can add new data in the case of absence of the monitoring devices in the building.

The forecasting of heat energy demand can be made with the help of the classical (statistical) method and using the opportunities of the artificial intelligence theory. While forecasting the impact of the weather conditions (temperature, air pressure, wind speed, direction etc.), fluctuations of the needs in heat energy in depending on the season must be considered. In forecasting of heat consumption in public sector it's necessary to take into consideration the periods of non-working time (weekends, holidays). On these days the heat demand is lower than in working days. The forecasted schedule should take into account the meteorological forecast (Chramcov, Dostál, & Baláte, 2003). These weather environmental conditions can be obtained by the data extraction from one of the specialized web-sites as well as to combine forecasted information from various sites using artificial neural networks for a more accurate forecasting for a year-round, for 7 days in advance and for the next day.

The forecasting subsystem is built with the usage of the neural networks. Neuro-fuzzy networks are widely used to predict the heat demand of the building at different times from one day to the entire heating season (Chramcov, Dostál, & Baláte, 2009; Chramcov, Baláte, & Princ, 2012; Grzenda, 2012).

The architecture of the neural network implemented in the HeatCAM system is the nonlinear autoregressive neural network architecture (NARX) with 16 hidden nodes and one output node. The forecasting is based on the previous data of the forecast variable and exogenous inputs.

Networks inputs X (t) and outputs Y (t) are listed in Table 6.

The data of the inputs of the neural network are the data collected from a special site of the weather conditions. The training sample is formed from the heat energy consumption data for each building for the previous three years, which are stored in a database on the server of the monitoring system.

Education of the neural networks is based 70% on the samples of training data, 20% on testing and 10% on the validation. Neural network is trained on Levenberg-Marquardt algorithm. The forecasting subsystem was tested for determining the heat demand of a building in the public sector. Whole data set for neural network consists of measured data for the past 3 years.

To summarize about "HeatCAM" system description it should be noted that testing of this system for energy management of a concrete building allows reducing the heat consumption approximately on 15%.

Thus, the short-term forecasting of the heat energy demand for a separate building provides support for making decisions about the mode of heating the building in the next few days, which enables to evaluate the RES demand for a certain heating period.

FUTURE RESEARCH DIRECTIONS

According to the present overview of the possibilities of using the renewable energy sources in Ukraine, there is a problem with the network construction with the RES meeting three basic requirements at a technical level: the availability of the resources, regional energy consumption, and the existing capacity of the grid, which can be increased at the expense of remodeling or building of a new grid.

Therefore, the future information model may be used in a decision support system, which allows calculating the load in the real time, to issue recommendations on the use of the RES, depending on the region of the grids construction. Before construction, it is necessary to determine the existing approaches that can be used and functionality.

The functional modules of the system should include:

- The monitoring system of regional weather conditions and geographical data, which allows conducting the operational control of the environmental change: speed measurements and the direction of wind, temperature, relative humidity and the atmospheric pressure.
- The common information database. Systematic, orderly brought in a total database the data allow us to give better recommendations for the construction of networks, as well as for power supply regulation.
- The calculation of the chosen design of the network and the source of the renewable energy.
- Planning and forecasting of the development of the information-analytical system of the integrated power supply.
- DSS also has to be supplemented by a simulation model of a distributed power grid to calculate the losses.

CONCLUSION

In accordance with the analysis of the current situation of using the RES in Ukraine, this Chapter describes the features of the RES integration into the Ukrainian power grid. It is the solid source of comprehensive reference material in the related research field. The questions about the RES potential of Ukraine, barriers of its implementation into Ukrainian power grid and the legislation for the regulation of this process were considered in it.

The urgent problems in the field of energy management using the information modeling of the electrical power grids and district heating were studied. The analysis of the existing methods of the power management, which can be used in the DSS was carried out, the problems of the short-term prediction of the heat demand were considered. Also it was found that according to the analysis of the existing methods that can be applied in the problems of the information modeling there are many disparate ways of solving the problems. Thus, further research in the field of construction of the power grids with the renewable energy is aimed to simulate the smart grids with different parameters of the renewable energy to achieve their optimal combinations.

The research results can serve as the basis for work on the construction of a Decision Support System, which allows simulating the work processes and receiving summary information about the systems, recommendations on their design, etc.

REFERENCES

AnyLogic. (n.d.). Retrieved April 20, 2014, from http://www.anylogic.ru

Babaev, V. M., Malyarenko, V. A., & Orlova, N. O. (2012). Development and implementation of increase the efficiency of communal energetics. *Energy Saving - Energetics - Energy Audit, 4*(98), 9-22.

Bivand, R., & Neteler, M. (2000, August). Open source geocomputation: Using the R data analysis language integrated with GRASS GIS and PostgreSQL data base systems. In *Proceedings of the 5th International Conference on GeoComputation*. Academic Press.

Chaly, V. V. (2013). Problemy obespechenia energeticheskoy bezopasnosti v Ukraine. *Upravlinnya rozvytkom, 22*, 144 – 146.

Chramcov, B. (2010). Heat demand forecasting for concrete district heating system. *International Journal of Mathematical Models and Methods in Applied Sciences, 4*(4), 231–239.

Chramcov, B., Dostál, P., & Baláte, J. (2003). Prediction of the heat supply daily diagram via artificial neural network. In *Proceedings of the 4th International Carpathian Control Conference*. Zittau, Germany: Academic Press.

Chramcov, B., Dostal, P., & Balate, J. (2009, June). Forecast model of heat demand. In *Proceedings of the 28th International Symposium on Forecasting*. Hong Kong, China: Academic Press.

Collegium of the Ministry Energy and Coal Industry of Ukraine. (2012). *Updates energy strategy of Ukraine till 2030*. Retrieved from http://energetyka.com.ua/normatyvna-baza/384-energetichna-strategiya-ukrajini-na-period-do-2030-roku

Cormas: Natural Resources and Agent-Based Simulations. (n.d.). Retrieved from: http://cormas.cirad.fr

Dagdouguia, H., Minciardia, R., Ouammia, A., Robbaa, M., & Sacilea, R. (2010, July). A dynamic optimization model for smart micro-grid: integration of a mix of renewable resources for a green building. In *Proceedings of the 5th Biennial International Congress on Environmental Modelling and Software Society: Modelling for Environment's Sake (ICEMSS'10)*. Academic Press.

NEPLAN Desktop Overview. (n.d.). Retrieved April 20, 2014, from http://www.neplan.ch

Ernst & Young. (2014). *Renewable energy country attractiveness index*. Ernst & Young Global Limited.

Geletukha G. G., & Zheliezna T. A. (2013). Barriers to bioenergy development in Ukraine. *Promislova teplotekhnika, 35*(5), 43-47.

Geletukha, G. G., Zheliezna, T. A., Kucheruk, P. P., & Oliinyk, Y. M. (2014). *State of the art and prospects for bioenergy development in Ukraine*. Retrieved from http://www.uabio.org/img/files/docs/position-paper-uabio-9-en.pdf

Geletukha, G. G., Zheliezna, T. A., & Oliinyk, Ye. M. (2013). *Opportunity of heat generation from biomass in Ukraine*. Retrieved from http://www.uabio.org/img/files/docs/position-paper-uabio-6-ua.pdf

Georgilakis, P. S. (2006). State-of-the-art of decision support systems for the choice of renewable energy sources for energy supply in isolated regions. *International Journal of Distributed Energy Resources, 2*(2), 129–150.

GeoServer. (n.d.). Retrieved April 20, 2014, from http://geoserver.org/

Grzenda, M. (2012). Consumer-oriented heat consumption prediction. *The Journal of Control and Cybernetics, 41*, 213–240.

gvSIG. (n.d.). Retrieved April 20, 2014, from: http://www.gvsig.com/

Hase, Y. (2007). *Handbook of power system engineering*. John Wiley & Sons. doi:10.1002/9780470033678

Inter P. S. S. Community. (n.d.). Retrieved from: http://www.interpss.org/

Java Agent Development Framework (JADE). (n.d.). Retrieved from http://jade.tilab.com/

Kalenska, S., Rahmetov, D., Kalenskiy, V., Yinik, A., & Kachura, I. (2012). Bioresource potential of Ukraine in settling of production and energy security. *International Scientific Electronic Journal Earth Bioresources and Life Quality, 1*. Retrieved from http://gchera-ejournal.nubip.edu.ua/index.php/ebql/article/view/20/pdf

Konechenkov, A. (2013). *Renewable energy. Focusing: Ukraine vision 2050*. Retrieved from http://www.inforse.org/europe/pdfs/VisionUA_ppt.pdf

Kosmo. (n.d.). Retrieved April 20, 2014, from http://www.opengis.es/

Kuznetsova, A., & Kutsenko, K. (2010). Biogas and "green" tariffs in Ukraine – A profitable investment? *German–Ukrainian Policy Dialogue in Agriculture. Institute for Economic Research and Policy Consulting, 26*, 1–38.

Law of Ukraine on Electricity. (n.d.). Retrieved April 20, 2014, from http://zakon2.rada.gov.ua/laws/show/575/97-%D0%B2%D1%80

Lazarou, S., Oikonomou, D. S., & Ekonomou, L. (2012). A platform for planning and evaluating distributed generation connected to the hellenic electric distribution grid. In *Proceedings of the 11th WSEAS International Conference on Circuits, Systems, Electronics, Control & Signal Processing*. Montreux, Switzerland: WSEAS Press.

Makarovsky, E.L. (2004). Energetichesky potential netradicionnyh i vozobnovlyaemyx istochnikov Ukrainy. *Integrovani technologii ta energozberezennya, 3*, 75-83.

MapGuide Project Home. (n.d.). Retrieved April 20, 2014, from http://mapguide.osgeo.org/

Massel, L. V., & Bakhvalov, K. S. (2012). Open integrated environment InterPSS as a basis of smart grid it-infrastructure. *Vestnik ISTU, 7*, 10–15.

MatLAB. (n.d.). Retrieved April 20, 2014, from http://uk.mathworks.com/

Ministry of Energy and Coal Industry of Ukraine. (n.d.). Retrieved from http://mpe.kmu.gov.ua

Monti, A., Ponci, F., Smith, A., & Liu, R. (2009). A design approach for digital controllers using reconfigurable network-based measurement. In *Proceedings of the International Instrumentation and Measurement Technology Conference 2009*. Piscataway, NJ: IEEE. doi:10.1109/IMTC.2009.5168415

Muslih, I. M., & Abdellatif, Y. (2011). Hybrid micro-power energy station: Design and optimization by using HOMER modeling software. In *Proceedings of the 2011 International Conference on Modeling, Simulation and Visualization Methods*. Pittsburgh, PA: ACTAPRESS.

National Report on the Man-Made and Natural Security in 2012. (2012). Retrieved from http://www.mns.gov.ua/files/prognoz/report/2012/3_4_2012.pdf

Network Simulation (OPNET Modeler Suite). (n.d.). Retrieved April 20, 2014, from http://www.riverbed.com/products/performance-management-control/network-performance-management/network-simulation.html

Nissen, M. E. (2001). Agent-based supply chain integration. *Journal of Information Technology Management, 2*(3), 289–312. doi:10.1023/A:1011449109160

On the Basic Parameters of the Boiler and Heating Networks in Ukraine by 2013. (2014). *The statistical bulletin* (Pub. No. 03.3-5/178-14). Kyiv: The State Statistics Committee of Ukraine.

OpenMap. (n.d.). Retrieved April 20, 2014, from http://openmap.bbn.com

Parfenenko, Yu. V., Okopnyi, R. P., & Nenja, V. G. (2012). Creation of control instruments for buildings' heating. *The Journal of the National Technical University NTU KPI. Series Information Systems and Networks, 34*, 93–97.

Parfenenko, Yu. V., Shendryk, V. V., Nenja, V. G., & Okopnyi, R. P. (2013). Information-analysis system for monitoring and prediction of heat-supply buildings. *Visnyk of East Ukrainian National University named after V. Dahl, 743*, 38–43.

Pavliuk, S. (2011). *The efficient national energy efficiency policy – Basis modernization of housing and communal services*. Retrieved from http://www.eu.prostir.ua/files/1341335344039/energy%20effeciencyPP_engl.pdf

PostGIS. (n.d.). Retrieved April 20, 2014, from http://postgis.refractions.net

Product Suite, P. S. S. (n.d.). Retrieved April 20, 2014, from http://www.simtec.cc/sites_en/sincal.asp/

Quantum, G. I. S. (n.d.). Retrieved April 20, 2014, from http://qgis.org/ru/site/

Ramachandra, T. V. (2009). RIEP: Regional integrated energy plan. *Renewable & Sustainable Energy Reviews, 13*(2), 285–317. doi:10.1016/j.rser.2007.10.004

Ramirez-Rosado, I., Garcia-Garrido, E., Fernandez-Jimenez, L., Zorzano-Santamaria, P., Monteiro, C., & Miranda, V. (2008). Promotion of new wind farms based on a decision support system. *Renewable Energy, 33*(4), 558–566. doi:10.1016/j.renene.2007.03.028

Renewable Energy Institute NAS of Ukraine. (n.d.). Retrieved April 20, 2014, from http://www.ive.org.ua/

Roncero, J. R. (2008). Integration is key to smart grid management. *Smart Grids for Distribution, CIRED Seminar, 9*, 1–4.

Rosenberger, C. (2012). Die policy of Ukraine in the power industry field (KAS Policy Paper 18). Retrieved from http://www.kas.de/wf/doc/kas_33444-1522-1-30.pdf?130206154511

Ryasnoy, D. (2010). *Energy becomes green: The introduction of «green» tariffs gave rise the development of alternative power generation in Ukraine*. Retrieved from http://www.ges-ukraine.com/maininfo_15-9.html

Safe Mode the Work of DEC Ukraine. (n.d.). Retrieved April 20, 2014, from http://www.ukrenergo.energy.gov.ua/ukrenergo/control/publish/category

Shevtsov, A. I., Barannik, V. O., Zemlyaniy, M. G., & Ryauzova, T. V. (2010). *State and perspectives of reforming the heat supply system in Ukraine*. Retrieved from http://www.db.niss.gov.ua/docs/energy/Teplozabezpechennya.pdf

SmartGrids Strategic Deployment Document Finalized. (n.d.). Retrieved April 20, 2014, from http://www.smartgrids.eu

State Agency on Energy Efficiency & Energy Saving of Ukraine. (n.d.). Retrieved April 20, 2014, from http://saee.gov.ua/en/

Strategic Environmental Review. Renewable Energy Scenarios. Interconnection and Transmission Considerations in Renewable Energy Development in Ukraine. (2013). Retrieved from http://www.uself.com.ua/fileadmin/uself-ser-en/3/E%20-%20Transmission.pdf

Sustainable Energy for All (SE4ALL). (2013). Retrieved April 20, 2014, from http://data.worldbank.org/data-catalog/sustainable-energy-for-all

Takai, M., Martin, J., & Bagrodia, R. (2001). Effects of wireless physical layer modeling in mobile ad hoc networks. In *Proceedings of the 2nd ACM International Symposium on Mobile Ad Hoc Networking & Computing*. New York, NY: ACM. doi:10.1145/501426.501429

The NS-3 Consortium. (n.d.). Retrieved April 20, 2014, from https://www.nsnam.org/

The Official Website Company Bioresurce Ukraine. (n.d.). Retrieved April 20, 2014, from http://bioresource.pro/index.php/uslugi

The Official Website Consulting Company NIKO Analytical Section. (2013). Retrieved from http://www.kua.niko.ua/wp-content/uploads/2010/07/Sector_Review_Energy_2011_full-v_ENG.pdf

The QualNet Communications Simulation Platform. (n.d.). Retrieved April 20, 2014, from http://web.scalable-networks.com/content/qualnet

Tiba, C., Candeias, A. L. B., Fraidenraich, N., Barbosa, E. M. S., & de Carvalho Neto, P. B. (2010). A GIS-based decision support tool for renewable energy management and planning in semi-arid rural environments of northeast of Brazil. *Renewable Energy*, *35*(12), 2921–2932. doi:10.1016/j.renene.2010.05.009

Tisue, S., & Wilensky, U. (2004). NetLogo: A simple environment for modeling complexity. In *Proceedings of the International Conference on Complex Systems*. Boston: NECSI Knowledge Press.

Varga, A., & Hornig, R. (2008). An overview of the OMNET++ simulation environment. In *Proceedings of the 1st International Conference on Simulation Tools and Techniques for Communications, Networks and Systems*. Marseille, France: ACM Digital Library. doi:10.4108/ICST.SIMUTOOLS2008.3027

Yi, Ch.-Y., Lee, J.-H., & Shim, M.-P. (2010). Site location analysis for small hydropower using geo-spatial information system. *Renewable Energy*, *35*(4), 852–861. doi:10.1016/j.renene.2009.08.003

Zerkalov, D. V. (2012). *Energy saving in Ukraine*. Kyiv: Osnova.

ADDITIONAL READING

Alanne, K., & Saari, A. (2006). Distributed energy generation and sustainable development. *Renewable & Sustainable Energy Reviews*, *10*(6), 539–558. doi:10.1016/j.rser.2004.11.004

Yumrutaş, R., & Ünsal, M , (2000). Analysis of solar aided heat pump systems with seasonal thermal energy storage in surface tanks. *Energy*, *25*(12), 1231–1243. doi:10.1016/S0360-5442(00)00032-3

Cavallaro, F., & Ciraolo, L. (2005). A multicriteria approach to evaluate wind energy plants on an Italian island. Energy Policy, 33(2), 235-244 , doi:10.1016/S0301-4215(03)00228-3

Dubovoy, V. M., Kabachiy, V. V., & Panochishin Yu. M. (2005). Monitoring and Control of District Heating Networks. Vinnitsa: Universe-Vinnitsa.

Finogeev, A. G., Dilman, V. B., Maslov, V. A., & Finogeev, A. A. (2010).. The remote monitoring system of the urban heating based on wireless sensor networks, *Proceedings of higher education institutions. The Povolzhye region. Technical sciences*, *3*(15), 27–36.

Haiyan, L., & Valdimarsson, P. (2009). District Heating Modeling and Simulation. In *Proceedings of Thirty-Fourth Workshop on Geothermal Reservoir Engineering*. Retrieved from https://pangea.stanford.edu/ERE/pdf/IGAstandard/SGW/2009/lei.pdf

Hledik, R. (2009). How Green Is the Smart Grid? *The Electricity Journal*, *22*(3), 29–41. doi:10.1016/j.tej.2009.03.001

Lasseter, R. (2007). Microgrids and Distributed Generation. *Journal of Energy Engineering, 133, Special Issue: Distributed Energy Resources-Potentials for the Electric Power Indsutry,* 144–149.

Lejeune, P., & Feltz, C. (2008). Development of a Decision Support System for setting up a wind energy policy across the Walloon Region (southern Belgium). *Renewable Energy*, *33*(11), 2416–2422. doi:10.1016/j.renene.2008.02.011

Lubchenko, V. Ya., & Pavluchenko, D. A. (2009). Reactive Power and Voltage Control by Genetic Algorithm and Artificial Neural Network. [IJTPE]. *International Journal on Technical and Physical Problems of Engineering, 1*(1), 23–26.

Malinovsky, P., & Ziembicki, P. (2006). Analysis of District Heating Network Monitoring by Neural Networks. *The Journal of Civil Engineering and Management, XII*(1), 21–28.

Nefedova, I., Finogeev, E.A., & Finogeev, A.G. (2013). Analysis of the Problems of Creating a Single System for Monitoring of DH Networks in the Region. *The International Research Journal. Engineering, 6 (13-1),* 74-75.

P, érez-Lombard, L., Ortiz, J., & Pout, C. (2008). A review on buildings energy consumption information. *Energy and Building, 40*(3), 394–398. doi:10.1016/j.enbuild.2007.03.007

Parfenenko, Yu. V., & Nenja, V. G. (2010). The Information Technology of Monitoring of the DH system with High Reliability. *The East European Journal of Advanced technologies, 4/9 (46),* 22-25.

Parfenenko, Yu. V., Nenja, V. G., & Ponomarenko, O. I. (2010). The Analysis of the Functioning of the District Heating System as the Control Object. *Visnyk of the National Technical University. KHPI, 57,* 264–268.

Ringel, M. (2006). Fostering the use of renewable energies in the European Union: The race between feed-in tariffs and green certificates. *Renewable Energy, 31*(1), 1–17. doi:10.1016/j.renene.2005.03.015

Russel, S., & Norvig, P. (2002). Artificial Intelligence: aModern Approach. International Edition. Englewood Cliffs, New Jersey: A Simon & Schuster Company.

Shendryk, V. V., & Vashchenko, S. M. (2011). The system of Data Collection, Allocation and Analysis. *Visnyk of the National University. Lviv Polytechnic., 715,* 343–353.

Vasek, L., & Dolinay, V. (2010). Simulation Model of Heat Distribution and Consumption in Municipal Heating Network. *The International Journal of Mathematical Models and Methods in Applied Science, 4*(4), 240–248.

Voronovskii, G. K, Klepikov, V. B, Kovalenko, M. V., & Makhotilo, K.V. (2000) Connected Neural Network Model of Heat and Electricity to Large Residential District of the City. *The Bulletin of the Kharkov State Polytechnic University. A series of Electrical engineering, electronics and electric drive, 113,* 363-366.

Voropay, N. I. (Ed.). (2010). *System Research in Energy: Retrospective Research Directions in the Energy System Studies Carried Out at the Institute of Energy Systems named by L. A. Melentyev.* Novosibirsk: Nauka.

Wernstedt, F. (2005). *Multi-Agent Systems for Distributed Control of District Heating Systems.* Retrieved from http://www.sea-mist.se/fou/Forskinfo.nsf/Sok/51e3dfb98bb6ba6bc1257107002f6d29/$file/Wernstedt_diss.pdf

Zhigang, Z., Pinghua, Z., Haoxuan, T., Xiumu, F., & Wei, W. (2006). Research on a Heat-supply Network Dispatching System Based on Geographical Information System (GIS). In *Proceedings of 6th Energy System Laboratory International Conference for Enhanced Building Operations.* Retrieved from http://repository.tamu.edu/handle/1969.1/5255

KEY TERMS AND DEFINITIONS

District Heating: An economic activity related to supplying heat produced centrally in one or several locations to a non-restricted number of customers.

Energy Saving: The reduction of the amount of energy used while achieving similar outcome of the end use.

Monitoring: An act of observing something to regulate and control processes in the observed system.

Terminal: A special device for the real-time acquisition of monitoring data.

The "Agent": The metaphor is used in agent-based systems, which are the result of the synthesis of the technology of object-oriented programming and the artificial intelligence.

The "Green" Tariff: Special rates at which one can purchase electricity from facilities that use RES (solar, wind, geothermal, wave and tidal energy, hydro power (installed capacity less than 10 MW), biomass, biogas, natural gas, etc.).

The Decision Support System: An interactive system, that can produce data and information, in some cases, contribute to the understanding relating to this subject area in order to provide useful assistance in solving complex and ill-defined problems.

The Geographic Information System: A computer system for the input, storage, maintenance, management, retrieval, analysis, synthesis and the output of geographic or location-based information.

The HeatCAM System: An information system for monitoring and forecasting of building heat consumption.

The Smart Grid: The concept of a fully integrated, self-regulating and self-healing power grid, which has a network topology and includes all the sources of generation, transmission and distribution networks and the all types of electricity consumers, managed into a unified network of information and control devices and systems in real time.

Section 2
Bordering Topics about Asset Management and Green Energies

Chapter 8
A System Safety Analysis of Renewable Energy Sources

Warren Naylor
Independent Researcher, USA

ABSTRACT

This chapter is focused solely on whether renewable energies can be implemented safely and if they are safer than the technologies they are replacing or supplanting albeit in small quantities at the current pace of implementation. Renewable or sustainable energy sources are necessary due to the ultimate erosion of traditional energy sources and the harmful effects they introduce into the environment and negatively affect our health. Regardless of how you personally feel concerning renewable energy sources, they are here and here to stay. With that simple understanding, we should ensure these systems are safe. This chapter evaluates the hazards associated with renewable energies and compares and contrasts them to those hazards posed by the traditional or legacy fossil fuel energies. The advantages of renewable energies are palpable and discussed in great detail in the other chapters of this book. This chapter focuses specifically on the safety of the renewable energy systems.

INTRODUCTION

Renewable energies are certainly controversial depending on your political persuasion, proposed financial gains or loss et cetera which makes objective research difficult. Renewable energy technologies are considered cleaner sources of energy possessing a much smaller environmental impact than conventional energy technologies. Notice I said smaller not nonexistent. Their acceptance is growing exponentially all over the globe as conflicts are prevalent in almost all fossil fuel producing countries with the continual fear that a war could shut down major supplies of petroleum. Regardless of how you personally feel concerning renewable energy sources, they are here and here to stay. With that simple understanding, we should ensure these systems are as safe as reasonably practical for the general public.

This chapter focuses specifically on the safety of renewable energy systems using a MIL-STD-882 DoD Standard Practice for System Safety or its commercialized counterpart ANSI/GEIA-STD-0010 Standard Best Practices for System Safety Program Development and Execution approach to identifying and proposing mitigations

DOI: 10.4018/978-1-4666-8222-1.ch008

A System Safety Analysis of Renewable Energy Sources

to the hazards identified within these pages. It should be understood that all systems that control or harness energy will possess hazards as well as residual risk.

This analysis is further focused on wind turbine and solar renewable energy technologies since they are at the time of this writing the most prominent. Additionally solar energy is divided into individual units (solar panels) and commercial distributed systems.

BACKGROUND

This chapter was written with an open mind. There are no intentional prejudices or presumptions prior to performing the research to write this paper. The author does not work in the energy field and is not a lobbyist, manufacturer, or solicitor of either conventional energies such as oil or natural gas. This is a simple system safety analysis of the various energy sources, identifying potential hazards and mitigations resulting in what are hopefully an honest assessment of the safety of wind and solar energy systems as compared to the more traditional energy sources. Not being in the industry obviously makes it difficult to gather actual safety data and a lot of the data that is available on the internet is either spun or outright wrong to fit ones political and or economic leanings. This analysis has tried to determine fact from fiction and spin from reality to form an honest and unbiased assessment. Validation of the analysis will be determined by the hoped for criticisms from both sides of the debate.

WIND TURBINES

Before being able to conduct a system safety analysis, the analyst must understand the environment in which the system, in this case wind turbines, is to operate for normal, abnormal, and emergency conditions.

Wind turbines operate in an extremely complex and ever changing environment commonly referred to as the weather. Weather variations are understood by most; however wind turbines are not exclusively built for specific geographic weather conditions. Wind turbines must be built for all credible weather environments ranging from extreme dry heat (i.e., Sahara Desert) to extreme cold (i.e., Artic) or hot humid Ocean environments (i.e., Northern Indian Ocean), and for tornados and extreme electrical storms in the Mid-Western United States, flooding, et cetera. Thoughtful consideration of weather drives the design of wind turbines to consider all credible weather conditions. A universal wind turbine designed to meet the environmental challenges of temperature extremes, humidity and moisture concerns (corrosion), friction and chaffing, health, and environmental concerns, and stability concerns. Each of these hazards, complete with potential mitigations, will be discussed as follows:

- **Temperature Extremes:** Temperature extremes can lead to several hazards including:
  - Equipment wear-out,
  - Metal and structural fatigue, and
  - Over and under-temperature to name just a few of the major hazards associated with weather. Mitigations include but not limited to the following:
    - Build in effective heating and cooling systems. The cooling system would likely contain Ethylene Glycol (EGW), Propylene Glycol (PGW), or Polyalfaolefin (PAO). All are toxic if ingested, however this is usually only an issue with animals and pets due to their respective sweet taste. Additionally, cooling systems using these compounds are quite common

and are used in automobiles, electronic systems, et cetera.
- Ensure to insulate against temperature excesses. Again, most insulation contains materials/chemicals that introduce occupation hazards, however again most are in common use in everyday life.
- Implement a closed loop temperature monitor system that fails safe should extreme temperatures be detected. Fail safe is the key point here. If power is lost the heating and cooling systems will fail. The easiest mitigation is to have backup generators that are adequately vented to ensure the safety of the turbine and any personnel involved in operations and maintenance activities.
- Design and build the system to the specific environment in which it is to be used. This mitigation is most likely the most expensive since it requires multiple designs with different attributes, maintaining individual configurations etc. but it is certainly feasible.

- **Humidity and Moisture:** Moisture such as humidity, condensation, rain, and flooding could wreak havoc on both the mechanical and electrical components of the wind turbine. Fortunately, the mitigations are well known and have become standard practice; however they are included to ensure that an untrained reader will understand the concerns and methods of mitigation. Mitigations include but not limited to the following:
  ◦ Develop a corrosion prevention plan that is adhered to during system design, development, installation, and full lifecycle maintenance to ensure against:
    - The use of dissimilar metals. Stated simply, dissimilar metals are metals that possess different electrode potentials. When two or more dissimilar metals come into contact in an electrolyte, one metal acts as anode and the other as cathode resulting in galvanic corrosion. Galvanic corrosion over time will result in mechanical and potentially electrical failures.
    - Failure to provide adequate moisture protection methods resulting in corrosion (rust, etc.). Corrosion as a result of moisture is probably the most understood and easiest to mitigate hazard. There are numerous coatings/platings (i.e., cadmium, nickel, zinc) are used to mitigate corrosion (Wilhelm, 1988).
  ◦ High Temperatures also pose corrosion concerns. High temperature environments (excessive heat) and its effects the corrosive coatings can be a concern and could lead to mechanical and potentially electrical failures when exposed to high temperature oxidizing environments such as flue gases, air or steam. Electroless nickel coatings excel in this environment up to their melting point (1630°F or 890°C).

- **Friction and Chaffing:** Friction and chaffing concerns will lead to reliability and safety concerns over the intended lifecycle

of a wind turbine. The wind turbine contains a rotary device that is subject to friction concerns such as, bearing damage over time. Mitigations include but not limited to the following:
- Include using proper lubricants that can withstand local environmental conditions such as salt, humidity, heat, et cetera.
    - Performing scheduled maintenance to ensure lubricants are applied and seals are maintained on a schedule to reduce the harmful effects of friction.
    - Control the internal environment; minimize moisture, humidity, and extreme temperatures as discussed previously.
- **Health:** Health concerns is one of the most controversial of all of the hazards regarding wind turbines and appears to be more related to politics/economics of energy rather than medical science. Putting the politics aside, there do appear to be some minor health and environmental concerns that can certainly be mitigated fairly easily and reasonably and in most cases are no longer even a concern (Layton, 2013). "Infrasound is the primary issue for those concerned about wind-turbine syndrome. They also say that audible sound and vibrations contribute to the health problems reported by some people who live close to wind farms. Symptoms of wind-turbine (Summary Report: Literature Search on the Potential Health Impacts Associated with Wind-to-Energy Turbine Operations, 2008) syndrome might include:
  - Headaches;
  - Sleep problems;
  - Night terrors or learning disabilities in children;
  - Ringing in the ears (tinnitus);
  - Mood problems (irritability, anxiety);
  - Concentration and memory problems;
  - Issues with equilibrium, dizziness and nausea.

These symptoms have been observed and documented by a limited number of scientists studying small groups of people, and the scientific community hasn't concluded whether wind-turbine syndrome exists. There are also mixed opinions on whether wind turbines emit infrasound and if the amount is any more than that emitted by diesel engines or waves crashing on the beach. But we do know that at high speeds, wind turbines can produce an audible hum and vibration that can be carried through the air (Layton, 2013). Researchers studying wind-turbine syndrome also recommend a larger buffer zone around wind farms to protect people from any ill effects. There are various recommended differences however none have produced solid scientific evidence to support their respective stand off recommendations.

Again placing politics aside, and assuming the concerns are real (still uncertain- but standard practice for system safety to assume worse case), one has to weigh these concerns against the health concerns of traditional energy sources such as nuclear, coal, and even natural gas. Given that, no catastrophic or critical injuries have been attributed to renewable energy sources. Mitigations include but not limited to the following:

- Include placing wind farms in rural communities with buffers from residential areas. Increase the distance from populated developments to reduce any negative impacts on humans. The actual safe buffer distances are controversial and still to be determined, but can certainly be achieved.
  - **Environmental:** Environmental concerns are also a politically charged issue. The primary concern with wind turbines is the killing of birdlife, including bats. Again, traditional energy sources also kill birds (i.e., Exxon

Valdez, The Gulf of Mexico Deep Water Oil spill,) and contribute to far more serious environmental concerns (i.e., air pollution, global warming, etc.), but again, this is more of a political/economic issue than a sincere science concern. Mitigations include but not limited to the following:

- Perform site surveys for wind farms prior to selection and construction to ensure the proposed site is not within or in near proximity to a bird migratory path, or
- Avoid locations near large colonies of bats as it is theorized bat's lungs are more sensitive to the pressure changes brought about by wind turbines resulting in the bats being more sensitive to barotrauma.

Note: One particularly highly publicized wind farm, Altamont Pass in California, has been a lightning rod of controversy because of the impact poor planning has had on the bird population. According to the Center for Biological Diversity, as many as 1,300 eagles, falcons, hawks and other predatory species are killed each year because the wind turbines were constructed along a critical migration route.

Again, given the severity of traditional energy environmental mishaps, the concerns stated above for renewable sources palls in comparison.

- **Stability/Structural:** There appears to be little published research concerning the structural integrity and health of wind turbine structures in in all respective challenging environments taking into consideration their projected life cycle. This is surprising as they are growing larger to increase capacity and are being installed in more challenging environments/climates. For example, there appears to be an impetus to locate them off shore in an ocean corrosive ocean environment making them more prone to corrosion (rust) and more difficult to repair. Obviously ocean installations are attractive due to the predictability of prevailing winds. Corrosion is a concern primarily due to the high humidity and moisture environment. Additionally to install in this environment requires the ability to withstand and operate in high wind conditions that could include weather conditions such as typhoons, hurricanes, tornados, water spouts, severe thunder storms, and high seas.

These weather conditions pose monumental risks, yet little concern appears to exist. For example, the Claire Carter wind turbine was destroyed by high winds on Scrabster Hill, Scotland. In this accident, 2 blades were ripped off an 18m wind turbine when hit by 40 mph gales in a school yard as reported by The Telegraph, 11 Jan 2014. Fortunately there were no catastrophic or critical injuries, but there were certainly lessons to be learned from this accident. Most significant, a school yard is probably not a safe nor logical place to install and operate a wind turbine and one would certainly expect that a wind turbine would be designed to withstand at least an 150 mph gale or greater. This design failed at only a 40 mph gale! So what is the lesson learned in Scrabster Hill? (Carter, 2014). This particular wind mill was poorly installed and did not meet minimal margins of safety.

In February 2008, a 10-year-old Vestas turbine with a total height of less than 200 feet broke apart in a storm. Large pieces of the blades flew as far as 500 meters (1,640 feet) -- more than 8 times its total height. (Safe setbacks: how far should wind turbines be from homes?, 2008). Fortunately there were no injuries reported.

These isolated incidents are not nor should it be an indictment on the safety of wind turbines in general, however they do indicate that proper set-back standards need to be applied to ensure safety for the future. Mitigations Include but not limited to the following:

- Designs need to accommodate worst credible weather conditions with adequate margins of safety (minimum 25%) built in. The designs must be able to withstand corrosion, high rotation in rates in high winds, degradation due to friction, high seas, et cetera.
- Explore the potential of developing a designed safe mode. For example, developing a means for folding or constraining the blades down to mitigate or at least minimize the potential effects of winds/seas making direct contact with the blades.
  ◦ **Wind Turbine Conclusion:** Wind energy currently accommodates only a small percentage of the world's energy needs; however the potential is significant and absolutely necessary to assist in making the world less fossil fuel dependent. The criticisms appear to be overstated for covert agenda reasons, however in performing a safety assessment one has to look at worse credible case so I took these criticisms at face value. Even at face value, objectively speaking; the end result is wind energy is far less harmful to humans and the environment including bird life than fossil fuels and should be pursued with vigor and without prejudice.

The following is an objective system safety analysis regarding the generation of electricity from photovoltaic (PV) solar panels, otherwise known as solar energy. Again, the criticisms are many; however this assessment is mainly looking at the safety impacts of solar energy. There is one major associated with solar power that must be addressed in this analysis and that is solar energy only produces energy in sunlight. Night operations require a battery backup system. Batteries, although getting safer do possess some inherent safety hazards among them being overheating, environmental hazards, disposal concerns, et cetera.

Currently there exist two (2) separate and distinct types of solar energy systems. They are Distributed Systems (most common) that are frequently used as single user or limited user systems. Distributed systems are most commonly associated in localized settings (i.e., free standing Solar panels or solar panels installed on your homes roof, etc.). The second type is call Centralized Solar Systems (less common) which are those designed to produce large quantities of what we will refer to as community energy to be shared on the grid. The differences regarding safety between the two types of solar energies are substantive and differ as much as their individual designs. Due to these design differences the safety of each will be discussed separately.

SOLAR

The potential of environmental, health, and safety hazards associated with Distributed Systems is focused on the solar panels. (Health and Safety Concerns of Photovoltaic Solar Panels, n.d.) The analysis assessed how Solar Panels were manufactured, installed, maintained, handled, stored, and environmental hazards (including mining of hazardous materials) and ultimately the disposal of the solar panels. This analysis considered the entire projected life cycle of the PV Panels.

In order to understand the environmental effects of distributed PVs, one must become aware that solar panels comprise a system of solar cells which are solid state semiconductor devices that convert sunlight into electricity. Solar cells are no different than any other semiconductor device. They use the same raw materials and possess many of the same hazards, with similar environmental effects, likelihood, and potential consequences as most other similar electrical components. To state this differently, the risk levels are more than acceptable by almost any definition and no different than any other semiconductor device.

Centralized systems also contain PVs, however the sunlight is concentrated with heat being the major hazard. In a practical sense, the hazards associated with centralized solar systems are similar to those discussed previously with wind turbines. Their excessive heat can and will harm wild life and in particular birds. Fortunately, the mitigations are also similar in that the locations of solar farms should be strategically planned to ensure they do not interfere or are not located in migratory bird paths from a safety perspective and in areas with substantive and reliable solar energy for performance and efficiency reasons.

- **Environmental:** To fully comprehend the environmental hazards associated with solar energy, one must consider the entire life cycle commencing during the mining and manufacturing phases, including operations, and ultimately, disposal or disposition. The environmental and occupational health hazards listed for solar energy, in the following paragraphs, will be addressed by life cycle phase.
- **Raw Materials:** The raw materials used in the manufacture of solar cells are considered hazardous, however these materials are also used in many other systems due to their unique properties. These raw materials include:
  ○ **Crystalline Silica:** Used for the manufacture of monocrystalline solar panels. Potential health effects include being a carcinogen (lung cancer), chronic obstructive pulmonary disease, rheumatoid arthritis, scleroderma Sjogern's syndrome, lupus, and renal disease. All serious hazardous materials with catastrophic consequences, if handled incorrectly (Yassin, Yebessi, & Tingle, 2005).
    - **Mitigations:** Monitor air quality, implement the automation of processes to limit human exposure, implement dust suppression measures, and define and provide personal protective devices for workers, such as, respirators.
  ○ **Metallurgical Grade Silicon:** Metallurgical grade silicon is used in the manufacture of metal alloys such as aluminum and steel, chemical silicones for use in lubricants, and epoxies, as well as, high purity polysilicon for the manufacture of semiconductors, including solar panels. This process involves multiple potentially hazardous materials and byproducts that, without proper safeguards, can pose a significant risk to human and environmental health. Chlorosilanes and hydrogen chloride are toxic and highly volatile, reacting explosively with water. Chlorosilanes and silane can also spontaneously ignite and under some conditions, explode.[6]
    - **Mitigations:** Monitor air quality, implement the automation of processes to limit human exposure, implement dust suppression measures, and define and provide personal protective devices for workers, such as respirators, when exposure occurs in a confined space with limited oxygen flow.
- **Solar Energy Conclusion:** Solar energy is the most understood and is generally accepted as a renewable energy sources, however I would argue that solar cells and solar panels wear out and do possess some

hazardous materials as discussed previously in this section. Regardless solar energy even with its imperfections is broadly accepted as providing safe and green energy and possesses a considerably less carbon footprint.

FUTURE RESEARCH DIRECTIONS

Currently, the use of the use of solar and wind renewable technologies comprise only a small percentage of our energy needs. The percentage of energy provided by renewable energy sources must grow exponentially overtime to ensure their viability and enhance their respective economic contributions to the world economy. To grow they must become more efficient and effective, however at the current moment the hopes of going solely to renewable solar and wind turbine technologies is rather optimistic at best. Additionally, as these technologies evolve, the manufacturers in concert with the system safety community must also insure the safety of these systems does not degrade.

Additionally, there are other renewable technologies that must be assessed as they evolve including but not limited to hydro, geothermal, tidal, and wave, all of which could provide additional niche power at a minimum in certain geographical locations which ultimately could further reduce global dependence upon fossil fuel or nonrenewable energies. Additionally, renewable energy sources such as wind and water are widely available leaving little opportunity for a hostile Government to restrict or withhold the forces of nature.

The bottom line is there are more opportunities to enhance the use of renewable energies; all that is needed is the realization of economic benefits that will spur the growth. The economic benefits will not be realized if the renewable energy sources are unstable, unsafe, or harmful to the environment. There are still safety concerns that must be addressed; however these concerns can be overcome resulting in an efficacious outcome. For example, wave energy possess tremendous power potential and are always present, however there are still concerns. Some concerns are:

- Potentially unsightly if implemented near to shore;
- Hazards for shipping (collision);
- Effects on fish populations;
- Corrosion and reliability;
- Source of pollution (lubricants, sealants, etc.).

Again, these technologies possess tremendous potential; however analyses must be performed to ensure their respective safety.

Finally, society at large must learn to use less energy. This can be accomplished either through personal sacrifice or through making our tools and toys more energy efficient. There is tremendous research being performed in this realm largely due to consumer demand. For example, I just updated my heating and air conditioning system for my home to the most efficient system I could find. Automobiles are becoming more fuel efficient and new technologies are being developed. Hybrid cars are not uncommon anymore with the movement to full electronic in the foreseeable future. The question is can the stored energy technologies be improved enough to meet the majority of needs of consumers? Can it be performed safely? Can they be performed without hazardous materials such as Lithium that must be mined, manufactured, and disposed of safely with minimal to no harm to the environment?

SUMMARY

As stated previously renewable energies are certainly controversial depending on your political persuasion, proposed financial gains or lose et cetera which makes objective research difficult. In conducting this analysis an open mind was maintained throughout. It can be honestly stated that personal feelings or political persuasion did not tarnish this research or the outcome of the analysis. With that simple understanding, the results of this analysis should be taken at face value. I can also emphatically state that much of the research was tainted or spun and when obvious was precluded.

There are some fairly minor and easily controllable safety concerns associated with renewable and sustainable energies. Most of these safety concerns have been effectively addressed, with future mitigations being actively pursued. Given the risks of current fossil fuel energy sources the safety risks are minimal and mostly easily mitigated. A few examples of fossil fuel mishaps include the grounding of Exxon Valdez (transportation mishap), The British Petroleum Gulf of Mexico Mishap (pumping), and the Texas City Mishap (handling) have resulted in literally billions of dollars of environmental, facility damage, and most importantly a significant loss of life. Additionally, how often do you hear of a train derailment (2013 Ontario, Canada) or just a plain home explosion? In reality, all systems' possess safety hazards and concerns, but fossil fuel usage, manufacture, and transport have generated what some would view as catastrophic mishaps. Whereas renewable energy sources mishaps are not even comparable in either occurrence or severity (GEIA-STD-0010, 2008).

The end result is rather simple when based solely on weighing the pros and cons based on historical accidents as demonstrated in this chapter. Given the likelihood of occurrence and the potential consequences of using sustainable energy sources (wind and solar), renewable energies should continue to evolve and displace as much of the traditional energy sources as feasible. Stated differently, there are definitively no safety hazard that cannot be controlled to prohibit the growth of renewable energy sources as they are definitively our future and the safer alternative!

REFERENCES

Carter, C. (2014, January 11). Wind turbine destroyed by high winds. *The Telegraph*.

GEIA-STD-0010. (2008, October). *Standard best practices for system safety program development and execution*. Academic Press.

Health and Safety Concerns of Photovoltaic Solar Panels. (n.d.). Retrieved from http://www.oregon.gov/odot/hwy/oipp/docs/life-cyclehealthandsafetyconcerns.pdf

Layton, J. (2013). Do wind turbines cause health problems?. *HowStuffWorks*.

Report, S. (2008, March). *Literature search on the potential health impacts associated with wind-to-energy turbine operations*. Retrieved from http://www.healthy.ohio.gov/~/media/ODH/ASSETS/Files/eh/HAS/windturbinehealthimpactreport.ashx

Safe Setbacks: How Far Should Wind Turbines be from Homes?. (2008, July 30). [Web log comment]. Retrieved from http://kirbymtn.blogspot.com/2008/07/safe-setbacks-how-far-should-wind.html

Wilhelm, S. (1988). Galvanic corrosion caused by corrosion products. In H. Hack (Ed.), *Galvanic corrosion*. Philadelphia: ASTM. doi:10.1520/STP26189S

Yassin, A., Yebesi, F., & Tingle, R. (2005). *Occupational exposure to crystalline silica dust in the United States: 1988–2003*. Academic Press.

ADDITIONAL READING

Fthenakis, V. "National PV environmental research center: Summary review of silane ignition studies." Retrieved from http://www.bnl.gov/pv/abs/abs_149.asp (September 2014)

IARC. (1987). *Silica and some silicates. IARC monographs on the evaluation of carcinogenic risk of chemicals to humans* (Vol. 42). Lyon, France: International Agency for Research on Cancer.

MineEx Health and Safety Council of New Zealand. (2008) (n.d.). "Guideline for the control dust and associated Hazards in surface mine and quarries." Retrieved from http://www.minex.org.nz/pdf/SUR_Dust_Mar08.pdf (September 2014)

The Fraunhofer Institute for Solar Energy Systems. (n.d.). Retrieved from http://www.fraunhofer.de/en/research-topics/energy-living/wind-power.html (October 2014)

Wind power problems, alleged problems and objections (n.d.). Retrieved from http://ramblingsdc.net/Australia/WindProblems.html (September 2014)

Yassin, A., Yebesi, F., & Tingle, R. (2005). Occupational exposure to crystalline silica dust in the United States: 1988–2003. *Environmental Health Perspectives*, *113*(3), 255–260. http://ehp.niehs.nih.gov/members/2004/7384/7384.pdf doi:10.1289/ehp.7384 PMID:15743711

KEY TERMS AND DEFINITIONS

Chaffing: Repeated relative motion between wiring system components, or between a wiring system component and structure or equipment, which results in a rubbing action that causes wear which will likely result in mechanical or electrical failure during the aerospace vehicles specified service life.

Corrosion: The gradual destruction of materials (usually metals) by chemical reaction with their environment.

Hazard: A real or potential condition that could lead to an unplanned event or series of events (i.e. mishap) resulting in death, injury, occupational illness, damage to or loss of equipment or property, or damage to the environment.

Hazardous Material: Any item or substance that, due to its chemical, physical, toxicological, or biological nature, could cause harm to people, equipment, or the environment.

Mishap: An event or series of events resulting in unintentional death, injury, occupational illness, damage to or loss of equipment or property, or damage to the environment. For the purposes of this Standard, the term "mishap" includes negative environmental impacts from planned events.

Mitigation: Action required eliminating the hazard or when a hazard cannot be eliminated, reducing the associated risk by lessening the severity of the resulting mishap or lowering the likelihood that a mishap will occur.

Photovoltaics (PV): A method of generating electrical power by converting sunlight into direct current electricity using semiconducting materials that exhibit the photovoltaic effect.

Residual Risk: The remaining risk associated with a possible event after all mitigation steps have been taken.

Safety: Freedom from conditions that can cause death, injury, occupational illness, damage to or loss of equipment or property, or damage to the environment.

System Safety: The application of engineering and management principles, criteria, and techniques to achieve acceptable risk within the constraints of operational effectiveness and suitability, time, and cost throughout all phases of the system life-cycle.

System: A bounded physical entity that achieves in its domain a defined objective through the interaction of its parts.

Wind Turbine: A device that converts kinetic energy from the wind into electrical power.

Chapter 9
Predictive Maintenance for Quality Control in High Precision Processes

María Carmen Carnero
University of Castilla – La Mancha, Spain

Rafael González-Palma
University of Cádiz, Spain

Carlos López-Escobar
Aluminium Company of America (ALCOA), Spain

Pedro Mayorga
Electrical Technology Institute (ITE), Spain

David Almorza
University of Cádiz, Spain

ABSTRACT

In external grinding processes, vibrations induced by the process itself can lead to defects that affect the quality of the parts. The literature offers models that cannot include all process variables in the analysis. This research applies theoretical models and experimental analysis to determine their suitability for predicting the chatter profile of parts in a plunge grinding process. The application of variance analysis to overall vibration value induced by grinding wheel-workpiece contact allows us to show that high frequency displacements vibration are sensitive to the process setup as well as to the quality of the products manufactured. The final statistical analysis has provided a determination of the spectral bands of the process in which the vibrations causes by grinding wheel-workpiece contact influence the existence of flaws in the workpieces. The methodology described can contribute to increasing the environmental sustainability of an industrial organization.

INTRODUCTION

Maintenance has a significant impact on the environmental sustainability of an organization, by, for example using more efficient and less wasteful machines and devices, using energy-saving candle bulbs or light bulbs, incorporating sensors into devices to limit or reduce water consumption, installing movement sensors to control lights outside working hours, recovery usable parts from unusable machines, recycling paper and cardboard used in normal operations or removing the need for documents based on paper and replacing them with computer-based files and procedures, and

DOI: 10.4018/978-1-4666-8222-1.ch009

especially, increasing the useful life of machines and devices by better maintenance, such as predictive maintenance.

Predictive Maintenance is a maintenance policy based on measuring and recording intermittently or continuously (on-line) certain physical parameters, including vibration, temperature, etc., associated with a working machine to obtain data and information through which failures can be detected and the future state of the machine can be determined as a function of the load to be applied to the equipment or process, that is, a prediction of the remaining life of the machine (Rao, 1996).

There exist a number of predictive diagnostic techniques, such as (Carnero, 2013; Mobley, 2001a; Scheffer & Girdhar, 2004):

- **Vibration Analysis:** This is especially suitable for rotary and reciprocating machines, although it can be used effectively on all electromechanical equipment too. It consists of periodically recording the vibration level of the machine, as it increases when there are anomalies such as unbalance, misalignment, damaged bearings and gears, looseness, etc. Alert and alarm levels are set which require the frequency of the vibration checks to be increased. The overall vibration level can be used as an indicator of the seriousness of a failure and spectral analysis can diagnose the specific type of fault. It is the most widely used due to its high diagnostic capacity. Furthermore, vibration analysis technology, the instrumentation required and the associated software packages have evolved faster than other predictive techniques, leading to more effective predictive maintenance programs.
- **Process Parameters:** Controlling the process parameters, such as pressure and temperature, can give information about the presence of anomalies or flaws. Also, controlling these parameters gives information about the efficiency of the machine, which has a significant impact on plant productivity and profitability.
- **Thermography:** Any object with a temperature above absolute zero emits radiation and infrared radiation has the shortest wavelength of all radiated energy. The intensity of infrared radiation emitted by an object is related to its surface temperature. Thermography, therefore consists in obtaining, remotely and without contact, an infrared image of a machine, to measure and record surface temperature accurately; this technique detects sources of thermal anomalies in high and low tension electric cables and installations, control panels, connections, spark plugs, fuses, transformers, electric motors, generators, etc., although it can be used to monitor structures, plant machinery and systems too.
- **Visual Inspection:** This consists of visually inspecting the machine, although at limes more sophisticated instruments are used, such as endoscopes and stroboscopic lamps. It can detect fissures, cracks, wear, changes in colour, etc. Despite its simplicity, this technique allows anomalies to be detected which would otherwise pass unnoticed by the use of other predictive techniques.
- **Tribology:** This consists of techniques to analyse lubricants and wear particles. Lubricant analysis controls the loss of protective function of lubricants due to use, by performing physical and chemical analysis of the lubricant in use such as viscosity, acid content, flash point, etc.; this gives information about the need to change the lubricant. Contamination of the lubricant through the presence of chemical substances of particles that were

not in the original composition of the lubricant and which alter its characteristics and efficient working can also be identified. Wear particle analysis provides information about the wearing condition of the machine. Particle size, shape, quantity and composition are analysed to determine the machine's condition. The presence of wearing particles, produced by an anomaly, can be detected from the first stages of production

- **Current Spectrum:** This consists in taking the intensity spectrum of the phases of the motor. It allows the effects of damaged rods, cracks in short circuit rings, poorly soldered joints, gaps caused by melting, static and/or dynamic irregularities of the air gap, etc., to be detected.
- **Ultrasound:** An apparatus produces a wave of ultrasound (with a frequency range between 20,000 and 100 kHz), which propagates between the emitter and the material to be inspected. From the difference between the emitted and received, or echo, waves, it is possible to determine the presence and size of defects and at what distance they are to be found. It allows the detection and characterization of discontinuities, the measurement of the width, breadth and degree of corrosion, and especially leak detection.
- **Penetrating Liquids:** A special staining fluid is applied to the clean surface to be analysed so that it penetrates into the possible flaws. After a period of time to allow the liquid to penetrate into surface openings, the stainer is cleaned and an absorbent or revealing liquid is used that absorbs the liquid that has penetrated into the cracks or surface flaws. The areas where there is absorbent liquid are those with surface discontinuities such as cracks.

Other nondestructive testing techniques may be seen by consulting National Aeronautics and Space Administration (2000) and in the nondestructive testing handbook series of the American Society of Nondestructive Testing (ANST).

Table 1 shows the typical applications for each predictive technique.

The benefits obtainable through the introduction of a Predictive Maintenance Program are (Carnero, 2005; Carnero, 2012; Charray, 2000; Jardine, Lin, & Banjevic, 2006; Mobley, 2001a):

- It gives a diagnosis of anomalies and faults.
- Maintenance actions are only carried out in machines which show signs of developing a fault.
- Failures are detected in their early stages of development and so there is enough time to plan the corrective or preventive actions via programmed stops under controlled conditions which minimize downtime and the negative effect on production.
- Predictive maintenance considerably reduces the level of application of other maintenance policies, such as corrective and preventive maintenance, as well as contributing to their more rational application.
- Reduction in insurance rates as the plant is now protected against interruptions in production.
- Increase in the availability of the plant, by increasing the Mean Time Between Failures (MTBF), and reducing the Mean Time To Repair (MTTR).
- Capacity to carry out checks on the quality of the maintenance actions, both in-house and outsourced.
- It facilitates optimum programming of maintenance actions.
- It allows effective programming of spare parts and labour.
- It is used in the design of machines via modal analysis.

Predictive Maintenance for Quality Control in High Precision Processes

Table 1. Applications of predictive maintenance techniques

Predictive Maintenance Techniques	Pumps	Electric Motors	Diesel Generators	Condensers	Heavy Equipment/Cranes	Circuit Breakers	Valves	Heat Exchangers	Electrical Systems	Transformers	Tanks, Piping
Vibration analysis	X	X	X		X						
Lubricant analysis	X	X	X							X	
Wear particle analysis	X	X	X		X						
Bearing, temperature/analysis	X	X	X		X						
Performance monitoring	X	X	X	X				X		X	
Ultrasonic noise detection	X	X	X	X			X	X		X	
Ultrasonic flow	X			X			X	X			
Thermography	X	X	X	X	X	X	X	X	X	X	
Non destructive testing (thickness)				X				X			X
Visual inspection	X	X	X	X	X	X	X	X	X	X	X
Insulation resistance		X	X			X			X	X	
Motor current spectrum		X									
Motor circuit analysis		X				X			X		
Polarization index		X	X						X		
Electrical monitoring									X	X	

(National Aeronautics and Space Administration, 2000).

- It reduces the number of hours dedicated to maintenance activities and the consumption of parts, since, as it eliminates catastrophic failures, fixing industrial equipment is cheaper than when they cease to function completely.
- It improves the image of the company by meeting delivery dates.
- Reduction in the cost of spare parts and labour.
- Machines may remain operative while predictive techniques are applied, and so the measuring process does not directly influence the availability of the equipment.
- A complete record of the history of the machines, which allows the reliability parameters to be more easily determined.
- It is easier to obtain and maintain the certificates ISO 9000, QS 9000, etc. Although ISO certification does not mention any requirements for predictive maintenance, its introduction considerably increases the probability of certification and will ensure long-term compliance with that standard.

However, the relationship between predictive maintenance and environmental sustainability should be especially highlighted, as follows:

- The increase in quality of the product allows the consumption of raw materials to be reduced because the quantity of defective product decreases.
- Machines working without failures are more productive and therefore there is better use of energy; additionally, some anomalies or faults produce an increase in energy consumption.
- It contributes to gaining and keeping certificate ISO 14001. This is because predictive maintenance improves an organisation's efficiency by reducing the consumption of resources and reduces the risk of pollution incidents and other damage to the environment.
- It reduces the consumption of spare parts since the components of a machine are replaced only when needed, and it therefore provides a lower consumption of raw materials.
- Failures are detected in their early stages and there is less influence of faulty components on other components, increasing the useful life of machines.
- Actions are carried out only in machines that show signs of developing a fault and so again there is a more efficient use of resources.
- It increases environmental protection because it can prevent leaks of gas, fuel, used lubricant, etc.

Further information about predictive maintenance can be seen in Rao (1996), Mobley (2001a), Mobley (2001b), Levitt (2003), Scheffer and Girdhar (2004), and Carnero (2012).

The use of predictive techniques, and specifically vibration analysis, to improve quality, have been applied in very limited fashion in companies (Brown, 1996), although excessive vibration may seriously affect the quality of the final product; thus, the condition of the machine should be an integral part of the quality control system of a company.

In the auto-part industry and above all in the manufacture of bearings for vehicles, which is the focus of this research, there is a direct relationship between the final quality of the finished bearing and each of the components on its own. And so, the final vibration of the assembled bearing is one of the quality parameters analysed to ensure that the life of the bearing, working noise levels and driving comfort are as desired. Although this research is focused on bearings, it should be borne in mind that it can be applied to other products and machine processes.

Vibration can be shown to play two different roles. On the one hand, it is the evaluation parameter for the end quality of the product, and on the other it is responsible for many of the manufacturing problems of the subcomponents of the assembled bearing.

Besides the vibration of the machines themselves, the processes also produce vibrations which frequently combine with the mechanical behaviour of the machine. Nevertheless, analysis of process-induced vibrations is little studied in the literature.

In external grinding processes vibrations appear which are generated by the process itself and not by flaws in the machine tool. These vibrations can cause defects which affect the quality of the workpieces produced. In the literature there are models and quantitative analyses which justify the instability of the behaviour of process-induced vibrations. These models show discrepancies with the real case, which are explained by the appearance of other sources of vibration not included in the models, because of the complexity of the process under consideration. And so the existing theoretical models have been updated to determine their applicability to plunge grinding. The technical parameters used to simulate the solution to the theoretical model are from a machining process for raceways in automotive vehicle bearings. Experimental tests were carried out to verify the model under normal operating conditions in an automotive wheel bearing manufacturing plant. The theoretical instability in the quality obtained from the process is then analysed; specifically, the theoretical lobing profiles are obtained from a combination of the process parameters.

The complex behaviour observed in the theoretical model shows the complications associated with an orbital grinding model which considered the magnetic grip strength, wearing, quality of input parts, mechanical adjustments, hardness of grinding wheels, quality of coolant and its affects, uniformity of previous thermal treatments, constant cut speed algorithms, etc. Therefore, the choice of working parameters which guarantee the quality of the process cannot be carried out by a theoretical model.

It is necessary, then, to apply methods of identification for the process parameters which cause the appearance of specific vibration frequencies, and thus affect the quality of the workpieces, by designing experiments and statistical analyses of vibration data obtained from the overall values of vibration and from the spectra.

The overall vibration value is a widely-used parameter in machine maintenance, since the appearance of electrical and mechanical problems tends to be associated with a significant increase in the overall value. However, the global vibration value is insufficient for quality control of the workpieces that the machine is making, as insignificant increases in global vibration values can show up as serious quality problems. For this reason, vibration spectra are applied to distinguish problematic vibration frequencies from those that are not. In the case of spectra it is almost impossible to find references for maximum permitted values and their frequencies in the literature, and so intercomparison of spectra between similar machines and studies of the change over time of spectra or waterfall plots are used. This concept is used because of the availability of identical machines in the plant that is subject to experimentation.

Spectral intercomparison between identical machines removes the problem of knowing maximum permitted values, as it is the machines themselves which report the existence of a problem when a spectrum comparison is performed.

This method considers the possible of establishing separate vibration limits in the machining of components so as to guarantee the quality of the final assembled product. These vibration limits as a guarantee of quality may be different from and more restrictive than the maximum permissible limits from the point of view of maintenance.

The methodology described broadens the use of predictive technologies, to use it where there are situation of risk of no quality in the machine processes. The conclusions of this research show how a combination of preventive techniques and statistical tools allows the innovative concept of predictive statistical quality control to be established. The methodology exposed can contribute to increase the environmental sustainability of an industrial organization due that contribute to a more efficient use of resources.

The rest of this chapter is structured as follows. First, there is a review of the literature about chatter detection; next it describes the characteristics of automotive bearings and then the theoretical models and the experimental and statistical models are presented. The paper ends with future lines of research and conclusions.

BACKGROUND

There is an extensive literature on chatter detection. For example, Govekar et al. (2002) present a method for automatically detecting chatter in outer-diameter grinding by monitoring the course-grained entropy rate, which is calculated from the fluctuations in the normal grinding force. The normalized values of this index can be used to automatically detect the presence of chatter. Gradišek et al. (2003) use entropy and the course-grained information rate to detect chatter in outer-diameter plunge feed grinding. These authors test their methods using signals of the normal grinding force and RMS acoustic emission. Khalifa and Densibali (2006) analyze images from chatter-free and chatter-rich turning processes. They use several parameters (average of gray level G, intensity histograms, and the variance, mean, and optical roughness parameter of the intensity distributions) to differentiate between chatter-free and chatter-rich processes.

Models of grinding processes can use either a frequency or a time domain, but the former can only consider linear phenomena and nor can quantitative values of relations between parameters be obtained from them. Snoeys and Wang (1968), Snoeys and Brown (1969), and Snoeys (1971) models use the time domain and closed-loop block diagrams to represent the grinding process.

Snoeys and Brown (1969) theoretically analyze how the grinding parameters affect the stability of the grinding process. They demonstrate the existence of a theoretical correlation between machine characteristics, the workspeed, the compliance of the grinding wheel-workpiece contact area, and the appearance of wheel regeneration. Snoeys (1971) studies the stability of self-induced vibrations, and proposes a mathematical grinding model that analyzes the forces and self-induced vibrations.

The number of studies that model and simulate grinding processes has increased considerably (Brinksmeier et al., 2006) due to the industry's need for models that can predict the results of grinding processes with the intention of improving the capacity of machine tools (Biera, Viñolas, & Nieto, 1997). The most significant are as follows. Tonshoff et al. (1992) compare different typologies of grinding process models (grinding topography models, chip thickness models, energy and force models) and use simulation to show the benefits and limitations of each model. Orynski and Pawlowski (2004) produce a model of a cylindrical plunge grinder to simulate the influence of the grinding process on the forced vibration damping of the wheelhead. Li and Shin (2006a) develop a model that simulates infeed centerless grinding processes. Their model considers the complete two-dimensional kinematics, dynamics, surface profiles, and the geometrical interactions of the workpiece with the grinding wheel, regulating wheel, and supporting blade to predict the coupled chatter and lobing process. They test their model by comparing the predicted chatter and lobing occurrences with experimental data. Li and Shin (2007) modify this model to handle the grinding force-dependent structure dynamics. Li and Shin (2006b) build a model that simulates surface grinding processes and predicts their regenerative chatter characteristics. The model incorporates aspects like the possibility of interrupted grinding on a series of surfaces and step-like wheel wear along the axial direction due to crossfeed. They test the model by comparing the simulated chatter frequencies and thresholds with experimental data. Brinksmeier et al. (2006) present the state of the art in the modeling and simulation of grinding processes. They discuss physical models (analytical and numerical models), empirical models (regression analysis and artificial neural net models) and heuristic models (rule-based models). Their models include the process characteristics (grinding force and temperature) as well as results like surface topography or integrity. These authors discuss the capabilities and limitations of each model and simulation approach. Choi and Shin (2007) develop a system that models grinding processes using analytical models, experimental data and heuristic information coming from experts. Choi et al. (2008) present generalized models of cylindrical plunge grinding processes. These models maintain the same functional forms regardless of the workpiece-wheel combinations used and hence can be readily applied to various grinding set-ups by determining their coefficients through a small number of designed experiments.

Snoeys and Brown (1969) and later Biera et al. (1994b) develop models and quantitative analyses in specific cases. The theoretical models cannot include all the process variables and sources of vibration observed in the manufacturing process taking into account the analytical complexity of its consideration. Thus, the authors extend the existing theoretical models to predict the chattered profile of the parts produced in a plunge grinding process. The model considered assumes the inexistence of movement between the workpiece axes and the workhead revolution axis although it is not the case of orbital grinding where relative movement part-workhead exists. Following the results of Snoeys and Brown (1969), and Biera et al. (1994b) the authors show the theoretical instability of the quality profiles obtained in the process analyzed. Specifically, the authors develop the theoretical chattering profiles obtained through a combination of process parameters. This work is a continuation of the contributions of Viñolas et al. (1997), who use the simulation software Simulink to model grinding processes including nonlinear phenomena and using the machine, grinding wheel and process parameters as simulation parameters, and Biera et al. (1997), who present a model of the external plunge grinding process. The model is built using a block-based simulation tool that facilitates the incorporation of the process variations and machine parameters. The time domain of the model permits the inclusion of nonlinear effects, so the model can be used to explain why although some process parameters lead to unstable conditions, they can produce workpieces of an acceptable quality.

In this article the authors apply existing theoretical models to determine its ability to how applicable they are to a plunge grinding process. The paper attempts to demonstrate the unpredictability of the self-induced vibrations and the resulting chatter when the problem is analyzed from the theoretical perspective. In this way the authors stress the need for experimental analysis of the vibrations generated in the process and their relation to workpiece quality (chatter).

DESCRIPTION OF AUTOMATIVE BEARINGS

This research is applied to integral vehicle wheel bearings (see Figure 1). These comprise two fully- or partially- hardened metallic pieces which contain the external and internal raceways. Between these two elements are the ball or roller bearings; they are also subjected to a heat treatment to guarantee their durability. The rolling elements are usually separated by light, tough pieces called ball cages, made of nylon, bronze, brass, etc. The system is lubricated by grease or oil. Contact between the balls and the raceways may be radial or angular depending on the loads to which the piece will be subjected. Automotive

Figure 1. Automative bearings

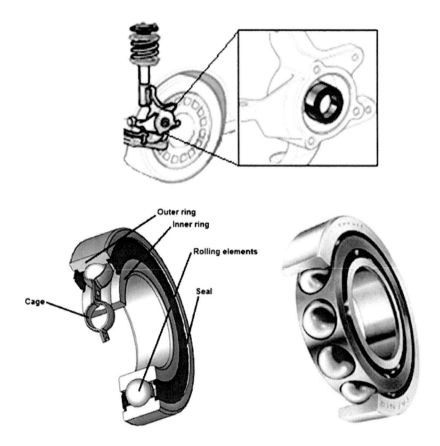

bearings use the angular type to absorb both axial and radial loads produced by tyre-road contact; the absorption capacity for axial loads is greater the greater the contact angle, although it is normally no greater than 40°.

Bearings are precision parts subject to rigorous inspection processes during manufacture, both during the preparation of individual pieces and at assembly, which gives a long life to the component if it is used according to the manufacturer's recommendations. These controls guarantee the non-existence of potential catastrophic flaws in the bearing, of a kind that could be a risk to the safety of the occupants of the vehicle.

When a failure takes place in a bearing, the reason is, probably, one of the following (Hutchings & Unterweiser, 1980):

- Reasons related to the manufacturing process:
    - Excessive preload during the assembly.
    - Incorrect lubrication.
    - Interior pollution.
    - Problems in the manufacturing of components (roundness and/or wrong finish of tracks, etc. This is the type of fault on which the present study is focused).
- Reasons related to incorrect use:
    - Overloads.
    - Functioning outside design limits.
    - Impacts.
    - Conditions of excessive temperature.
    - Incorrect assembly.

These reasons motivate the premature failure of the bearing in different ways that depend on the predominant reason; the best-known are fatigue, fractures, indentation of solids, corrosion, etc.

VIBRATION ANALYSIS

The period of the vibration is defined as the amount of time required to complete one full cycle of the vibration. Vibration frequency is the number of complete cycles that occur in a specified period of time such as cycles per minute (cpm). Frequency is the inverse of the period of the vibration. Although vibration frequency may be expressed in cycles per minute or second, the common practice is to use Hertz (Hz).

Amplitude is defined as the intensity or maximum value of the vibration. The amplitude of the vibration may be given in different formats: peak-peak, zero-peak and root mean square (rms) (see Equation 1). The rms is generally applied because it gathers historical information and information about the energy content of the vibration signal, and therefore it is correlated with the damaging power of the vibration.

$$rms = \sqrt{\frac{1}{T}\int_0^t x^2(t)dt} \qquad (1)$$

Vibrations can be obtained in displacement $(x(t))$, velocity $(v(t))$ and acceleration $(a(t))$. These magnitudes may be plotted against time and against frequency; in the latter case they are spectra. The mathematical expression of these magnitudes in their change with time is similar to that which defines simple harmonic vibration.

$$x(t) = X\sin(\omega t + \varphi) \qquad (2)$$

$$v(t) = \frac{dx(t)}{dt} = \omega X \cos(\omega t + \varphi) \qquad (3)$$

$$a(t) = \frac{d^2x(t)}{d^2t} = -\omega^2 X \sin(\omega t + \varphi) \qquad (4)$$

where $\omega = 2\pi f$ and φ is the phase angle with respect to the time origin.

The use of displacement, velocity or acceleration to gather vibrational signals depends largely on the revolutions of the machine to be analysed. The aim is to use the magnitude which gives the most uniform spectrum (Harris, 1995). Displacement is used in slow-speed machines or with frequencies below 10 Hz (600 cpm). Movements below 10 Hz gives few vibrations in the acceleration magnitude, moderate vibration in velocity and a lot in displacement, which is why the vibrational data are collected in displacement up to 10 Hz. If velocity is used, the vibrational amplitude measured is independent of the frequency for the range 10-1,000 Hz (60 kcpm), and so velocity is a good indicator of the severity of vibrations; it is the magnitude in which most mechanical faults show up, and since the majority of rating machinery operates in the 10-1,000 Hz range, the velocity is the magnitude which is generally used in vibration analysis. Acceleration is used for vibration measurement at frequencies above 1,000 Hz.

Fourier's theorem states that any periodic signal may be expressed as a combination of pure sinusoidal signals, whose frequencies are multiples of the fundamental frequency, and constitute a frequency spectrum (Muszynska, 1992). The Fast Fourier Transform algorithm (FFT) transforms N samples in the time domain to N/2 or N/2.56 spectral samples. The equations which show the relation in the time and frequency domains are respectively:

$$S_x(f) = \int_{-\infty}^{\infty} x(t)e^{-i2\pi ft}dt = F\big[x(t)\big] \qquad (5)$$

$$x(t) = \int_{-\infty}^{\infty} S_x(f)e^{i2\pi ft}df = F^{-1}\big[S_x(f)\big] \qquad (6)$$

where $x(t)$ is the time signal, and $S_x(f)$ is the Fourier transform of the signal $x(t)$ (Harris, 1995).

The FFT is applied to continuous signals, and the discrete Fourier transform (DFT) is applied to discrete signals, which are the result of gathering the signal in digital systems.

The FFT therefore allows a time vibration signal to be transformed into components, each with an amplitude, a phase, and a frequency (see Figure 2). The fundamental frequency of a spectrum also called the first harmonic (1×rpm) is the rotational speed of a rotor. The second harmonic (2×rpm) is twice the fundamental frequency and the third harmonic (3×rpm) is three times the fundamental frequency and so on; therefore, the harmonics are frequency components at a frequency that is an integer multiple of the fundamental frequency (1×rpm).

By associating the frequencies with machine characteristics, and looking at the amplitudes, it is possible to diagnose faults very accurately. Tables have been created which identify the fault by the presence of specific harmonics and subharmonics in the spectrum (Goldman & Muszynska, 1999; Rao, 1996).

Spectrum comparison is the most reliable technique for fault detection; it should be used when similar operating conditions exist in the machine.

The overall vibration value is a single value obtained from a spectrum and measured in displacement, velocity or acceleration of the total vibration amplitude over a filtered bandwidth, or the sum of all vibrations, measured within a specified frequency range. It is expressed in rms in velocity and acceleration and peak-peak values for displacement in predictive maintenance. The overall vibration value (OVV) in a frequency range can be calculated from Equation (7).

$$OVV = \sqrt{\frac{\sum_{i=1}^{N} x_i^2}{1.5}} \qquad (7)$$

where x_i is the amplitude in the ith line of resolution in the frequency range, and N is the number of lines of resolution in the frequency range.

The overall vibration value is controlled by a trend study, from a value associated with an acceptable state of the machine until a value considered dangerous (alarm) to the condition of the machine is passed. The application of this analytical tool is very widespread in standards for vibrational severity; also, it can be individualized for each machine, as it can include the influence of the bench setting, transmission of vibrations from other machines, etc, and is simple to apply in practice. Its application, however, does not give reliable results for the condition of the machine (Randal, 1987; Harris, 1995). This is due to the fact that the tiny modifications in the vibrational level which produce flaws in the initial stage are drowned by higher amplitude effects, such as

Figure 2. Time waveform (left) and resulting spectrum (right)

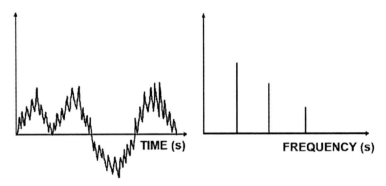

imbalance, which exist in all rotational machinery. Trend analysis requires the use of a logarithmic scale, where a fault can be seen as an exponential trend over time (Harris, 1995). Once the fault has been identified by this technique, the intersection of the vibration trend curve with the given alarm value gives an idea of the remaining lifetime of the machine.

Waterfalls consist of a representation of successive vibration spectra as a function of the variation in velocity of the machine. To this end, the vibration analyser creates a three-dimensional graph, with frequency, amplitude and velocity on the x, y and z axes, respectively, and obtains a register over a given time interval.

In this way the trend of vibration can be seen in the 1×rpm values and their harmonics (Bently Nevada, 1997). A fixed trend in the maximum values and their harmonics, as velocity varies across successive spectra, suggests these components are independent of velocity, while the orders (multiples and submultiples) of operating velocity can be seen on the diagonal. This technique is very useful when starting and stopping, and when the machines experiences changes in its operating conditions (Valverde, 1994).

A full description of the general procedures of measurement and evaluation of mechanical vibrations by machine type can be seen in standard ISO 18016. A more detailed description of how to do vibration analysis can be seen in Harris (1995), Shreve (1995), Brown (1996), Mobley (2001b), Harris and Piersol (2002), and Scheffer and Girdhar (2004).

THEORETICAL MODEL

The technical parameters used to simulate the response of the theoretical model are those of the manufacture of the inner and outer ball race of automotive bearings. The experimental tests to verify the model were carried out in an automotive bearing manufacturing plant during the normal manufacturing process. The authors then analyzed the theoretical instability of the quality obtained in the process analyzed; specifically, they developed the theoretical chattering profiles obtained through the combination of process parameters.

Geometric Relations

The model identifies the relations among the different forces between the workpiece and the grinding wheel and their effects on the different elements directly affected by the process (machine, workpiece and grinding wheel). To define the magnitudes the authors establish the coordinate origins for each as well as the positive directions. The starting point is that of contact between workpiece and grinding wheel without any force between them and without any wear in either. The positive direction of the cutting force $F_c(t)$ is that of compression of workpiece and grinding wheel, in other words, the direction that favors the cut. The grinding wheel infeed is defined as u(t). The function u(t) defines the workpiece to grinding wheel approximation. The instantaneous wear of workpiece $\delta_w(t)$ and of grinding wheel $\delta_g(t)$ is defined as the difference between the initial radius and the current radius at the point of contact for each of them.

For the workpiece:

$$\delta_w(t) = R_{w0} - R_w(t) \qquad (8)$$

For the grinding wheel:

$$\delta_g(t) = R_{g0} - R_g(t) \qquad (9)$$

where R_{w0} and R_{g0} are the initial radii of the workpiece and the grinding wheel, respectively.

The machine deformation $Y_M(t)$ is defined as the deflection of the machine by the instantaneous effects of the grinding process; the positive direction is opposite to the one favoring the cut. The deformation of the contact $Y_f(t)$ between workpiece and grinding wheel is caused by the effects of the force on the workpiece and the grinding wheel; its positive direction is opposite to the one that favors the cut.

Equation 3 shows the relation between the different magnitudes:

$$\delta_w(t) + \delta_g(t) = u(t) - Y_M(t) - Y_f(t) \qquad (10)$$

Wear of Workpiece

The increase in wear of the workpiece per revolution $\Delta\delta_w(t)$ is proportional to the cutting force $F_c(t)$:

$$F_c(t) = K_w \Delta\delta_w(t) \qquad (11)$$

where K_w is the cutting stiffness and $\Delta\delta_w(t)$ is the increase in wear at the point of contact between the previous and current revolutions.

$$\Delta\delta_w(t) = \delta_w(t) - \delta_w(t)(t - T_w) \qquad (12)$$

where T_w is the workpiece's period of revolution on its axis. The cut parameter K_w is proportional to the width of the grinding wheel (b).

$$K_w = K_{w'} b \qquad (13)$$

Snoeys and Brown (1969) show that the parameter $K_{w'}$ experimentally has values ranging from 2 to 20 N/µm-mm.

Wear of Grinding Wheel

As for the workpiece, a direct relation exists between the cutting force and the increase in wear of the grinding wheel per revolution (see Equation 14).

$$F_c(t) = K_s \Delta\delta_g(t) \qquad (14)$$

where $\Delta\delta_g(t)$ is the increase in wear of the grinding wheel at the point of contact between the previous and current revolutions.

$$\Delta\delta_g(t) = \delta_g(t) - \delta_w(t - T_g) \qquad (15)$$

where T_g is the grinding wheel's period of revolution on its axis.

The grinding wheel wear resistance K_s depends on the type of grinding wheel, but tends to be between 100 and 10,000 times bigger than K_w for normal grinding wheels according to Snoeys and Brown (1969), and up to 100,000 times bigger for cubic boron nitride grinding wheels (Biera, Nieto, Iturriza, Viñolas, & Goikoetxea, 1994a).

Deformation of Machine

For the deformation of the machine the authors consider the simplest model possible in the frequency domain. The dynamics of the machine is a second-order system $H(s)$, so the transfer function between the cutting force $F_c(s)$ and the deformation of the machine $Y_M(s)$ is as shown in Equation 16.

$$Y_M(s) = H(s) F_c(s) \qquad (16)$$

where

$$H(s) = \frac{\omega_n^2}{K_m\left(s^2 + 2A_m \omega_n s + \omega_n^2\right)} \qquad (17)$$

In this transfer function in the frequency domain $H(s)$, ω_n corresponds to the first resonance frequency, which is 5 Hz, K_m is the modal stiffness of vibration mode m, which is approximately 100 N/mm, and A_m is the relative damping of vibration mode m, which ranges from 0.2 to 0.5.

Deformations of Contact

This deformation depends on many variables: the composition of the grinding wheel, the grinding wheel dressing setup conditions, the workpiece and grinding wheel diameters, and so on.

$$Y_f(t) = F_c(t) / K_c \qquad (18)$$

The new parameter K_c is the workpiece-grinding wheel contact stiffness (Biera et al., 1994b). Snoeys and Brown (1969) provide the empirical equations to calculate the contact stiffness and indicates that it can be of the same order of magnitude as the cutting stiffness K_w.

SOLUTIONS AND RECOMMENDATIONS

Given the above equations, and the stiffness constants defined in the previous section, Equations 19-21 follow:

$$Y_M(s) = H(s)F_c(s) \qquad (19)$$

$$F_c(t) = K_s \Delta \delta_g(t) \qquad (20)$$

$$F_c(t) = K_w \Delta \delta_w(t) \qquad (21)$$

The authors introduced these equations into a model built in the Simulink environment of the Matlab software. Figure 3 shows a diagram of the resulting linear model. From this linear model the results shown in Figure 4 were obtained. The figure shows that the process becomes unstable. It is also clear that the deformation has no memory, which means that the workpiece can have less wear at the end of a revolution than it had at the end of the previous revolution. This is physically impossible since it would mean material that has been previously cut adhering to the workpiece. This defect in the response of the model led the authors to elaborate a nonlinear model that avoids this physical impossibility.

The main problem with the linear model is that it allows the generation of both positive and negative cutting forces, and so the increase in wear can be negative, in other words the workpiece can increase in diameter. To avoid this, it is necessary to include in the model the condition that the increase in wear must always be positive. This problem appears both for the workpiece and for the grinding wheel, but to simplify things this condition is only applied for the workpiece. This is coherent with the results of the real process, in which the diameter of the grinding wheel remains practically unchanged from revolution to revolution. Thus the equation for the cutting force becomes:

$$\Delta \delta_w(t) = F_c(t) / K_w \ if \ F_c(t) \geq 0 \qquad (22)$$

$$\Delta \delta_w(t) = 0 \ if \ F_c(t) < 0 \qquad (23)$$

Putting the above equations in the nonlinear model leads to the block diagram of the nonlinear grinding model shown in Figure 5.

Figure 6 shows the results of using this model for the same parameters as from the linear model. If the results of the two previous models are superimposed, the curve shown in Figure 7 is obtained.

Figure 7 shows that at the start of the machining, when the most material is eliminated from the surface of the workpiece and the forces are

Figure 3. Block diagram of linear grinding model

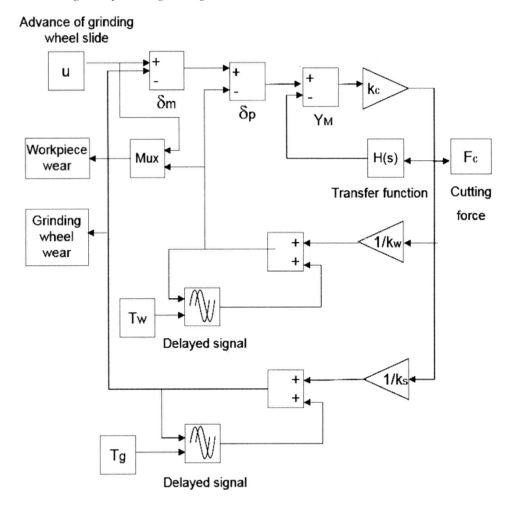

stronger (rough grinding), the linear and nonlinear models coincide. But when the process moves on there may be no contact between workpiece and grinding wheel due to the lack of interference between successive revolutions. The model should consider this possibility for it to behave logically.

Although Snoeys' model considers the contact stiffness, other authors such as Biera et al. (1994b) ignore this parameter in their models, whether linear or nonlinear, because they consider that the contact between workpiece and grinding wheel is not constant. In the current paper, the authors include the contact stiffness in their analysis.

The initial variables and constants used to define the operation are:

1. **Geometric:**
   a. Grinding wheel working width (b) = 53×10^{-3} m.
   b. Workpiece speed = 100 (revolutions per minute (rpm)).
   c. Grinding wheel speed = 1500 rpm.
   d. Workpiece initial diameter = 40×10^{-3} m.
   e. Grinding wheel initial diameter = 580×10^{-3} m.

Figure 4. Simultaneous representation of workpiece wear (unstable) and advance of grinding wheel slide over time, assuming linear model; mechanical recuperation of system means workpiece's deformation must be maintained or delayed with respect to movement of grinding wheel slide.

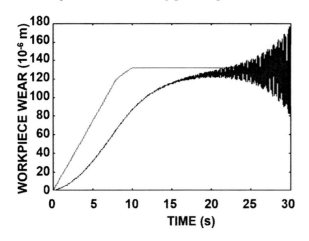

2. **Kinematics of Machining:**
   a. Rough grinding cycle time = 8 s.
   b. Finishing grinding cycle time = 10 s.
   c. Spark-out time = 15 s.
   d. Rough grinding wheel feed = 15×10^{-6} m/s.
   e. Finishing grinding wheel feed = 6×10^{-6} m/s.

Thus for the workpiece and grinding wheel the time it takes to make one revolution is T_w = 60/workpiece speed and T_g = 60/grinding wheel speed, respectively.

Given the above parameters, the result is the wear behavior of the workpiece and grinding wheel with respect to the advance of the grinding wheel slide shown in Figure 8a and 8b respectively. The interference phenomenon was modeled for the workpiece wear but it has not been considered for the grinding wheel wear as the grinding wheel's behavior, applying the nonlinear model to the workpiece wear, shows no incoherence and is monotonically increasing and without fluctuations.

The analysis of the cutting force reveals that this function contains numerous commutations due to the interference phenomenon, in other words, contact-no contact between tool and workpiece.

To compare the profile of forces of Figure 9 with the habitual ones from the linear models, the authors filtered the cutting force. For this, they applied a third-order low-pass filter on its mean square value, with the three poles identical and at -0.15 rad/sec. Figure 10 shows the result. The force reaches a maximum in the rough grinding stage, and then drops progressively to 0 at the end of the spark-out period. This is consistent with what occurs in the real process.

To obtain the theoretical profiles of the workpieces the authors again ran the simulation using the Simulink software.

To be able to compare the effects of these phenomena on the final form of the workpiece the authors ran a series of simulations, modifying the rotational speed of the workpiece (V_w) (4 levels) and the grinding wheel (V_g) (5 levels). Table 2 shows the result obtained in terms of non-quality (chatter) of the theoretical profiles obtained. Table 3 shows the profiles obtained in the matrix of theoretical instability from Table 2.

Looking at the results matrix, the conclusion is that increasing the workpiece's frequency of revolution does not translate into an increase in the chatter frequency. The grinding wheel rotational speed does not have a significant effect on the theoretical model in terms of number of lobes. This conclusion is not consistent with the reality of the process on the shopfloor, where a strong sensitivity to this parameter in the chattering effect is observed. The lobe amplitude sinks to a minimum at value 3 in the workpiece's revolution speed (see Table 2). Subsequently, when the workpiece's revolution speed increases, average amplitude lobes reappear.

From the results of the theoretical models it can be seen that the choice of working parameters to guarantee the quality of the process cannot be

Figure 5. Block diagram of nonlinear grinding model; inclusion of switch makes $F_c(t)$ always positive and workpiece diameter decrease during grinding process.

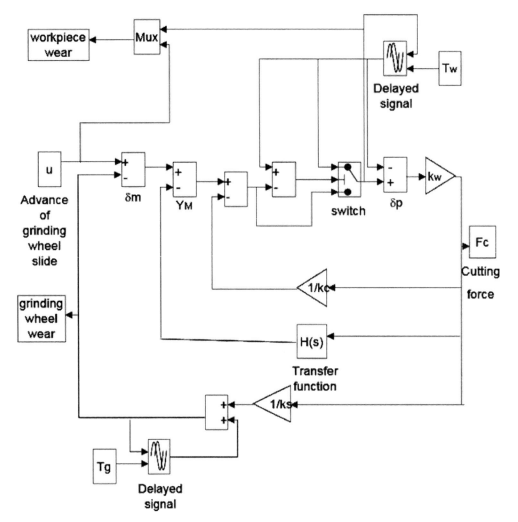

Figure 6. Simultaneous representation of workpiece wear and advance of grinding wheel slide assuming nonlinear model; diameter of workpiece converges in this case toward final position of grinding wheel slide while forces generated by advance of tool during initial rough grinding stage and subsequent finishing stage decline.

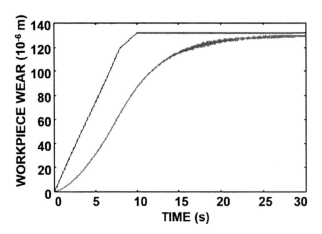

Figure 7. Superimposing workpiece wear linear and nonlinear models; models coincide perfectly up to middle of cycle, and separate at the end due to instability of linear model. Nonlinear model is perfectly stable.

Figure 8. a) Workpiece wear with respect to advance of grinding wheel slide; small fluctuations evident, which will affect final workpiece chatter; b) grinding wheel wear, expressed in 10^{-6} m; no fluctuations evident, with monotonically increasing behavior

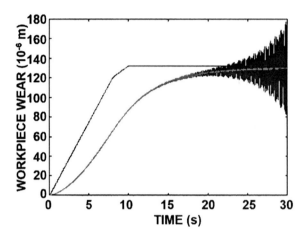

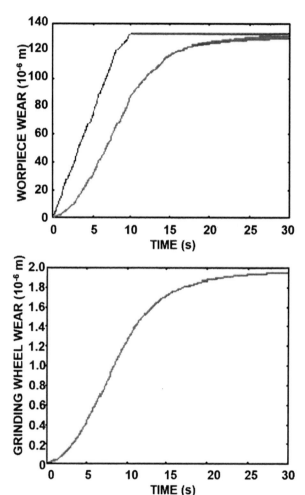

carried out by theoretical analysis. The limitations of the theoretical control of parameters affecting quality require methods for identifying important parameters of the process which cause given vibration frequencies to appear, and which, as a result, affect the quality of manufactured parts, via statistical conclusions based on real experiments.

EXPERIMENT DESIGN AND HYPOTHESIS

A series of prior experiments was carried out (Carnero, Gónzalez-Palma, Almorza, Mayorga, & López-Escobar, 2010) to determine the variables which influence the spectra and global values of the process (chatter) and quality (harmonics). The final set of variables considered is:

- **Cause Variables:** Rotational speed of the grinding wheel, rotational speed of the workpiece, diameter of the grinding wheel, magnetism, machine used and interactions between all the previous variables.
- **Alert Variables:** Overall vibration values during the process (displacement (m), velocity (m/s) and acceleration (g = 9.8 m/s^2)) and spectra of process vibration;
- **Effect Variables:** Level of quality (frequency and amplitude of ground lobes).

Figure 9. Evolution of cutting force during grinding cycle; strongly fluctuating behavior at end (spark-out)

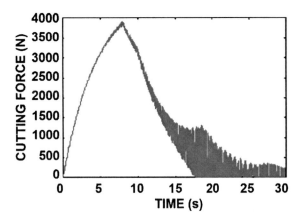

Figure 10. Filtered profile of cutting force; tool-workpiece contact almost inexistent at end of spark-out stage

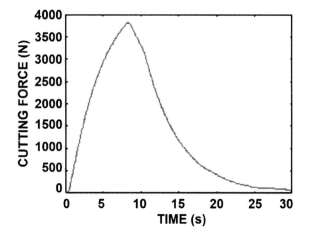

Experimental Design

The experimental process to be carried out includes a large number of process variable and interactions between them, and so it is necessary to use an experimental design that reduces the number of experiments and facilitates data analysis.

The use of the Taguchi, Shainin and classical methods was considered. The use of the Taguchi method has the weakness that it does not consider the effect of interactions between variables with the same weight as that of individual variables. In this research, this could be disadvantageous since it has been noted that such interactions can be the most important cause of no quality situations in our process.

Table 2. Chatter matrix; represents theoretical variation in severity of defect obtained (number and amplitude of lobes) when varying workpiece and tool's revolution velocities

		Chatter Amplitudes				
		Grinding Wheel Speed (V_g)				
		Value 1	Value 2	Value 3	Value 4	Value 5
Workpiece Speed (V_w)	Value 1	High	High	High	High	High
	Value 2	High	High	High	High	High
	Value 3	Low	Low	Low	Low	Low
	Value 4	Medium	Medium	Medium	Medium	Medium
		Chatter Frequencies				
		Grinding Wheel Speed (V_g)				
		Value 1	Value 2	Value 3	Value 4	Value 5
Workpiece Speed (V_w)	Value 1	6	6	6	6	6
	Value 2	4	4	4	4	4
	Value 3	3	3	3	3	3
	Value 4	2	2	2	2	2

Table 3. Chatter profiles

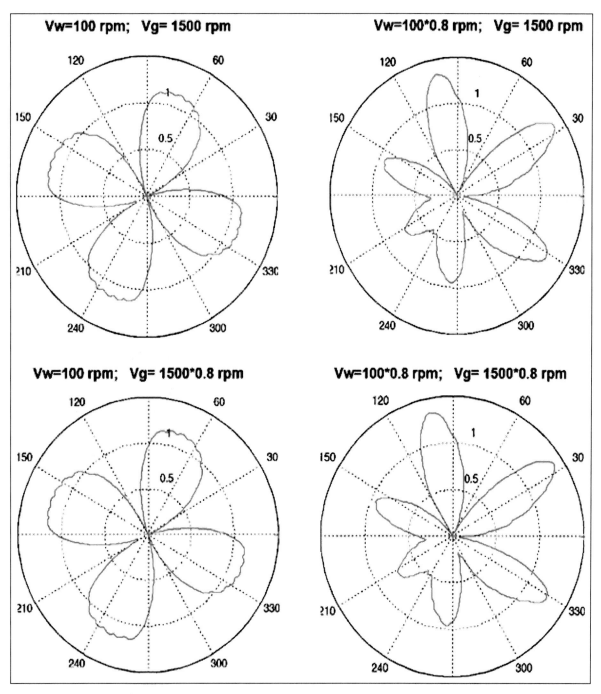

continued on following page

Table 3. Continued

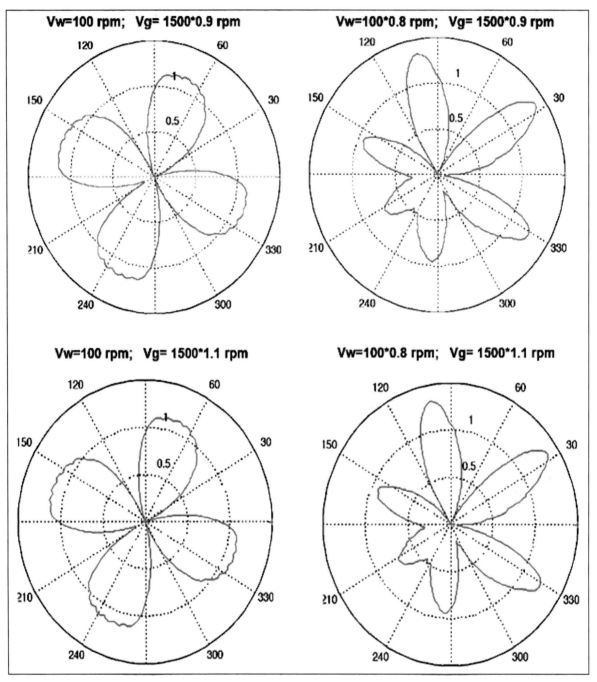

continued on following page

Table 3. Continued

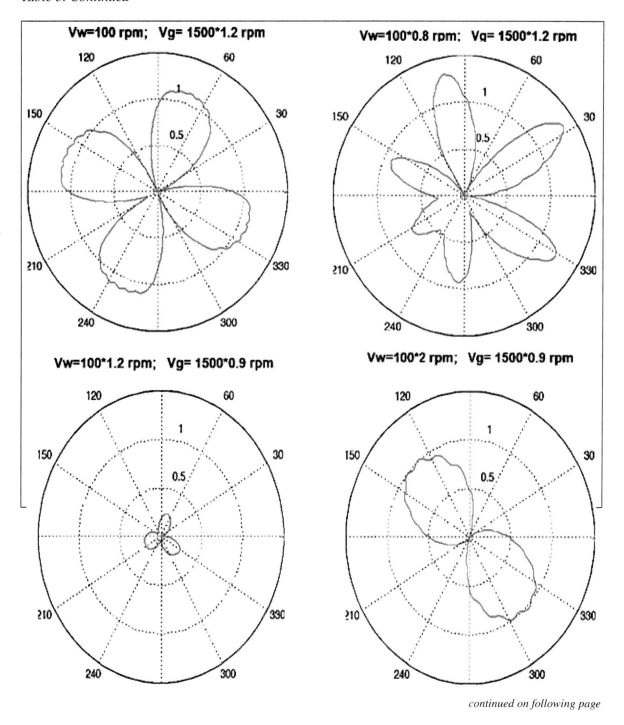

continued on following page

Table 3. Continued

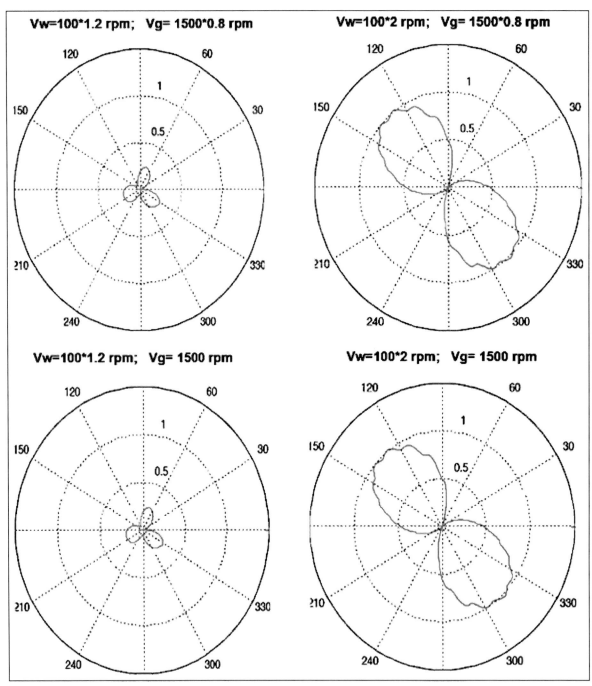

continued on following page

Table 3. Continued

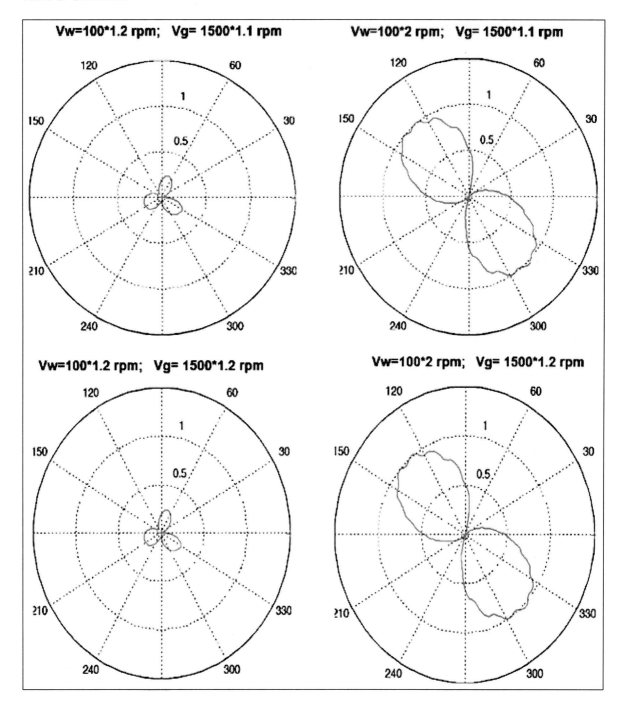

This research intends to explore more deeply the mechanical behaviour of the processes and the effect of each variable and its interaction on the chatter generated, without ignoring the effect on this behaviour of any of the chosen variable; therefore, the Shainin approach is not suitable. A classical approach will thus be used, considering all the interactions and systematically applying variance analysis. This approach allows a large number of variables to be included, which means performing a large number of experiments but it includes the effect of interactions between different variables.

Hypothesis

For a given machine the following controllable cause variables are considered as members of the control variable matrix of our process: rotational speed of the grinding wheel (rpm), rotational speed of the workpiece (rpm) and magnetism (gauss).

Each of these variables is assigned two levels of possible variation (maximum and minimum allowed according to the specifications of the process). Because of time constraints on data acquisition, through-speed will be held constant, without distinguishing between the subprocesses of polishing and finishing.

The grinding wheel diameter is considered a controllable factor at the experimental stage.

Although there are other variable such as quality and quantity of coolant in the process, quantity of material to be ground, distribution of through-speeds, external vibrations, etc., their influence on the quality obtained during manufacture is less important than those analysed in the research.

Each experiment has provided:

1. Overall vibration values for displacement, velocity and acceleration of the process vibration and lobing of the ground pieces.
2. Process vibration spectra.
3. Quality characteristics of the workpiece (roundness and number of lobes). Breakdown into harmonics. In acquiring spectra, the machine used is included as a cause variable, to be able to apply spectral intercomparison criteria for the process between identical machines. The set of experiments carried out will provide data to assess the following hypotheses:

a. The overall vibration values (displacement, velocity and acceleration) show a tendency associated with the type of process and dependent on its control variables. This trend is repeated cycle by cycle if the process conditions are maintained.
b. The overall vibration values (displacement, velocity and acceleration) during contact of the grinding wheel with the workpiece is shown to depend on the process variables; these variables are used in the spectral analysis subsequently carried out.
c. The rotating speed of the grinding wheel and workpiece affect the overall vibration value.
d. Interactions between the variables considerably affect process vibrations. It is possible to determine the most important interactions with respect to process vibrations and quality.
e. The important interactions should correspond to the actual practice of the technicians with regard to accessible process variables when there are quality problems.
f. The trend in overall vibration values during grinding wheel - workpiece contact is not sufficient to detect quality problems.
g. The spectral behaviour of the process vibrations is a function of the process variables. Action on the process variables causes changes in the distribution of frequency and amplitude of the vibration spectrum, and in the quality of the product.

h. The process spectra for displacement, velocity and acceleration show repetitive behaviour sensitive to fluctuations in quality of the workpieces obtained, that is, there is spectral identity between the processes.
i. The amplitude of the spectral changes seen in displacement, velocity and acceleration when quality problems occur is very small compared to the overall vibration value.
j. There are spectral vibration bands forbidden by quality constraints. The appearance of spectral activity in these bands shows that the process is in an area of non-controlled quality
k. To determine operating behaviour outside safe quality zones, it is possible to use spectral intercomparison between processes although the machines may be physically different.

OVERALL VIBRATION VALUE ANALYSIS

To analyse the relationship between the process mechanics and the resulting quality it is necessary to assess the behaviour of the overall vibration value during work cycles and to analyse the effect of the process variables.

In order to measure the quality of the manufactured workpieces a dial gauge was used with precision of 10^{-6} m which compares values by indirect measurement of displacement of a spherical point of contact when the device is fixed to a support. These devices can detect two types of chattering:

- **High Frequency Chattering (Ch A):** Very fast movements of the dial indicator upon rotating the workpiece.
- **Low Frequency Chattering (Ch B):** Slow movements of the dial indicator upon rotating the workpiece.

Furthermore, three levels for each type of chattering (low and high frequency) were established:

Level 0: Null or practically imperceptible displacements of the needle of the dial gauge (null chattering).
Level 1: Slight displacements of the needle of the dial gauge (intermediate chattering).
Level 2: Significant displacement of the needle of the dial gauge (severe chattering).

The experiments gathered data on the profile of the changes in magnitude over the operating cycles, including the loading and unloading cycles of the workpieces, the took-workpiece distance, polishing, finishing and sparkout. It should be noted that this last stage of the process, in which the tool and the workpiece are in contact but with no relative movement is considered by some studies to be important to the final lobing profile of the workpiece.

The aim of the analysis of variance of overall vibration values in displacement, velocity and acceleration is also to distinguish which should be used for the later spectral analysis.

For a given machine four sizes will be considered for the grinding wheel diameter (ϕ): ϕ_{max}, $\phi > \phi_{medium}$, $\phi < \phi_{medium}$, ϕ_{min}. For each grinding wheel diameter experiments will be carried out on all possible combinations of maximum and minimum permitted values of the parameters: rotational speed of the grinding wheel, rotational speed of the workpiece and magnetism (see Table 4).

Analysis of Variance 1

The variables used in the analysis of variance are:

- **Controllable Cause Variables:** Rotational speed of the grinding wheel (revolutions per minute (rpm)), rotational speed of the workpiece (rpm), magnetism (gauss).
- **Uncontrollable Cause Variable:** Diameter of the grinding wheel.

Table 4. Experimental data

Experiments in Overall Vibration Value									
Diameter of the Grinding Wheel (10^{-3}m)	Rotational Speed of the Grinding Wheel (rpm)	Rotational Speed of the Workpiece (rpm)	Magnetism (G)	HFD (10^{-6} m)	LFD (10^{-6} m)	Velocity (10^{-3} m/s)	Acceleration (g)	Ch A	Ch B
595	45	100	50	1.5	6	0.37	0.26	1	0
595	45	100	60	1.5	7	0.32	0.21	0	0
595	45	120	50	2	6	0.34	0.23	0	0
595	45	120	60	1.4	5	0.29	0.21	0	0
595	60	100	50	2	6	0.44	0.24	1	0
595	60	100	60	2	10	0.36	0.2	0	0
595	60	120	50	2	7	0.43	0.22	1	0
595	60	120	60	2.2	6	0.36	0.18	0	0
575	45	100	50	1.7	7	0.43	0.27	2	2
575	45	100	60	1.7	7	0.4	0.23	2	2
575	45	120	50	1.7	7	0.44	0.23	2	2
575	45	120	60	1.4	7	0.32	0.19	0	0
575	60	100	50	2.5	6	0.32	0.13	0	0
575	60	100	60	2.5	6	0.33	0.17	1	1
575	60	120	50	3	7	0.3 5	0.19	0	0
575	60	120	60	2.6	7	0.46	0.15	1	0
529	45	100	50	2.9	7	0.35	0.24	0	1
529	45	100	60	2.7	18	0.32	0.27	0	2
529	45	120	50	2.6	12	0.37	0.25	0	2
529	45	120	60	2.7	9	0.42	0.18	0	2
529	60	100	50	3.8	7	0.52	0.19	0	0
529	60	100	60	3.5	8	0.41	0.14	1	0
529	60	120	50	3.1	8	0.38	0.14	0	0
529	60	120	60	3.5	7	0.35	0.14	0	0
495	45	100	50	2.6	10	0.34	0.19	0	0
495	45	100	60	2.6	7	0.33	0.18	0	0
495	45	120	50	2.3	6	0.32	0.21	0	0
495	45	120	60	2.5	9	0.32	0.22	1	0
495	60	100	50	2.2	6	0.38	0.2	1	0
495	60	100	60	2.3	10	0.41	0.17	0	0
495	60	120	50	2.4	8	0.37	0.17	0	0
495	60	120	60	2.2	7	0.37	0.16	0	0

continued on following page

Table 4. Continued

Experiments in Spectral Analysis						
Machine-Tool	Diameter of Grinding Wheel (10^{-3} m)	Peripheral Velocity of Grinding Wheel (m/s)	Overall Vibration Value (HFD) (10^{-6} m)	Displacements in the High-Frequency Band (2,640-11,250 cpm)	Displacements in the Low-Frequency Band (0-2,640 cpm)	Quality
1	566	60	0.4	0.04	0.4	0
1	566	45	0.2	0.03	0.2	0
1	554	50	0.3	0.05	0.3	0
1	554	55	0.3	0.1	0.3	1
1	554	60	0.4	0.05	0.4	0
1	547	45	0.3	0.06	0.3	1
1	547	50	0.4	0.06	0.4	1
1	547	60	0.4	0.07	0.4	1
1	509	45	1.7	0.09	1.7	0
1	509	50	1.5	0.08	1.5	2
1	509	55	0.9	0.16	0.9	1
1	509	60	0.6	0.18	0.6	1
1	496	45	0.7	0.12	0.7	1
1	496	60	1	0.14	1	2
1	488	60	1.2	0.04	1.2	0
1	488	60	1.2	0.03	1.2	0
1	488	45	1.8	0.04	1.8	0
2	566	45	0.2	0.04	0.2	0
2	566	50	0.2	0.04	0.2	0
2	566	55	0.2	0.08	0.2	0
2	566	60	0.2	0.04	0.2	0
2	560	45	0.2	0.04	0.2	0
2	560	50	0.3	0.04	0.3	1
2	560	55	0.2	0.07	0.2	1
2	560	60	0.3	0.04	0.3	0

- **Alert Variables:** High frequency displacements (HFD) (10^{-6} m), low frequency displacements (LFD) (10^{-6} m), velocity (10^{-3} m/s) and acceleration (g).
- **Quality Variables:** High frequency chattering (Ch A) and low frequency chattering (Ch B).

With these experimental data for the overall vibration value, shown in Table 4, a variety of multifactor variance analyses are carried out, using the controllable variables and analysing their effect on the alert and quality variables. The results obtained are:

1. HFD is affected by the grinding wheel diameter and the rotational speed of the grinding wheel (p-value = 0.0000 and p-value = 0.0002 respectively, and in both cases the values are below 0.05).

2. The overall vibration acceleration is affected by the rotational speed of the grinding wheel and by magnetism (p-value = 0.000 and 0.0252 respectively).
3. The overall vibration displacement and velocity are not significantly affected by the process variables.
4. Ch A and Ch B are affected by the grinding wheel diameter with a 95% confidence interval; furthermore, Ch B is influenced by the rotational speed of the grinding wheel.

Then a variance analysis is performed, centred on HFD, with respect to the controllable cause variables and their possible interactions. As a result it can be seen that the interaction diameter - rotational speed of the grinding wheel (p-value = 0.0130) is more important that rotational speed of the grinding wheel alone (p-value = 0.0553).

The overall vibration value in HFD generated by the process increases as the diameter is reduced, and decreases as the rotational speed of the grinding wheel is reduced. This statistical observation agrees with what is in fact observed, since problems of lobing tend to appear when the life of the tool is exhausted, that is, with small grinding wheel diameters.

The results of the variance analysis are subjected to Chi-squared, Shapiro-Wilk, symmetry and Kurtosis tests, to check whether they satisfy the normality conditions. All the experiments give a p-value less than 0.32, and so it can be guaranteed with a 90% confidence level that HFD behaves as a normal distribution.

It should also be checked that the HFD variable satisfies the homoscedasticity condition with respect to the controllable variables. This is done using the trials of Cochran, Harlett and Hartley, with the null hypothesis being to consider that the variances of HFD for each level of the controllable variables are the same. Applying this criterion shows that the hypothesis of equal variance is satisfied for HFD with respect to rotational speed and the diameter of the grinding wheel.

However, the homoscedasticity hypothesis is not confirmed when analysing the effect of the interaction grinding wheel diameter-rotational speed of the grinding wheel on the HFD. As a consequence of the non-fulfilment of homoscedasticity the Kruskal-Wallis test is applied, showing significant differences between the means with a 95% confidence level. There are also significant differences in the diameter, which also fails to confirm homoscedasticity. For this reason the null hypothesis, that the variation of the means for the four diameter levels is zero, is checked. Again, the confidence level for obtaining different means for the distinct stages of the life of the grinding wheel is over 95%

Then new analyses of variance, incorporating the interaction grinding wheel diameter – rotational speed of the grinding wheel are carried out, and their effect on the quality obtained is analysed. The following conclusions are found:

1. The HFD are affected by the grinding wheel diameter - rotational speed of the grinding wheel interaction.
2. The LFD are affected by the grinding wheel diameter - rotational speed of the grinding wheel interaction.
3. Acceleration is affected by the grinding wheel diameter - rotational speed of the grinding wheel interaction.
4. The Ch B is affected by the grinding wheel diameter - rotational speed of the grinding wheel interaction.

Ch A and Ch B are discrete variable and so they do not satisfy the requirements of normality and homoscedasticity necessary to apply variance analysis; contingency tables have therefore been applied to analyse the independence of these variables with respect to each of the controllable variables, and including the interaction.

Each of these contingency tables analyses the frequency of appearance of each of the three degrees of chattering (0, 1 and 2) for each value of the controllable variable whose effect needs to be analysed. Then the Chi-squared test was used to determine whether or not there was statistical dependence between the variables or on the contrary they are not related. The p-values resulting from applying the contingency tables give the following results:

1. Ch A and Ch B are independent of the plate speed and magnetism.
2. Ch A and Ch B are dependent on the grinding wheel diameter and the rotational speed of the grinding wheel.
3. Ch A is dependent on the interaction.
4. Ch B is independent in the set of experiments carried out.

Now spectral analysis is carried out to determine which vibration frequencies are damaging to the process.

SPECTRAL ANALYSIS

In the previous analysis in can be seen that there is no clear relationship between the mechanical behaviour measured in the overall vibration value and the quality, and so it is necessary to perform another study, using spectral analysis in order to determine the vibrational energy contained in different vibration bands, and its possible relationship to quality.

This analysis first considers the existence of spectral intercomparison between different machines which perform the same process. A set of experiments is therefore applied to different machines to determine whether the vibrations induced by the process itself allow spectral comparisons to be made that would suggest which operating conditions lead to no quality.

The analysis of variance 1 shows the particular sensitivity of HFD to the controllable process parameters, and so this is the variable that is held in the analysis. The process parameters will be the grinding wheel diameter and the rotational speed of the grinding wheel since these have been shown to have an effect on the overall vibration values and the quality.

Spectral Identity of the Process

When a machine is operating with no load, it is natural to find specific frequencies dominating the spectrum. These frequencies and their harmonics depend on the mechanical characteristics of the machines, such as the number of teeth on the gears, length and diameter of pulleys, and the characteristics of the bearings.

When a machine tool begins a cycle, the quality of its operation also depends on other mechanical sources from the process itself. These sources of excitement in the process should show a characteristic spectral behaviour, or spectral identity, as the behaviour in quality of a machine has specific characteristics if there is no disturbance (López-Escobar, González-Palma, Almorza, Mayorga, & Carnero, 2012).

The chattering of a workpiece can be decomposed into given frequencies and their corresponding amplitudes, and so it can also be represented as a Fourier series.

Spectral identity allows the spectral behaviour of a process to be known, and it makes it easier to relate the process spectrum with the quality spectrum obtained, which in turn allows the phenomenon of quality transfer from the machine to the workpiece through the process to be described.

To determine the degree of repetition of the process-induced vibrations a number of workpieces have been produced consecutively, from which vibrational data have been derived with respect to displacement, velocity and acceleration.

The waterfalls plots in Figure 11 represent the spectral characteristics in displacement, velocity and acceleration in the chosen frequency ranges for the external raceway grinding process. It is seen that the frequency of appearance of the vibrational peaks is maintained, showing that the process retains spectral identity if the process variables are not modified. From the spectral identity obtained it is possible to define vibration bands which allow the vibration content of each to be quantified.

Quality Bands

Now we determine which of the vibration frequencies are prejudicial to the quality of the processes. The study focuses on HFD and its variation with respect to influential variables. The analysis is applied to two machines performing identical operations.

At the experimental stage spectra are taken over the lifetime of the grinding wheel, as well as data on the development of chattering in the manufactured components. Figure 12 shows that low frequency vibration is greatly influenced by grinding wheel diameter. The existence of random, high frequency spectral activity can also be seen. We therefore choose two vibration bands:

- **Overall Band:** From 60 cycles per minute (cpm) to 15,000 cpm.
- **High Frequency Band:** From 2,640 cpm to 60,000 cpm.

Experiments were then carried out in spectral analysis, as shown in Table 4.

The high frequency measurements appear to be random, which suggests a new statistical analysis to explore further the behaviour of this process-induced mechanical parameter.

It is hypothesized that HFD activity is related to no quality in the processes.

Analysis of Variance 2

Predictive statistical spectral analysis focuses on the search for a relation between the band values of HFD and the appearance of chattering, taking into account two different machines.

Of the experimental data from the two machines shown in Table 4 it can be seen that spectral activity of HFD shows subtle differences between the different grinding cycles and the two machines used.

The Shapiro-Wilk test confirms that HFD in overall vibration value does not follow a normal distribution. The median test (p-value = 0.063) confirms with a 95% confidence level that overall vibration value in HFD does not permit the detection of quality problems in grinding processes.

The HFD in the high frequency band does not satisfy the normality conditions; the subsequent median test confirms that there are no significant differences between the two machines with a confidence level of 95%.

The median test carried out on HFD and quality rejects the null hypothesis that the medians of HFD are the same between quality levels (p-value = 0.01) with a confidence level of 95%.

The values of the HFD in the high frequency band for Ch A = 0 and Ch B = 0 or when the workpiece presents chattering do not follow a normal distribution. Although the median test applied to HFD in the high frequency band between the two machines indicates that there are no significant differences between them, running the median test again to find the relationship between HFD in the high frequency band with Ch A= 0 and Ch B = 0 or Ch A $\neq$ 0 and Ch B $\neq$ 0, it can be seen that there is a difference between the medians, with a confidence level of 95%.

The direct conclusion from this result is the possibility of establishing a surveillance band, in this case between 2,640 cpm and 15,000 cpm, where HFD in the high frequency band guarantees the stability of the process with respect to quality. This statement is valid even in two different machines, which allows the definition of generic control bands, that is, dependent on the process and independent of the specific machine.

Once the maximum permissible vibration displacement values for high frequency vibration are known, limit spectra can be defined, controlled cycle by cycle, or even connected to the permanent

Figure 11. Waterfalls plots of the external raceway grinding process in displacement, velocity, and acceleration (from left to right)

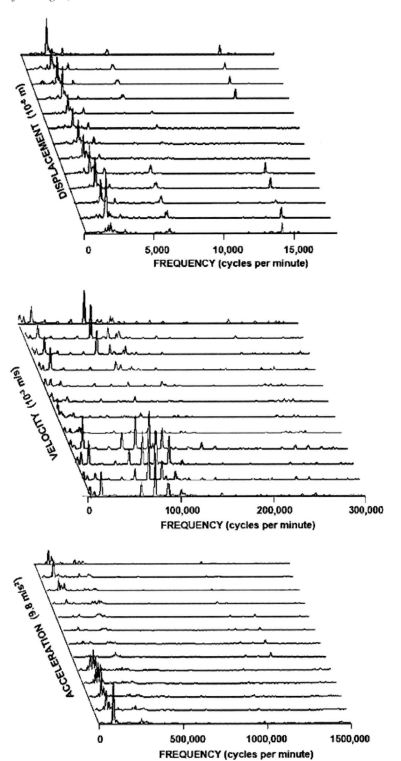

Figure 12. Waterfalls of HFD; definition of spectral bands

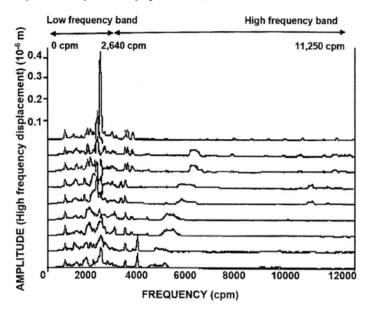

monitoring panel of the machines, so that they can modify the sensitive parameters identified at the overall vibration value analysis stage.

FUTURE RESEARCH DIRECTIONS

This research is intended to broaden the field of application of predictive techniques to other maintenance functions within a company, such as quality control. The concept can also be extended to other areas such as safety. In this way it would be possible to achieve the integration of company functions based on predictive techniques, which could lead to the concept known as Total Predictive Assurance.

Total Predictive Assurance would reduce the company's activities according to results obtained by predictive techniques; and so activities related to maintenance, quality and safety would be carried out only when recommended by a predictive indicator, minimizing preventive activity in these areas. Although this concept is applied to the maintenance of machines, it has not been extended to other areas.

Total Productive Maintenance (TPM) strengthens the participation of the operator in the condition of the machines, recognising the importance and usefulness of the information obtained from the people who are constantly around the machines. TPM is usually focused on cleaning of machines and inspections, as well as on simple maintenance tasks such as restocking lubricants, checking pressure gauges, etc. Total Predictive Assurance broadens the participation of the operator, allowing simple predictive controls (the simplest predictive technique, little used, is the impression of the operator about the behaviour of the machines and process he/she is in charge of), not only aimed at maintenance, but also taking in parameters indicative of developing quality and safety problems. These first level predictive inspections may include checks of temperature, pressure, suspicious trends in the processes, early wear in tools, etc.

Also, when applying Total Predictive Assurance it would be necessary to use advanced predictive technologies which require a highly qualified crew of maintenance workers.

Spectral intercomparison between identical machines and waterfall analysis at specific points are key areas for achieving a powerful system of problem prediction based on vibration analysis, and they are the foundation for success when training maintenance staff. These tools allow problems to be predicted more accurately and without the need for a detailed interpretation of the frequency of appearance of peaks. The worker assigned to collection and analysis of predictive data only has to look out for a deviation from spectral identity in a specific point of the machine and to make comparisons with equivalent points in identical machines. This simple technique is very efficient and can be applied to simple mechanical adjustments as well as to solving complex quality problems. Each problem has different associated peak appearance spectra, and it is not necessary to explain them in each case, only their trend. This method rarely leads to unjustified interventions in machines or processes.

The method previously described minimizes the need for action by more qualified staff and is applicable to any predictive technology, however complex it is, which means putting Total Predictive Assurance into action is considerably simplified.

Thus, Total Predictive Assurance gathers information about the trends in predictive indicators from two separate sources: TPM and predictive technologies to develop procedures to guarantee adequate uptime, controlled process quality and even improvements in the safety of personnel.

CONCLUSION

Grinding processes cause induced vibrations which depend on the process variables. These self-induced vibrations are independent of the input variables, the amplitudes and the frequencies of the vibrations and the chattering found in the workpieces. This conclusion is seen both in the theoretical model and in the experimental results. However, the complexity of the phenomenon and the huge number of variables involved prevent the same quantitative results from being obtained when the two models are compared. In practice, the theoretical model for determining lobing is of very limited applicability.

The linear model does not produce coherent results due to the constraints in the real process about the sign of the forces and of the tool wear. The nonlinear model shows the appearance of self-induced vibrations that may originate the final defects in the workpieces manufactured. Increasing the spark-out time does not eliminate these defects, the origin of which cannot be attributed to forces outside the process because they are not considered when building the model. The complex behavior observed in this model makes it clear that it would be extremely difficult to produce a complete orbital grinding model including factors like the grasp magnetic force, wear of elements, starting workpiece quality, mechanical adjustments, hardness of grinding wheel, quality of refrigerant and its effect, uniformity of previous thermal treatments, constant cutting velocity algorithms, and so on.

The amplitudes and frequencies of the modelled lobes show a fluctuating theoretical behavior, as can be seen from the chatter matrix. This is the case with a simplified model, in other words one considering only some of the variables that may affect how the process quality behaves.

The selection of operating parameters that guarantee process quality (acceptable chattering levels) is not possible using a theoretical analysis. Given the difficulty in estimating all the variables considered in the model, and the fact that not all phenomena affecting the appearance of defects have been considered, the authors suggest exploring quality control systems on the basis of the control of characteristics induced by the machining process itself, through measurement of real process variables, such as vibrations induced during the tool-workpiece contact. The nature of these vibrations and their dominant frequencies are the symptoms of the generation of quality problems.

The theoretical analysis of the parameters impacting the quality has its limitations, so it is necessary to establish methods to identify the process parameters influential in the appearance of certain vibration frequencies. These, in turn, can be analyzed in real experiments generating statistical results that show their influence on the quality of the workpieces being manufactured. This suggests the need to use (overall vibration value analysis and spectral analysis) vibration analysis to determine the process conditions where non-quality conditions appear in grinding processes.

There are spectra with particular characteristics associated with the vibrations produced by grinding wheel-workpiece contact; this is known as spectral identity, and these spectra are repeated when the process variables are maintained. Thus, for each process quality warning bands can be defined. The spectral identity of self-induced vibrations leads to the concept of spectral intercomparison.

Statistical analysis to determine the relationship between the overall vibration value and the quality of the workpieces when modifying the variables does not lead to relevant conclusions. In high-precision grinding processes the overall vibration value is not sufficient to show no-quality. This conclusion differs from those in the current literature.

The HFD are greatly influenced by the grinding wheel diameter - rotational speed of the grinding wheel interaction.

The HFD should be used to perform the spectral intercomparison which allows problems of quality in grinding processes to be detected. In each process a high-frequency band can be defined in which the displacements should be controlled to ensure operation with consistent quality. In this way, its use together with the concepts of spectral identity and intercomparison between spectra allow surveillance bands to be defined in which the spectral activity suggests the imminent appearance of quality problems.

The LFD have little effect on quality.

The maximum permissible vibration limits which guarantee the quality of the processes differ from the alert values used in predictive maintenance to detect breakdowns. Thus, in order to use vibration analysis for quality control the procedure described in this chapter should be used, firstly applying overall vibration value analysis, and then defining bands using spectral identity. Analysis of the high-precision grinding process has established that when the vibrations amplitudes are above $1 \cdot 10^{-7}$ m in a band between 2,640 cpm and 15,000 cpm, it is possible to predict the lack of quality in the ground workpieces.

The methodology described can contribute to increasing the environmental sustainability of manufacturing organizations with a more efficient use of resources, as the quality of pieces can be guaranteed; therefore, the energy used in manufacture, the material used to make the workpieces and the effort involved in the process of remanufacture is avoided.

REFERENCES

Bently Nevada. (1997). *Machinery diagnostics course*. Warrington, UK: Bently Nevada.

Biera, J., Nieto, F. J., Iturriza, I., Viñolas, J., & Goikoetxea, P. (1994a). Characterisation of the behaviour of a CBN wheel when grinding at high speed. In *Proceedigns of the X Congress de Investigatión*. San Sebastián, Spain: Diseño y Utilizatión de Máquinas-Herramientas.

Biera, J., Nieto, F. J., Viñolas, J., & Goikoetxea, P. (1994b). *Proccedings of the XI National Congress of Mechanical Engineering*. Valencia, Spain: Academic Press.

Biera, J., Viñolas, J., & Nieto, F. J. (1997). Time-domain dynamic modelling of the external plunge grinding process. *International Journal of Machine Tools & Manufacture, 37*(11), 1555–1572. doi:10.1016/S0890-6955(97)00024-2

Brinksmeier, E., Aurich, J. C., Govekar, E., Heinzel, C., Hoffmeister, H. W., Klocke, F., . . . Wittmann, M. (2006). Advances in modelling and simulation of grinding processes. *CIRP Annals - Manufacturing Technology, 55*(2), 667-696.

Brown, R. (1996). Profile. *Brüel & Kraer, 4*(3), 10–12.

Carnero, M. C. (2005). Selection of diagnostic techniques and instrumentation in a predictive maintenance program: A case study. *Decision Support Systems, 38*(4), 539–555. doi:10.1016/j.dss.2003.09.003

Carnero, M. C. (2012). *Programas de mantenimiento predictivo: Análisis de lubricantes y vibraciones*. Saarbrücken, Germany: LAP LAMBERT Academic Publishing GmbH & Co.

Carnero, M. C. (2013). Mantenimiento predictivo en pequeña y mediana empresa. *DYNA Management, 1*(1). doi:10.6036/MN5790

Carnero, M. C., Gónzalez-Palma, R., Almorza, D., Mayorga, P., & López-Escobar, C. (2010). Statistical quality control through overall vibration analysis. *Mechanical Systems and Signal Processing, 24*(4), 1138–1160. doi:10.1016/j.ymssp.2009.09.007

Charray, C. (2000). Mantenimiento predictivo: Una técnica que reduce o elimina averías inesperadas. *DYNA, 75*, 28–34.

Choi, T., & Shin, Y. C. (2007). Generalized intelligent grinding advisory system. *International Journal of Production Research, 45*(8), 1899–1932. doi:10.1080/00207540600562025

Choi, T. J., Subrahmanya, N., Li, H., & Shin, Y. C. (2008). Generalized practical models of cylindrical plunge grinding processes. *International Journal of Machine Tools & Manufacture, 48*(1), 61–72. doi:10.1016/j.ijmachtools.2007.07.010

Goldman, P., & Muszynska, A. (1999). Application of full spectrum to rotating machinery diagnostics. *Orbit (Amsterdam, Netherlands), 20*(1), 17–21. PMID:12048694

Govekar, E., Baus, A., Gradišek, J., Klocke, F., & Grabec, I. (2002). A new method for chatter detection in grinding. *CIRP Annals - Manufacturing Technology, 51*(1), 267-270.

Gradišek, J., Baus, A., Govekar, E., Klocke, F., & Grabec, I. (2003). Automatic chatter detection in grinding. *International Journal of Machine Tools & Manufacture, 43*(14), 1397–1403. doi:10.1016/S0890-6955(03)00184-6

Harris, C. M. (1995). *Shock and vibration handbook*. New York, NY: McGraw-Hill.

Harris, C. M., & Piersol, A. G. (2002). *Harris' shock and vibration handbook*. New York, NY: McGraw-Hill.

Jardine, A., Lin, D., & Banjevic, D. (2006). A review on machinery diagnostics and prognostics implementing condition based maintenance. *Mechanical Systems and Signal Processing, 20*(7), 1483–1510. doi:10.1016/j.ymssp.2005.09.012

Khalifa, O. O., Densibali, A., & Faris, W. (2006). Image processing for chatter identification in machining processes. *International Journal of Advanced Manufacturing Technology, 31*(5-6), 443–444. doi:10.1007/s00170-005-0233-4

Levitt, J. (2003). *The complete guide to preventive and predictive maintenance*. New York, NY: Industrial Press Inc.

Li, H., & Shin, Y. C. (2006a). Wheel regenerative chatter of surface grinding. Transactions of ASME. *Journal of Manufacturing Science and Engineering, 128*(2), 393–403. doi:10.1115/1.2137752

Li, H., & Shin, Y. C. (2006b). A time-domain dynamic model for chatter prediction of cylindrical plunge grinding processes. *Journal of Manufacturing Science and Engineering, 128*(2), 404–416. doi:10.1115/1.2118748

Li, H., & Shin, Y. C. (2007). A study on chatter boundaries of cylindrical plunge grinding with process condition-dependent dynamics. *International Journal of Machine Tools & Manufacture, 47*(10), 1563–1572. doi:10.1016/j.ijmachtools.2006.11.009

López-Escobar, C., González-Palma, R., Almorza, D., Mayorga, P., & Carnero, M. C. (2012). Statistical quality control through process self-induced vibration spectrum analysis. *International Journal of Advanced Manufacturing Technology, 58*(9-12), 1243–1259. doi:10.1007/s00170-011-3462-8

Mobley, R. K. (2001a). *An introduction to predictive maintenance*. Woburn, MA: Butterworth-Heinemann.

Mobley, R. K. (2001b). *Plant engineer's handbook*. Woburn, MA: Butterworth-Heinemann.

Muszynska, A. (1992). Vibrational diagnostics of rotating machinery malfunctions. Course Rotor Dynamics and Vibration in Turbomachinery, Belgium.

National Aeronautics and Space Administration. (2000). *Reliability centered maintenance guide for facilities and collateral equipment*. Washington, DC: National Aeronautics and Space Administration.

Orynski, F., & Pawlowski, W. (2004). Simulation and experimental research of the grinder's wheelhead dynamics. *Journal of Vibration and Control, 10*(6), 915–930. doi:10.1177/1077546304041369

Rao, B. K. N. (1996). *Handbook of condition monitoring*. Oxford, UK: Elsevier.

Scheffer, C., & Girdhar, P. (2004). *Machinery vibration analysis & predictive maintenance*. Oxford, MA: Elsevier.

Snoeys, R. (1971). Principal parameters of grinding machine stability. *Ingenieursblad, 40*(4), 87–95.

Snoeys, R., & Brown, D. (1969). Dominating parameters in grinding wheel, and workpiece regenerative chatter. In *Proceedings of the 10th International MTDR Conference, Advance in Machine Tool Design and Research* (pp. 325-348). Manchester, UK: Academic Press.

Snoeys, R., & Wang, I. C. (1968). Analysis of the static and dynamic stiffnesses of the grinding wheel surface. In *Proceedings of the 9th MTDR Conference* (pp. 1133-1148). Birmingham, UK: MTDR.

Tonshoff, H. K., Peters, J., Inasaki, I., & Paul, T. (1992). Modelling and simulation of grinding processes. *CIRP Annals - Manufacturing Technology, 41*(2), 677-688.

Valverde, A. (1994). *Análisis de la disponibilidad de los equipos dinámicos y su incidencia en el mantenimiento en plantas industriales*. (Thesis). UNED, Madrid, Spain.

Viñolas, J., Biera, J., Nieto, J., Llorente, J. I., & Vigneau, J. (1997). The use of an efficient and intuitive tool for the dynamic modelling of grinding processes. *CIRP Annals - Manufacturing Technology, 46*(1), 239-252.

ADDITIONAL READING

Abouelatta, O. B., & Madl, J. (2001). Surface roughness prediction based on cutting parameters and tool vibrations in turning operations. *Journal of Materials Processing Technology, 118*(1-3), 269–277. doi:10.1016/S0924-0136(01)00959-1

Al-Najjar, B. (1996). Total quality maintenance: An approach for continuous reduction in costs of quality products. *Journal of Quality in Maintenance Engineering, 2*(3), 4–20. doi:10.1108/13552519610130413

Al-Najjar, B., & Alsyouf, I. (2000). Improving effectiveness of manufacturing systems using total quality maintenance. *Integrated Manufacturing Systems, 11*(4), 267–276. doi:10.1108/09576060010326393

Bisu, C. F., Darnis, P., Gérard, A., & K'nevez, J.-Y. (2009). Displacements analysis of self-excited vibrations in turning. *International Journal of Advanced Manufacturing Technology, 44*(1-2), 1–16. doi:10.1007/s00170-008-1815-8

Bouguerriou, N., Haritopoulos, M., Capdessus, C., & Allam, L. (2005). Novel cyclostationarity-based blind source separation algorithm using second order statistical properties: Theory and application to the bearing defect diagnosis. *Mechanical Systems and Signal Processing, 19*(6), 1260–1281. doi:10.1016/j.ymssp.2005.07.007

Brinksmeier, E., Heinzel, C., & Meyer, L. (2005). Development and application of a wheel based process monitoring system in grinding. *CIRP Annals - Manufacturing Technology, 54*(1), 301-304.

Cardi, A. A., Firpi, H. A., Bement, M. T., & Liang, S. Y. (2008). Workpiece dynamic analysis and prediction during chatter of turning process. *Mechanical Systems and Signal Processing, 22*(6), 1481–1494. doi:10.1016/j.ymssp.2007.11.026

Cho, D. W., & Eman, K. F. (1988). Pattern recognition for on-line chatter detection. *Mechanical Systems and Signal Processing, 2*(3), 279–290. doi:10.1016/0888-3270(88)90024-6

Choudhury, S. K., Goudimenko, N. N., & Kudinov, V. A. (1997). On-line control of machine tool vibration in turning. *International Journal of Machine Tools & Manufacture, 37*(6), 801–811. doi:10.1016/S0890-6955(96)00031-4

Devillez, A., & Dudzinski, D. (2007). Tool vibration detection with eddy current sensors in machining process and computation of stability lobes using fuzzy classifiers. *Mechanical Systems and Signal Processing, 21*(1), 441–456. doi:10.1016/j.ymssp.2005.11.007

Drew, S. J., Mannan, M. A., Ong, K. L., & Stone, B. J. (2001). The measurement of forces in grinding in the presence of vibration. *International Journal of Machine Grinding wheels and Manufacture, 41*(4), 509-520.

Ema, S., & Marui, E. (2003). Theoretical analysis on chatter vibration in drilling and its suppression. *Journal of Materials Processing Technology, 138*(1-3), 572–578. doi:10.1016/S0924-0136(03)00148-1

González, F. (2013). Maximization of process tolerances using an analysis of setup capability. *International Journal of Advanced Manufacturing Technology, 67*(9-12), 2171–2181. doi:10.1007/s00170-012-4638-6

Inasaki, I., Karpuschewski, B., & Lee H. S. (2001). Grinding chatter – origin and suppression. *CIRP Annals - Manufacturing Technology, 50*(2), 515-534.

Liao, Y. S., & Shiang, L. C. (1991). Computer simulation of self-excited and forced vibrations in the external cylindrical plunge grinding process. *ASME Journal of Engineering for Industry, 113*(3), 297–304. doi:10.1115/1.2899700

Lu, C. (2008). Study on prediction of surface quality in machining process. *Journal of Materials Processing Technology, 205*(1-3), 439–450. doi:10.1016/j.jmatprotec.2007.11.270

Lu, C., & Costes, J. P. (2008). Surface profile prediction and analysis applied to turning process. *International Journal of Machining and Machinability of Materials, 4*(2-3), 158–180. doi:10.1504/IJMMM.2008.023192

Mechanalysis, I. R. D. Inc. (1995). Signal processing for effective vibration analysis. Columbus, OH: Shreve, D. H.

Rahnama, R., Sajjadi, M., & Park, S. S. (2009). Chatter suppression in micro end milling with process damping. *Journal of Materials Processing Technology*, *209*(17), 5766–5776. doi:10.1016/j.jmatprotec.2009.06.009

Rubio, E., & Jáuregui-Correa, J. C. (2012). A wavelet approach to estimate the quality of ground parts. *Journal of Applied Research and Technology*, *10*(1), 28–37.

Ruxu, D., Elbestawi, M. A., & Ullagaddi, B. C. (1992). Chatter detection in milling based on the probability distribution of cutting force signal. *Mechanical Systems and Signal Processing*, *6*(4), 345–362. doi:10.1016/0888-3270(92)90036-I

Salgado, D. R., Alonso, F. J., Cambero, I., & Marcelo, A. (2009). In-process surface roughness prediction system using cutting vibrations in turning. *International Journal of Advanced Manufacturing Technology*, *43*(1-2), 40–51. doi:10.1007/s00170-008-1698-8

Tarng, Y. S., Kao, J. Y., & Lee, E. C. (2000). Chatter suppression in turning operations with a tuned vibration absorber. *Journal of Materials Processing Technology*, *105*(1-2), 55–60. doi:10.1016/S0924-0136(00)00585-9

Tobias, A. (1961). Machine tool vibration research. *International Journal of Machine Tool Design and Research*, *1*(1-2), 1–14. doi:10.1016/0020-7357(61)90040-3

Tsang, A. H. C. (2002). Strategic dimensions of maintenance management. *Journal of Quality in Maintenance Engineering*, *8*(1), 7–39. doi:10.1108/13552510210420577

Tumer, I. Y., Wood, K. L., & Busch-Vishniac, I. J. (2000). Monitoring of signals from manufacturing processes using the Karhunen-Loeve transform. *Mechanical Systems and Signal Processing*, *14*(6), 1011–1026. doi:10.1006/mssp.1999.1278

Vafaei, S., Rahnejat, H., & Aini, R. (2002). Vibration monitoring of high speed spindles using spectral analysis techniques. *International Journal of Machine Grinding wheels & Manufacture*, *42*, 1223-1234.

Wang, L., & Liang, M. (2009). Chatter detection based on probability distribution of wavelet modulus maxima. *Robotics and Computer-integrated Manufacturing*, *25*(6), 989–998. doi:10.1016/j.rcim.2009.04.011

Wegerich, S. (2005). Similarity-based modeling of vibration features for fault detection and identification. *Sensor Review*, *25*(2), 114–122. doi:10.1108/02602280510585691

Yao, Z., Mei, D., & Chen, Z. (2010). On-line chatter detection and identification based on wavelet and support vector machine. *Journal of Materials Processing Technology*, *210*(5), 713–719. doi:10.1016/j.jmatprotec.2009.11.007

Zeng, Y., & Forssberg, E. (1994). Application of vibration signal measurement for monitoring grinding parameters. *Mechanical Systems and Signal Processing*, *8*(6), 703–713. doi:10.1006/mssp.1994.1050

Zhongqun, L., & Qiang, L. (2008). Solution and analysis of chatter stability for end milling in the time-domain. *Chinese Journal of Aeronautics*, *21*(2), 169–178. doi:10.1016/S1000-9361(08)60022-9

KEY TERMS AND DEFINITIONS

Analysis of Variance: Collection of statistical models and their associated procedures which allows a comparison to test whether the values of a set of numerical data are significantly different from the values of one or more other sets of data. It derives from the premise that the dependent variable is measured at least over an interval, the independence of the observations, the distribution of the dependent variable must be normal, and the homoscedasticity condition or the condition of homogeneity of variances must be satisfied.

Chattering: An effect generated in machining processes consisting in relative movement between the workpiece and the grinding wheel (tool) which leads to vibrations resulting in waves on the machined surface.

Overall Vibration Value: A scalar value which quantifies the vibrational energy in the wave emitted by the machine, and which is indicative of the destructive capacity of the wave and is used as a measure of severity.

Spectral Identity: Characteristic spectrum associated with the vibrations generated by the tool-workpiece contact, which tends to be repeated when the process conditions are not altered and which is in turn sensitive to the influence of multiple variables.

Spectral Intercomparison: Comparison of spectra in different machines carrying out the same process and with similar operating variables.

Spectrum: The Fourier theorem states that any periodic signal can be expressed as a combination of pure sinusoidal signals, with frequencies which are multiples of the fundamental frequency. The series of signals at different frequencies makes up the frequency spectrum.

Waterfalls: Representations of successive vibration spectra as a function of the variation of the machine velocity, via a three-dimensional graph, with frequency, amplitude and velocity of the X, Y and Z axes respectively, and taking a reading over a given time interval.

Chapter 10
Retrospection of Globalisation Process and the Sustainability of Natural Environment in Developing Countries

Shahul Hameed
Te Wananga o Aotearoa, New Zealand

ABSTRACT

Globalization is an inevitable integrating process and vital to the world economy but it generates many challenges towards the integration of "economic independence" of the nation states like (a) economic integration through investment/trade and capital flow, (b) initiating multilateral political interaction between the countries, and (c) diffusion of dominant cultural values and beliefs over other cultures. globalization accelerates structural change, which alters the industrial structure of host countries, for instance the excessive use of natural resources and contributes to the physical environmental deterioration. Further, globalization transmits and magnifies market failures and policy distortions if not properly addressed. The chapter attempts to (a) identify the key links between globalization and environment deterioration, (b) identify some issues in multilateral economic agreements in trade, finance, investments, and intellectual property rights that affect environmental sustainability, (c) identify and review priority policy issues affecting multilateral economic agreements on environment issues.

INTRODUCTION

Globalization is a multifaceted eco-politico-social phenomenon that tends to be a centre of controversy due to its nature of processes and output in the host countries. Whilst looking in terms of its actual benefits incurring to the host economies, the concept of globalization does become debatable and contested by the stakeholders. This may be its due to the complexity in its operation and processes being involved against the background of the chaotic nature of the host developing countries in regards to the sustainability of physical environment. The question generally arises whether the globalisation process serves the interest of all or only just benefiting few countries or corporates

DOI: 10.4018/978-1-4666-8222-1.ch010

/ group of individuals. This may be mainly due to the nature of definition of the concept of 'globalization' and while viewing across the various disciplines since each of this discipline offers a varying definition and interpretations of globalization process. Whilst in the past, the process of the globalization and its impact have been researched in the context of various empirical case studies but the definition of the globalization in these context tends to appear vague, elusive, and many times contradictory to its generic meaning. As Ardic (2009) comments that the term "globalization" is generally used to refer to a 'single phenomenon' but in reality it may not be a single phenomenon but rather a kind of unified process involving various stakeholders. Sometimes it becomes hard and there seems to be no 'universal' accepted definition describing the globalization process but as a matter of fact this process results and involves a high degree of interconnectivity, integration, interdependence and openness features stand out as used interchangeably by all the scholars in an attempt to give meaning to the concept (Mimiko, 2010). Take for instance, the arguments put forward towards the definition of globalization by De and Pal (2011) that the globalization can be defined as the process of opening up the host economies to the outside world in order to facilitate trade, reduction in physical and other barriers and to increase its mobility of goods and factors of production (land, labour and capital) in the host country. There is no doubt that the globalization process eventually creates a process of integration of economies through economic, social and political processes. But the lack of an essential definition may have to be contributed, at least partially, to keep globalization as a highly contested subject (Acosta & Gonzalez, 2010). Whilst keeping the pace with the main title of the book being on "sustainability practices through energy engineering and asset management" I did attempted to explore the process of globalization process and its sustainability of physical environment of the host country.

On the other hand while looking at the massive trade liberalization being witnessed and experienced by the world economy due to the process of globalization process. And followed and re-enforced by the complex global financial integration of the countries has created a kind of inter-dependency among the Nation States. No doubt this process of interdependency does have huge impact on the physical environment, on culture, on political systems, on economic development and prosperity, and on human physical wellbeing in the societies around the World (Kefela, 2011). Whilst looking from the environmental sustainability, the process of globalization has made huge impact in the host developing countries due to its chaotic nature of political economy and resulting urban environmental destruction. Take for instance, the case of the environmental deterioration seen in the urban city, towns, and the growth of proliferation of the ghetto type 'living environment squatters' generally are being associated with the haphazard growth pattern of industrialization of the developing countries. The working industrial section of the society have no other option but to live in close quarters sharing common sanitation and toilet facilities, Further there is no proper waste management planning facilities due to the defunct nature of civil amenities in general in the developing countries. The surroundings physical environments are being highly polluted and drainage, ponds and stream being choked with the day to day waste of humans as well the industrial toxic chemicals. This physical environment affects the residents in general and in particular in terms of health and living conditions around the manufacturing units induced by the globalization process. One wonders whether these scenarios is to do with the anthropogenic factors alone which further add up and plays a major role in the urban megacities in the developing countries effecting the natural environment but also presents risks to highly concentrated population which has potentials to trigger floods, mudslides, tsunamis, earthquakes and draught

(United Nations, 2009b) Further these snapshots of the urban physical environment deterioration are linked to the development process linked to the globalization process in the host developing countries. But one must also remember that this may be due to the economic development and planning swayed due to the institutional corruption at different levels within the host developing countries. As a matter of fact, the erratic haphazard compulsive nature of the developing countries rushing towards to integrate with the rest of the world economy through globalizations results and strains the fragile existing environment. While recalling the arguments put forward by the various academicians / policy makers / politicians and other experts who argue that the globalization process which does offers both opportunities and risks especially to the developing countries unless proper functional domestic policies are in place. And in case of the host developing countries it is mainly the risks being involved due to their poor management of their physical environmental assets towards sustainability.

On one hand while, exploring the beneficial solutions in the developing countries in terms of resolving the environmental issues, one is always reminded about the existence of the harsh 'economic rationality' prevailing which compels them to get involved into the globalization process. This economic rationality results in pushing the environmental concerns at the backburner of the priority agenda of the developing country. Whilst realizing the fact that the globalization does have potentials for creating many economic advantages and will eventually benefit to the society. But the reality seems that not every country or society has been benefited from its optimal attainment of economic welfare and wellbeing after being involved in the globalization. This results in creation of huge gaps between the developing and developed countries in terms of economic growth and human development and the standard of living. Whilst looking in the case of the rising economic powers such as China, India, Brazil, Mexico there is no doubt that millions have been lifted out of poverty cycle, but often incurring at a high per environmental cost in terms of their health and wellbeing. Looking at the United Nation Environmental Programme report which states that "The economic growth of recent decades has been accomplished mainly through drawing down natural resources, without allowing stocks to regenerate, and through allowing widespread ecosystem degradation and loss" (United Nations Environment Programme (UNEP), 2011, p.10). Further not the denying the fact that the strong economic growth in many developing countries, their level of Gross Development Product (GDP) per capita increased substantially, particularly in the last decade (80% since 1992, 45% since 2002) was due to industrial growth and development.

Due to the limitation of the study in this chapter, I did attempt to explore some of the specific channels through which such impacts of globalization process are transmitted. Whilst looking at the background of the current growing demand for large volume of economic activities witnessed around the world which results in the rise the aggregate level of natural resource utilization towards its depletion. Further, the inherent lack of political willingness witnessed in the host developing countries towards formulating appropriate domestic environmental policies and its feeble enforcement does exacerbate and degrades the physical environment. Whilst looking from the technological perspective, the developing countries do not have access to the current technology nor capital nor labour (skilled) to contain the sustainability of environment. And in case to purchase the technological knowhow or transfer of technology, the developing host countries are reluctant since these technological knowhow have huge price tag on them. To make the matter worse, the growth of high population witnessed in the developing countries results the tendency to encroach its natural resources (forest, pasture lands, marshy lands, public lands) resulting in the exploitation of physical natural environment like (forests, land, water bodies, mangroves) . Whilst looking from trade perspective, the globalized trade regimes in the developing countries do facilitates the diffusion of products, technologies and processes across borders but eventually generates huge industrial waste such as toxic chemicals, hazardous waste, endangered species and disease bearing pests which has to be regulated by robust domestic environmental policies or multilateral trade agreements. In this chapter, I have extensively used the notion of 'developed and developing countries' dichotomy since it has no doubt as UNDESA (2009) states that has been reflected on the developmental process of various socio-economic and political domains like the education and health care sectors as well as on the environment itself and its sustainability.

REVIEW OF LITERATURE

The impact of the globalization process and the sustainability of the socio-cultural and economic factors within the countries in the past have been argued from a wide spectrum of interpretations in social sciences research. Take for instance, that some of the authors concentrate specifically on the economic factors like Leitao (2011), Rao, Tamazian and Vadlamannati (2011), and Rao and Vadlamannati (2011) and whilst others scholars do address the impacts of the globalization process in the context of political issues (Bjornskov & Potrafke, 2011; Chang & Berdiev, 2011; Dreher, Sturm & Vreeland, 2009). Many other scholars did conceptualize and investigate many other various aspects such as human development and growth like De and Pal (2011), human welfare (Tsai, 2007) and life satisfaction, water pollution (Lamla, 2009), military interstate disputes (Choi, 2010). Whilst looking at the literature in terms of in-depth research being conducted and sponsored by the United Nations (UN) and its centres on the impact and effects of the globalization in terms of development and sustainability is very useful.

Apart from the above literature, there exists a wide array of case studies on the other aspects of globalization at macro and micro level done in the academia circle. Whilst some research interpretations are driven from the neo-liberal economic theories whilst other from the socialist perspectives. Keeping aside the literature on the globalization process, the reality is that the developing host countries adopt a more open arm towards globalization process. As Obadan (2010) argues that the countries which tend to enjoy higher growth rates than those that close their economies to international trade are generally globalized

The fact being that the host developing countries do realize that being market friendly will eventually initiate trade being vital 'integral part and engine' of the globalization process. And the attempt to open the domestic markets through trade liberalization process along with structural adjustments, the host country tends to integrate to the world economies system. There exists option for the infusion of new technology to the domestic economy which enhances a better productive capacity for domestic manufacturing industries through technical collaboration. As Akinmulegun (2011) argues that the opening an economy to foreign competition may stimulate efficiency in domestic production especially in the developing countries. The basic arguments in the above literature on globalization concludes that being market friendly and opening the economy will initiate the domestic producers to strive and survive in the competitive world market in comparative terms. Whilst the literature from economic perspective argues that the current globalisation process may be regarded as a new phase of world integration with increased density and frequency of international or global social interactions relative to local or national ones. In these new dynamics, Nation States are influenced by transnational processes occurring on multiple economic, political and cultural levels (Walby, 2009). And the current integration of the world economies through the initiation of the globalization has resulted in the phenomenal growth of trade between the countries. And this may be strongly linked to increasing population numbers and the need for shelter, food and an improved standard of living towards economic sustainability (UNTAD, 2012c)

Globalisation and Environment and Its Sustainability in the Host Developing Countries

Whilst the main features of the globalization process is to initiate the economic integration within the countries with the infusion and transmission bodies of knowledge, technology and labour. Not denying the fact that the globalization process has helped the host developing countries to liberalize its economics by creating a global market place in which all nations must participate directly or indirectly. This undoubtedly led to growing web of economic activities and clout and power of international financial investors mainly driven by the Multinational Corporations (MNCs) (Jaja, 2010). Generally, the scholars do focus the globalization process and its impact on the economic dimensions in terms of employment, transfer of technology, Gross Domestic product etc. But the globalization processes also exacerbates environmental problems which are not taken seriously or addressed adequately. The negative effects of globalization seem to be more prevalent and noticed in the developing countries due to its inherent institutional issues. And the attempts to integrate with rest of the world inevitable results in 'cascading consequences' towards the deterioration of the fragile natural physical environment. Whilst looking at the current scenario of 'Integration of Economies' initiated through the trade agreements at the bi-lateral and multi-lateral level, foreign direct investment, short-term capital flows and the International movement of workers (skilled and unskilled) across the countries demands for a plethora sustainability of policies to contain the situation. Take for instance; one may notice that many regions in Asia may have got quickly integrated into the global economy. But in the case of African continent, the story of globalization and its impact depicts a different story apart from existence of high level of political instability, poverty and endemic corruption plaguing the region. There are many showcases of the Africa region and its apathy towards the environmental degradation and the constant intra conflicts between the countries make the environmental sustainability worse. While the interdependence on the global economy and inter-transfers and flows of trade, capital and investment between and among countries of the world is expected to have a far reaching effects on the economic growth of a participating nation through its multiplier and accelerator effect. But despite the over whelming theoretical evidence of causality between globalization and economic growth, interpretations empirical studies done in the past are divergent. But the scholars in the past reiterate that the globalization process brings along with its benefits as well as pains in terms of socio-cultural, economic, health and or political (Steger, 2009) and these pains experienced by the developing host countries are illustrated below:

Environmental Debt

During the industrialization of the developed countries last few centuries, for instance, the excessive use of natural resources was exploited like coal, wood, fossil oil especially in Europe. The realization that the excessive utilization of resources being limited and not sustainable in the long run in Europe, triggered colonization of the developing countries by force during the last centuries. Thus the depletion of physical environmental resources in the developed countries known as 'ecological debt' was merely the outcome of the disequilibrium use of physical resources between the countries: the majority who over-exploited the global resources are the developed countries who owe an ecological debt to those in possession of resources being the developing countries. Whilst looking at the ecological debt in the developing countries, which were not using even a small portion of their legitimate share of the global physical resources due to lack of technology knowhow and political instability was insignificant in comparative terms with developed countries.

1. The resultant outcome being that the 'environmental debt' being accumulated by the developed countries currently is transferred to the developing countries by the method of involving the developing host countries through international trade which is trigged by globalization. Currently, the developing countries indiscriminately exported agricultural products and other consumer goods by the means of exploitation of their natural resources. Further, we are aware historically over the centuries, and
2. The developed countries have polluted the atmosphere and the land mass without limits during their industrialization process and colonization of the developing countries in the past which again is the ecological debt.

And in order to address the environmental debt at global level the developing countries led by the initial multi-lateral discussions on the ecological debt took place around 1990 by the initiative from Latin American Non-Governmental Organisations (NGOs) and then backed by Friends of the Earth International in the developing countries. And in 1992, during the Rio Summit, the idea of a 'Debt Treaty' was proposed, which introduced the notion of an ecological debt in contraposition to the external debt. But the developed countries opposed the proposal and the notion of debt treaty. Another global attempt was made in 2009 by the Centre for Sustainable Development (CDO) at Ghent University (Paredis, Goeminne, Vanhove, Maes & Lambrecht, 2009) proposed as a working definition:

1. The ecological damage caused over time by a country in other countries or to ecosystems beyond national jurisdiction through its production and consumption patterns;
2. The exploitation or use of ecosystems (and its goods and services) over time by a country at the expense of the equitable rights to these ecosystems by other countries.

The ecological debt concept focuses on the lack of political power of poor regions and countries. Further the debt arises from:

1. Exports of raw materials and other products from relatively poor countries or regions being sold at prices which do not include compensation for local or global externalities;
2. Rich countries or regions making disproportionate use of environmental space or services without payment (for instance, to dump carbon dioxide).

The notion of the ecological debt raises some difficult political and ethical questions to the countries concerned in terms of injustice caused to the present generation that have to pay for the debts of past generations. Theoretically, it may be possible to put a money value on ecological debt by calculating the value of the environmental and social externalities associated with historic resource extraction and adding an estimated value for the share of global pollution problems borne by poor countries as the result of higher consumption levels in rich ones. This includes efforts to value the external costs associated with climate change. Such monetary accounts Goemmine and Paredis (2010) argue might be useful to address the ecological debt campaigners from civil society. The ecological debt concept, therefore, casts a new light on our understanding of 'sustainable development', not just by adding a historical dimension but by bringing power and justice to centre stage, to reveal control over resources and pollution burdens as an issue of power relations. The concept of the ecological debt has the potential to help the implementation of sustainability and to fight environmental injustices. Similarly, whilst exploring the fallout of the globalisation process in the South developing countries, there seems to be a kind of transfer of 'environmental currency' from developed North to Developing South countries known as 'Ecological footprints' (A measure of the impact humans have on the environment is called an ecological footprint.). The environment is now gaining major significance and being a 'common heritage of mankind'. There are deliberate attempt at the global level through Multi-lateral agreements to address the environmental problems due to its cross-border effects and being impossible for just one country to deal with the issues.

Deforestation/Export of Cash Crops

Whilst looking at the five factors that strain the biodiversity of the planet earth namely

1. Loss namely the habitat change,
2. Overexploitation,
3. Pollution,
4. Invasive alien species, and
5. Climate change (CBD 2010).

The deforestation for the export of the corps through displacement, destruction and removal of ecosystems like the mangrove, marshy lands, forest to accommodate the growing population poses a big challenge for mankind. For instance, the massive deforestation has been occurring in worldwide with the logging industries being fuelled by the need for wood products. On the other hand the commercialization of agriculture and preferring the growth of cash crops instead of traditional crops in the developing countries amongst smallholders and big holders being initiated during the colonial times onwards in the developing countries. For instance, in the case of the African continent, the cash crops products were groundnuts in present day Senegal, the Gambia and Guinea Bissau and palm oil in Sierra Leone, south -eastern Nigeria and northern Cameroon (Austin, 2009). In the Latin America the growth of cash crops like (banana, coco, cotton, groundnuts, soya etc.). In the case of Brazil, twenty years henceforth, much of the forest will be cleared and replaced by soy fields and cattle pastures. Further followed by severe droughts during the 2005 and 2010 increased the frequency of fire, and have reinforced concerns that the Amazon is reaching a tipping point where large areas of forest could be replaced by a more savannah-like ecosystem (Lewis, 2011; Malhi et al., 2009).

Another instance, to meet the demand of growing global interest in renewable energy has led to a tripling of global biofuel production during the period 2000-2007 and it is expected to double again in 2007 – 2011 (Molony & Smith, 2010). The Biofuels have been produced for decades in sub-Saharan Africa, e.g. jathropa crop grown in Mali and Tanzania, although it has not been considered highly profitable until recently. A growing number of African countries, e.g. Malawi, Mali, Mauritius, Nigeria, Senegal, South Africa, Zambia and Zimbabwe are now enacting new pro-biofuel national strategies and the number of joint ventures with other countries and private enterprises is growing (Molony & Smith 2010). These crops are largely being raised for the production of ethanol (sugar cane), export for livestock fodder (soybeans) and ingredients for food and drug products or other biofuel production (palm oil). In many or most cases they are grown on industrial-scale farms established by the clear-cutting or burning of vast areas of tropical forest (International Tropical Timber Organization) (ITTO, 2011); Looking from the sustainability of the environment, however; more information on biofuels is raising new concerns about their production and use. Among these are direct environmental and social impacts from land-clearing and conversion, the introduction of potentially invasive species, the overuse of water, along with related consequences for the global food market. A major reason for concern is the trend of numerous wealthy countries to buy or contract for land in other, typically developing and sometimes semi-arid countries, in order to produce food and often biofuels. This trend may have potentially serious impacts on fossil and renewable water resources, as well as the local food security. Biodiversity in the tropics is dramatically declining, by 30% since 1992, indicating the ecosystem's severe degradation due to high deforestation rates of primary forest and transformation into agricultural land and pasture (World Wildlife Fund, 2010).

Air Pollution

While over half of the world population now lives in urban areas, they also account for 75% of global energy consumption (United Nations-Habitat, 2009) and 80% of global carbon emissions at least when viewed from a consumption perspective (Satterthwaite, 2008). On the other hand, the top 25 cities in the developed countries in the world create more than half of the world's wealth. This on-going rapid urbanisation indicates that long-term investments addressing the associated vulnerabilities are critically needed. "The urgency is acute considering that 30-50% of the entire population of cities in developing countries live in settlements that have been developed in environmentally fragile areas, vulnerable to flooding or other adverse climate conditions, and where the quality of housing is poor and basic services are lacking" (UNDESA, 2009b). Industrial facilities release greenhouse gas (GHG) emissions, particulate matter, sulphur dioxide, nitrogen dioxide, lead and chemicals. These emission of toxic air not only accelerate the climate change and but also intensifies the atmospheric pollution, but they also degrade ecosystems and cause health risks. For instance, the Manufacturing industries account for up to 17 per cent of air pollution-related health problems. Pollution also has an economic effect and incurs economic costs: Estimates of gross air pollution damage range from 1 to 5 per cent of global gross domestic product (GDP) (UNEP, 2011a). Take for instance the case of India, as Lim et al. (2013) study sums up that the Indian urban cities today are among the most polluted in the world and it is estimated that outdoor air pollution leads to approximately 670,000 deaths annually. This may be due to the globalization process involving host countries towards the contribution of the air pollution.

E-Electronic Waste

Another accusation of the developing countries is that the developed countries exploit the environmental regulatory regimes in the former colonies. This may be due to the different factors where in the developed countries waste treatment has become more stringent and the overhead costs of domestic safe disposal increased. There is tendency of the developing countries to allow the developed countries to dump their industrial as well other toxic waste. Take for instance, the clandestine operations being carried towards the 'toxic disposal' waste produced in the developed countries towards the 'willingly recipient developing countries'. Most of the amounts of e-waste that are still being imported to these countries, both legally and illegally, it is evident that the problem is exploding, with many dangers for human health and the environment (McCarthy, 2010). The multi- lateral agreement known as Basel Convention controls the Trans boundary movement of hazardous wastes and their disposal, and is the most significant multilateral environmental agreements (MEAs) in relation to tackling the issues surrounding e-waste and its management. And as on September 2010, the Convention had 178 signatories (Basel Convention) (b). However, the US, being a major actor, has not yet ratified Basel convention. Take for instance the case of the Global output in the chemicals industry has grown from US$ 170 billion in 1970 to over US$ 4.1 trillion today, with a steady shift in the production, use and disposal of chemical products from developed countries to emerging and developing economies, where safeguards and regulations are often limited. Poisonings from industrial and agricultural chemicals are among the top five leading causes of death worldwide, contributing to over 1 million deaths annually and 14 million Disability Adjusted Life Years (UNEP, 2013) which poses big challenge towards sustainability of environment caused by globalization process.

Another challenge the developing host countries faces due to the globalization process in the consumption of the electronic goods creates has created huge stock pile of electronic waste known as e-waste due to mass production and new technology driven gadgets. 'Electronic waste' or 'e-waste' may be defined as all secondary computers, entertainment device electronics, mobile phones, and other items such as television sets and refrigerators, whether sold, donated, or discarded by their original owners. This definition includes used electronics which are destined for reuse, resale, salvage, recycling, or disposal (Omatek Ventures, 2011). This e-waste need to be properly disposed but finds its way to the developing countries for its recycling of its parts and extraction of the precious metals .It has adverse environmental and health implications as developing countries face economic challenges and lack the infrastructure for sound hazardous waste management, including recycling, or effective regulatory frameworks for hazardous waste management (Strategic Approach to International Chemicals Management, 2009). This is because of the fact being that most of the South (developing host countries) that do not have robust regulatory environmental policies in place to deal with the disposal of the electronic waste which are high in toxic content. For instance, the challenge of globalization in Nigeria, China and India is proper method of disposal of the electronic waste (e-waste) which adds to the sustainability of their environment.

Environmental Sustainability

Whilst looking from the Race-to-the-bottom" theory in regards to the sustainability of environment. The theory race to bottom states that pollution-intensive economic activity will tend to migrate to those jurisdictions where costs related to environmental regulation are lowest (Lepawsky & McNabb, 2010) There is apparent lax attitude towards the enforcement of regulations or having no appropriate regulations in place in the developing host countries. This may be directly linked to the "race-to-the-bottom" theory since the high polluting intensive goods tend to migrate to countries with weaker environmental regulations. As a resultant the polluting industries concentrate in developing countries with low environmental standards having lax environmental regimes. Take for instance, the case of India, despite the fact of having environmental legislations, there is no specific laws or guidelines for e-waste disposal of management. The regulations banning the importation of hazardous waste for disposal are weak and imported e-waste still finds its way into the country. Furthermore, although various laws regulating areas such as importation, the environment, labour and factories could have an impact upon e-waste in India, virtually none apply to the informal sector. The provision of environmental protection is delegated among India's various states, also, and this is known to give rise to lax enforcement. The provision of environmental protection is delegated among India's various states, also, and this is known to give rise to lax enforcement. In addition to a lack of funds and government capacity for enforcement, this situation also originates in low awareness of issues, low literacy levels, poverty, highly bureaucratic structures and attitudinal issues (Sinha, Mahesh, Donders & Van Breusegem, 2011).

Some of the Initiatives Influencing Globalization and Sustainability of Environment: Multilateral Agreements, Finance, Investments, and Intellectual Property Rights

Multilateral Environmental Agreements (MEAs) Concerning the Developing Countries

In 1992, the first United Nations Conference on Sustainable Development, otherwise known as the Rio Earth Summit, was convened in Rio de Janeiro by the member countries and Brazil was given the privilege to address the state of the environment and sustainable development during the deliberations of the conference. There was no doubt that the Earth Summit was the result due to the initial several important agreements by the member countries of United Nations and one being the "Agenda 21", a plan of action agreed and ratified by 178 member countries in order to address human impact and effects on the natural physical environment. This was a culmination of the agreements being carried at local, national and global levels, and review of key treaties on climate change, desertification and biodiversity. Further this was followed by the second multilateral International conference held at Johannesburg in 2002 known as 'World Summit on Sustainable Development' and the respective governments agreed on the Johannesburg Plan of Implementation, reaffirming their commitment to Agenda 21. Agenda 21 was a kind of global plan towards achieving sustainable development at local, national and global level which recognises the essential importance of the participation of the member countries. Whilst during 2012, the Nations Conference on Sustainable Development, or Rio+20 Earth Summit focused on the 'Green Economy' in the context of sustainable development, poverty eradication, and the institutional framework for sustainable development.

The main objectives figured were to renew the political initiative and commitment towards the sustainable development, review progress and identify implementation gaps, and address new and emerging challenges (UNEP, 2011b) As Pring and Pring (2009) argues that the awareness towards the environmental laws generates demand for new and expanded environmental and resource management agencies to administer and enforce the environmental laws among member countries. Further the enforcement of environmental law, including by environmental litigation, generates a demand for courts and tribunals with specialist environmental expertise and jurisdiction. And the decisions of these environmental courts and tribunals in turn develop environmental jurisprudence (Preston, 2012) towards the sustainability of environment.

There has been a growth in international institutions responsible for administering MEAs, including the United Nation's specialised agencies and programs such as the United Nations Environment Programme (UNEP'), the United Nations Development Programme (UNDP') and the Food and Agriculture Organisation of the United Nations (FAO'), which have played a leading role in setting law-making agendas, providing negotiating forums and expertise, as well as assisting countries in implementing international environmental law and in capacity building. Regional bodies, such as the United Nations Economic Commission for Europe (UNECE') and the European Union (EU') have contributed significantly to the development and implementation of MEAs. Transnational corporations, industry associations and development banks, such as the World Bank and the Asian Development Bank, have helped drive the development of common standards and some nascent environmentally sound business practices throughout the developing world, particularly with respect to environmental impact assessment. The 1992 Rio Conference on Environment and Development (UNCED) and the negotiation of the Conventions on Biological Diversity, Climate Change13 and Ozone Depletion had unprecedented numbers of country participants. First, there has been an increase in the number of MEAs, as well as an increase in the scope of issues and topics addressed by international environmental law (Anton, 2013).

Markets for Organic Products and Eco-Labelling

There appeared attempts in terms of investment in the environmental friendly goods and services by the countries. And this may be the fact that the products being manufactured around the world has increased the awareness and concern about environmental issues and the need to address them with some working rational solutions. As Birnie, Boyle and Redgwell (2009) argues that this may be partly a product of the use of consensus negotiating procedures, which have a greater potential for securing general acceptance of negotiated texts. Take for instance, the consumer demand for goods that are produced in a sustainable way has boosted the tag of certification and eco-labelling, like the Forest Stewardship Council (FSC) and the Programme for the Endorsement of Forest Certification (PEFC) for forest products, the Marine Stewardship Council (MSC) for fish products, and 'bio' or 'organic labels' for many agricultural products including coffee, tea and dairy products. Of course there may appears different questions regarding the labelling of the product, like 'what does a label measure?' may generate different interpretations. Further as Saunders (2009) argues that competition between label types provides solutions to a variety of needs that exist in the market? While a consolidation of Eco labels is likely to occur, there will remain a variety to address the diversity of consumer concerns. Rainforest Alliance Certified differs from the USDA Organic certification in that it allows for some limited uses of agrochemicals, requiring continual reduction of agrochemical use and encouraging the use of biological alternatives whenever possible. It also has standards for wildlife conservation and worker welfare, which the USDA Organic labels do not (Rainforest Alliance, n.d.).

Intellectual Property Rights

Whilst exploring the environmental regulations in the developing countries as well at the on-going multi-lateral conventions on environment sustainability, the sections and clauses have highlighted restricting the use of harmful technology but do they do not propose the options for the host developing countries to buy innovative beneficial technology towards sustainability. No doubt the intellectual property laws are well entrenched in the system for promoting invention and facilitating commercial development but this seems to occur only in the Research and Development (RD) laboratories situated in the developed countries. And the accessibility to this technology know how is very expensive for the host countries to purchase. The fact being that the application of intellectual property rights, they play a vital catalyst role towards invention and innovation of environmental technology. But the Intellectual property law can be applied to improve environmental sustainability in several ways which will benefit the whole of the planet. But in reality this does not occur, take for instance the existing intellectual property rights which do not generally distinguish between environmentally harmful or beneficial technology. Further most of the environmental laws do not consider generally 'intellectual property' incentives into account whilst seeking to protect the environment sustainability. Historically there seems to be no voluntary move to transfer the environmental efficient technology to the South through the multi-lateral agreements on environment. Take for instance in the case of Exxon Valdez oil spill initiated growing pressure on industry to audit environmental compliance and impact of the environmental damage towards environmental audit in assessing the risks created by Exxon Valdez Company to audit compliance and impact. The number as well as the total quantity of oil from accidental oil spills from tankers (including combined carriers and barges) has decreased significantly since 1992. Although the vast majority of spills are relatively small (i.e., less than seven tonnes) (International Tanker Owners Pollution Federation (ITOPF), n.d.), the accumulated amount is nearly one million tonnes since 1992.The fact being that most of the spills from tankers result from routine operations such as loading, discharging and bunkering which normally occur in ports or at oil terminals. Thus the current environmental technology management lacks the criteria for technology audit to be considered in the multilateral as well domestic agreements.

Carbon Trading and Other Environmental Market Tools

One of the major initiatives by the countries at the global level was placing a monetary value on greenhouse gas emissions and creating a market for trade in carbon is a new and increasingly utilized concept to address climate change. Further other initiatives relating to the market frameworks which include biodiversity offset and compensation programs, habitat credit trading and conservation banking, with a goal toward reducing biodiversity loss and mainstreaming impacts into economic decisions. This has resulted for instance, at least 45 compensatory mitigation programs and more than 1100 mitigation banks now exist (UNDP, 2011) to address the climatic change. Latest estimates show that global CO_2 emissions accumulated to 30 600 million tonnes in 2010 (International Energy Agency, 2011). But on the other hand the prospect of limiting the Global Increase in Temperature to 2°C is seems to be bleak. Since whilst we see at the gravity of the situation of the global increase in temperature, the increased air travel and goods transport is the additional emissions of CO_2 as well as particulates, nitrogen oxides (NOx) and water vapour, which can have more than twice the warming effect of the carbon dioxide alone (Environmental Transport Agency, n.d.). Nevertheless large differences exist between regions and countries in terms of emissions, for instance, around 80% of the global CO_2 emissions being generated by 19 countries mainly those with high levels of economic development which are the developed countries and large density populations like china and India. Whilst the total emissions of CO_2 in developed countries increased by nearly 8%, and although per capita emissions declined steadily by 18%, they are still 10 times higher than those of developing countries. In addition, many developed countries profited from a significant shift of production to developing countries, thus leading to declining domestic emissions, but nevertheless increasing consumption-based emissions (Peters, Minx, Weber, & Edenhofer, 2011)

RECOMMENDATION AND SOLUTIONS

The developing host countries do have to more prioritize its economic growth and development with proper enforcement of domestic policies towards environmental sustainability.

1. The better access to more efficient technologies at affordable price.
2. To better manage and regulate the environmental governance through enforcement of multi-lateral agreements and domestic environmental rules.
3. The Green economy initiative may be viable option and a pathways for fundamentally shifting economic development to be a low-carbon, climate resilient, resource efficient, and socially inclusive, as well as for valuing ecosystem services, are now being proposed widely and increasingly pursued. The degree of flexibility inherent in the formulation and application of green economy policies is expressed through the consideration that

such policies provide options for policymaking but should not be a rigid set of rules rather acknowledge the need for supporting developing countries through technical and technological assistance.
4. At the same time, it remains crucial for developing countries to build or reinforce R&D capacity, particularly related to environmentally sound technologies. In addition, technological advancements and spill overs facilitated by international trade can lead to further specialisation in the production of more energy- and resource-efficient goods and services. What economies worldwide need is absolute decoupling of the environmental pressure associated with resource consumption from economic growth. This will be easier to achieve to the extent that resource use itself becomes more efficient" during their trade initial agreements (WTO-UNEP, 2009).
5. As far the part of sustainable of the trade practices: Having the 'trade in certified products' or 'environmental goods and services' creates awareness before its consumption. This may prove to be an important move towards mainstreaming sustainability in production and trade at the global level. Further, trade, when accompanied by appropriate regulation, can facilitate the transition to a green economy by fostering the exchange of environmentally friendly goods and services (including environmentally sound technologies) and by increasing resource efficiency and generating economic opportunities and employment.

CONCLUSION

Finally to sum up the globalization process and its impact on the sustainability in the host developing countries, one may have to consider overall nature of the political economy of the concerned developing country in terms of its political and economic stability. Further one may also have to consider towards the formulation of the robust domestic environmental policies and the willingness to ratify the multi-lateral policies towards sustainability by the developing country taking in account the Sovereignty of the member country. The role of the developed countries does play vital part towards formulation of the multilateral agreements in trade and environment and can eventually yield better results towards environmental protection. The developed countries are better positioned to make huge difference during the multilateral agreements through participation whole heartedly. They must exhibit a more positive attitude and being more realistic in negotiations with the developing countries concerns and issues and be willing to share some of the technological knowhow towards sustainability. The fact being that the developed countries cannot ignore that they were the initial perpetuators of the deterioration of the environment at the early stages of industrialization and are currently involved in huge per capita consumption of goods and services affecting the environmental sustainability. The developing countries do have the capacity of being immense economic powers and must be willingly to share those technologies which will enhance environmental sustainability. There must be more cohesive integrated environmental and trade policies agreements towards better sustainability at the domestic as well at the International level against the background of globalization. And this integration may take the form of institutionalization of environmental issues in future bilateral, multilateral and regional trade agreements also the private and public sector and various other Non-Governmental Organisations (NGOs).

The major challenges that the host developing countries confront is to manage the process of globalization in such a way that it promotes environmental sustainability and equitable human development. Globalization process must accommodate the interest of both people and the environment as part of social responsibility. There is no doubt that the Globalization process changes the "balance of power" between markets, national governments and international collective action. But it also enhances the influence of domestic markets on economic, social, and environmental outcomes and reduces the degree of freedom and unilateral management capabilities of Nation State. There appears to be a necessity for developed as well developing countries to cooperate both in the management of the global commons and the coordination of domestic policies and multi-lateral agreements. Further realizing the generic nature of globalization process which do creates market driven political pressures to gain or maintain competitiveness and this forces premature and not necessarily appropriate convergence of environmental policy at the multi-lateral and domestic level. Whilst in the presence of diversity of environmental endowments, assimilative capacities and preferences efficient environmental management requires sensitivity to local ecological and social conditions which cannot be ignored towards having viable environmental sustainability.

REFERENCES

Acosta, O., & Gonzalez, J. (2010). A thermodynamic approach for emergence of globalization. In K. Deng (Ed.), *Globalization: Today, tomorrow* (pp. 38–49). Sciyo. doi:10.5772/10223

Akinmulegun, S. O. (2011). *Globalization, FDI and economic growth in Nigeria 1986-2009.* (Unpublished doctoral thesis). Adekunle Ajasin University.

Alliance, R. (n.d.). *Standards for sustainable agriculture.* Retrieved from http://www.rainforest-alliance.org/agriculture.cfm?id=san

Anton, D. K. (2013). Treaty congestion in contemporary international environmental law. In S. Alam, J. Bhuiyan, T. Chowdhury, & E. Techera (Eds.), *Routledge handbook of international environmental law* (pp. 651–665). London: Routledge.

Ardıc, N. (2009). Friend or foe? Globalization and Turkey at the turn of the 21st century. *Journal of Economic and Social Research, 11*(1), 17–42.

Austin, G. (2009). Cash crops and freedom: Export agriculture and the decline of slavery in colonial West Africa. *International Review of Social History, 54*(1), 1–37. doi:10.1017/S0020859009000017

Birnie, P., Boyle, A., & Redgwell, C. (2009). *International law and the environment* (3rd ed.). Oxford, UK: Oxford University Press.

Bjornskov, C., & Potrafke, N. (2011). Politics and privatization in Central and Eastern Europe: A panel data analysis. *Economics of Transition, 19*(2), 201–230. doi:10.1111/j.1468-0351.2010.00404.x

Chang, C. P., & Berdiev, A. N. (2011). The political economy of energy regulation in OECD countries. *Energy Economics, 33*(5), 816–825. doi:10.1016/j.eneco.2011.06.001

Choi, S. (2010). Beyond Kantian liberalism: Peace through globalization. *Conflict Management and Peace Science, 27*(3), 272–295. doi:10.1177/0738894210366513

De, U. K., & Pal, M. (2011). Dimensions of globalization and their effects on economic growth and human development index. *Asian Economic and Financial Review, 1*(1), 1–13.

Dreher, A., Sturm, J., & Vreeland, J. R. (2009). Development aid and international politics: Does membership on the UN Security Council influence World Bank decisions? *Journal of Development Economics, 88*(1), 1–18. doi:10.1016/j.jdeveco.2008.02.003

Environmental Transport Association. (n.d.). *Air travel's impact on climate change*. Retrieved from http://www.eta.co.uk/env_info/air_travel_climate_change

Goeminne, G., & Paredis, E. (2010). The concept of ecological debt: Some steps towards an enriched sustainability paradigm. *Environment, Development and Sustainability, 12*(5), 691–712. doi:10.1007/s10668-009-9219-y

International Energy Agency. (2011). *Prospect of limiting the global increase in temperature to 2°C is getting bleaker*. Retrieved from http://www.iea.org/newsroomandevents/news/2011/may/name,19839,en.html

International Tanker Owners Pollution Federation. (n.d.). *Oil tanker statistics 2014*. Retrieved from http://www.itopf.com/information-services/data-and-statistics/statistics/

International Tropical Timber Organization. (2011). *Survey of world's embattled tropical forests reports 50% increase in areas under sustainable management since 2005*. Retrieved from https://www.google.co.nz/url?sa=t&rct=j&q=&esrc=s&source=web&cd=1&cad=rja&uact=8&ved=0CCAQFjAA&url=http%3A%2F%2Fwww.itto.int%2Fdirect%2Ftopics%2Ftopics_pdf_download%2Ftopics_id%3D2642%26no%3D4%26disp%3Dinline&ei=eA7lVMzMDuH2mQWHyYHoBw&usg=AFQjCNEKAP9h8pxnH3l7FKiJS9w1qVmcNA&sig2=r3Pu4JB2vnvNU0V0ghJl1A

Jaja, J. M. (2010). Globalization or Americanization: Implications for sub-Saharan Africa. In K. Deng (Ed.), *Globalization: Today, tomorrow* (pp. 138–140). Sciyo.

Kefela, G. T. (2011). Driving forces of globalization in emerging market economies developing countries. *Asian Economic and Financial Review, 1*(2), 83–94.

Lamla, M. (2009). Long-run determinants of pollution: A robustness analysis. *Ecological Economics, 69*(1), 135–144. doi:10.1016/j.ecolecon.2009.08.002

Leitao, N. C. (2011). Foreign direct investment: Localization and institutional determinants. *Management Research and Practice, 3*(2), 1–6.

Lepawsky, J., & McNabb, C. (2010). Mapping international flows of electronic waste. *Canadian Geographer, 54*(2), 177–195. doi:10.1111/j.1541-0064.2009.00279.x

Lewis, A. (2011). *Europe breaking electronic waste export ban*. Retrieved from http://www.bbc.co.uk/news/world-europe-10846395

Lim, S. S., Vos, T., Flaxman, A. D., Danaei, G., Shibuya, K., Adair-Rohani, H., & Blyth, F. et al. (2012). A comparative risk assessment of burden of disease and injury attributable to 67 risk factors and risk factor clusters in 21 regions, 1990–2010: A systematic analysis for the global burden of disease study 2010. *Lancet, 380*(9859), 2224–2260. doi:10.1016/S0140-6736(12)61766-8 PMID:23245609

Malhi, Y., Aragão, L., Galbraith, D., Huntingford, C., Fisher, R., Zelazowski, P., ... Meir, P. (2009). Exploring the likelihood and mechanism of a climate-change-induced dieback of the Amazon rainforest. *Proceedings of the National Academy of Sciences of the United States of America, 106*(49), 20610-20615. doi:10.1073/pnas.0804619106

McCarthy, M. (2010). *The big question: How big is the problem of electronic waste, and can it be tackled?* Retrieved from http://www.independent.co.uk/environment/green-living/the-big-question-how-big-is-the-problem-of-electronic-waste-and-can-it-be-tackled-1908335

Mimiko, N. O. (2010). *Swimming against the tide: Development challenge for the long-disadvantaged in a fundamentally skewed global system: An inaugural lecture delivered at the Oduduwa Hall, Obefemi Awolowo University, Ile-Ife, Nigeria, on Tuesday October 12, 2010*. Ile-Ife, Nigeria: Obafemi Awolowo University Press.

Molony, T., & Smith, J. (2010). Biofuels, food security, and Africa. *African Affairs, 109*(436), 489–498. doi:10.1093/afraf/adq019

Obadan, M. I. (2010). *Globalization: Concepts, opportunities and challenges for Nigeria development*. Paper presented at the Training Programme on Economic Reform.

Paredis, E., Goeminne, G., Vanhove, W., Maes, F., & Lambrecht, J. (2009). *The concept of ecological debt: Its meaning and applicability in international policy*. Gent, Belgium: Academia Press.

Peters, G., Minx, J., Weber, C., & Edenhofer, O. (2011). Growth in emission transfers via international trade from 1990 to 2008. *Proceedings of the National Academy of Sciences of the United States of America, 108*(21), 8903–8908. doi:10.1073/pnas.1006388108 PMID:21518879

Preston, B. J. (2012). Benefits of judicial specialization in environmental law: The land and environment court of New South Wales as a case study. *Pace Environmental Law Review, 29*(2), 396–440.

Pring, G., & Pring, C. (2009). *Greening justice: Creating and Improving environmental courts and tribunals*. Washington, DC: The Access Initiative.

Rao, B. B., Tamazian, A., & Vadlamannati, K. C. (2011). Growth effects of a comprehensive measure of globalization with country specific time series data. *Applied Economics, 43*(4-6), 551–568. doi:10.1080/00036840802534476

Rao, B. B., & Vadlamannati, K. C. (2011). Globalization and growth in the low income African countries with the extreme bounds analysis. *Economic Modelling, 28*(3), 795–805. doi:10.1016/j.econmod.2010.10.009

Satterthwaite, D. (2008). Cities' contribution to global warming: Notes on the allocation of greenhouse gas emissions. *Environment and Urbanization, 20*(2), 539–549. doi:10.1177/0956247808096127

Saunders, J. (2009). *The war over eco-labels*. Retrieved from http://www.greenbiz.com/blog/2009/04/28/war-over-eco-labels

Sinha, S., Mahesh, P., Donders, E., & Van Breusegem, W. (2011). *Waste electrical and electronic equipment: The EU and India: Sharing best practices*. Available: http://eeas.europa.eu/delegations/india/documents/eu_india/final_e_waste_book_en.pdf

Steger, M. (2009). *Globalization*. New York, NY: Sterling.

Strategic Approach to International Chemicals Management. (2009). *Background information in relation to the emerging policy issue of electronic waste*. Paper presented at the SAICM/ICCM.2/INF36 Disseminated at the International Conference on Chemicals Management, Geneva, Switzerland.

Tsai, M. (2007). Does globalization affect human well-being? *Social Indicators Research, 81*(1), 103–126. doi:10.1007/s11205-006-0017-8

UN. (2009b). *UN-DESA policy brief no. 25*. United Nations Department of Economic and Social Affairs. Accessed on November 1, 2014 at http://www.un.org/esa/analysis/policy briefs/policy brief 25.pdf

UNEP. (2009c). Making Tourism More Sustainable: A Guide for Policy Makers.

UNEP. (2011b). *Keeping Track of our Changing Environment: From Rio to Rio+20.* UNEP.

UNEP. (2011b). *A guidebook to the green economy issue 1: Green economy, green growth, and low-carbon development – History, definitions and a guide to recent publications division for sustainable development.* UNDESA. Retrieved from https://sustainabledevelopment.un.org/content/.../GE%20Guidebook.pdf

United Nations Department of Economic and Social Affairs. (2009b). *UN-DESA policy brief no. 25: The challenges of adapting to a warmer planet for urban growth and development.* Retrieved from http://www.un.org/en/development/desa/policy/publications/policy_briefs/policybrief25.pdf

United Nations Department of Economic and Social Affairs. (2009b). *UN-DESA policy brief no. 25: The challenges of adapting to a warmer planet for urban growth and development.* Retrieved from http://www.un.org/en/development/desa/policy/publications/policy_briefs/policybrief25.pdf

United Nations Development Programme. (2011). *Human development report 2010: The real wealth of nations: Pathways to human development: Summary.* Retrieved from http://hdr.undp.org/sites/default/files/hdr_2010_en_summary.pdf

United Nations Environment Programme. (2011a). Keeping track of our changing environment from Rio to Rio + 20 (1992-2012). New York, NY: Author.

United Nations Environment Programme. (2011b). Towards a green economy: Pathways to sustainable development and poverty eradication. New York, NY United Nations Department of Economic and Social Affairs.

United Nations Environment Programme Paris UNCTAD. (2012c). *Investment policy framework for sustainable development.* Retrieved from 12 November 2014 from http://unctad.org/en/PublicationsLibrary/webdiaepcb2012d6_en.pdf

United Nations Environment Programme. (2013). *Global chemicals outlook: Towards sound management of chemicals: Trends and changes.* Retrieved from http://www.unep.org/chemicalsandwaste/Portals/9/Mainstreaming/GCO/Rapport_GCO_calibri_greendot_20131211_web.pdf

United Nations Environment Programme. (n.d.). Available at: http://www.unep.org/geo/pdfs/keeping_track.pdf

United Nations Environment Programme. (2009a). World population ageing 2009. New York, NY: Author.

Ventures, O. (2011). *Understanding e-waste.* Retrieved from: http://lasepa.org/E-waste%20Day%201/understanding%20ewaste%20final%20of%20all.pdf

Walby, S. (2009). *Globalization & inequalities: Complexity and contested modernities.* London: Sage.

World Wildlife Fund. (2010). *Living planet report 2010: Biodiversity, biocapacity and development.* Gland, Switzerland: Author.

WTO-UNEP. (2009). *Trade and climate change.* WTO-UNEP Report. Available at: http://unfccc.int/files/adaptation/adverse_effects_and_response_measures_art_48/application/pdf/part_iv_trade_and_climate_change_report.pdf

KEY TERMS AND DEFINITIONS

Bio-Diversity: Biodiversity is the degree of variation of life and a measure of the variety of organisms present in different ecosystems. This may happen through genetic variation, ecosystem variation, or species variation (number of species/plants) within an area, biome, or planet. Environmental deterioration causes mass extinctions of organism/plants within the Bio-diversity.

Cash Crops: A cash crop is an agricultural crop which is grown for sale to return profits thereby relegating substituting its native traditional corps. For instance, cotton, soybeans, coco, rubber, peanuts are some of the cash crops.

E- Waste: Electronic Waste compromises the physical parts or whole of secondary computers, entertainment device electronics, mobile phones, and other items such as television sets and refrigerators, whether sold, donated, or discarded by their original owners.

Ecological Debt: Ecological debt is the level of resource consumption and waste discharge by a population in excess of locally sustainable natural production and assimilative capacity. The term has been used since 1992 by some environmental organizations from the developing countries.

Ecological Footprints: A measure of the impact humans have on the environment is called an ecological footprint.

Ecosystems: An ecosystem is a community of living organisms (plants, animals and microbes) in conjunction with the non-living components of their environment (things like air, water and mineral soil), interacting as a system.

GDP (Gross Domestic Product): GDP is defined by the Organisation for Economic Co-operation and Development (OECD) as "an aggregate measure of production equal to the sum of the gross values added of all resident, institutional units engaged in production (plus any taxes, and minus any subsidies, on products not included in the value of their outputs. GDP estimates are commonly used to measure the economic performance of a whole country or region.

Globalization: The current wave of globalization has been driven by policies that have opened economies domestically and internationally. Driven by the policy and technological developments (internally and externally) which have spurred increases in cross-border trade, investment, and migration skilled labour towards the integration of the world economy and its development.

Multilateral: Multilateral or Multilateralism is that where in multiple countries working in concert on a given issue. Multilateralism was defined by Miles Kahler as "international governance of the 'many,'" and its central principle was "opposition [of] bilateral discriminatory arrangements that were believed to enhance the leverage of the powerful over the weak and to increase international conflict."

Multilateral Environmental Agreements: Multilateral Environmental Agreements (MEAs) have been established or entered into force in the last two decades to address emerging global environmental issues, including the United Nations Framework Convention on Climate Change (UNFCCC), the Convention on Biological Diversity (CBD), agreements related to chemicals (Basel, Rotterdam and Stockholm Conventions), Convention to Combat Desertification (UNCCD) Initiated by United Nations member countries by the consensus of the member participating countries.

Nation States: Nation State is a geographical area that can be identified as deriving its political legitimacy from serving as a sovereign nation. A state (country) is a political and geopolitical entity, while a nation is a cultural and ethnic one.

Physical Environment: Environment is all of physical and social conditions that surround a person and can influence that person's health. Physical environment includes both your outdoor and indoor surroundings. The quality of air you breathe and the water you drink are important to your health. For instance, Water (rivers/seas/oceans), Natural Vegetation, Landform and rocks, weather and climate Natural resources are found here Examples: rivers, seas, ocean, mountains, rocks, volcanoes, tornadoes are the physical environments.

Rio + 20 Earth Summit: Rio + 20 Earth's 'Green Economy' in the context of sustainable development, poverty eradication, and the institutional framework for sustainable development initiated by United Nations and rectified by many of its member countries.

Sustainability: Sustainability is how biological systems remain diverse and productive. For instance, the long-lived and healthy wetlands and forests are examples of sustainable biological systems. Further, sustainability is the endurance of systems and processes. The organizing principle for sustainability is sustainable development, which includes the four interconnected domains: ecology, economics, politics and culture.

Trade Liberalization: An attempt of any country (willing) towards its removal / reduction / restrictions/ barriers to integrate in International trade. This includes the removal or reduction of both tariff (duties and surcharges) and non-tariff obstacles (like licensing rules, quotas and other requirements).

UNDESA: The United Nations Department of Economic and Social Affairs (UN DESA) is part of the United Nations Secretariat and is responsible for the follow-up to the major United Nations Summits and Conferences, as well as services to the United Nations Economic and Social Council. UN DESA assists countries around the world in agenda-setting and decision-making with the goal of meeting their economic, social and environmental challenges.

UNDP: United Nations Development Programme being part of the United Nation initiative which advocates for change and connects countries to knowledge, experience and resources to help people build a better life. It provides expert advice, training, and grant support to developing countries, with increasing emphasis on assistance to the least developed countries.

UNEP: The United Nations Environment Programme (UNEP) is an agency of the United Nations that does coordinates its environmental activities, assisting developing countries in implementing environmentally sound policies and practices. Its activities cover a wide range of issues regarding the atmosphere, marine and terrestrial ecosystems, environmental governance and green economy.

Chapter 11
Clean Technology Industry:
Relevance of Patents and Related Service Providers

Liina Tonisson
Fraunhofer MOEZ, Germany

Lutz Maicher
University of Jena, Germany & Fraunhofer MOEZ, Germany

ABSTRACT

Many clean technology transfer barriers have been associated to Intellectual Property (IP) rights. The objective of this chapter is to give insights to the types of IP rights services the clean technology industry needs to overcome. By conducting in-depth qualitative interviews with a convenience sample of 25 clean technology companies in 2012, most outsourced intellectual property-related services were discovered. The clean technology companies specified the following top three IP services required from service providers: legal services for IP protection, legal services for IP transactions, and IP consultancy (i.e. IP portfolio analyses). The companies investigated outsource IP-related processes to service providers. That leads to the conclusion that outsourcing patent-related activities is an efficient management decision for the clean technology industry. Outsourcing tasks to competent service providers who are familiar with foreign technology and legal markets were found to be especially useful against infringement threats from developing countries.

1. INTRODUCTION

The degree to which intellectual property rights (please see "Key Terms and Definitions" at the end of the chapter) are important varies over the comprehensive range of clean technology companies (Wright & Shih, 2010). Clean technologies cover a broad range of fundamentally different types of innovation, including alternative energy resources, technologies retaining alternative energy sources, energy storage, distribution and management technologies, recycling and waste technologies, industrial processes, and technologies for capture, storage, and sequestration or disposal of greenhouse gases (Popp, 2010). The core technologies behind clean innovations vary

DOI: 10.4018/978-1-4666-8222-1.ch011

greatly, and range from high-tech innovations to low-tech innovations. These technologies differ in other ways as well, for example the fixed costs of innovation, the patenting and adoption involved and their applicability across industries (Hall & Helmers, 2010).

A patent on a basic invention with no substitutes may allow its holder to block follow-on inventors who would be willing to invest in R&D to create socially useful applications. That kind of invention stands out from the variety of low – or high tech inventions because as an example, in the case of an owner who has patents covering fundamental clean technology inventions essential for advancing follow-on research, such as research tools in biotechnology areas, his/her patents do not allow other inventors to access those patented clean technologies in reasonable conditions. Any innovation processes in these cases could be harmed, and thus public benefit could be decreased considerably. To address this issue, some public authorities develop guidelines for patent licensing in certain technology fields closely related to the public interest for the purpose of advancing further research by facilitating the diffusion of patented technology. The OECD Council, for example, adopted in 2006 Guidelines for the Licensing of Genetic Inventions that outlines principles and best practices for the licensing of genetic inventions used for purposes of human health care. These licensing guidelines are also applied to genetic inventions used for the purpose of human health care. Nevertheless, no such guidelines for environmental technologies exist (OECD, 2006a, 2006b, 2006c).

For almost all of the various technologies that promote sustainable development managing industrial property rights successfully is a key to business sustainability. The clean technology industry in general is patenting and therefore the companies must address the issue. Inventions in some technological fields such as biotechnology rely to a large extent on patenting, other technologies such as irrigation systems, construction techniques or more service-related sectors may be largely free of patenting (Smithers & Blay-Palmer, 2001). For some technologies, intellectual property rights protect only components of a technology. For others, intellectual property rights protect the end product. One example is the agricultural sector where the need for local adaptation of clean technologies to extremely heterogeneous local conditions may also limit the importance of intellectual property rights (Hall & Helmers, 2011). The number of renewable energy patents issued in the U.S. ballooned from 200 per year between 1975 and 2000 to 1,000 per year between 2000 and 2009, according to a study released in 2013 by the Massachusetts Institute of Technology and the Santa Fe Institute. Most of the growth has come from solar and wind technologies, whose yearly growth rates have approached or exceeded the growth rates of new patents for other technologies such as semiconductors and digital communications, according to the study (Bettencourt, Trancik & Kaur, 2013). The renewable patent explosion has been driven by a raise in the R&D investment as well as government programmes designed to promote renewable energy development, and intellectual property service providers do not expect the push for new clean technology patents to subside anytime soon. That has made legal service providers speculate that the increasing rate of new clean energy patents will soon spur a rush to the courthouse by patent holders looking to protect their increasingly profitable innovations leading the way to patent wars (Goldberg, 2013).

In order to assist renewable energy companies with managing industrial property right in times when the patenting rates are growing fast several innovative intellectual property service providers have emerged in order to help technology companies and make sure they can focus on their core business – building and developing innovations (Yanaglsawa & Guellee, 2009). Over the recent years, the market of "traditional" industrial property services, like legal services, has grown in the past and have tendencies towards specialization and professionalization. A number of business

Clean Technology Industry

models for providing intellectual property related services and thus making money out of intellectual property have emerged (Millien & Laurie 2007, 2008). Service market focusing on industrial property has grown fast. Statistical evidence shows the increase in patent reassignments as well as the number of supporting service providers (Chesbrough, 2006). Intellectual property rights service providers have emerged to facilitate more efficient market transactions of technologies, technical knowledge, and particularly patents by developing new models like patent auctions and patent portfolio funds. The current role of intellectual property related service providers is diverse (U St Gallen & Fraunhofer MOEZ, 2012) but surely technology companies are choosing to turn to them for assistance with managing their industrial property better and more efficiently.

This research investigates what are most important intellectual property related services for clean technology industry. It does not solely focus on one or few clean technology fields, but rather looks on the entire clean technology sector by investigating a random sample of clean technologies companies to get a broad overview on the clean technologies sector. This chapter investigates the relevance of patents, patenting behaviors and intellectual property related service needs for companies from a range of clean technologies. Managing industrial property rights is very important for the clean technology company directors according to the underlying research done and it is also considered important according to related literature. The chapter provides overview of clean technology industry dependency on intellectual property rights related services to give insights into how clean technology companies see and value their intellectual property. It will not answer the question of whether intellectual property rights harm or boost innovation in the clean technology sector, but rather will give a better understanding of how important are intellectual property rights for the clean technology sector internally. The chapter is organized as follows:

This section introduced the clean technology industry from intellectual property perspective. The next section will discuss intellectual property relevance for the industry and will focus on recent developments concerning patenting and intellectual property market developments. The chapter is organized into 10 sections:

1. Introduction.
2. Importance of industrial property rights in clean technology transfer.
3. Methodology.
4. Internally recognized value of intellectual property rights for clean technology industry.
5. Patenting and related services for clean technology sector.
6. Outsourcing intellectual property rights related services.
7. Clean technology sector specifics.
8. Implications for developing countries.
9. Future research directions.
10. Conclusion.

2. IMPORTANCE OF INDUSTRIAL PROPERTY RIGHTS IN CLEAN TECHNOLOGY TRANSFER

The intellectual property system provides the regulatory framework where most commercially valuable technologies work and get developed. Global climate change challenges mitigation will involve the development and dissemination of a large number and variety of innovative technological solutions. This worldwide challenge has led to an increased interest in the mechanisms that encourage the development and adoption of new technologies. Recent rapid economic growth indicators in several large developing countries has focused policy attention on the technology transfer and technology development in countries that are not generally on the technology frontier in facilitating the use of clean technologies. Raising the standard of living in third world countries to similar levels enjoyed in the West without a great deal of energy and environment related innovation would have negative consequences for global warming, it is also true that prompt growth means a great deal of new investments, and new investments are an opportunity for substantial progress for technologies (Hall & Helmers, 2010).

Intellectual property rights and the necessary processes associated with the patenting processes for clean technologies do not differ from other technologies, although the need to promote the clean energy industry compared to other industries is evident. Global discussions on private sector and governmental levels over whether intellectual property rights promote innovation and encourage implementation of clean technologies or stand as an obstacle to their development and deployment are currently in focus.

In some research, the intellectual property rights system has been associated with a several barriers to access and dissemination of technologies (Barton, 2007c). The most important ones are: high transaction costs of collecting information, difficulties with negotiating and acquiring intellectual property protected technologies and a lack of clarity when defining what is protected by a patent and what is not. Consequently, these limitations create the basis for potential market failures related to asymmetric information (Barton, 2007a; Newell, 2009).

Therefore, both the UN Secretariat and many developing country parties have recommended weakening or even removing intellectual property rights in clean technologies (Rimmer, 2011). There have been numerous recent partnerships, joint ventures and licensing engagements among clean technology firms despite the presence of noteworthy intellectual property rights (Maboye, 1995; Lane, 2011). For environmental purposes, it would make sense to disseminate clean technologies to the maximum number of end users. The maximum number would be achieved when it would be distributed for either no or at minimum costs. Giving away the technology for minimum or no cost, however, undermines the incentives that private companies have in order to build up innovative technologies. Free usage, or in other words no enforceable property rights, is reasonable for individuals, but it results in depriving the whole community of its users. So it is as well with intellectual property rights: free usage of knowledge results in societies that produce too little new knowledge (Thurlow, 1997). Therefore, it can be argued that the removal of intellectual property rights when dealing with clean technology transfer is the way forward to adapt more to green technologies (Lane, 2011).

Indeed the intellectual property system, and in particular the World Trade Organization Agreement on Trade Related Intellectual Property Rights, does not provide for any particular treatment or flexibilities for access and distribution of environmental sound technologies as it does in the field of health or nutrition (Barton, 2007b). That might be due to limited understanding and evidence that this particular treatment might really prove to be useful. The existing evidence using patent counts to describe the geographical distribution of inventors of green technologies and their international transfer are by construction incomplete in terms of insights they can provide. While useful in improving our understanding of the patenting distribution in clean technologies across sectors, countries and time, these studies cannot provide an answer to the question of whether patents help or hinder the development and transfer of clean technologies. Furthermore, to-date there is no consensus with regard to the definition of relevant clean technology classes as identified by patents' intellectual property codes (IPCs). However, any interpretation of results hinges crucially on this definition. Furthermore existing evidence suggests that there are two groups of developing countries. In the first group are emerging economies, such as Brazil, China, India, and Mexico, and in the second group a larger number of less developed countries. Similar to the conclusion drawn from the general literature on technology development and transfer, the evidence on green technologies suggests that a strengthening of intellectual property rights for the group of emerging economies will most likely have a positive impact on the domestic development of technology and its transfer from developed economies. The current research does not allow drawing a similar conclusion in the case of less developed countries (Hall & Helmers, 2010). Given the huge range of clean technologies related to the problems discussed above, the question of the role of intellectual property rights protection in promoting the development and transfer of technology is not clear.

It is clear that under the present law system and regulatory framework, clean technology companies need to deal with intellectual property rights – at least to a certain extent. Intellectual property rights can play key role in clean technology transfer and is therefore an important topic in clean technology development in general. As the issues of intellectual property is important and vital both on micro and macro level, intellectual property related service providers can assist with their specific knowhow, experience and expertise. Currently, it is not apparent how extensively and which intellectual property rights related services clean technology companies' use. The existing literature addressing the link between intellectual property rights importance and the development of climate change-related technologies is rather small. There is also little or almost no literature on usage of intellectual property rights in clean tech industry. On the whole, the existing evidence on the role of intellectual property rights in promoting the development and diffusion of climate change-related technologies does not provide sufficient insights to reach any substantial conclusion on the role of industrial property rights in clean technology transfer (Hall & Helmers, 2010). This calls for additional efforts in investigating the relationship between intellectual property rights and clean technology companies.

Intellectual property rights and related services have an essential role in employing and transferring clean technologies under the current law scheme. Unless drastic policies are not introduced that would eliminate the industrial property rights in clean technology transfer the corresponding patenting will experience strong growth worldwide hand–in-hand with corresponding clean technologies investments. The global growing energy demand is the driver for increase investments that in turn drives up the patenting rates. The percentage of global clean technologies patents rose by over 2% to 4% from 2000 to 2008 (Arvanitis et al., 2011). In general clean technology patenting has been on rise globally hand in hand with heavy R&D investments and therefore it is important to understand the role of industrial property rights from micro level (Bettencourt, Trancik & Kaur, 2013). Following there is an illustration of the acceleration of clean technology patenting activity, where the drop after 2007 is due to lags in data. Interestingly Figure 1 also shows the positive effect of Kyoto Protocol for clean technology patenting rates.

Overall, clean-energy technologies represent a very small share of total patents, less than 1% over the period (1988-2007), indicating that the creation of new clean technologies is still in its early stages. However clean-energy technology patent numbers are growing fast. Historically up to the mid-1990s, clean technologies patent growth rates were stagnant. In relative terms we can state that the patent growth rates were even declining because the general patenting activity grew. Since the late 1990s, the clean-energy technology patenting numbers have been growing. There is an ascending trend that holds steady when compared to the traditional energy sectors (fossil fuels and nuclear). The traditional energy sectors have been on a descending trend since 2000. Going more technology specific we see that solar PV, wind and carbon capture patenting rates have shown the most patenting activity. Biofuels patenting rates

Figure 1. Acceleration of global clean technology patenting; the drop since 2007 is due to data lags. Counts are measured in terms of Patent Cooperation Treaty applications worldwide, shown as a three-year moving average, indexed to 1997 = 1.0, and are taken from the OECD Patent Database.
Source: Renssen, 2011.

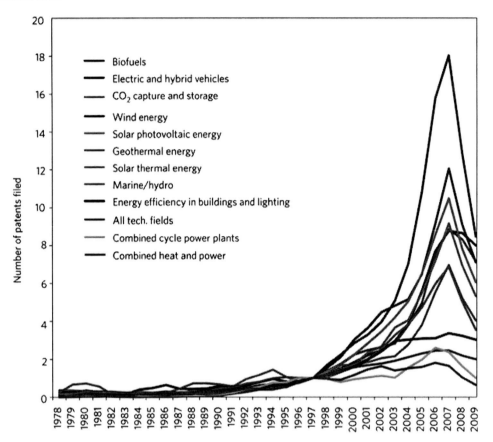

have shown an upwards trend only recently. Most probably due to the early stage of development IGCC and solar thermal and geothermal are not yet taking off. Therefore the empirical evidence shows that intellectual property rights when it comes to clean technology is gaining importance (Veugelers, 2011).

Moving to a competitive low-carbon economy by 2050, the European Commission (2011) recognizes the crucial parts that will be played by the development and deployment of new technologies. It stresses the importance of the competitiveness of European clean energy technology companies on world markets. In case climate change targets are to be met in a cost-effective manner, European clean technology industry needs to prosper. The European Union's Strategic Energy Technology Plan, which has become central to the achievement of the European Union's ambitions has been developed that highlights some curtail intellectual property matters as well.

European clean technology market is perfect for setting the example for third world countries. Many technology transfer initiatives have been created, for example Germany's Federal Ministry of Education and Research and Brazil's Research Ministry jointly held the German-Brazilian Year of Science, Technology and Innovation 2010 and

2011. The initiative was kicked off in São Paulo on 12 April 2010. Various activities took place in both countries between April 2010 and April 2011 under the slogan sustainable: innovative with the aim to jointly find sustainable and innovative solutions to the great challenges of our time. More recent similar initiative was the German-South African Year of Science 2012 and 2013. The aim was to strengthen cooperation between both countries and to create new networks among emerging young scientists. Targeting German and South African research and higher education institutions, as well as businesses, upon dealing with various regarding the global climate challenge: how can we adjust to climate change without slowing down our growth? What will the cities of the future look like? How can we use the resources we have sustainably? How can science and research help? Nevertheless, none of the initiatives including German-Chinese one has specifically looked at the role of industrial property rights.

To sum up intellectual property remains an important issue for clean technology industry developments. This research will investigate clean technology companies and their relationship to patenting and industrial property management. As issue that is highly important on industry and government levels. This chapter will give overview of which different intellectual property services offered are used the most by clean technology companies.

3. METHODOLOGY

In the spring of 2012, in-depth telephone interviews were conducted with 25 clean technology companies. Qualitative analysis was applied within this study. Quantitative methods were not sufficient to draw conclusions on patenting importance due to lack of relevant data. There is no statistical proof of required patent related services for clean technology companies or of patent related activities outsourcing preferences.

Respondents were identified from Fraunhofer MOEZ previous research partners and from internet search. Clean technology company owners, CEOs or intellectual property managers of various clean technology companies, such as from recycling business to environmentally sound transportation providers were recruited. Clean technology producers, clean service related processes providers as well as clean technology machinery producers were included into the sample investigated. Respondents included one founder not operating as current CEO, 11 CEOs, 2 CFOs, 6 R&D managers, 3 intellectual property managers, 2 operations managers. The types of clean technology companies included were: 2 fuel cell systems producers, 5 energy efficient vehicles/motor producers, 4 smart home/building companies, 2 waste management firms, 10 clean energy converters/producers, 2 green material (plastic, wood) producers and one semiconductor company. The company representatives where asked to elaborate on the following questions:

1. On scale from 1 to 10 how important are intellectual property rights for their organization/specific department?
2. Which intellectual property related services are they currently using the most?
3. Which intellectual property related services are they currently missing (if any)?
4. Why are they missing some intellectual property related services, are there any barriers to use any?
5. Which of the services named in intellectual property services catalogue (Appendix) do they use? Do they feel that some function is missing for them?

6. What do they consider in intellectual property services perspective to be clean technology industry specifics? How do they differ from other industries (in case they do differ)?

Prior to the interviews the respondents were asked to get acquitted with intellectual property intellectual property service taxonomy developed by Fraunhofer MOEZ Competitive Intelligence group (please see Appendix) The results are described below.

4. INTERNALLY RECOGNIZED VALUE OF INTELLECTUAL PROPERTY RIGHTS FOR CLEAN TECHNOLOGY INDUSTRY

Industrial property is considered very important for the clean technology industry. The average score from 1 (lowest) to 10 (highest) for the sample was 7,56. That indicates that under the current law scheme, intellectual property rights play an important role and cannot be overlooked when it comes to clean technology development despite the ongoing discussions on various levels. The value of patents was well recognized by most of the interview partners. Patenting was seen as a process that might harm companies' activities due to disclosing valuable information that can be discovered by competitors. Small versus big technology companies' pitfall in court systems was brought up in several cases. The challenge comes in after filing if a large or several large companies are potentially infringing your intellectual property and have significant resources to extensively fight in various court systems with the smaller firm until the smaller clean technology firm has lost all its resources and might be facing bankruptcy - even if the larger firm is in the wrong. Infringement threats from third world countries were mentioned in several cases and industrial property enforcement was seen as an effective tool to fight with these threats. In one of the cases the interview partner mentioned that it had proven to be useful to email the infringing company by saying that they will take action. The outcome of that email was surprisingly positive, the company stopped infringing and started paying license fees without involving going to court.

An important takeaway should be that although industrial property rights are important for clean technology companies, they require very well thought through strategy for their management. Enforcing industrial property rights is seen as the most useful tool for fighting infringements under this law system. The intellectual property strategy should always be in line with the company's overall strategy and should be integrated into clean technology company financial plan as well. In the research done more than 95% percent of clean technology companies decide to patent and consider patents very important for them.

5. PATENTING AND RELATED SERVICES FOR CLEAN TECHNOLOGY SECTOR

Although almost all patent their technologies in clean technology sector attitudes toward patenting and the intellectual property system differ for clean technology companies. For start-ups, patenting is seen as costly and going to court is very resource demanding. These results are somewhat consistent to literature as the questions concerning if to patent (Vance, 2012), when to patent (whether to first patent and then build the technology or vice versa) (Wilkof, 2012), how to patent (narrow or wide patents, process related patents or technology patents) (OECD, 2004) are still on-going in media, literature and mostly in industry. As it became clear during the interviews managers have to face the industrial property strategy related questions on daily bases. Among the interview partners very different patenting practices were evident. Although clean technology startups might not have the necessary resources to deal with the industrial property management in firms' early stages, the need to do it was clearly recognized. For them outsourcing the patent related activities instead on hiring experts in house has proven to be resource efficient. Successful intellectual property management is somewhat linked to overall business growth and success in all size clean technology firms.

Clean technology companies are not aware of the extent of intellectual property related services that are currently offered on the intellectual property market. They do not fully know what can be outsourced. That can be concluded from their replies when answering which intellectual property services they are aware of. All 25-interview partners found that the intellectual property services taxonomy (Appendix) used as basis for the interviews was comprehensive and covered all known or even yet unknown, currently available intellectual property related services. Additionally the taxonomy before the interview made them aware of many intellectual property services that they did not know about.

Many of interview partners did agree that the European intellectual property services market is comparable to a "black-box". Most services get acquired to personal networks and connections. In the sample, there was one highly qualified intellectual property expert with more than 25 years of experience in the intellectual property field. During the interview, it became obvious that having such a person in-house makes outsourcing some intellectual property tasks to intellectual property service providers unnecessary. For the rest outsourcing industrial property related services seemed like an easier way of getting the necessary things done. Many clean technology companies have emerged intellectual property management tasks with other responsibilities and thus it can be the CEO, R&D manager or even sales person who is made responsible for intellectual property. That is mainly due lack of resources and therefore intellectual property rights can be left without proper attention. Therefore outsourcing patent related services to reliable service providers can be seen as a beneficial strategic decision for the technology company. Especially when clean technology companies are dealing with developing countries where the underlying legal systems might be different and furthermore, technology diffusion requires working in a different language, outsourcing some of the intellectual property related tasks, might prove to benefit all parties. Hiring experts in-house to handle the technology diffusion and management are resource costly, money and time wise. Having the opportunity to outsource technology transfer related tasks and concentrating on technology companies' core business – building innovation, might save time and money.

6. OUTSOURCING INTELLECTUAL PROPERTY RELATED SERVICES

One of the key findings from the conducted interviews was that almost all clean technology companies need assistance with legal steps of patent proceeding and processes. Even when having an in-house lawyer, is it still necessary in most cases to consult a patent attorney on legal aspects. All of the companies interviewed outsource legal services at least to some extent. That is clear proof that nowadays under current legal system outsourcing patent related activities is current practice. For the less developed countries it can set an example of technology management could be handled more efficiently. Our example had EU origin technology companies that have somewhat proven to be successful (named among top innovators, performers or received a reward in last

5 years for their practice). This current practice has therefore proven to be successful and could be suggested for starting clean technology firms working on less developed markets where there are even fewer resources to successfully develop innovative technologies.

Managing the intellectual property portfolio is usually done in house according to our interview partners. That is suggested by literature as well (Tietze & Herstatt, 2010). The final decisions on what to patent and if to patent is up to the companies who own the corresponding technological innovations. Different stages for coming up with these kind of decisions might be outsourced. Many of the interview partners have used various consulting services, and some have been disappointed in the results due to lack of technology-specific knowledge, whereas others have received valuable help for making strategic decisions related to industrial property rights management.

Interestingly, clean technology companies who produce their own technologies might have unused intellectual property that they would be willing to license out. Licensing out unused intellectual property rights would earn extra revenues for clean technology companies for advancing their businesses. The intellectual property rights licensing service was missed by 3 interview partners, furthermore that service was falsely assumed to be missing from current intellectual property market from one of the interview partners. That proves that clean technology companies are not aware of their current possibilities and they do not have good access to information regarding intellectual property intellectual property services and service providers. A wind turbine producer, for example, mentioned that their turbine technology intellectual property could be easily used in submarine industry, which is not their competitor. Therefore licensing out to a different industry would create relatively risk free (it would be bad business strategy to license to competitors, but not to other industries) extra income (licensing revenues). Consequently for clean technology sectors, service providers related to licensing activities might see an increase in demand for their services. On the other hand, few of the interview partners stated that they do not see licensing out as an extra source of revenue; intellectual property exchange and licensing were in some ways perceived as "a dirty game" (e.g. ChromoGenics, Nualight). Therefore the mindsets for some of the clean technology companies might have to be changed first.

Several clean technology companies mentioned that they are currently missing some intellectual property related services, and the most common barrier to using some of the services they would like to outsource was the information barrier. Clean technology companies simply were not aware of their opportunities, and once the service they felt they are missing was named, it was rather easy to find that kind of service or services from the taxonomy. That means these services do exist, and clean technology companies are just not aware of their intellectual property service outsourcing opportunities.

7. CLEAN TECHNOLOGY SECTOR SPECIFICS

An important observation on clean technology specifics related to intellectual property is that some very broad patents exist in clean technology market. Therefore some clean technology managers felt that the number of infringement is assumed to increase in the future. Most of the interview partners believe that the number of infringements will increase in the future, but if this is purely clean technologies specific, the exact number of litigations cannot be predicted. Few of the interviewed felt strongly that it is not just clean technologies-specific phenomenon.

Nevertheless, service providers that help clean technology companies with court proceedings and detecting cases on infringements might become more popular. This occurrence might take place due to several reasons, mainly due to very broad patents already granted in the past and clean technology companies having to patent narrower and narrower nowadays. Narrow patenting is also taking place because of strategic reasons. Decision to keep some of the technical details and knowhow as a secret, is due to safety issues. Patent filing process makes your technology easily discoverable for competitors and that makes patenting less desirable (Bar-Gill & Parchomovsky, 2003). This patenting behavior among companies could possibly increase infringement possibilities and cases in near future (some interview partners gave and approximate 5 years' timeline). About a third of the interview partners did not see any difference between clean technology and other industries when it comes to intellectual property related services. Due to very wide sample selection of clean technology companies, industry specifics were not evident. Companies interviewed were active on various fields, and therefore had each different insight. Additional aid from governments was expected in various cases as clean technology was considered to have great social benefits, and therefore it was assumed to be in the interest of governments to promote and support the industry. According to the expertise of experienced intellectual property managers interviewed, clean technology does not stand out in great extents when it comes to intellectual property related services. The intellectual property -related service market in general is complex to grasp. Companies, not being aware of the full extent of the intellectual property service market, are consequently not well aware of the market specifics for their industry. Interview partners seemed open minded and willing to use reasonably useful intellectual property services. Therefore in the future the percentage of outsourced intellectual property related tasks is predicted to increase foe clean technology sector.

8. IMPLICATIONS FOR DEVELOPING COUNTRIES

This chapter based on evidence gathered from 25 European clean technology companies active globally. It is believed that clean technology developers in third world countries face the same difficulties as the ones elsewhere. The issues apply to all with some exceptions.

China for example was seen as a country by the interview partners where industrial property rights are not getting the respect they deserve. Many company managers interviewed where afraid of infringement cases from the east. In this case outsourcing industrial property rights protection to language and law competent intellectual property service providers in east was seen as one of the solutions. The third world countries should respect and try to work together with vast range of various patent systems in the west. Local industrial property law systems in developing countries need to achieve mature and stable levels of intellectual property acknowledgment, culture and respect towards the intellectual property rights in general. Co-operation with the more developed, or old(er) intellectual property systems will be the second step. Meanwhile it can be a good idea for the clean technology companies in developing countries to outsource some of the intellectual property related tasks to experts who have the knowhow. It can be assumed that clean technology companies in developing countries are even less aware of the outsourcing possibilities when it comes to intellectual property related task than they are in Europe and in the USA. Governments can improve the situation by supporting initiatives that dedicate themselves to spreading the intellectual property related knowledge.

9. FUTURE RESEARCH DIRECTIONS

It is suggested to further examine detailed patenting patters of clean technology companies.

Notably this must be done under non-disclosure agreement due to confidentiality risks. Focus should be taken on how to improve the patenting patterns. Focus should be taken on how to improve the patenting patterns, for example by examining ways for the EPO to launch instruments that are similar to UK IPOs "fast track" for green patents. Additionally, the increase of infringements phenomenon on intellectual property landscape should be investigated. Policy instruments to aid and services to assist SMEs and start-ups with patenting and intellectual property rights protection should be considered and further examined with a focus on third world countries. Several interview partners from UK mentioned the benefits for having a "fast track" for green patents in UK. Such instruments benefits could be investigated for developing countries as well. Literature on incentives of big technology companies acquiring smaller companies is scarce or non-existent. Big technology companies eliminating competition by buying up start-ups who might become their competitors in the future might hinder faster and further development among green technology companies. Policies and research should target the issue as soon as possible.

Another problem that needs examining is how to provide services that aid small technology companies that are fighting in courts with big industry companies. SMEs can be driven into bankruptcy with extensive court cases involving big industry players due to their lack of resources, even if the infringers are big companies. Thus several pitfalls exist for clean technology companies when it comes to patenting and especially for small companies. All the issues mentioned above and discussed over by interview partners are suggested for further research.

To sum up, because intellectual property continues to be a dynamic and constantly evolving field, closely tied to technological, economic, political and social changes, companies involved in the intellectual property market must update themselves about the changes frequently (ICC, 2012). Obviously, when having fewer resources (which is the case with developing countries and start-ups and young clean technology companies in general), they do not have enough resources to keep themselves up to speed with intellectual property -market updates, and that was confirmed by our interview partners. Therefore, efforts making the intellectual property market globally more transparent are highly appreciated by interview partners.

10. CONCLUSION

Addressing climate change, rousing the development and deployment of advanced and innovative clean technology are an innovation imperative. The International Energy Agency (IEA) report from 2008 states that clean technology innovation must rise by a factor between two and ten to meet global climate change goals (IEA, 2008). The needed investments are estimated to be EUR 0.84 trillion per year (in real terms) through 2050, or around 1.1% of global GDP. Due to increased investments in the clean technology sector, vast developments for clean technology companies are expected. As the world struggles with global warming, we are expected to face rising demands to take action in terms of technological changes. Economies may have to acquire or develop more and more new technologies. Therefore, the roles of clean technology transfer and the related intellectual property rights are receiving much attention from industry and policy makers (Mytelka, 2007; Newell, 2010).

Protecting intellectual property by patenting their innovations is one of the main means by which clean technology companies can capture value related to developing new clean technologies. Supporting intellectual property protection, encouraging and promoting technology transfer, implementing environmental policies, and strengthening public-private partnerships are important steps in global clean technology

Clean Technology Industry

development (Lane, 2009). All these activities are currently supported according to the interviews by intellectual property related service providers. Patents are considered very important for the clean technology industry. The average score from 1 (lowest) to 10 (highest) was 7.56 given the research done. That indicates that under current law system intellectual property plays an important role and cannot be overlooked. Many of the clean technology companies investigated are outsourcing intellectual property related services.

Patenting patterns of the companies examined differ to a great extent, and the rationale behind it is different based on specific industry. Clean technology companies are not well aware of the intellectual property services provided. In general, the use of intellectual property related services is limited, and therefore mostly legal aspects - the basic ones of patenting are outsourced. Clean technology companies interviewed were not fully aware of their possibilities when it comes to receiving aid with intellectual property management. The information barrier limits organizations from recognizing possible services that they might be missing from their management strategy. Recent market developments force clean technology firms to continuously innovate. However, clean innovation is a cumulative process that increasingly requires the arrangement of internally developed technologies with externally acquired ones, including external assistance – particularly for increasingly complex products. The emergence of new intellectual property service providers can be seen as one of the outcomes of this on-going specialization. A forward fragmentation of the technology owners who outsource certain services to intellectual property service providers can be expected. This new division of work between technology owners, buyers and intellectual property service providers on the clean technology market level has an impact on the governance structures of intellectual property transactions on the micro level of the firm, more precisely, on how firms manage intellectual property transactions (Tietze & Herstatt, 2010).

The survey results in this chapter summarized that the 25 clean technology companies interviewed use intellectual property service providers quite actively. Patent drafting and legal/administrative matters surrounding patent acquisition are 100% outsourced according to interview results. About half are or have used some sort of intellectual property consultancy. That is signal for using outsourced intellectual property services for efficient clean technology company management.

The top three clusters of intellectual property services required from service providers for the sample were:

1. Legal services, namely intellectual property protection (patent and trademark searches, applications and renewals of intellectual property, representation at industrial property office)
2. Legal services, more precisely intellectual property rights contracting (due diligence and administrative assistance with intellectual property transactions)
3. Intellectual property consultancy, i.e. intellectual property portfolio analyses (legal and quality assessment, commercial value assessment, intellectual property portfolio landscaping)

Intellectual property portfolio management-related topics are handled in-house, according to the interviews. This is consistent with literature by Stigler (1951) and Tietze, Herstatt (2010). Companies preferably out-source tasks that do not represent increasing returns or diminishing costs. Tasks which might need further governance, coordination and communication of the intellectual property rights transaction process can be out-sourced. Initial decisions about intellectual property sales/licensing are always done in-house.

Clean technology companies tend to patent narrowly. Such kind of patenting is taking place for strategic or safety reasons (i.e. to keep some of the technical details and knowhow secret) and as well as due to industry specifics (wide patenting has taken place ages ago and innovators are going narrower and narrower nowadays). Because patent filing process makes your technology easily discoverable for competitors (Bar-Gill & Parchomovsky, 2003), the safety issues were the number one reason for narrow patenting.

Closer investigation on clean technology specifics related to intellectual property rights highlighted that indeed some very broad patents exist already, and narrow patenting is taking place more and more. Consequently, the number of infringement cases is assumed to increase in the future. The reasoning behind it is that wide patents might be stated to cover some of the narrow patents being granted nowadays. Nevertheless, some interviewees felt many old patents would have difficulties in current court systems and might not prove to be valid anymore. Most of the interview partners believe that the number of infringements will increase in the future, but cannot be predicted as it specifically relates to clean technologies. Few of those interviewed felt strongly that the issue of infringements is not just clean technologies specific phenomenon. Therefore for emerging countries the patent validity should be one of the key issues. Patent infringement cases are harmful for technology development because they delay the production processes according to one of our interview partners. Therefore emerging countries should avoid patenting narrow and having strong valid patent could save costs for the clean technology industry. That could be one of the main takeaways from this chapter for developing countries. It is important to have a strong trusted underlying legal system on which a patent system that produces strong valid patents is built on. That can help developing countries in disseminating and developing clean technologies more efficiently.

To sum up outsourcing industrial property matters to trusted external expert is evident in current clean technology industry. It is a accepted management practice among clean technology companies. It saves money and time for companies. Mainly legal activities are being outsourced at the moment. That might be due to the information barriers that exist on intellectual property service markets.

ACKNOWLEDGMENT

This research project would not have been possible without the support of many people. The authors wish to express their gratitude to the competitive intelligence team at Fraunhofer MOEZ institute, especially to Pirjo Jha, Fabian Bartsch, and Michael Prilop.

REFERENCES

Arvanitis, S., Ley, M., Soltmann, C., Stucki, T., Wörter, M., & Bolli, T. (2011). *Potenziale für Cleantech im Industrie- und Dienstleistungsbereich in der Schweiz, KOF Studien, 27*. Zurich: KOF Swiss Economic Institute.

Bar-Gill, O., & Parchomovsky, G. (2003). *The value of giving away secrets*. Harvard Law and Econ Discussion Paper No. 417. Cambridge, MA: Harvard Law School.

Barton, J. H. (2007a). *IP and access to clean energy technologies in developing countries: ICTSD biofuel and windtechnologies issues paper*. Geneva: ICTSD. Retrieved from http://www.wipo.int/wipo_magazine/en/2008/01/article_0003.html

Barton, J. H. (2007b). *Intellectual property and access to clean energy technologies in developing countries: An analysis of solar photovoltaic, biofuel and wind technologies.* Geneva: ICTSD. doi:10.7215/GP_IP_20071201

Barton, J. H. (2007c). *New trends in technology transfer.* Geneva: ICTSD.

Bettencourt, L. M. A., Trancik, J. E., & Kaur, J. (2013). Determinants of the pace of global innovation in energy technologies. *PLoS ONE, 8*(10), e67864. doi:10.1371/journal.pone.0067864 PMID:24155867

Chesbrough, H. (2006). *Open business models: How to thrive in the new innovation landscape.* Boston, MA: Harvard Business School Press.

European Commission. (2011). *Roadmap for moving to a competitive low-carbon economy by 2050* [Policy paper]. European Commission.

Goldberg, K. (2013, October 30). Clean energy patent boom sets stage for IP wars. In *Law360*. New York, NY: Portfolio Media, Inc.

Hall, B. H., & Helmers, C. (2010). *The role of patent protection in (clean/green) technology transfer.* NBER Working Papers 16323. Cambridge, MA: National Bureau of Economic Research.

Hall, B. H., & Helmers, C. (2011). Innovation and diffusion of clean/green technology: Can patent commons help? *Journal of Environmental Economics and Management, 66*(1), 33–51. doi:10.1016/j.jeem.2012.12.008

Intellectual Property Office UK. (2009). *UK green inventions to get fast-tracked through patent system* [Press release]. Retrieved September 10, 2014, from www.ipo.gov.uk/about/press/press-release/press-release-2009/pressrelease-20090512.htm

International Chamber of Commerce (ICC). (2012). *ICC intellectual property roadmap: Current and emerging issues for business and policymakers 2012.* Paris: ICC.

International Energy Agency (IEA). (2008). *Energy technology perspectives 2008.* Paris: IEA. Retrieved from http://www.iea.org/Textbase/npsum/ETP2008SUM.pdf

Lane, E. (2009). Clean tech reality check: Nine international green technology transfer deals unhindered by intellectual property rights. *Santa Clara Computer and High-Technology Law Journal, 26,* 533.

Lane, E. (2011). *Clean tech intellectual property: Eco-marks, green patents, and green innovation.* Oxford, UK: Oxford University Press.

Maboye, B. (1995). *Technology transfer overlooked in GEF solar project.* Stockholm: Stockholm Environmental Institute.

Millien, R., & Laurie, R. (2007). *A summary of established & emerging IP business models.* Paper presented at the Sedona Conference, Phoenix, AZ.

Millien, R., & Laurie, R. (2008). *Meet the middlemen.* London: Globe White Page Ltd.

Mytelka, L. (2007). *Technology transfer issues in environmental goods and services.* ICTSD Trade and Environment Papers No. 6. Geneva: ICTSD. Retrieved from http://www.ictsd.org/sites/default/files/research/2008/04/2007-04-lmytelka.pdf

Newell, R. G. (2010). International climate technology strategies. In J. Aldy & R. Stavins (Eds.), *Post-Kyoto international climate policy: Implementing architectures for agreement.* Cambridge, UK: Cambridge University Press.

OECD. (2004). *Patents and innovation: Trends and policy challenges.* Paris: OECD Publishing.

OECD. (2006a). *OECD science, technology and industry outlook 2006.* Paris: OECD Publishing.

OECD. (2006b). *Biotechnology update.* Paris: OECD.

OECD. (2006c). *Guidelines for the licensing of genetic inventions* [Policy guideline]. Retrieved from http://www.oecd.org/sti/biotechnology/licensing

Popp, A. (2000). Swamped in information but starved of data' information and intermediaries in clothing supply chains. *Supply Chain Management, 5*(3), 151–161. doi:10.1108/13598540010338910

Renssen, S. (2011). Driving technology transfer. *Nature Climate Change, 1*(6), 289–290. doi:10.1038/nclimate1193

Rimmer, M. (2011). *Intellectual property and climate change: Inventing clean technologies*. Cheltenham, UK: Edgar Elgar Publishing. doi:10.4337/9780857935885

Smithers, J., & Blay-Palmer, A. (2001). Technology innovation as a strategy for climate adaptation in agriculture. *Applied Geography (Sevenoaks, England), 21*(2), 175–197. doi:10.1016/S0143-6228(01)00004-2

St. Gallen, U., & Fraunhofer, M. O. E. Z. (2012). *Creating a financial marketplace for IPR in Europe: Final report for European Commission to EU tender No 3/PP/ENT/CIP/10/A/NO2S003*. European Commission.

Stigler, G. J. (1951). The division of labor is limited by the extent of the market. *Journal of Political Economy, 59*(3), 185–193. doi:10.1086/257075

Thurlow, L. C. (1997). Needed a new system of IP rights. *Harvard Business Review, 75*(5), 94–103. PMID:10170334

Tietze, F., & Herstatt, C. (2010). *Technology market intermediaries and innovation*. Paper presented at DRUID Summer Conference, London, UK.

Vance, A. (2012, August 9). Patents - Startups' new creed: Patent first, prototype later. *Bloomberg Businessweek*. Retrieved from http://www.bloomberg.com/bw/articles/2012-08-09/startups-new-creed-patent-first-prototype-later

Veugelers, R. (2011). Europe's clean technology investment challenge. *Bruegel Policy Contribution*. Retrieved from http://www.bruegel.org/download/parent/561-europes-clean-technology-investment-challenge/file/1408-europes-clean-technology-investment-challenge/

Wilkof, N. (2012, August 21). So which is it for a start-up: A patent or a proto-type? [Blog post]. Retrieved from http://ipfinance.blogspot.de/2012/08/so-which-is-it-for-start-up-patent-or.html

Wright, B., & Shih, T. (2010). *Agricultural innovation*. NBER Working Paper 15793. Cambridge, MA: National Bureau of Economic Research.

Yanaglsawa, T., & Guellee, D. (2009). *The emerging patent marketplace*. OECD Science, Technology and Industry Working Paper, 2009/9. Paris: OECD Publishing.

ADDITIONAL READING

Barrett, S. (2009). The coming global climate-Technology revolution. *The Journal of Economic Perspectives, 23*(2), 53–75. doi:10.1257/jep.23.2.53

Brunneimer, S., & Cohen, M. (2003). Determinants of environmental innovation in US manufacturing industries. *Journal of Environmental Economics and Management, 45*(2), 278–293. doi:10.1016/S0095-0696(02)00058-X

Dechezleprêtre, A., Glachant, M., Hascic, I., Johnstone, N., & Ménière, Y. (2011). Invention and transfer of climate change mitigation technologies on a global scale: A study drawing on patent data. *Review of Environmental Economics and Policy*, *5*(1), 109–130. doi:10.1093/reep/req023

Fischer, C., & Newell, R. (2008). Environmental and technology policies for climate mitigation. *Journal of Environmental Economics and Management*, *55*(2), 142–162. doi:10.1016/j.jeem.2007.11.001

Geroski, P. (1995). Markets for technology: Knowledge, innovation, and appropriability. In P. Stoneman (Ed.), *Handbook of the Economics of Innovation and Technological Change* (pp. 90–131). Oxford, UK: Blackwell Publishers.

Jaffe, A. B. (1986). Technological opportunity and spillovers of R&D: Evidence from firms' patents, profits and market value. *The American Economic Review*, *76*(5), 984–1001.

Jaffe, A. B., & Lerner, J. (2001). Reinventing public R&D: Patent policy and the commercialization of national laboratory technologies. *The Rand Journal of Economics*, *32*(1), 167–198. doi:10.2307/2696403

Jaffe, A. B., & Palmer, K. (1997). Environmental regulation and innovation: A panel data study. *The Review of Economics and Statistics*, *79*(4), 610–619. doi:10.1162/003465397557196

Johnstone, N., Hascic, I., & Popp, D. (2009). Renewable energy policies and technological innovation: Evidence based on patent counts. *Environmental and Resource Economics*, *45*(1), 133–155. doi:10.1007/s10640-009-9309-1

Keller, W. (2004). International technology diffusion. *Journal of Economic Literature*, *42*(3), 752–782. doi:10.1257/0022051042177685

Lanjouw, J. O., & Mody, A. (1996). Innovation and the international diffusion of environmentally responsive technology. *Research Policy*, *25*(4), 549–571. doi:10.1016/0048-7333(95)00853-5

Lovely, M., & Popp, D. (2008). Trade, technology and the environment: Why do poorer countries regulate sooner? *Journal of Environmental Economics and Management*, *61*(1), 16–35. doi:10.1016/j.jeem.2010.08.003

Pakes, A. (1985). On patents, R&D, and the stock market rate of return. *Journal of Political Economy*, *93*(2), 390–409. doi:10.1086/261305

Popp, D. (2006). International innovation and diffusion of air pollution control technologies: The effects of NOX and SO2 regulation in the U.S., Japan, and Germany. *Journal of Environmental Economics and Management*, *51*(1), 46–71. doi:10.1016/j.jeem.2005.04.006

Rose, A., Bulte, E., & Folmer, H. (1999). Long-run implications for developing countries of joint implementation of greenhouse gas mitigation. *Environmental and Resource Economics*, *14*(1), 19–31. doi:10.1023/A:1008396829502

Snyder, L. D., Miller, N. H., & Stavins, R. N. (2003). The effects of environmental regulation on technology diffusion: The case of chlorine manufacturing. *The American Economic Review*, *93*(2), 431–435. doi:10.1257/000282803321947470

Söderholm, P., & Klaassen, G. (2007). Wind power in Europe: A simultaneous innovation-diffusion model. *Environmental and Resource Economics*, *36*(2), 163–190. doi:10.1007/s10640-006-9025-z

KEY TERMS AND DEFINITIONS

Clean Technology: New products, processes and services which improve current energy efficiency, therefore minimising pollution and the environmental footprint compared to conventional technologies.

Industrial Property: Intangible property such as inventions, industrial designs, trademarks, which is afforded protection under national and international intellectual property laws. Within this chapter in mainly refers to patents.

Intellectual Property Rights: The term "Intellectual Property Rights" refers to the legal rights granted with the aim to protect the creations of the intellect. These rights include Industrial Property Rights (e.g. patents, industrial designs and trademarks) and Copyright (right of the author or creator) and Related Rights (rights of the performers, producers and broadcasting organizations). Within this chapter the term has been mainly used as a synonym for patents.

Intellectual Property Services: All the services that are mentioned in the Appendix.

Intellectual Property: Refers to the ownership of intangible and non-physical goods. This includes ideas, names, designs, symbols, artwork, writings, and other creations. Within this chapter in mainly refers to patents and has been used as a synonym for patents.

APPENDIX

This is the comprehensive list of various intellectual property rights related services. The list was developed by conducting desk research on existing intellectual property service providers.

- **100 IP-Related Finance Services:** IP private equity and venture capital firms raise funds from institutional investors such as companies, banks, governments or high net worth individuals, as well as private equity fund managers themselves. Here are services dealing with resource allocation as well as resource management, acquisition and investment. Services similar to traditional venture capital (VC) or private equity firms services, but specializing in spinning out promising non-core IP which has become "stranded" within larger technology companies, or creating joint ventures between large technology companies to commercialize the technology and monetize the associated IP.
- **110 Investment Products based on Royalty Liquidation/Streams:** Services related to the counsel, assistance and/or providing capital to patent owners performing IP securitization financing transactions (which resemble the more common mortgage-backed securities).
- **120 Financing IP and Innovation Processes:** Providing capital for IP creation and aggregation. Includes loan based (backed by IP) financing.
  - **121 Private Financing:** Related to service of providing private financing for IP owners, either directly or as intermediaries, usually in the form of loans (debt financing), where the security for the loan is either wholly or partially IP assets (i.e., IP collateralization).
  - **122 Public Funding:** Similarly to private funding (see 131), government funding to develop further specific technology areas or promote certain technologies.
  - **123 PPP Financing:** Similarly to private funding (see 131), composition of public and private funding for IP creation.
- **130 IP Litigation Funding:** Litigation funders are interested in providing financial means for IP litigation and particularly patent litigation cases for a fixed fee or % on the amount gained from infringing party.
- **140 IP Insurances:** Intellectual Property Insurance service protects companies for copyright, trademark or patent infringement claims arising out of the company's operation. It pays the defence costs and any judgment up to the policy limits.
  - **141 IP Litigation Insurances for Inventors:** Insurances focused on inventors that cover legal fees for claiming and litigating own intellectual property rights. IP coverage helps pay the legal expenses of suing an individual or firm that has violated your intellectual property rights.
  - **142 IP Litigation Insurances for Third-Parties:** Insurances that cover legal fees related to IP litigation. Third party coverage protects you if you are sued for infringing on another party's intellectual property rights and it funds your legal defence.
- **200 Matchmaking & Trading:** Services related to arrangement of intellectual property rights related development needs of companies with available resources. Trading involving exchange of ownership.
- **210 Matchmaking:** Service of linking IP (development) needs with available resources (including researchers).

- - **211 Onsite Matchmaking Services:** Desktop-based matchmaking, conferences or forums created for purpose of connecting IP (development) needs with available resources.
  - **212 Online Matchmaking Platforms:** Web-based platforms for services connecting IP (development) needs with available resources.
- **220 IP Brokerage:** Services related to assisting patent owners in finding licensees, buyers for their IP. Service includes negotiating IP related contracts, IP purchases, - or sales in return for a fee or commission.
- **230 IP Scouting:** Specific services that help you to find necessary IP. It is a team of IP and technology experts or an expert who observes and recommends promising IP for acquiring.
- **240 IP Auctions:** A public sale in which intellectual property or IP portfolios are sold to the highest bidder.
  - **241 Onsite IP Auctions:** Live IP auctions.
  - **242 Online IP Auctions:** Web-based IP auctions.
- **250 IP Exchanges:** Traded exchanges like IPXI (whether physical or online locations) similar to the NYSE and NASDAQ where yet-to-be created IP-based financial instruments would be listed and traded much like stocks are today.
- **260 IP Sharing:** Services dedicated to various forms of IP sharing.
  - **261 Defensive Publishing:** Defensive publishing or platforms where inventions are made public. Disclosing an enabling description and/or drawing of the product, apparatus or method so that it enters the public domain and becomes prior art.
  - **262 (Online) IP Pools for Public Use:** Platforms for sharing IP for free.
- **270 IP Pooling/Aggregation:** The process of scouting and acquiring existing patents.
  - **271 Offensive IP Aggregation:** The purchasing of patents in order to assert them against companies that would use the inventions protected by such patents (operating companies) and to grant licenses to these operating companies in return for licensing fees or royalties.
  - **272 Defensive IP Aggregation:** This service was introduced by RPX Corporation. The purchasing of patents or patent rights to keep such patents out of the hands of entities that would assert them against operating companies.
- **280 IP-Driven M&A Advisory:** Services similar to traditional investment banking services where a percentage fee is received. Services advising technology companies in their merger and acquisition (M&A) activities and earning fees based on the value of the entire deal (or apportioned according to the value of the IP within the deal).
- **290 Purchases and Sale of IP:** Services that provide assistance with actions that involve exchange of IP ownership.
- **300 IP and Portfolio Processing:** Various services related to creation of IP portfolios and partial management processes of the portfolio related to creating revenues out of IP.
- **310 Patent Document Processing:** Services related to assisting with the documentation of patent process / patenting itself.
  - **311 Patent & Design Drawings:** Services creating visuals of inventions for better marketing and patenting purposes.
  - **312 IP Translations:** Services related to assistance of translations of IP documentation.
- **320 IP Portfolio Management:** Services related to outsourcing the activities related to your patent portfolio.

- **330 IP Portfolio Administration:** Updating the patents the portfolio consists of as well as collecting royalty rates and dealing with licensing.
- **340 IP Augmentation:** IP creation, either creating new technologies by cooperating with other institutions and as an results being the owner (or co-owner) of the patents created out of that process; or developing new technologies and getting patents on them in-house, using own R&D resources.
    - **341 IP Augmentation Through In-House Labs:** Developing patents within the institution in order to develop technologies or IP portfolios.
    - **342 IP Augmentation Through Outsourcing:** Services related to IP creation for organizations by third parties.
- **350 Licensing IP:** Services of licensing and advising for licensing, e.g. done by Licensing Agents. An authorization (by the licensor) to use the licensed material (by the licensee).
    - **351 Carrot Licensing:** Services executing carrot licensing involve bringing together licensing partners voluntarily. A carrot patent licensing approach is appropriate when the prospective licensee is not practicing the patented invention and is under no compulsion to take a license.
    - **352 Stick Licensing:** Services pursuing stick licensing involve to some degree infringement. A stick patent licensing approach is applied when the prospective licensee is already using your patent technology and, thereby, infringing your patent.
- **400 Legal Services:** Services involving legal or law related matters like issue of patents, preparation of patent filing documents and litigation processes.
- **410 IP Legal Protection:** Process of assuring legal rights to the objects of IP (e.g. inventions, literacy and artistic works, images, designs).
    - **411 Patent and Trademark Searches:** Prior art search and investigation and comparison of existing intellectual property rights and applications regionally and worldwide.
    - **412 Applications and Renewals of IP:** Applications for IP protection and renewals of IP protection at industrial property offices (e.g. EPO, DPMA, USPTO, JPO).
    - **413 Representation at Industrial Property Office:** Official representation of the IP owners at industrial property offices (e.g. in patent grant and litigation proceedings).
    - **414 Patent Drafting:** Services related to the drafting of a patent application. The service is a multifaceted activity. Patent drafting refers to the process of writing the patent description and claims, which is the core of any patent application, and in due course, if allowed, of the granted patent specification.
- **420 IP Contracting:** The branch of legal services dealing with assisting with formal IP related agreements between parties.
    - **421 Due Diligence:** IP related due diligence services prior to IP transactions (e.g. licensing, acquisition, sale).
    - **422 IP Transaction Support:** Negotiations for and draft of IP contracts (e.g. licensing, acquisition, sale of IP rights), and development of legal strategies for IP protection and use.
- **430 IP Litigation:** A legal proceeding in a court or a judicial contest to determine and enforce IP rights.
    - **431 Non-Judicial Proceedings:** Legal services lying outside the proceedings in the court (e.g. determination of possible infringement cases, negotiations for extrajudicial settlements).

- **432 Judicial Proceedings:** Legal services associated with the protection of IP in the court (e.g. representation in civil and criminal proceedings of IP owner or alleged infringer of IP rights).
- **433 Arbitration And Mediation:** Legal services covering the arbitration and mediation proceedings (e.g. preparation of claims, and representation of IP owners or alleged infringer of IP rights).
- **440 IP-Granting Authority:** Industrial property offices that grant and renew legal rights to the objects of IP (EPO, DPMA, USPTO, JPO).
- **450 Standardization Authorities:** Legal and regulatory services related to IP standards setting.
- **460 Anti-Trust Authorities:** Services opposing or intended to regulate business monopolies, such as trusts or cartels, especially in the interest of promoting competition.
- **470 Competition Law Authorities:** Service that seeks to maintain market competition by regulating anti-competitive conduct by companies.
- **500 IP Consultancy:** Advisory services related to various IP aspects providing professional or expert advice in a particular area such as market specifics for precise industry for patenting, technology and IP roadmaps, and various analyses.
- **510 IP Portfolio Analysis:** Services for assessment of patents.
  - **511 Legal Quality Assessment:** Examining the legal strength of patent.
  - **512 IP Valuation:** Determination or estimation the market value for patents or the underlying technology. Includes valuation of patent portfolios and technology.
  - **513 IP Portfolio Landscaping:** Assessment services that comprise mapping technology fields and existing patents according to the given patent portfolio and thus estimating its market position.
- **520 IP Strategy Development:** Consulting services for examining the best solutions of IP usage and further development. Includes strategic planning of technology trajectories/technology paths and IP portfolio development.
- **530 Competitive Intelligence:** Collection and analysis of IP related data. It is the service of defining, gathering, analysing, and distributing intelligence about IP, IP holders, IP portfolios and any aspect of the IP environment needed to support executives and managers in making strategic IP decisions for an organization.
  - **531 Industry Analyses:** Examining existing competitors and companies involved in IP market.
  - **532 Technology Analyses:** Examining patented technologies, their technical details and – requirements for patenting purposes.
  - **533 Patent Analyses:** Services related to examining existing patents and drawing conclusions on patenting related information/activities.
- **540 Crowd-Sourcing Platform for Prior Art Search:** Service that allows an organization or an individual to collaborate with a community to find out if a specific technology exists/is patented.
- **550 Fighting Infringement, Counterfeiting, & Piracy:** Services specialized on detecting and interfering IP infringements.
  - **551 Infringement Intelligence:** Services for searching and demonstrating IP infringements.
  - **552 Technical Infringement Analysis (Software/Circuits):** Services that comprise the technical detection of infringements (e.g. reverse engineering).

- **553 Crowd-Sourcing Platform for Infringement Search:** Service that allows an organization or an individual to collaborate with a community to find out if an organization or inventor has been involved in litigations or not.
  - **554 Collaboration with Customs/Police/Standards/Trademark Offices:** Assistance in searching and actively blocking infringed products through cooperation with customs.
  - **554 Technology Developments:** Services that support building technological solutions or technology developments that make it difficult to counterfeit.
- **560 Internationalisation Support:** Services for supporting internationalization and trade of IP. Includes assistance in finding investors and business partners abroad and also offering any advice in legal, strategic or politic topics for certain countries (e.g. local patent laws, local technologies and clusters, societal and environmental issues).
- **600 Media & People:** Publications, journals, blogs and educational materials on IP topic as well as unions and IPinterest groups.
- **610 IP Related Education and Publications:** Services based on specialized education and coaching in IP related topics and non-academic publishers specialized on IP topics.
  - **611 IP Related Education:** Services based on specialized education and coaching in IP related topics.
  - **612 IP Related Publications:** (Online) Journals focusing on IPR related topics. Includes internet blogs. Excludes IP related scientific publications from 680 IP-related scientific research.
  - **613 E-Learning Solutions for IP:** Internet-based education and online courses about intellectual property rights and related issues.
- **620 IP Software:** Various gadgets and instructions and data stored electronically and created for evaluating patents and IP related features.
  - **621 IP Portfolio Management Software:** Software for Managing IP Portfolio (e.g. Licensing and collecting of royalties, Application and Renewal support, IP decision management or IP portfolio related business intelligence solutions).
  - **622 IP Portfolio Management Software for Attorneys:** Specialised IP Portfolio Management software for patent attorneys.
  - **623 IP Valuation Software:** Software that valuates or supports valuation of patents and/or portfolios.
  - **624 IP Search Software:** Software or web-based platforms for searching patent databases (EPO, DPMA, USPTO, JPO). Includes further examining and monitoring of patent databases and providing patent information.
  - **625 Patent-Based Public Stock Indexes:** Stock indexes that are based on aggregated patent and technology value.
- **630 IP Databases:** Service related to organized collection of IPR related data, today typically in digital form. The data are typically organized to model relevant aspects of patents, intellectual property, and protected technology in a way that supports processes requiring patent related information.
  - **631 Patent Document Data:** Services related to collecting data on patents.
  - **632 IP Litigation Database:** Services related to collecting data on patent law cases.

- ○ **633 Online Databases of Industrial Property and Trademark Offices:** Official design, trademark and patent databases.
- **640 IP-Centric HR Services:** Headhunting and scouting services specialised on persons in the field of intellectual property. It includes services that help to recognize outstanding inventors among other IP community members, HR recruitment platforms and conferences on IP related topics for HR people for networking purposes.
  - ○ **641 IP Job Platforms:** Online platform posting IP expert vacancies.
  - ○ **642 IP HR Agencies/Services:** Headhunting services for finding IP experts.
- **650 Interest Group:** Organisations with IPR related political or legal strategies as main topic. Excludes associations of IP professionals.
- **660 Conferences and Meetings Specialized On IP Topics:** Gatherings or meetings for IP consultation, exchange of IP related information, or discussion, especially ones with a formal agenda on IP related topics.
- **670 Association of IP Professionals:** Networks and associations of professionals with business or academic interest in IP. Includes academic research groups and bar associations.
- **680 IP Related Scientific Research:** Scientific research and publications in the fields of intellectual property (mostly in an economic or legal perspective).

Chapter 12
Mathematical and Stochastic Models for Reliability in Repairable Industrial Physical Assets

Pablo A. Viveros Gunckel
Universidad Técnica Federico Santa María, Chile

Fredy A. Kristjanpoller
Universidad Técnica Federico Santa María, Chile

Adolfo Crespo Márquez
Universidad de Sevilla, Spain

Rene W. Tapia
RelPro SpA, Chile

Vicente González-Prida
Universidad de Sevilla, Spain

ABSTRACT

Generally, assets present a varied behaviour in their life cycle, which is related directly to the use given and consequently related to the technical assistance traditionally known as maintenance or maintenance policies. It can be of a diverse nature: perfect, minimum, imperfect, over-perfect, and destructive as appropriate. This feature requires the application of advanced techniques in order to model the behaviour of assets life, adapting ideally to each reality of use and wear out. In this chapter, the stochastic models PRP, NHPP, and GRP are explained with their conceptual, mathematic, and stochastic development. For each model, the conceptualization, parameterizing, and stochastic simulation are analysed. Additionally, complementing the analysis and resolution pattern, these models are concluded with a numeric application that allows one to show step by step the mathematic and stochastic development as appropriate.

1. INTRODUCTION

The model and analysis of repairable equipments are of great importance, mainly in order to increase the performance oriented to reliability and maintenance as part of the cost reduction in this last item. A reparable system is defined as:

A system that, after failing to perform one or more of its functions satisfactorily, can be restored to

DOI: 10.4018/978-1-4666-8222-1.ch012

fully satisfactory performance by any method other than replacement of the entire system. (Ascher & Feingold, 1984)

A system that, after failing in order to develop an activity properly, is possible to restore satisfactorily its functioning by some method. (Ascher & Feingold, 1984)

Depending on the type of maintenance given to an equipment, is possible to find 5 cases (Veber et al., 2008):

1. **Perfect Maintenance or Reparation:** Maintenance operation that restores the equipment to the condition "as good as new".
2. **Minimum Maintenance or Reparation:** Maintenance operation that restores the equipment to the condition "as bad as old".
3. **Imperfect Maintenance or Reparation:** Maintenance operation that restores the equipment to the condition "worse than new but better than old".
4. **Over-Perfect Maintenance or Reparation:** Maintenance operation that restores the equipment to the condition "better than new"
5. **Destructive Maintenance or Reparation:** Maintenance operation that restores the equipment to the condition "worse that old".

For a perfect maintenance, the most common developed model corresponds to the Perfect Renewal Process (PRP). In it, we assume that repairing action restores the equipment to a condition as good as new and assumes that times between failures in the equipment are distributed by an identical and independent way. The most used and common model PRP is the Homogeneous Processes of Poison (HPP), which considers that the system not ages neither spoils, independently of the previous pattern of failures. That is to say, it is a process without memory. Regarding case b), "as bad as old" is the opposite case to what happens in case a) "as good as new", since it is assumed that the equipment will stay after the maintenance intervention in the same state than before each failure. This consideration is based that the equipment is complex, composed by hundreds of components, with many failure modes and the fact that replacing or repairing a determined component will not affect significantly the global state and age of the equipment. In other words, the system is subject to minimum repairs, which does not cause any change or considerable improvement. The most common model to represent this case is through Non –Homogeneous Processes of Poison (NHPP), in this case the most used model to represent NHPP is called "Power Law". In this model, it is assumed a Weibull distribution for the first failure, that later it is modified over time.

Although the models HPP and NHPP are the most used, they have a practical restriction regarding its application, since a more realistic condition after a repairing action is what we find between both: "worse than new but better than old". In order to find a generalization to this situation and not distinguish between HPP and NHPP it was necessary to create the Generalized Renewal Process (GRP) (Kijima & Sumita, 1986), which establishes an improvement ratio. Unfortunately, the incorporation of this variable can complicate the analytic calculation of parameters and adjustments of probability. Therefore, its applicability in mathematic terms is complex. For this reason, it has been considered solutions through the Monte Carlo simulation (MC) being one of the most validated methods according the proposal developed by author Krivstov (2000) where time series of good functioning are generated through the use of the inverse function of the probability distribution (pdf) that has as a base a random variable.

Understanding the importance and applicability of methods PRP, NHPP and GRP this paper introduces the conceptual, mathematical and stochastic development for each one, as explained and presented briefly in the previous paragraphs. Each model is explained and developed in the following way: conceptualization, parameterising and stochastic simulation. In addition, each model will be complemented with a numerical application that allows to show step by step the mathematical and stochastic development as appropriate. The applied cases correspond to 3 pulp pumps (water, copper concentrate and inert material) used in the mining industry of Chile, which suffer different levels of erosion due to use intensity, geographic height and of course according to the maintenance type developed in its life, planned or not planned.

2. DEFINITION OF THE STOCHASTIC MODELS

2.1 Perfect Renewal Process (PRP)

The Perfect Renewal Process model describes the situation in which a repairable system is restored to a state "as good as new" and the times between failures are considered independent and identically distributed. This process assumes that the equipment restores to an identical condition to the original, as if it is replaced.

The graph of the failure rate depending on the time elapsed for an equipment with growing failure rate, considering the general case of a Weibull distribution, would be according to Figure 1.

As proved in Figure 1, the evolution of the failure rate is reset after each failure, evidently due to fact that the equipment remains in perfect conditions. The PRP model possesses main application over those equipments that have a complete maintenance over all its components or if it a 100% replacement of the equipment.

In mathematical terms: Be t_i the functioning time between failures i-1 and the i-th. Then, under a PRP model, any time t_i will obey to the same probability distribution with inalterable parameters in time, for example for the two-parameter Weibull case:

$$f(t_i) = \begin{cases} \dfrac{\beta}{\alpha}\left(\dfrac{t}{\alpha}\right)^{\beta-1} e^{-\left(\frac{t}{\alpha}\right)^{\beta}} & t \geq 0 \\ 0 & t < 0 \end{cases}$$

Being α and β continuous and inalterable in time. The origin of this type of distribution is in order to consider increasing or decreasing failure rates along the time from the last repair. Being α and β the scale and form parameters respectively.

Figure 1. Failure rate $\lambda(t)$ in Perfect Renewal Process

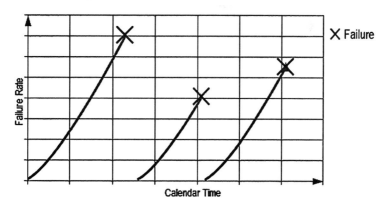

Parameterising of the PRP Model

Exponential Distribution

The density function of exponential probability correspond to the distribution that model the time between 2 random events described by a distribution of Poisson. The pdf correspond to Equation 1.

With statistical (shown in Equation 2).

The exponential function is applied commonly over the time of good functioning, specifically for those equipment that are found in useful life phase, since this allows working with a mean rate of constant failure.

Linking these parameters with mean time to failure (MTTF) and reliability R(t) these are defined as shown in Equation 3.

Equation 1. Pdf for exponential distribution

$$f(t) = \lambda e^{-\lambda t}; t \geq 0$$

Equation 2. Statistical of exponential distribution

$$E(t) = 1/\lambda, \quad \sigma^2 = 1/\lambda^2$$

Equation 3. MTTF and reliability for exponential distribution

$$MTTF = 1/\lambda, \quad R(t) = e^{-\lambda t}$$

Table 1. Time between failures of the pump 1 in hours

Number of Failure	Time between Failures [h]	Number of Failure	Time between Failures [h]
1	434.09	16	375.05
2	226.04	17	496.48
3	266.67	18	649.41
4	681.33	19	540.68
5	1127.85	20	491.59
6	634.69	21	947.31
7	474.84	22	718.78
8	38.72	23	953.63
9	31.89	24	182.2
10	711.52	25	391.55
11	726.9	26	327.71
12	574.15	27	986.04
13	1043.54	28	352.53
14	336.54	29	631.86
15	771.23	30	680.52

Table 2. Time between failures of pump P2 in hours

Number of Failure	Time between Failures [h]	Number of Failure	Time between Failures [h]
1	270.19	16	135.21
2	89.42	17	173.39
3	451.74	18	4,53
4	30.69	19	40.36
5	17.00	20	23.05
6	176.96	21	38.15
7	42.67	22	31.61
8	78.72	23	27.65
9	42.84	24	34.05
10	214.63	25	0.25
11	0.96	26	20.56
12	80.70	27	127.85
13	8.54	28	11.15
14	5.79	29	52.26
15	104.35	30	26.52

Mathematical and Stochastic Models for Reliability

Table 3. Time between failures of the pump P3 in hours

Number of Failure	Time between Failures [h]	Number of Failure	Time between Failures [h]
1	860.05	13	367.41
2	1608.24	14	2757.98
3	1134.24	15	355.50
4	2703.12	16	1084.39
5	645.38	17	855.52
6	95.15	18	280.52
7	1278.48	19	490.48
8	605.34	20	945.55
9	344.33	21	105.32
10	1054.68	22	127.33
11	680.57	23	61.85
12	405.38	24	326.30

The reliability for the case of an exponential distribution is possible to derive it of the following way, understanding that failure rate is constant:

$$\frac{dR(t)}{dt} = -\lambda R(t)$$
$$\frac{dR(t)}{R(t)} = -\lambda dt$$
$$\int \frac{dR(t)}{R(t)} = \int -\lambda dt$$
$$\ln(R(t)) = -\lambda t$$
$$R(t) = e^{-\lambda t}$$

Is important to indicate that the failure rate concept corresponds to the rate or probability that elements fail and that have survived until a determined time t.

In order to obtain the mean life or MTTF according to the most appropriate term, is possible to reach it through the following analysis:

$$R(t) = e^{-\lambda t}$$
$$F(t) = 1 - R(t) = 1 - e^{-\lambda t}$$
$$f(t) = F'(t) = \lambda e^{-\lambda t}$$
$$MTTF = E[t] = \int_0^\infty \left(\lambda e^{-\lambda t} \times t\right) dt = \frac{1}{\lambda}$$

The data fitting or parameterising for an exponential function, using the method of Likelihood Maximum, is expressed in the following way:

$$P(x_i \text{ in } [x_i, x_i + dx] \forall i \in \{1,...,n\}) = \prod_{i=1}^{n} f(x_i; \theta)$$

$$L(\theta) = \prod_{i=1}^{n} f(x_i; \theta)$$

Resolving we have:

$$L = \left[\prod_{i=1}^{n} \left\{\lambda e^{-\lambda t_i}\right\}\right]$$

$$L = \lambda^n e^{\left(-\lambda \sum_{i=1}^{n} t_i\right)}$$

Then, it is applied the natural logarithm in order to simplify the expression:

$$\ln(L) = \ln(\lambda^n) + \ln\left\{e^{-\lambda \sum_{i=1}^{n} t_i}\right\} =$$

$$n \ln(\lambda) + -\lambda \sum_{i=1}^{n} t_i$$

And finally, it is derived with respect to the adjusted parameter.

$$\frac{\partial \ln(L)}{\partial \lambda} = 0$$

$$\frac{\partial}{\partial \lambda}\left(n\ln(\lambda) + -\lambda\sum_{i=1}^{n} t_i\right) = 0$$

$$\left(n\frac{1}{\lambda} + -\sum_{i=1}^{n} t_i\right) = 0$$

$$\Rightarrow \lambda = \frac{n}{\left(\sum_{i=1}^{n} t_i\right)}$$

Then, the parameter L is equal to the multiplicative inverse of the average of times observed, that is to say the *mean rate* in which occurs the failure events.

Weibull Distribution

From a practical point of view, it is characterized by having 2 parameters, where β corresponds to the form parameter linked to the well-known bath curve and to the respective phase of life cycle of the asset, and the parameter α known as the scale parameter, which is linked directly with the variability and dispersion of the life data that the asset analyzed has. The probability density function of failure (pdf) corresponds to Equation 4.

In this case the failure rate is defined as shown in Equation 5.

The mean time to fail (MTTF) and reliability R (t) correspondingly are shown in Equation 6.

Note that Γ corresponds to the gamma function (Equation 7).

The case of the three-parameter Weibull distribution also exists. The addition of the third parameter to the Weibull distribution allows adjusting the location of the function, setting a minimum of lifetime γ in units of time, of course.

The addition of this parameter modifies the previous equations to the following form (Equations 8, 9, and 10).

The addition of this third parameter allows that the adjustment to the function of density of probability being more precise, achieving a major goodness of adjustment. The Weibull function for itself constitutes a generalization for the exponential function since for values of β=1 we have $\alpha = 1/\lambda$ and $\gamma = 0$

Equation 4. Pdf for a two-parameter Weibull distribution

$$f(t) = \frac{\beta}{\alpha}\left(\frac{t}{\alpha}\right)^{\beta-1} e^{-\left(\frac{t}{\alpha}\right)^{\beta}}$$

Equation 5. Failure rate for a two-parameter Weibull distribution

$$\lambda(t) = \frac{\beta}{\alpha}\left(\frac{t}{\alpha}\right)^{\beta-1}$$

Equation 6. MTTF and reliability for a two-parameter Weibull distribution

$$MTTF = \alpha \cdot \Gamma\left(1 + \frac{1}{\beta}\right), \quad R(t) = e^{-\left(\frac{t}{\alpha}\right)^{\beta}}$$

Equation 7. Gamma function

$$\Gamma(t) = \int_{0}^{\infty} x^{t-1} e^{-x} dx$$

Mathematical and Stochastic Models for Reliability

Equation 8. Pdf for a three-parameter Weibull distribution

$$f(t) = \frac{\beta}{\alpha}\left(\frac{t-\gamma}{\alpha}\right)^{\beta-1} e^{-\left(\frac{t-\gamma}{\alpha}\right)^{\beta}}$$

Equation 9. Failure rate for a three-parameter Weibull distribution

$$\lambda(t) = \frac{\beta}{\alpha}\left(\frac{t-\gamma}{\alpha}\right)^{\beta-1}$$

Equation 10. MTTF and reliability for a three-parameter Weibull distribution

$$MTTF = \gamma + \alpha \cdot \Gamma\left(1 + \frac{1}{\beta}\right), \quad R(t) = e^{-\left(\frac{t-\gamma}{\alpha}\right)^{\beta}}$$

The adjustment through Weibull distribution allows representing the state of equipment in any of the 3 phases of the bath curve, through the adjustment of the parameter β that acts as a factor to the time *t* in the failure rate function.

For the parameterising, the function of likelihood maximum is expressed as:

$$P(x_i \text{ in } [x_i, x_i+dx] \forall i \in \{1,...,n\}) = \prod_{i=1}^{n} f(x_i;\theta)$$

$$L(\theta) = \prod_{i=1}^{n} f(x_i;\theta)$$

Nevertheless, resolving it is:

$$L(\theta \mid t_1,...,t_n) = \left[\prod_{i=1}^{n}\left\{\frac{\beta}{\alpha}\left(\frac{t_i}{\alpha}\right)^{\beta-1} e^{-\left(\frac{t_i}{\alpha}\right)^{\beta}}\right\}\right]$$

$$L(\theta \mid t_1,...,t_n) = \left(\frac{\beta}{\alpha^{\beta}}\right)^{n}\left[\prod_{i=1}^{n}\left\{t_i^{\beta-1} e^{-\left(\frac{t_i}{\alpha}\right)^{\beta}}\right\}\right]$$

Then, it is applied the natural logarithm (shown in Box 1).

Then, it is derived partially with respect of each parameter and equalize to 0. This is:

$$\frac{\partial}{\partial \theta} \ln\left[L(\theta \mid t_1,...,t_n)\right] = 0$$

Derived partial over α (Box 2).

Box 1.

$$\ln\left[L(\theta \mid t_1,...,t_n)\right] = \ln\left[\left(\frac{\beta}{\alpha^{\beta}}\right)^{n}\right] + \sum_{i=1}^{n}\left[\ln\left[t_i^{\beta-1} e^{-\left(\frac{t_i}{\alpha}\right)^{\beta}}\right]\right]$$

$$\ln\left[L(\theta \mid t_1,...,t_n)\right] = n\ln\left[\left(\frac{\beta}{\alpha^{\beta}}\right)\right] + \sum_{i=1}^{n}\left[\ln\left[t_i^{\beta-1}\right] - \left(\frac{t_i}{\alpha}\right)^{\beta}\right]$$

$$\ln\left[L(\theta \mid t_1,...,t_n)\right] = n\left(\ln[\beta] - \beta\ln[\alpha]\right) + \sum_{i=1}^{n}\left[\ln\left[t_i^{\beta-1}\right]\right] - \sum_{i=1}^{n}\left[\left(\frac{t_i}{\alpha}\right)^{\beta}\right]$$

Box 2.

$$\frac{\partial}{\partial \alpha}\left(n\left(\ln[\beta] - \beta\ln[\alpha]\right) + (\beta-1)\sum_{i=1}^{n}\left[\ln[t_i]\right] - \sum_{i=1}^{n}\left[\left(\frac{t_i}{\alpha}\right)^j\right]\right) = 0$$

$$n\left(\frac{-\beta}{\alpha}\right) - \beta\alpha^{-(j+1)}\sum_{i=1}^{n}\left[t_i^{\,j}\right] = 0$$

$$\frac{-n}{\alpha^{-j}} = \sum_{i=1}^{n}\left[t_i^{\,j}\right]$$

$$\boxed{\alpha = \left(\frac{-\sum_{i=1}^{n}\left[t_i^{\,j}\right]}{n}\right)^{1/j}} \quad (1)$$

And then the partial derivative over β (shown in Box 3).

With this, it is used the equivalence of (1) and it is replaced in the former equation to obtain finally:

$$\boxed{\frac{n}{\beta} + \sum_{i=1}^{n}\left[\ln[t_i]\right] = \left(\frac{n\sum_{i=1}^{n}\left[t_i^{\,j}\ln(t_i)\right]}{\sum_{i=1}^{n}\left[t_i^{\,j}\right]}\right)} \quad (2)$$

For the resolution of this type of adjustment, there are specialized softwares. One of these is RelPro (2014) which disposes of advanced and efficient algorithm for the resolution of this kind of problems. The advantages of this tool is the flexibility of analysis of different stochastic models, the potential analysis post parameterising, the usage and fast learning of the user and evidently the capacity to model, analyze and simulate simple and complex systems. RelPro solves the equation (2) for random search over the parameter β and then obtaining this parameter, it is replaced in the equation (1) and it is obtained the parameter a.

Box 3.

$$\frac{\partial}{\partial \beta}\left(n\left(\ln[\beta] - \beta\ln[\alpha]\right) + (\beta-1)\sum_{i=1}^{n}\left[\ln[t_i]\right] - \sum_{i=1}^{n}\left[\left(\frac{t_i}{\alpha}\right)^j\right]\right) = 0$$

$$n\left(\frac{1}{\beta} - \ln[\alpha]\right) + \sum_{i=1}^{n}\left[\ln[t_i]\right] = \left(\alpha^{-j}\left(\ln(\alpha)\sum_{i=1}^{n}\left[t_i^{\,j}\right] - \sum_{i=1}^{n}\left[t_i^{\,j}\ln(t_i)\right]\right)\right) = 0$$

$$n\left(\frac{1}{\beta} - \ln[\alpha]\right) + \sum_{i=1}^{n}\left[\ln[t_i]\right] = \left(-n\ln(\alpha) - \alpha^{-j}\sum_{i=1}^{n}\left[t_i^{\,j}\ln(t_i)\right]\right) = 0$$

Simulation for the PRP Model

The simulation of functioning times in equipment represented by the PRP model is relatively simple, and it is only necessary to develop the resolution process of the inverse function.

It is obtained the Weibull function.

$$F(t) = \int_0^x f(t) = \int_0^x \frac{\beta}{\alpha}\left(\frac{t}{\alpha}\right)^{j-1} e^{-\left|\frac{t}{\alpha}\right|^j} dt$$

$$F(t) = 1 - e^{-\left|\frac{t}{\alpha}\right|^j} = u$$

The accumulated distribution function is evidently distributed between 0 and 1.

The resolution of the random variable t in function of a variable evenly distributed could be:

$$F(t) = u \sim U[0,1]$$

$$1 - e^{-\left|\frac{t}{\alpha}\right|^j} = u$$

$$e^{-\left|\frac{t}{\alpha}\right|^j} = 1 - u$$

$$\ln\left(e^{-\left|\frac{t}{\alpha}\right|^j}\right) = \ln(1-u)$$

$$\left(\frac{t}{\alpha}\right)^j = -\ln(1-u)$$

$$\frac{t}{\alpha} = \left(-\ln(1-u)\right)^{1/j}$$

$$\boxed{t = TBF = \alpha\left(-\ln(1-u)\right)^{1/j} \wedge u \in U[0,1]}$$

In this way, from the creation of a random variable μ, is possible to generate the respective values of t. Moreover, in this case as this is a complete renewal process, any t will be equivalent to a new time of simulated good functioning (TTF) knowing TTF as the time that the equipment operates without anomalies, according the applied usage profile.

Numeric Exemplification: PRP Case

The numeric case corresponds to the analysis of a pulp pump 1 (P1). According to Table 1 contains the time between failures records.

Then, in order to develop the respective adjustment of probability distribution, it is implemented step by step:

Step 1: Tendency Test

According the information shown in Table 1, for the pump 1 has been recorded 30 times between failures, in hours and in order of occurrence.

According to literature, it is necessary to develop a tendency test that allows determining quantitatively if the times of functioning of a system show a significate tendency, which can be an improvement or a degradation tendency. In this case it has been chosen the tendency test of Laplace (Nist/Sematech, 2014) optimum method to distinguish between "there is not tendency" or "there is tendency" following the model of exponential law NHPP. Among other tests there is the *Test of inverse order"* and the *"Test of military book of EEUU* (Nist/Sematech, 2014) both of them do not apply due to fact that they are for cases such as application of Duane Model (Duane, 1964).

The Laplace Test gives as result a statistician z that must be compared with the standard normal distribution, due to it approaches to a random variable. The statistician z is calculated as:

$$z = \frac{\frac{\sum_{i=1}^{n} t_i}{n} - \frac{t_o}{2}}{t_o \sqrt{\frac{1}{12n}}}$$

where

t_o: total time lapsed,
n: total number of times,
t_i: total time lapsed to the event i.

It is assumed that for statistician z:

$z = 0$; without tendency

$z > 0$; tendency to spoilage

$z < 0$; tendency to improvement

Evaluating the expression, we obtain a z value=-0.0953.

From the standard normal distribution, considering a significance level equal to 0.10, the critical value is equal to 1.645. If -1.645<z<1.645 then, it is failed refusing the hypothesis that there is no tendency. Therefore, having a value z equal to -0.0953 is not possible to say that there is tendency in the data, therefore PRP modelling is applicable.

Step 2: Distribution Adjustment

Once the PRP modelling is appropriately justified (there is not tendency) it is proceeded to apply the distribution adjustment under the two-parameter Weibull distribution.

According to the method of maximum plausibility previously presented, the values of parameters $\{\alpha, \beta\}$ that give the maximal likelihood, are those that satisfy the following equations:

$$\hat{\alpha} = \left(\frac{\sum_{i=1}^{n} (t_i)^{\hat{\beta}}}{n} \right)^{1/\hat{\beta}}$$

$$\frac{\sum_{i=1}^{n} (t_i)^{\hat{\beta}} \ln(t_i)}{\sum_{i=1}^{n} (t_i)^{\hat{\beta}}} - \frac{1}{\hat{\beta}} = \frac{1}{n} \sum_{i=1}^{n} \ln(t_i)$$

Solving the equations through the software RelPro we have that the parameters are $\alpha=632.04$ and $\beta=1.99$ respectively.

2.2 Non-Homogeneous Poisson Process

NHPP is a Poisson process with a parametric model used to represent events with an occurrence of evolutional failure in time and always with the same tendency. This case applies specially for those equipments that are composed by many components where the replacement of one of them does not affect the global reliability: consider an equipment composed by hundreds of component that work in series, if one of them fails, this component is replaced and the equipment continues working but with a level of waste almost identical to previous one. For this reason the NHPP model applies for the called "minimum maintenance"

Next, in Figure 2 is presented the graph for the behaviour of the failure rate in time, being this completely accumulative between one and other failure.

As we appreciate in the former graphic, for the case of NHPP, the failure rate remains dependent on total time elapsed.

Figure 2. Failure rate λ(t) in NHPP model

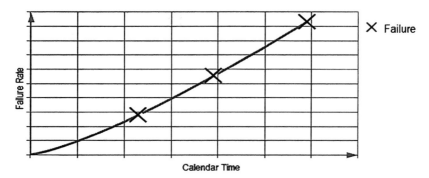

In the case of NHPP, the functions of the reliability and failure probability are expressed according to Figure 3.

Having as a base the former graphic, let's consider that one equipment has a failure in a t_1 time. After being repaired, the functioning is restarted and begins to work in that same point. Then, the reliability function from t_1, for a t time that represents the elapsed time beyond t_1, will be gives as:

$$R(t \mid t > t_1) = \frac{R(t)}{R(t_1)} = 1 - F(t \mid t > t_1)$$

This is called by various authors (Bebbington et al., 2009) as "Mission Time", where t corresponds to elapsed calendar time. For a Weibull distribution, from the previous equation, is obtained:

$$R(t_i) = e^{-\left(\frac{t_i}{\alpha}\right)^j} \wedge R(t_{i-1}) = e^{-\left(\frac{t_{i-1}}{\alpha}\right)^j}$$

$$R(t_i \mid t_i > t_{i-1}) = \frac{e^{-\left(\frac{t_i}{\alpha}\right)^j}}{e^{-\left(\frac{t_{i-1}}{\alpha}\right)^j}} = 1 - F(t_i \mid t_i > t_{i-1})$$

$$\Rightarrow F(t_i \mid t_i > t_{i-1}) = 1 - \exp\left\{\left(\frac{t_{i-1}}{\alpha}\right)^j - \left(\frac{t_i}{\alpha}\right)^j\right\}$$

Figure 3. R(t) and f(t) on NHPP

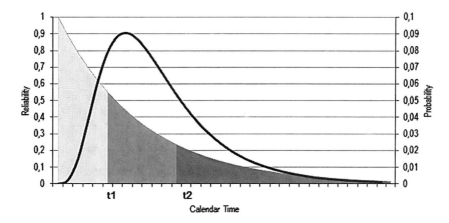

where t_{i-1} corresponds to the *total time* elapsed until the last failure and t_i the total time (calendar) elapsed after generate the failure i-th. From this is possible to conclude the following pdf:

$$f(t_i \mid t_i > t_{i-1}) = \frac{\beta}{\alpha}\left(\frac{t_i}{\alpha}\right)^{j-1} \exp\left\{\left(\frac{t_{i-1}}{\alpha}\right)^j - \left(\frac{t_i}{\alpha}\right)^j\right\}$$

Parameterising of the NHPP Model

In order to obtain the parameters α and β, the lineal regression is not a choice. It is ideally made an adjustment by maximum likelihood.

$$P(x_i \text{ in } [x_i, x_i + dx] \;\forall i \in \{1,...,n\}) = \prod_{i=1}^{n} f(x_i;\theta)$$

$$L(\theta) = \prod_{i=1}^{n} f(x_i;\theta)$$

where Θ correspond to the vector of the parameters of distribution to which obey the $f(t)$. Moreover x_i corresponds to the element i-th of the sample. As it is wished to obtain maximum likelihood between the data and one pdf: $f(t;\Theta)$, the values of the vector Θ are adjusted with the aim to reach that maximum.

Conceptually, parameters are searched in order to better fit to a sample $x_1,...,x_n$ in such way that the probability of the series of values that be presented in a random sample be maximal:

$$P(x_i \text{ in } [x_i, x_i + dx] \;\forall I \in \{1,...,n\})$$

The parameters Θ that give maximum likelihood are those which fulfil with:

$$\frac{\partial L}{\partial \theta} = 0, \quad i = 1, 2, ... m$$

In this way, for the actual case, the likelihood function is shown in Box 4.

Box 4.

$$L = \underbrace{\left[\frac{\beta}{\alpha}\left(\frac{t_1}{\alpha}\right)^{j-1} e^{-\left(\frac{t_i}{\alpha}\right)^j}\right]}_{f(t_1)} \cdot \prod_{i=2}^{n}\left\{\left(\frac{\beta}{\alpha}\right)\left(\frac{t_i}{\alpha}\right)^{j-1} e^{\left[\left(\frac{t_{i-1}}{\alpha}\right)^j - \left(\frac{t_i}{\alpha}\right)^j\right]}\right\}$$

Box 5.

$$L = \underbrace{\left[\frac{\beta}{\alpha}\left(\frac{t_1}{\alpha}\right)^{j-1} e^{-\left(\frac{t_i}{\alpha}\right)^j}\right]}_{f(t_1)} \cdot \left[\left(\frac{\beta}{\alpha}\right)^{n-1} \prod_{i=2}^{n}\left\{\left(\frac{t_i}{\alpha}\right)^{j-1} e^{\left[\left(\frac{t_{i-1}}{\alpha}\right)^j - \left(\frac{t_i}{\alpha}\right)^j\right]}\right\}\right]$$

$$\rightarrow L = \underbrace{\left[\frac{\beta}{\alpha}\left(\frac{t_1}{\alpha}\right)^{j-1} e^{-\left(\frac{t_i}{\alpha}\right)^j}\right]}_{f(t_1)} \cdot \left[\left(\frac{\beta}{\alpha}\right)^{n-1} e^{\sum_{i=2}^{n}\left[\left(\frac{t_{i-1}}{\alpha}\right)^j - \left(\frac{t_i}{\alpha}\right)^j\right]} \prod_{i=2}^{n}\left(\frac{t_i}{\alpha}\right)^{j-1}\right]$$

Mathematical and Stochastic Models for Reliability

Simplifying (Box 5).

After applying partial derivatives and equal to 0, the result of the approaches for NHPP are the following:

$$\hat{\alpha} = \frac{t_n}{n^{1/j}}$$

$$\hat{\beta} = \frac{n-1}{\sum_{i=1}^{n-1} \ln\left(\frac{t_n}{t_i}\right)}$$

where t_i corresponds to the elapsed time until the failure i-th and t_n the elapsed time until the last failure.

Simulation for NHPP Model

The lifetime simulation for the equipment modelled by NHPP, having the Weibull case as a base, will be as follows:

$$F(t_i) = 1 - \exp\left\{\left(\frac{t_{i-1}}{\alpha}\right)^j - \left(\frac{t_i}{\alpha}\right)^j\right\}$$

$$F(t_i) = u \in U[0,1]$$

$$1 - \exp\left\{\left(\frac{t_{i-1}}{\alpha}\right)^j - \left(\frac{t_i}{\alpha}\right)^j\right\} = u$$

$$\exp\left\{\left(\frac{t_{i-1}}{\alpha}\right)^j - \left(\frac{t_i}{\alpha}\right)^j\right\} = 1 - u$$

$$\ln\left(\exp\left\{\left(\frac{t_{i-1}}{\alpha}\right)^j - \left(\frac{t_i}{\alpha}\right)^j\right\}\right) = \ln(1-u)$$

$$\left(\frac{t_{i-1}}{\alpha}\right)^j - \left(\frac{t_i}{\alpha}\right)^j = \ln(1-u)$$

$$\left(\frac{t_i}{\alpha}\right)^j = \left(\frac{t_{i-1}}{\alpha}\right)^j - \ln(1-u)$$

$$\frac{t_i}{\alpha} = \left(\left(\frac{t_{i-1}}{\alpha}\right)^j - \ln(1-u)\right)^{1/j}$$

$$t_i = \alpha\left(\left(\frac{t_{i-1}}{\alpha}\right)^j - \ln(1-u)\right)^{1/j}$$

$$t_i = \left(\alpha^j \left(\frac{t_{i-1}}{\alpha}\right)^j - \alpha^j \cdot \ln(1-u)\right)^{1/j}$$

$$\boxed{t_i = \left((t_{i-1})^j - \alpha^j \cdot \ln(1-u)\right)^{1/j} \wedge u \in U[0,1]}$$

In this case, t_i is the total elapsed time, so the time between failures will be expressed as:

$$TBF_i = t_i - t_{i-1}$$

$$\boxed{TBF_i = \left(t_{i-1}^j - \alpha^j \cdot \ln(1-u)\right)^{1/j} - t_{i-1}}$$

Like for PRP, it is here suggested a previous tendency test, commonly developed in order to identify if the equipment has or not modelled behaviour by NHPP.

Numeric Exemplification: NHPP Case

The numeric case corresponds to the analysis of a pulp pump 2(P2). According to, Table 2 contains the time between failures records.

Then, in order to develop the respective adjustment of probability distribution, it is developed step by step:

Step 1: Tendency Test

According to the information shown in Table 2, for the pump 2 it has been recorded 30 times between failures, in hours and in occurrence order.

Again, it is necessary to develop a tendency test, where (as in the previous case) it has been chose the Laplace tendency test, which gives as a result a statistician z that must be compared with the standard normal distribution, as it approaches to a random variable. The statistician z is calculated according to:

$$z = \frac{\frac{\sum_{i=1}^{n} t_i}{n} - \frac{t_o}{2}}{t_o \sqrt{\frac{1}{12n}}}$$

t_o: Total accumulated time,
n: Number of times,
t_i: Total accumulated time until element i.

This model is strictly valid for n>3.

$z = 0$; there is no tendency

$z > 0$; increasing tendency

$z < 0$; decreasing tendency

Evaluating the previous expression we obtain a value $z = 3.3559$. From the standard normal distribution, considering a significance level equal to 0.10, the critical value is equal to 1.645. If $-1.645 < z < 1.645$ then is failed when refusing the hypothesis that there is no tendency. Therefore, having a value z equal to 3.3559 is possible to say that there is tendency in the data, therefore, it is applicable the NHPP model.

Step 2. Distribution Adjustment

Once the NHPP model is justified appropriately (there is tendency), we proceed to apply the distribution adjustment under two-parameter Weibull distribution.

According to maximum likelihood method previously shown, the values of the parameters Θ now $\{α, β\}$ that give the maximum likelihood are those which satisfy the following equations already shown:

$$\hat{\beta} = \frac{n-1}{\sum_{i=1}^{n-1} \ln\left(\frac{t_n}{t_i}\right)}$$

Solving the equations through the software RelPro® we have that the parameters are $α = 632.04$, $β = 1.99$ respectively.

2.3 Generalized Renewal Process

The traditional models already shown are only able to model 2 types of maintenance: the completely perfect and the completely imperfect. GRP model is the generalization for any level of perfection that has the maintenance, including the both mentioned.

GRP adds a new parameter, called "virtual age". The parameter A_n represents the age of the system at the immediate instant when the n-th repair is carried out. In this way, if $A_n = y$, the element has a time of functioning associated to a probability distribution conditioned for this age y. That is to say, all the failure times have different probability distributions as the time passes by.

Graphically, Figure 4 represents the change of failure rate after each failure.

The way to incorporate this variable is considering that equipment begins to operate with certain waste, which is reflected in the reliability function.

Mathematical and Stochastic Models for Reliability

Figure 4. Failure rate on GRP

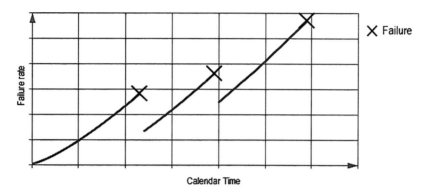

In this manner, the accumulated reliability and probability distribution for t_{n+1} is:

$$F(t \mid A_n = y) = \frac{F(t+y) - F(y)}{1 - F(y)}$$

$$R(t \mid A_n = y) = \frac{R(t+y)}{R(t)}$$

By this way, it is clear that this "virtual age" is the age of waste in which the equipment begins to work again. The reliability function remains similar to the "Mission Time" only that this does not correspond to a real time elapsed, but to an equivalent.

Be x the i-th time of good functioning and T_n the total accumulated time elapsed until failure n-th. As follow:

$$T_n = \sum_{i=1}^{n} x_i \quad (1)$$

On the other hand, the parameter A_n is given by:

$$A_n = A_{n-1} + q \cdot x_n \quad (2)$$

Using (1) remains that:

$$A_n = qT_n = q \sum_{i=1}^{n} x_i \quad (3)$$

where q is the parameter that decides the ineffectiveness of the repair, in this way q=0 implies that $A_n = 0$, that is to say virtual age equal to 0. Therefore q=0 correspond to a perfect repair case, that is to say is completely effective. In the case it was $q=1$ then it begins to operate in the same part of the reliability function where the equipment failed. This would be:

$0 < q < 1$: GRP

$q = 0$: PRP (HPP)

$q = 1$: NHPP

Plotting the existing relation between real life and virtual age that evolves, it is possible to generate a comparative graphic for PRP, NHPP and GRP. Details in Figure 5.

Figure 5. Real age against virtual age on PRP, NHPP, and GRP

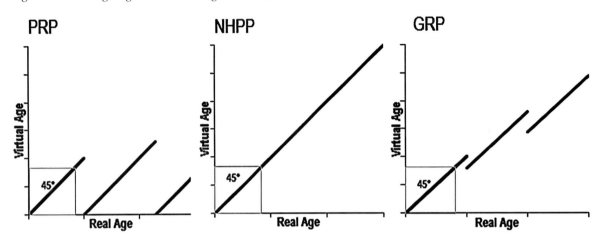

As in NHPP, it is determined the conditioned reliability and the respective pdf.

$$R(t_i \mid t_i > q \cdot t_{i-1}) = e^{\left(\frac{q \cdot t_{i-1}}{\alpha}\right)^{\beta} - \left(\frac{t_i}{\alpha}\right)^{\beta}}$$

$$F(t_i \mid t_i > q \cdot t_{i-1}) = 1 - e^{\left(\left(\frac{q \cdot t_{i-1}}{\alpha}\right)^{\beta} - \left(\frac{t_i}{\alpha}\right)^{\beta}\right)}$$

$$f(t_i \mid t_i > q \cdot t_{i-1}) = \frac{\beta}{\alpha}\left(\frac{t_i}{\alpha}\right)^{\beta-1} e^{\left(\left(\frac{q \cdot t_{i-1}}{\alpha}\right)^{\beta} - \left(\frac{t_i}{\alpha}\right)^{\beta}\right)}$$

Parameterising of the GRP Model

The adjustment developed is on the basis of a pdf with two-parameter Weibull (α, β), and adding the parameter q, so then we have 3 parameters to determine. The most common approach for parameters corresponds by maximum likelihood to the following expression (Box 6).

With the partial derivative in each variable, it is obtained a set of 3 equations with 3 unknown quantities, these are: α, β, q.

Then, there is a set of 3 equations, which are:

$$\frac{\partial[\ln(L)]}{\partial \alpha} = 0, \frac{\partial[\ln(L)]}{\partial \beta} = 0, \frac{\partial[\ln(L)]}{\partial q} = 0$$

Developing each partial derivative, there is (Box 7).

Generalizing the main expressions, there is (Box 8).where: $X_i = t_i + t_{i-1}(q - 1)$ ^ $Y_i = q \cdot t_{i-1}$

The parameters, α, β, q are the 3 values to identify. Searching these parameters is a very exhaustive procedure, as it requires more precision in the procedure, generally is a long process, so it is suggested to use the Monte Carlo simulation.

The searching of α, β, q starts with the simulation of q and β, repeated by even distributions.

Box 6. Likelihood function

$$L = \underbrace{\left[\frac{\beta}{\alpha}\left(\frac{t_1}{\alpha}\right)^{\beta-1} e^{-\left(\frac{t_1}{\alpha}\right)^{\beta}}\right]}_{f(t_1)} \cdot \left[\prod_{i=2}^{n}\left\{\left(\frac{\beta}{\alpha}\right)\left(\frac{t_i}{\alpha}\right)^{\beta-1} \exp\left(\left(\frac{q \cdot t_{i-1}}{\alpha}\right)^{\beta} - \left(\frac{t_i}{\alpha}\right)^{\beta}\right)\right\}\right]$$

Box 7.

$$\frac{\beta}{\alpha^{\beta+1}}\left[\sum_{i=2}^{n}\left[\left(t_i+t_{i-1}(q-1)\right)^{\beta}-\left(q\cdot t_{i-1}\right)^{\beta}\right]\right]+\frac{\beta}{\alpha}\left[\left(\frac{t_1}{\alpha}\right)-(n)\right]=0 \quad (1)$$

$$\left[\frac{(n)}{\beta}+\ln(t_1)-(n)\ln(\alpha)-\left(\frac{t_1}{\alpha}\right)^{\beta}\ln\left(\frac{t_1}{\alpha}\right)\right]+$$

$$\sum_{i=2}^{n}\left[\begin{array}{l}\ln\left(t_i+t_{i-1}(q-1)\right)-\left(\dfrac{t_i+t_{i-1}(q-1)}{\alpha}\right)^{\beta}\\ \ln\left(\dfrac{t_i+t_{i-1}(q-1)}{\alpha}\right)+\left(\dfrac{q\cdot t_{i-1}}{\alpha}\right)^{\beta}\ln\left(\dfrac{q\cdot t_{i-1}}{\alpha}\right)\end{array}\right]=0 \quad (2)$$

$$(\beta-1)\sum_{i=2}^{n}\left[\frac{t_{i-1}}{t_i+t_{i-1}(q-1)}\right]+\frac{\beta q^{(\beta-1)}}{\alpha^{\beta}}\sum_{i=2}^{n}(t_{i-1})^{\beta}-$$

$$\frac{\beta}{\alpha^{\beta}}\sum_{i=2}^{n}\left(t_i+t_{i-1}(q-1)\right)^{\beta-1}(t_{i-1})=0 \quad (3)$$

Box 8.

$$\frac{\beta}{\alpha^{\beta+1}}\left[\sum_{i=2}^{n}\left[X_i^{\beta}-Y_i^{\beta}\right]\right]+\frac{\beta}{\alpha}\left[\left(\frac{t_1}{\alpha}\right)-(n)\right]=0 \quad (1)$$

$$\left[\frac{(n)}{\beta}+\ln(t_1)-(n)\ln(\alpha)-\left(\frac{t_1}{\alpha}\right)^{\beta}\ln\left(\frac{t_1}{\alpha}\right)\right]+$$

$$\sum_{i=2}^{n}\left[\ln(X_i)-\left(\frac{X_i}{\alpha}\right)^{\beta}\ln\left(\frac{X_i}{\alpha}\right)+\left(\frac{Y_i}{\alpha}\right)^{\beta}\ln\left(\frac{Y_i}{\alpha}\right)\right]=0 \quad (2)$$

$$(\beta-1)\sum_{i=2}^{n}\left[\frac{t_{i-1}}{X_i}\right]+\frac{\beta q^{(\beta-1)}}{\alpha^{\beta}}\sum_{i=2}^{n}(t_{i-1})^{\beta}-\frac{\beta}{\alpha^{\beta}}\sum_{i=2}^{n}(X_i)^{\beta-1}(t_{i-1})=0 \quad (3)$$

$$q \sim U[0,1]$$
$$\beta \sim U[0,10]$$

The parameter q only can be between 0 and 1 and the parameter β hardly exceed the value 10 (Yañez et al., 2002). On the other hand, the value of α has a wide rank of possibilities that is why it is used the third equation of the system which allows expressing α from q and β. Similar to approach a for GRP Model remains:

$$\hat{\alpha} = \sqrt[\beta]{\frac{\sum_{i=2}^{n}\left[\left(t_i + t_{i-1}(q-1)\right)^{\beta} - \left(q \cdot t_{i-1}\right)^{\beta}\right] + t_1^{\beta}}{n}}$$

Then the values of α, β, q are given by those 3 values that satisfy in a better way the first 2 equations of the set. (The third one is obviated since it is used to obtain a).

Figure 6 represents the procedure for the GRP adjustment.

Figure 6. Diagram process for GRP adjustment

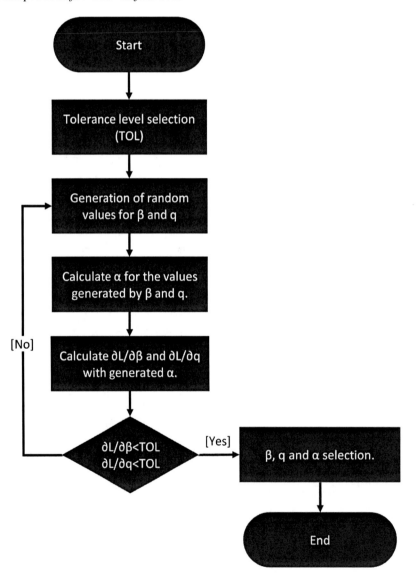

As far as the probability to find the values that grant the global maximum of the maximum likelihood function is virtually invalid through random search, it is necessary to define a value of tolerance for the partial derivatives ($\partial L/\partial B$ and $\partial L/\partial q$) equal to 0, being necessary to fix this value of tolerance "TOL" as an acceptable rank to consider that is found in a global maximum and in this way to accept the respective distribution adjustment.

Considering the adding of a new parameter, in this case q, always the adjustment GRP will give a higher likelihood than a PRP or NHPP adjustment. Nevertheless, in order to consider the existence and applicability of these cases, it is necessary to count on selection criteria. This is applied after the adjustment through GRP once obtained the parameter q.

As q value is always a continuous value, the probability to be exactly q=1 or q=0 is practically null, therefore it is considered a new tolerance level, which has been called TQ.

This tolerance level corresponds to higher and lower percentage of the possibilities that value q has.

This goal of this value is to identify when would be more appropriate to consider a PRP or NHPP model. Therefore the practical expression corresponds to:

$0 + TQ < q < 1 - TQ$: GRP
$q \leq TQ$: PRP
$q \geq 1 - TQ$: NHPP

Simulation for GRP Model

For the simulation, the case is very similar to NHPP Model, due to fact that it is originated from the function of density of conditional probability that the equipment has survived until a given **t** time.

Considering (Box 9):

Then, as t_i corresponds to the total time elapsed, a simulated TBF value will be depicted by:

$$TBF_i = t_i - t_{i-1}$$

$$\boxed{TBF_i = \left((q \cdot t_{i-1})^\beta - \alpha^\beta \cdot \ln(1-u) \right)^{1/\beta} - q \cdot t_{i-1}}$$

Numeric Exemplification: GRP Case

The numeric case corresponds to the analysis of a pulp pump 3 (P3). According to, Table 3 contains the time between failures records.

Therefore, in order to develop the respective adjustment of probability distribution, it is developed step by step:

Step 1: Tolerance Level

The tolerance level is defined for the partial derivatives as TOL= 0.01 and tolerance for q value is TQ=5%

Step 2: Distribution Adjustment

Once defined the tolerance level for GRP adjustment it is proceed with respective adjustment.

Solving the equations through RelPro software, the parameters are:

α=1986.067; β= 2.026 and q =0.192

With these parameters we obtain:

$$\left| \frac{\partial[\ln(L)]}{\partial \beta} \right| = 1,76 \times 10^{-7} < TOL = 0,01$$

$$\left| \frac{\partial[\ln(L)]}{\partial q} \right| = 0,009833 < TOL = 0,01$$

Therefore, the distribution adjustment is suitable, due to fact that the tolerance level in the random search is completely appropriately.

Box 9.

$$F(t_i) = 1 - \exp\left\{\left(\frac{q \cdot t_{i-1}}{\alpha}\right)^\beta - \left(\frac{t_i}{\alpha}\right)^\beta\right\} \wedge F(t_i) = u \in U[0,1]$$

$$1 - \exp\left\{\left(\frac{q \cdot t_{i-1}}{\alpha}\right)^\beta - \left(\frac{t_i}{\alpha}\right)^\beta\right\} = u$$

$$\exp\left\{\left(\frac{q \cdot t_{i-1}}{\alpha}\right)^\beta - \left(\frac{t_i}{\alpha}\right)^\beta\right\} = 1 - u$$

$$\ln\left(\exp\left\{\left(\frac{q \cdot t_{i-1}}{\alpha}\right)^\beta - \left(\frac{t_i}{\alpha}\right)^\beta\right\}\right) = \ln(1-u)$$

$$\left(\frac{q \cdot t_{i-1}}{\alpha}\right)^\beta - \left(\frac{t_i}{\alpha}\right)^\beta = \ln(1-u)$$

$$\left(\frac{t_i}{\alpha}\right)^\beta = \left(\frac{q \cdot t_{i-1}}{\alpha}\right)^\beta - \ln(1-u)$$

$$\frac{t_i}{\alpha} = \left(\left(\frac{q \cdot t_{i-1}}{\alpha}\right)^\beta - \ln(1-u)\right)^{1/\beta}$$

$$t_i = \alpha\left(\left(\frac{q \cdot t_{i-1}}{\alpha}\right)^\beta - \ln(1-u)\right)^{1/\beta}$$

$$t_i = \left(\alpha^\beta \left(\frac{q \cdot t_{i-1}}{\alpha}\right)^\beta - \alpha^\beta \cdot \ln(1-u)\right)^{1/\beta}$$

$$\boxed{t_i = \left((q \cdot t_{i-1})^\beta - \alpha^\beta \cdot \ln(1-u)\right)^{1/\beta} \wedge u \sim U[0,1]}$$

Regarding q value there is:

$$TQ = 0,05 < q = 0,192 < 1 - TQ = 0,95$$

Therefore, when we find an adjustment solution by maximum likelihood with acceptable derivatives that guarantee the quality of the adjustment and a q parameter with a numeric value between the rank $0.05 < q < 0.95$, it is possible to affirm that the use of GRP model is suitable for the case.

3. CONCLUSION

The reliability model is an essential aspect for the management and optimization of physical industrial assets. The wide range and variability of their behaviour, demands the application of techniques of diverse complexity and depth, that allow adapting in a better way to each one of the realities.

The variable that defines and determines the use of the techniques is the state in which remains the assets after a repair. In this sense, the repair classifications are five: perfect, minimum, imperfect, over-perfect and destructive.

For a perfect maintenance, it is used and recommended the Perfect Renewal Process PRP through the Homogeneous Poisson Process HPP. In the case of a minimum repair, generally is depicted through Non-Homogeneous Poisson Process NHPP being the most used model the "Power Law".

In spite of the large application of the previous models, there are varied situations that are not covered due to the fact that the generality of the repair cases are between the perfect and the minimum. For these situations, it is headed and developed the "Generalized Renewal Process" GRP. Given the flexible structure that the GRP model has, that allows adapting to the different kinds of maintenance, its development is quite complicated; due to the partial incorporation of the parameter virtual age An and the weighting factor q. Moreover, as the probability to find values that grant the global maximum of the maximum plausibility function is practically null, through the random search, it is necessary to define a tolerance value for the partial derivatives ($\partial L/\partial \beta$ and $\partial L/\partial q$), that are equal to cero, establishing this tolerance value "TOL" as an acceptable rank to consider the distribution adjustment. It is for this reason that the simulation techniques (especially Monte Carlo simulation), arise as powerful alternatives for its resolution.

In order to learn in detail the step by step of each model is a fundamental task to apply effective and correctly each model. Diverse researches omit the process of resolution and only present final results indicating the use of a model and the use of some computed tool with integrated algorithm. It was showed in different researches, which motivated the research team to develop a specific conceptual pattern and resolution practical for each stochastic parametric model former mentioned. This is fundamental to recognize the value of this work and its contribution for future researchers who wish to learn and apply this knowledge. For this reason, this research becomes an analytic and explicative procedure about the definition, calculation methodology and criteria that must be considered to parameterise industrial assets under certain degradation level after maintenance, complementing in addition its analysis with a numeric application that allows demonstrating step by step the mathematic and stochastic development as appropriate. The practical cases chosen were developed in the mining industry of Chile.

REFERENCES

Ascher, H., & Feingold, H. (1984). *Repairable systems reliability: Modeling, inference, misconceptions and their causes*. New York, NY: Marcel Dekker.

Bebbington, M., Chin-Diew, L., & Zitikis, R. (2009). Balancing burn-in and mission times in environments with catastrophic and repairable failures. *Reliability Engineering & System Safety*, 94(8), 1314–1321. doi:10.1016/j.ress.2009.02.015

Duane, J. T. (1964). Learning curve approach to reliability monitoring. *IEEE Transactions on Aerospace*, 2(2), 563–566. doi:10.1109/TA.1964.4319640

Kijima, M., & Sumita, N. (1986). A useful generalization of renewal theory: Counting process governed by non-negative markovian increments. *Journal of Applied Probability*, 23(1), 71–88. doi:10.2307/3214117

Krivtsov, V. (2000). *Monte Carlo approach to modeling and estimation of the generalized renewal process in repairable system reliability analysis*. (Dissertation for the degree of Doctor of Philosophy). University of Maryland, College Park, MD.

NIST/SEMATECH. (2014). *e-Handbook of statistical methods*. Retrieved June 6, 2014. http://www.itl.nist.gov/div898/handbook/

RelPro®, Reliability and Production, Analysis and Simulation. (2014). *RelPro SpA*. Retrieved April 10, 2014. http://www.relpro.pro

Veber, B., Nagode, M., & Fajdiga, M. (2008). Generalized renewal process for repairable systems based on finite Weibull mixture. *Reliability Engineering & System Safety*, *93*(10), 1461–1472. doi:10.1016/j.ress.2007.10.003

Yanez, M., Joglar, F., & Modarres, M. (2002). Generalized renewal process for analysis of repairable system with limited failure experience. *Reliability Engineering & System Safety*, *77*(2), 167–180. doi:10.1016/S0951-8320(02)00044-3

ADDITIONAL READING

Damaso, V. C., & Garcia, P. A. (2009). Testing and preventive maintenance scheduling optimization for aging systems modeled by generalized renewal process. *Pesqui. Oper.*, *29*(3), 563–576. doi:10.1590/S0101-74382009000300006

Kumar, U., & Klefsjö, B. (1992). Reliability analysis of hydraulic systems of LHD machines using the power law process model. *Reliability Engineering & System Safety*, *35*(3), 217–224. doi:10.1016/0951-8320(92)90080-5

Lopera, C. M., & Manotas, E. C. (2011). Aplicación del análisis de datos recurrentes sobre interruptores FL245 en Interconexión Eléctrica. S.A. *Revista Colombiana de Estadística*, *34*(2), 249–266.

Malaiya, Y. K., Li, M. N., Bieman, J. M., & Karcich, R. (2002). Software reliability growth with test coverage. *IEEE Transactions on Reliability*, *51*(4), 420–426. doi:10.1109/TR.2002.804489

Mettas, A., & Zhao, W. (2005). Modeling and analysis of repairable system with general repair. IEEE 2005 Proceedings annual Reliability and Maintainability Symposium, Alexandaria, Virginia USA, January 24-27. doi:10.1109/RAMS.2005.1408358

Muhammad, M., Abd Majid, M. A., & Ibrahim, N. A. (2009). A case study of reliability assessment for centrifugal pumps in a petrochemical plant. In: *4th World Congress on Engineering Asset Management*, 28-30 September 2009, Athen, Greece.

Wang, P., & Coit, D. W. (2005). Repairable systems reliability trend tests and evaluation. Annual Reliability and Maintainability Symposium 20. 416-421 doi:10.1109/RAMS.2005.1408398

Weckman, G. R., Shell, R. L., & Marvel, J. H. (2001). Modeling the reliability of repairable systems in aviation industry. *Computers & Industrial Engineering*, *40*(1-2), 51–63. doi:10.1016/S0360-8352(00)00063-2

KEY TERMS AND DEFINITIONS

Failure Rate: Is the rate of failure occurrence per units of time, distance, cycles or other measure. Degradation Level: Effects of the wearing down over a determined element, given by: time, conditions of use, among other factors.

Imperfect Maintenance: Maintenance action where the intervened item is restored to a condition "worse than new but better than old".

Non Repairable System: A system that cannot be restored to an operational status after a failure, and have to be replaced to return the functionality.

Parameterization: Defining the parameters necessary to model or represent a specific variable.

Repairable System: A system that can be restored to an operational status after a failure.

Tolerance Level: Corresponds to a maximum value (near to zero) for differential equations that maximize a certain likelihood function, in which can be assumed that has been found an optimal solution.

APPENDIX: NOTATIONS

PRP: Perfect Renewal Process.
NHPP: Non-Homogeneous Poisson Process.
GRP: Generalized Renewal Process.
L(*t*): Failure Rate of an element in a given time t.
T_i: Function time between failure i-1 and i-th.
F(t): Probability density function of failure (pdf) of an element with operation time t.
F(t): Probability density function of accumulated failure of an element with operation time t.
G(t): Gamma function.
F(t;Θ): Probability density function of failure of an element with operation time t, with forma and scale parameters given by vector 0.
L(Θ): Likelihood function for parameters vector 0 in a pdf given.
MTTF: Mean time to failure.
MTTR: Mean time to repair.
â: Estimated value of the parameter *a* (applicable for all parameters).
A_n: Virtual age of system at the immediate moment of the repair of n-th.
T_n: Virtual age (of operation) at the immediate moment of the repair of n-th.
q: Parameter which establishes the defect of the repair.
TOL: Numeric grade to consider acceptable a distribution adjustment through the likelihood maximum.
TQ: Tolerance that corresponds to higher or lower percentage of the possibilities of value q.

Chapter 13
Challenges in Building a Green Supply Chain:
Case of Intel Malaysia

Yudi Fernando
Universiti Sains Malaysia, Malaysia

Kurtar Kaur
Universiti Sains Malaysia, Malaysia

Ika Sari Wahyuni-TD
Andalas University, Indonesia

ABSTRACT

Consumers today are focusing on products that are manufactured using sustainable, environmentally friendly methods. Profitability or even existence of an industry can be impacted by public opinion. Governments all over the world are also coming up with stricter regulations for industries to comply with on items like pollution, hazardous content, conflict minerals, child labor, exploitation, etc. A number of requirements have been set up by the semiconductor industry, and Intel worldwide is working on some of the current issues: (1) conflict-free minerals sourcing; (2) using green/sustainable energy; (3) reduction of water consumption/recycling of water; and (4) migrating to unleaded parts and halogen free parts. This chapter presents the Intel experiences and challenges in building a green supply chain at both the corporate and regional levels.

INTRODUCTION

The current environment in the semiconductor industry is influenced by governmental legislation, required industry certifications, public opinion, increasingly discerning customers and non-governmental organizations' agenda. The industry is facing a lot of pressure to stop pollution, to use electrical energy and water in a sustainable way, use ethically sourced material and usage of environmental friendly, non-hazardous (green) materials. Many of the semiconductors that are in Malaysia have their origins in the United States and need to adhere to both the US governmental rules

DOI: 10.4018/978-1-4666-8222-1.ch013

and regulations and the local laws. In addition, the industry itself has many standards that must be adhered to. Non adherence can cause harsh actions to be taken against any company and this can include hefty fines and loss of revenue through product boycotts or plant closures. In Intel Malaysia's case both the US and the Malaysian laws need to be adhered to – the more stringent one will apply. Intel has taken proactive steps in implementing sustainable use of natural resources and ethical sourcing from its founding. This chapter is useful for those who interested in area of green supply chain and conflict of conflict-free mineral sourcing in Semiconductor Company alike. The secondary data and information have been collected as the basis of discussion in a single case study.

GREEN SUPPLY CHAIN

A supply chain is considered sustainable or green when sustainability and sustainable development concepts are applied into the traditional operational principles, i.e., economic, environmental and social friendly practices are considered part of the operational procedures on a regular basis (Pusavec et al., 2010). There are many definitions being brought up by researcher on sustainable supply chain or green supply chain management (GSCM). Srivastava (2007) defines sustainable supply chain management or GSCM as the integration of the sustainable thinking that incorporated into the supply chain, including product design, material sourcing and selection, manufacturing processes, delivery of goods to the final products to the consumers as well as the product management after its useful life also known as reverse logistics (Vanalle et al., 2011). Zsidisin and Siferd (2001) argue that environmental supply chain management is a series of policies and actions on design, procurement, production, bulk dispatch, utilization, reutilization, and disposal, undertaken by businesses out of concern for the natural environment. Skjoett-Larsen (2000) suggested that green should include each link in the chain from initial manufacturer at the raw material stage to the end-user which including production, processing, packaging, shipping, handling, and so on. Supply chain strategies are developed with the purpose to improve effectiveness and efficiency of processes across the operation. Integration of environmental concern in supply chain or green supply chain is one strategy for operational improvisation (Hasan, 2013). Green supply chain management main concern is to integrate environmental interest into corporate supply chain practices that include purchasing, material handling, and logistics procedure. Thus, it is of vital importance that the upstream members of the supply chain, which are the suppliers to be aware of the focal company aspiration of establishing green supply chain management (Zhu et al. 2014). Besides, Zucatto et al. (2008) defines green supply chain management as a way of environmental improvement that can involve initiatives in purchasing, production, shipping and reverse logistics, including material suppliers, service contractors, salesmen, distributors and final users, all them working together to reduce or eliminate adverse environmental impact from their activities (Zucatto et al., 2008; Vanalle et al., 2011).

COMPANY BACKGROUND

Intel Corporation was formed by Gordon. E. Moore and Robert Noyce. Gordon Moore was as chemist and a physicist while Robert Noyce was a physicist. Robert Noyce is part of the duo who invented the integrated circuit. Both Gordon Moore and Robert Noyce left Fairchild Semiconductor Corporation and formed Intel. Legend has it that the new start-up company was to be named "Moore Noyce" but this was quickly discarded as this name would never do for a semiconductor company. They later bought the rights to the

Intel name. The first employee to join this new company was Andy Grove, a chemical engineer who also came from Fairchild and served in the capacity of president, CEO and later chairman of the board over different periods in time. Intel is the largest semiconductor business in the world today. The X86 series microprocessors that are the brain of many computers were invented by Intel. Currently, the company is helmed by Brian M. Krzanich (who was appointed CEO on May 16th 2013) and Renée J. James (president of Intel Corporation).

Intel Kulim started off as a manufacturer of motherboards in 1995 and has grown from one building to five buildings. Intel Kulim manufactures and assembles computer and communications components like micro-processors, chipsets, motherboards, wireless and wired connectivity products and integrated digital technology platforms. Intel Kulim is an important site for building mobile modules and also has a research and development unit which focuses on early development.

Target Market

The components manufactured are sold to original equipment manufacturers (OEMs), original design manufacturers (ODMs), industrial and telecommunication equipment manufacturers and lately mobile phone manufacturers. There are other customers who buy Intel products through distributors, resellers, retailers and other OEM channels around the world. Intel has three major customers (Hewlett Packard Company, Dell Inc. and Lenovo Group Limited) and many smaller direct and indirect customers.

Company Vision and Mission

The company mission is to "create and extend computing technology to connect and enrich the life of every person on earth in this decade (Intel CSR report, 2012)". To realize this mission a number of strategies are in place:-

- Grow PC and Datacenter business with new users and uses
- Extend Intel Solutions to win in adjacent market segments
- Create a continuum of secure, personal computing experiences
- Care for our people, the planet, and inspire the next generation

This is further internalized in each business unit. For the materials department, the goal is to accelerate a cool user experience with technology and supply chain solutions. Material's mission is: *Be the leader in materials strategies, sourcing, and supply chain solutions that strengthen Intel's competitive advantage.* This is further translated into a department goal of "Simple, Fast & Hassle-Free' – a very easy tagline to remember.

Size of the Business

Intel Corporation is a USD 53.3 billion revenue company today - revenue contribution by the major three customers was at 43% for 2012. Net income for the year ended 2012 was USD 11 billion. This represents a market share of 16.4% in 2012 which is 1% lower than 2011. However Intel still retained the market share position at number 1 for the 21st consecutive year (Pls refer to Table 1). As of Dec 29th 2012, Intel had 105,000 employees worldwide with 51% located in the US. Intel Malaysia has roughly nine thousand employees.

In Intel, a number of initiatives are being implemented that at a minimum must be met and if possible industry surpassed. Some of the main ones are:-

1. Conflict free minerals sourcing.
2. Using green energy (in US) & energy efficient lighting.
3. Reduction of water consumption.
4. Elimination of leaded parts and moving to halogen free parts.

Table 1. Top 10 semiconductor vendors by revenue, worldwide, 2012 (millions of dollars)

Rank 2011	Rank 2012	Vendor	2011 Revenue	2012 Revenue	2011-2012 Growth (%)	2012 Market Share (%)
1	1	Intel	50,669	49,089	-3.1	16.4
2	2	Samsung Electronics	27,764	28,622	3.1	9.5
6	3	Qualcomm	9,998	13,177	31.8	4.4
4	4	Texas Instruments	11,754	11,111	-5.5	3.7
3	5	Toshiba	11,769	10,610	-9.8	3.5
5	6	Renesas Electronics	10,650	9,152	-14.1	3.1
8	7	SK Hynix	9,388	8,965	-4.5	3.0
7	8	STMicroelectronics	9,635	8,415	-12.7	2.8
10	9	Broadcom	7,160	7,846	9.6	2.6
9	10	Micron Technology	7,643	6,917	-9.5	2.3
		Others	151,343	146,008	-3.5	48.7
		Total Market	**307,773**	**299,912**	**-2.6**	**100**

Source: Gartner (2013)

Most of these programs have either been implemented across all sites or are in the process of implementation. In Intel, there is terminology that is often used – "Copy Exactly" – whatever good practices implemented at one site are adopted across the virtual factories. This saves time and effort by avoiding wastage of resources on the same problem. Conflict- free minerals sourcing is being driven at a corporate level while the others are being driven at both the corporate and local levels.

CONFLICT-FREE MINERAL SOURCING

What is a conflict mineral? The semiconductor industry uses many raw materials in their processes and some of these raw material originates from mines in eastern part of the Democratic Republic of the Congo (DRC). The DRC has abundant natural resources in the form of mineral ores. Trade in these ores brings in money for those who own the mines or the trading routes. Armed conflict arises when different groups (both local and neighboring countries) try to gain control over the mines and the trading routes. The earnings from these mines is used to fund further weapon purchases which are then used to threaten and intimidate the local population. The main forms of violence are mass rapes, child labor and slavery like conditions. The current list of conflict minerals consists of 4 minerals ores: columbite-tantalite, cassiterite, wolframite and gold (this list may increase in future). These are then smelted to produce tantalum, tin, tungsten and gold. Tantalum is used in many different industries – in the semiconductor industry is mainly used for high performance capacitors. Tin is used to produce solder which is used to hold components on circuit boards and is also used in other industries. Tungsten is used in cell phones and gold is used as wire bonding material in Integrated circuits. Table 2 shows a list of products that use these minerals:-

The raw ores after mining are sent for processing to smelters in East Asia through neighboring countries. The mineral ores pass through many middlemen in the process of being shipped out of the DRC. Smuggling of the ores across national boundaries is a common occurrence. Traceabil-

Table 2. A list of products that use these minerals

Metal Ore	Mineral Extracted	Uses
Columbite-tantalite (Coltan)	Tantalum	Capacitors, hearing aids and pacemakers, airbags, GPS, ignition systems and anti-lock braking systems in automobiles, laptop computers, mobile phones, video game consoles, video cameras and digital cameras. In the carbide form used in jet engine/turbine blades, drill bits, end mills and other tools.
Cassiterite	Tin	Tin cans, solder for semiconductor industry, component of biocides, fungicides. As tetrabutyl tin/tetraoctyl tin, an intermediate in polyvinyl chloride (PVC) and high performance paint manufacturing.
Wolframite	Tungsten	Fishing weights, dart tips and golf clubs. Tungsten carbide used for metalworking tools, green ammunition and the vibration mechanism of cell phones.
Gold	Gold	Jewelry, electronics, dental products and semiconductor manufacturing processes.

Source: Ernst & Young (2012)

ity of the ores is a big concern and to take care of this the Dodd-Frank Wall Street Reform and Consumer Protection Act has been passed in 2010 (in the US) and it applies to materials originating (or claimed to originate) from the DRC and nine of its immediate neighbor countries. There are countries have impacted by the US conflict such as Democratic Republic of the Congo (DRC), Central Africa Republic, South Sudan, Zambia, Angola, The republic of the Congo, Tanzania, Burundi, Rwanda and Uganda. This law is effective from November 13, 2012 and the first report out to the SEC is in May 31st. The main thrust of the law is to ensure that American companies are able to trace the origin of the ores and that the traceability documentation is auditable. As Intel is a US based company, it has to follow US regulations. Intel's rule of thumb is to follow whichever law that is more stringent (US or local).

ACTIONS TAKEN BY INTEL CORPORATION

Intel has currently mapped more than 90% of the microprocessor supply chain and identified more than 140 unique smelters. Intel has also taken the lead to work towards a solution on the conflict minerals and eliminating these from semiconductor supply chains. Figure 1 shows the percentage of progress by multinationals on responsible sourcing of conflict material.

Intel's internal goal is to develop the first verified conflict-free microprocessor by end of 2013. Current only tantalum is used in the Intel microprocessor has been verified as conflict free. The DRC has 2 main sources of income – farming and mining. The dilemma was if Intel/ semiconductor stopped purchasing minerals originating from the DRC – it would eventually impact the DRC economy and its citizens. Intel teamed up with Hewlett Packard and Apple to work out some resolution to this problem. The Intel team that was working on this issue – let's call it A-Team - went through many brain storming sessions on what would lead to a win-win situation for both Intel/ industry and the DCR.

Upfront decision made was not to impact the DRC economy – more poverty will lead to escalation of violence and cause hardship for the DRC citizens. It was decided that the best solution was to have a traceability system which could be audited and more importantly getting the smelters certified as not using ores that originated from the 'conflict' mines. The smelters were the best place to begin – they were the ones that got the

Figure 1. Percentage of progress towards responsible sourcing of conflict minerals
Source: Lezhnev and Hellmuth (2012).

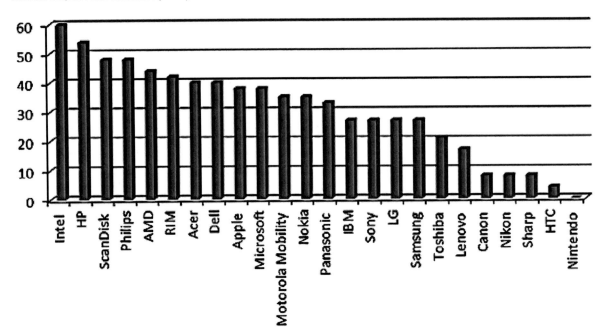

raw ore and sent it the minerals to manufactures after the extraction process. Therefore a need to look at the smelters – some of them did not have proper records or records were incomplete.

The A-team visited sixty smelters in twenty countries to provide education on conflict minerals and encouraged their participation in a conflict free smelter program. This certification is provided to smelters that are able to trace the raw mineral ores to their origins – show that it comes from non-conflict sources and pass an audit process by a 3rd party. A system to audit and verify conflict minerals was implemented. Those smelters that were willing to participate to get this certification were partially funded (50%) on the certification and audit costs. Intel co-chaired working groups on the conflict minerals and called for cross industry participation to implement systems that will weed out conflict minerals from their supply chains. The current system put in place is a tagging system – each bag of mineral is tagged and tack and trace details updated into a data base. The tag can be traced right to the mine that the ore originated from. Even this tagging system has weaknesses – some tags were found to have been sold illegally by certain parties but this was eventually traced and resolved within a week.

The other portion is due diligence by the end user. The SEC has set up a 3 –step compliance process:

Step 1: Each company evaluates if any if these four minerals are present in their products and define if the presence of this mineral is critical to its functioning. If these minerals are not used – there is no need to do anything.

Step 2: If these minerals are present / required, then the company will need to determine if the minerals are obtained from a non-conflict source by doing a RCOI (Reasonable Country of Origin Inquiry). The company will then need to disclose information and the process used for the RCOI to the SEC. This disclosure is to be in a form called Form SD (specialized disclosure).

Step 3: If the findings are that the mineral originated from a conflict country, the company will need to countercheck whether income from the minerals benefited any armed groups. This benefit could be direct or indirect. A Conflict Minerals Report will need to be submitted together as a supporting document for Form SD. This report will need to state the materials' origin and the steps involved in determining the COO (country of Origin). All minerals listed as conflict will need to be stated in the Conflict Minerals Report and must be audited by an independent auditor.

In addition to conflict free, Intel also has to do due diligence to ensure no issue of worker exploitation and child labor in upstream supply chains. When doing supplier selection, these criteria is thoroughly checked and audits are conducted regularly to ensure compliance by our suppliers.

ANALYSIS

There are three competitors of Intel in the marketplace such as Advanced Micro Devices, Inc, Texas Instruments Incorporated and Samsung Electronics Co., Ltd. The details of strength and weaknesses are discussed in Table 3.

Strengths

In 2012, In 2012 Intel, HP and GE come up with a program to pay half of the audit costs incurred by the smelters in getting themselves certified as a 'Conflict-Free Smelter'- (CFS). This provide an incentive for the smelters to go for the certification and the certification is published. Additionally, other companies will be more likely to approach the CFS smelters as their sources are known and auditable. The semiconductor industry is also actively mapping their own supply chains so as to eliminate conflict materials – either due to the US legislation or due to peer pressure from within the industry. Individual organisations procurement

Table 3. General strength and weaknesses of four companies

	Strength	Weaknesses
Intel Corporation	Advanced technology capabilities with strong durable semiconductor solutions. Most of the computers have x86 microprocessors and almost every brand which makes computer is using Intel products in it. invested heavy amount in Research and Development.	Technical strategy which is not easily understood. Don not has focus on the mobile market.
Advanced Micro Devices, Inc	Strong brand in Microprocessor and PC market. AMD operates in GPU (graphical processing unit) and Microprocessor market with only single competitor in each market, Nvidia and Intel respectively.	Lack of presence in the smartphone market. Weak performance of computing segment due to advent of tablets, handhelds
Texas Instruments Incorporated	worldwide customers and has a great brand name amongst college graduates which helps them attract the best talent	Dependence on third party contractors. More of a supplier than end retailer.
Samsung Electronics Co., Ltd.	Hardware integration with many open source OS and software. Innovation and design. Low production costs. Largest share in mobile phones.	Patent infringement. Too low profit margin. Lack its own OS and software. Focus on too many products.

teams are also auditing their own processes to comply with this rule. Investigation into the supply chain is also carried out with more urgency.

Opportunities

There is already an established supply chain in the DRC and the other neighboring countries – this can be utilized to improve tracking and tracing the ores. Governmental commitment – like in Rwanda will push the other countries authorities to improve conditions in their own countries. The multinationals themselves will have a reason to improve their supply line information due to the US government's legislation. The monitoring done by NGO's like "Enough" is will also force organisations to work on improvements in their supply chains and procurement activities.

Weaknesses

From the white paper published by Intel in February 2013 - many smelters did not have country of origin of the mineral ores. Infrastructure for tracing the ores throughout the supply chain was not available and this resulted in many suppliers having limited knowledge of their minerals country of origin. Companies impacted who are not able to trace their minerals to the origin would have to look for alternative sources and indirectly impact the livelihood of the people in the mining areas. Also some minerals may not come direct from the nine countries impacted by the Dodd-Frank Wall Street Reform and Consumer Protection Act, but from other sources. This may not be able to be traced to its origin – ways to take care of this need to be looked into. The tagging system also needs to be made more robust so as to prevent abuses.

Threats

The inability to trace the ores to their origin might result in organisations seeking for alternative sources and impacting the economy of countries like the DRC. If ores are still obtained from the DCR and its neighboring countries, the likelihood of funding armed conflict is always present. The inability of the various governments to put a stop to armed conflict and genocide might one day just lead customers to other sources of mineral ores. Governments also need to put in additional efforts to stop smuggling of the mineral ores.

Table 4. SWOT analysis

Strengths	Weaknesses
• Publishing list of smelters who comply with audit requirements. • Reimbursement of half of audit costs to smelters – early adopters • Early adoption of conflict free philosophy by many large multinationals. • Peer pressure from the early adopters • Self-audits conducted by individual companies	• Many smelters had ship from country Information but not origin country. • Infrastructure to trace 100% of material non existent • Tagging system – tags can be abused • Resold minerals / outsourcing supply lines not able to trace origin
Opportunities	Threats
• Variability in in the amount of information suppliers knew about their own supply chain. • Governmental participation. • Using existing communication and supply chain stakeholders in the DRC. • Monitoring of organisation's compliance by NGO's	• Difficulty in tracing minerals to their origin • Drive down demand for Central African mineral ores. • Profits funding armed conflict and violence • US government legalization • Smuggling activities

POTENTIAL SOLUTIONS

1. **Robust Track and Trace System:** This could begin with a simple database from the mines and fed into servers that are co-owned by the smelters or companies that will use these minerals. Multinational should take the initiative to do backwards integration of their supply chains. Systems are available – the focus should be on implementation and education on the usage of existing systems and also to build the necessary infrastructure to track the shipments throughout the supply chain. Not only should the bags be tagged – all transportation modes and routes used in the supply chain need to be mapped. The smelters also can provide documentation on the origin of the material by keeping accurate records of their inventory, incoming and outgoing information.
2. Monitoring and auditing of smelters on origin of ores by 3rd parties like Ernest and Young etc.
3. The Central African governments need to have the political will to end armed conflict especially when there is a possibility industries might source material from elsewhere.
4. All industries using these 4 minerals must be one voice in committing to conflict free supply lines and responsibly sourced minerals. The publishing of data (Table 3) by organizations like *'enough'* through their *"Raise Hope for Congo"* program will also provide the necessary push for the laggards in the industries involved to implement conflict free sourcing.

Green Energy and Reduction of Water Consumption

Intel worldwide is working on reducing its environmental footprint and reducing the use of natural resources by recycling and other sustainable initiatives. Intel's goal is to reduce energy consumption by 4% per year per unit of production in the manufacturing operations. The goal has been consistently exceeded for the past years by upgrading existing buildings, designing in energy efficient technology into new equipment and buildings. Intel has allocated funds (several millions) annually for all its plants for energy conservation. From the year 2002 – which Intel is using as its baseline - the company's energy use per unit of production has been reduced by 20% and higher. 500 million kilowatt-hours of electricity has been saved by investing in approximately 250 projects with an investment cost of $20 million. This is equivalent to power usage in 50,000 U.S. homes. Many of the new buildings are going for Leadership in energy and Environmental Design (LEED)'s certification. Current, Intel Malaysia is 100% Leeds certified. Intel Malaysia has almost 82 million kilowatt hours saved since year – this is enough to power up 361,000 households monthly (assuming average household electricity consumption of 227KWh/month). 150 gallons of water has been saved equivalent to providing drinking, cooking, sanitation and domestic use for 16,200 households per month (based on United Nations Water Statistics- average of 220 liters per household/day).

Intel also is working with its suppliers and SEMATECH (representation from semiconductor industries, universities, government and suppliers) to develop standards that will lead to energy efficient tools and processes. Water conservation and energy conservation education is being done for all employees. In Penang, there is a program to harvest rainwater and this water is used for the PG8 cooling towers.

ROHs Initiative

The industry is moving towards having lead–free parts as lead is a poison and is a neurotoxin that can accumulate both in the bones and soft tissue. Intel has actively replaced leaded parts with lead free versions from the year 2007. A lead free part is any part with less than a thousand parts per million of lead content. Most countries have legislation in place for reducing or eliminating lead content of products. Many of Intel's suppliers are already providing lead free / halogen free parts as this initiative is roughly 10 years in the pipeline. The manufacturers are in tune with industry trends and provide lead free / halogen free parts. However, there are some smaller manufacturer's that either do not have the expertise to make the changes or their processes may not be able cater for lead elimination or they might not be able to get an efficient replacement for the leaded component.

Current process:- Intel has a large numbers of parts from thousands of suppliers. In Intel, we have a database which contains details of all components used in our products. In this database, new part numbers can be added but older part numbers cannot be removed. This means that old parts that are leaded are still in the database and what is being done to eliminate usage of the leaded parts is to change the status of the leaded parts to *disapproved*. As this database feeds the part info into the procurement system – the disapproved part cannot be purchased. The procurement system will give a hard stop if a disapproved manufacturing part number is used and a purchase order cannot be generated.

A cross functional team has worked on implementing ROHs (Restriction of Hazardous Substances) and lead free initiative from many years ago and Intel's component part list is clean – all leaded parts are at disapproved status.

Current issue is when new part numbers are added by requestors - some of the requestor's do not do their due diligence to counter check with the suppliers or lack knowledge on what is required. This has led to a few cases of purchasing becoming policemen in catching the discrepancy. This is easy to catch as all new part numbers start with a different suffix value. The suppliers also act as a gate by informing when a certain request is not lead free or halogen free. They will then inform on the correct manufacturing part number and the buyers will work with the business units to update the database accordingly. This is very reactive in nature.

What is being done now – is a cross functional virtual team has been set up to look in to all aspects of the databases and resource requirements. The top item on the agenda is to identify all potential users for this database and communicate requirements for part set up. A formal training package is already available for all potential users and the user will get access upon completion of this training.

The next phase is to re-educate all potential users on required criteria to use when creating a new part and incorporate this into their desktop procedures. (Desktops refers to any set of process steps that are kept by an individual group and is used as a guide for their individual operation). The system could also be enhanced so that the requestor will need to input more values before a part can be set up.

Possible Solution

1. Virtual team to review and redefine training criteria
2. Data base cleanup on a regular basis – this should be automated so that any fallout encountered is reported out.
3. Re-education package which has to be reviewed on a periodic basis so that people are aware of latest requirement.

CONCLUSION

This chapter displayed the case of Intel's sustainability efforts, the various qualitative analysis techniques were employed in this chapter to achieve objective of the study. The results from the analysis conducted displayed that the green supply chain is very challenging issue in the industry. Intel is leading efforts in reducing its' environmental footprint, driving for sustainability and responsible use of natural resources in whichever country it operates in. The vision and mission of the corporation is internalized by the various business units and their departments. Alignment is done at all levels through virtual open forums, departmental meetings, Q &A's and through staff meetings. Employees are encouraged to submit ideas for sustainability and are provided funds to put those ideas into action. The conflict free mineral initiative is one way of fulfilling Intel's CSR and giving back to society. A lot of works still needs to be done and the monitoring and publishing of environmental and social KPI's will help in persuading reluctant organization to adopt industry best practices. The semiconductor industry has to adopt these strategies if they want to stay in business and remain profitable. Intel's suppliers are also in line with this initiative with most suppliers having statements/ commitment on their web site on these topics. The conflict free mineral initiative is one way of fulfilling Intel's CSR and giving back to society. Non-compliance will result in legal action, fines, a tarnished reputation and of course impact to the bottom line – profit of the company.

FUTURE RESEARCH DIRECTIONS

In order to operation sustainably in the future, it is of vital importance that companies give utmost priority to preserving the environment through practising green supply management. As such, more resources should be pooled in and expertise should be sought in order to develop the most practical green supply chain for the benefit of the current and future social and environmental wellbeing (Glover, 2013; Ibrahim et al. 2012). Practising respectable and environmentally ethical supply chain would consequently benefit the corporations by increasing the corporations' value and at the same time assist the corporations to achieve operational efficiency.

The current changing in environmental requirements that influenced manufacturing activities had increased attention in developing environmental management strategies for the supply chain (Beamon, 1999). Thus, the future study should explore further on the issue of sustainable supply chain integration. The concept of sustainable supply chain integration arises as a new systematic approach and becoming an important factor for business activities today. Zhu et al. (2010) claimed that sustainable supply chain can be regarded as an environmental innovation. By integrating the 'green concept and sustainability to the supply chain' concept, it has created a new research agenda where the supply chain will have a direct relation to the environment and economic.

REFERENCES

Beamon, B. M. (1999). Designing the green supply chain. *Logistics Information Management*, *12*(4), 332–342. doi:10.1108/09576059910284159

DesMarais, C. (2013). *Intel, Microsoft and Kohl's top EPA's green power list*. Retrieved December 1, from http://www.greenbiz.com/blog/2013/07/24/intel-microsoft-kohls-top-epa-lists-green-power

Ernst & Young. (2012). *Conflict minerals: What you need to know about the new disclosure and reporting requirements and how Ernst & Young can help*. Retrieved January 23, 2014, from http://www.ey.com/Publication/vwLUAssets/Conflict_minerals/$FILE/Conflict_Minerals_US.pdf

Gartner. (2013). *Worldwide semiconductor revenue declined 2.6 percent in 2012*. Retrieved January 23, 2014, from http://www.gartner.com/newsroom/id/2405215

Glover, J. l., Champion, D., Daniels, K. J., & Dainty, A. J. D. (2013). An Institutional Theory perspective on sustainable practices across the dairy supply chain. *International Journal of Production Economics*, *10*(5), 220–241.

Hasan, M. (2013). Sustainable supply chain management practices and operational performance. *American Journal of Industrial and Business Management, 3*(01), 42–48. doi:10.4236/ajibm.2013.31006

Ibrahim, A., Zailani, T., & Tan, A. C. (2012). Extending green practices across the supply chain: The impact of upstream and downstream integration. *International Journal of Operations & Production Management, 26*(7), 795–821.

Intel. (2010). *Intel water policy*. Retrieved November 24, 2013 from http://www.intel.com/content/www/us/en/policy/policy-water.html

Intel. (2012). *ESG corporate social responsibility*. Retrieved November 24, 2013 from www.intel.com/content/www/us/en/corporate-responsibility/inspire-the-next-generation.html

Intel. (2013). *2012 Annual report*. Retrieved November 24, 2013 from http://www.intc.com/intel-annual-report/2012/

Intel. (2013). *Conflict minerals whitepaper*. Retrieved November 25, 2013 from http://www.intel.com/content/www/us/en/policy/policy-conflict-minerals.html

Intel News Release. (2008). *Intel becomes largest purchaser of green power in the U.S.* Retrieved November 24, 2013 from http://www.intel.com/pressroom/kits/green/rec/?wapkw=green+energy

Lezhnev, S., & Hellmuth, A. (2012). *Taking conflict out of consumer gadgets*. Retrieved January 23, 2014 from http://www.enoughproject.org/files/CorporateRankings2012.pdf

Miller, J. (2012). *Intel, Apple praised for clean mineral efforts*. Retrieved November 17, 2013 from http://www.cbsnews.com/8301-501465_162-57494957-501465/intel-apple-praised-for-clean-mineral-efforts

OECD. (2013). *Helping companies source minerals responsibly*. Retrieved November 17, 2013 from http://www.oecd.org/daf/inv/mne/mining.htm

Pusavec, F., Krajnik, P., & Kopac, J. (2010). Transitioning to sustainable production – Part I: Application on machining technologies. *Journal of Cleaner Production, 18*(2), 174–184. doi:10.1016/j.jclepro.2009.08.010

Reuters. (2012). *Intel, Apple leading charge against conflict minerals -report*. Retrieved November 23, 2013 from http://in.reuters.com/article/2012/08/16/conflictminerals-enoughproject-idINL2E8JF7HR20120816

Skjoett-Larsen, T. (2000). European logistics beyond 2000. *International Journal of Physical Distribution & Logistics Management, 30*(5), 377–387. doi:10.1108/09600030010336144

Source Intelligence. (2012). *What are conflict minerals?* Retrieved November 19, 2013 from http://www.sourceintelligence.com/what-are-conflict-minerals

Srivastava, S. K. (2007). Green supply-chain management: A state-of-the-art literature review. *International Journal of Management Reviews, 9*(1), 53–80. doi:10.1111/j.1468-2370.2007.00202.x

Tysiac, K. (2012). *Build a strong team to comply with conflict minerals rule*. Retrieved November 23, 2013 from http://www.cgma.org/magazine/news/pages/20126875.aspx

Tysiac, K. (2012). *Highly scrutinized SEC conflict mineral regs. include new audit requirement*. Retrieved November 23, 2013 from http://www.journalofaccountancy.com/News/20126307.htm

Tysiac, K. (2013). *Conflict minerals rule poses compliance challenge*. Retrieved November 23, 2013 from http://www.journalofaccountancy.com/Issues/2013/Apr/20127083.htm

Vanalle, R. M., Lucato, W. C., & Santos, L. B. (2011). Environment requirements in the automotive supply chain: An evaluation of a first tier company in the Brazilian auto industry. *Procedia Environmental Science*, *10*, 337–343. doi:10.1016/j.proenv.2011.09.055

Zhu, Q., Geng, Y., Fujita, T., & Hashimoto, S. (2010). Green supply chain management in leading manufacturers: Case studies in Japanese large companies. *Management Research Review*, *33*(4), 380–392. doi:10.1108/01409171011030471

Zhu, Q., Sarkis, J., & Lai, K.-. (2014). Green supply chain management: Pressures, practices and performance within the Chinese automobile industry. *Journal of Cleaner Production*, *15*(11-12), 1041–1052. doi:10.1016/j.jclepro.2006.05.021

Zsidisin, G. A., & Siferd, S. P. (2001). Environmental purchasing: A framework of theory development. *European Journal of Purchasing and Supply Management*, *7*(1), 61–73. doi:10.1016/S0969-7012(00)00007-1

Zucatto, L. C. (2008). *Inovações em processos como uma forma de estruturar uma cadeia de suprimentos sustentável:são possíveis?* Rio de Janeiro: XXVIII Encontro Nacional de Engenharia de Produção.

KEY TERMS AND DEFINITIONS

Business Sustainability: A dynamic state that occurs when a company creates ongoing value for its shareholders and stakeholders.

Green Global Business: Across national borders business transaction is driven largely by the escalating deterioration of the environment.

Green Innovations: Reduce the environmental burden of the logistics, particularly in terms of pollution and greenhouse gas emissions.

Green Supply Chain Management: A supply chain management system that is concerned on ensuring all conventional supply chain deliveries while centralizing the environmental and social concern in every stage of supply chain to reduce product environmental footprint.

Supply Chain Management: The system used by an organization to improve logistics and planning efficiency, as well as material and information control internally and externally among the upstream and downstream members.

Supply Chain Sustainable Development: The companies must move beyond mere "green" that indicates relentless pursuit of short-term profitability towards long-term sustainability.

Sustainability: A key for any corporation to develop a sustained business operation as it suggests that businesses should grow and meet the needs of the present without compromising the ability of the future generations to meet their own needs.

Chapter 14
Low Carbon Footprint:
The Supply Chain Agenda in Malaysian Manufacturing Firms

Muhammad Shabir Shaharudin
Universiti Sains Malaysia, Malaysia

Yudi Fernando
Universiti Sains Malaysia, Malaysia

ABSTRACT

Malaysia has committed to a 40% reduction of carbon emissions by 2020. The government has encouraged industry, society, and non-government organizations to work together to achieve this objective. The government has provided incentives through several energy programmes such as energy efficiency, renewable energy, green technology, and green building. One key area that has been targeted is logistics and supply chain, which has been contributing to high carbon emissions in manufacturing industries. Scholars and practitioners have only recently begun to pay attention to creating a low carbon supply chain. Furthermore, Small Medium Enterprises (SMEs) have faced several challenges in adopting low carbon activities. SMEs are unable to take the advantage of energy initiatives because of a lack of knowledge, a shortage of funds, and inadequate facilities. Almost 90% of firms are in the service industry working with large manufacturing firms and some SMEs working in manufacturing industry are working closely with their supply chain networks; achieving low carbon targets is hampered by the readiness of the manufacturing itself. This chapter discusses the challenges and future agenda of creating low carbon supply chains in manufacturing in Malaysia. Possible solutions are provided at the end of the chapter.

INTRODUCTION

Malaysia is expecting to increase its gross domestic product, by concentrating on international trade. Many foreign investors have begun to set up production plants in the country due to favourable government support, political stability, and competitive labour costs, which help to lower the cost of production. With rich natural and talent resources, Malaysia is projected to produce a broad range of products to serve the international market. Yet, producing quality, low-cost products

DOI: 10.4018/978-1-4666-8222-1.ch014

accounting for environmental aspects has become a serious challenge. In 2013 Malaysia recorded unusual climate phenomena such as cold weather during the normally hot tropic season, rainstorms, and floods. As climate change has risen to the top of the agenda around the world, manufacturing firms have begun undertaking initiatives to reduce their carbon footprint in production and supply chains.

The term carbon footprint derives from the language of Ecological Footprinting (Wackernagel & Rees, 1998) and stands for gas emissions contributing to climate change due to human production or consumption activities. Implicitly, the term "footprint" has become a unit measurement (Wiedmann & Minx, 2008). As global warming became an issue, international representatives under a United Nations programme introduced the Kyoto Protocol in 1997. This international treaty requires countries to reduce the emissions of six important greenhouse gasses (GHGs), namely, carbon dioxide (CO_2), methane (CH_4), nitrous Oxide (N_2O), sulphur hexafluoride, perfluorocarbons and hydroflurocarbons (Pandey, Agrawal & Pandey, 2011). Stern (2006) stated that an annual loss of as much as 5% of world GDP could be expected if nations failed to take early preventive actions. As output of production for goods increases almost daily to cater to the seemingly insatiable needs and wants of consumers, manufacturing firms have been upon called to find solutions benefitting global society. A focus on emissions associated with physical processes or carbon reporting sometimes overlooked important factors emerging from the interaction among the multiple firms comprising elements of the supply chain. Nowadays, however, the supply chain has become a vital aspect contributing to carbon emission increases in the atmosphere. The argument has been made that production activities and logistics are using many resources and emitting high carbon dioxide. Thus, studying the carbon footprint from supply chain perspectives is a necessity.

Scholars have argued that the low carbon footprint could be achieved if the firms manage their supply chains. The lack of coordination among multiple firms within the supply chain has increased the overall carbon footprint (Benjaafar, Li & Daskin, 2013). Manufacturing firms should be able to design a supply chain focusing on aspects such as global sourcing and just-in-time deliveries (Halldórsson & Svanberg, 2013). From the logistics perspective, data about energy consumption is used to measure cost in transportation (Leonardi & Browne, 2010) and its helps to estimate carbon footprints (McKinnon, 2010). Although much has been written about the carbon footprint of supply chains, the research community in operations and supply chain management has only recently begun to pay attention to this area of study (Benjaafar et al., 2013).

This current work is written in response to Halldórsson and Kovacs's (2010) idea that a need exists to rethink energy efficiency and operational design. Dey, LaGuardia & Srinivasan (2011) also found that little has been work done to understand the role of logistics in a low carbon supply chain. Logistics are main contributors to carbon emissions of a typical firm, producing up to 75% of a firm's carbon footprint (The Council of Supply Chain Management Professionals, 2008). Moreover, in Malaysia, carbon emissions in the transport sector alone have increased from 35 billion kilograms in 2000 to 119 billion kilograms in 2010 according to the Malaysian Ministry of Energy, Green Technology and Water (KeTTHA) (2011).

This chapter has been developed to explore future directions for those who are interested in studying the energy aspect in supply chain management as well as to fill the literature gap in this topic. Even though many studies has looked at the carbon footprint, limited attention has been given to the modelling aspect of carbon emissions of a supply chain network. Optimizing carbon emissions in manufacturing and transportation offers a fragmented approach to green supply chain management (Shaw, Shankar, Yadav & Thakur, 2013).

Thus, the objective of this chapter is twofold: first, to review the current issues with respect to lowering the carbon footprint in the manufacturing sector in Malaysia, second is to discuss the carbon footprint agenda with respect to lowering carbon emissions in a supply chain.

CARBON FOOTPRINT

Understanding a carbon footprint requires a clear understanding of the term being used. Literally, definitions of carbon footprint available in literature come from industry, as academia seems reluctant to try generalizing the definition (Matthews, Hendrickson & Weber, 2008). However, a common understanding among academicians and businesses is that carbon footprint is an effort to record carbon emissions by an individual or firm. Firms involved in sustainable green practices have successfully contributed to definition of carbon footprint as shown in Table 1. However, the definitions are derived from what has been called grey literature because the definitions are not easily accessible or not available in any form of academic publication. The extant literature so far mostly defines carbon footprint from the perspective of the practitioner; thus, specific, technical and scientific terms are usually used when studying carbon footprints.

Definitions from Table 1 show that the driver of the definition is the same, which is to measure or record GHGs particularly carbon dioxide (CO_2) emitted from human activities including business activities that have negative impacts on social, environment and economy. Despite the fact that definitions of carbon footprint have been identified since 2006, the actual implementation and monitoring of a carbon footprint has not much gained attention from practitioners and industry. However, global societies have now realized the important of reducing carbon footprints and that research area needs to be considered.

Scholars also have defined carbon footprint. For example, Kitzes et al. (2008) and Matthews et al. (2008) have similar definition about carbon footprint as "the terms of ecological footprint" and carbon footprint are often used to quantify a person, region, business, or country's impact on the environment due to GHGs.

To guide firms in recording their carbon emission, a protocol needs to be followed. A protocol serves as guidance of scope of GHGs to be recorded. Many carbon footprint protocols have been used, but not all protocols are the same and usage depends on the capacity and availability of data. The two most widely used protocols in literature are from the World Resource Institute (WRI) and the World Business Council for Sustainable Development (WBCSD) (2004). Matthews et al. (2008) suggested the use of the WRI/WBCSD protocol, but add Tier IV, which includes total life cycle emissions for production and the delivery, use and end-of-life if such information is available.

Table 1. Definitions of carbon footprint from the grey literature

Source	Definition
British Sky Broadcasting (Sky) (Patel, 2006)	The carbon footprint was calculated by "measuring the CO2 equivalent emissions from its premises, company-owned vehicles, business travel and waste to landfill.
Carbon Trust (2012)	A carbon footprint is the total greenhouse gas (GHG) emissions caused directly and indirectly by an individual, organisation, event or product, and is expressed as a carbon dioxide equivalent (CO2e). A carbon footprint accounts for all six Kyoto GHG emissions
Grubb & Ellis Company Michael (Groppi & Burin, 2007)	A carbon footprint is a measure of the amount of carbon dioxide emitted through the combustion of fossil fuels. In the case of a business organization, it is the amount of CO2 emitted either directly or indirectly because of its everyday operations. It also might reflect the fossil energy represented in a product or commodity reaching market.

Table 2 shows two carbon footprint protocols that firms often use to record emissions. In Malaysia, carbon footprint measurement is using the WRI protocol to capture each stage of supply chain carbon emissions.

Once firms are able to understand and classify their carbon emissions, they then need to develop an instrument or metric to measure their carbon footprint. Many GHGs accounting standards are available as shown in Table 3. Using these standards, not only firm will be able to record their carbon emission but also to find a solution for reducing their carbon emissions. These guidelines share two common goals: to effectively measure a carbon footprint and to promote a low carbon footprint in production and management.

Several methods exist to document carbon emissions. Most widely used, and one of the oldest methodologies, is the Life Cycle Assessment (LCA) model in which carbon emissions are recorded at various life cycle stages (see WRI Table 2). Scholars who have used this model either to explain or measure carbon footprint are Lenzen (2000), Williams and Tagami (2002), Matthews et al. (2008) and Wakeley, Hendrickson, Griffin and Matthews (2009). The LCA model is useful

Table 2. Carbon footprint protocols

Protocol	Description	Reference
California Climate Action Registry (CCAR) or The Climate Registry (TCR)	Requires all firms to report all direct emissions from their facilities and company vehicles as well as purchases of electricity, steam, heating and cooling. Suggests that firms report all GHGs (optional) but it is mandatory to report all CO2 emissions.	California Climate Action Registry General Reporting Protocol, 2008
World Resources Institute (WRI) and World Business Council for Sustainable Development (WBCSD)	Scopes for footprint by classifying it into tiers: Tier I: direct emission of the firm Tier II: carbon emission of energy inputs used Tier III: other indirect activities (optional)	WBCSD/WRI. The Greenhouse Gas Protocol; World Business Council for Sustainable Development and World Resources Institute, 2004

Table 3. Resources of GHG accounting standards

Resource	Description	Reference
World Resource Institute/World Business Council on Sustainable Development (WBCSD)	There are two standards: (1) A Product Life Cycle Accounting and Reporting Standard and (2) Corporate Accounting and Reporting Standard.	(WRI/WBCSD 2004, 2005)
ISO 14064	ISO 14064 (parts 1 and 2): it is an international standard for determination of boundaries, quantification of GHG emissions, and removal. It also provides standard for designing of GHG mitigation projects.	(ISO 2006a, ISO 2006b)
Publicly Available Specifications 2050 (PAS2050), British Standard Institution (BSI)	Specifies the requirements for assessing the life cycle GHG emissions of goods and services.	(PAS, B., 2008)
2006 IPCC guidelines for National GHG inventories	All sources of GHG emissions are classified into four sectors—energy, industrial process and product use, agriculture, forestry and other land use, and waste.	(IPCC, 2006)

for recording carbon emissions at the firm level and along the supply chain; the model goes into great detail for each category or stage. However, obtaining participation from numerous firms is impossible, and it is difficult to benchmark with other firms in other industries due the large quantity of data needed. Furthermore, stakeholders often find this data difficult to understand as it involves technical data. However, supply chain members that document their carbon footprint can easily improve the performance of their supply chain by sharing and comparing carbon footprint data among members.

Another method that can be used to record carbon footprint is the Economic Input-Output (EIO) model. This method of documenting carbon emission on the sector or industry level means that the data can be generalized to entire sectors and easily used by firms and stakeholders to understand. Also, it allows easy benchmarking of carbon emissions for firm with industry standards. However, this method is insufficient for analysis of an individual firm and its supply chain members. Among researchers that used EIO methodology are Huang, Lenzen, Weber, Murray and Matthews, (2009a, 2009b), Minx et al. (2009), Wiedmann (2009) and Wiedmann, Lenzen and Berrett (2009). There are also hybrid methods combining both LCA and EIO and several new methods using mathematical calculation, and other quantitative engineering techniques. LCA measurement seems doable as the concern is to lower carbon across the supply chain and not directly looking at the measurement of carbon emissions.

Climate change concerns helped lead to the Kyoto protocol in 1997, which created three instruments to reduce GHGs emissions, especially for Annex 1 countries (countries that emit high quantities of GHGs). The first instrument is carbon-trading system whereby developed countries can trade GHGs internationally based on the "cap and trade system" (Amran, Zainuddin & Zailani, 2013). Basically, government imposes a limit to emissions allowed for firms (the cap) and firms will be limited in the quantity of GHGs they can emit. In order to exceed the limit, firms need to buy a permit (trade). To do so, firms can negotiate with other firms that produce lower carbon emissions to increase their limit. The seller gains economic benefit from this trade. The second instrument is collaboration between Annex 1 countries to take on emission reduction projects by sharing reduction credits. The third instrument is the Clean Development Mechanism (CDM), whereby Annex 1 countries conduct emission reduction projects in developing countries to earn emission reduction credits (Amran et al., 2013). Of all these instruments, Malaysia can participate in CDM to lower its own carbon emission.

LOW CARBON SUPPLY CHAIN

A low carbon supply chain is an extension of green supply chain practices. The short-range aim of a low carbon supply chain is eliminating or reducing waste of the processes in the firms and the members of supply chain (which is a GSCM objective) while monitoring and reducing carbon emissions throughout the supply chain. However, ElTayeb, Zailani & Jayaraman (2010) found that study was lacking in two key areas: the existence of green supply chains in Malaysia and the incentives for developing a green supply chain in Malaysia. In this context, an understanding of why Malaysian firms have difficulties in supporting low carbon supply chain practices becomes clearer. The difficulties in supporting low carbon supply chain practices is often because Malaysian firms do not well understand the initiatives present in the existing literature for developing green supply chain

The government must educate firms about green practices than more before by offering incentives for lowering their carbon footprints. Firms need to know that low carbon initiatives do not necessarily mean changing directly to green technology but that they need to reduce wastes that have negative impact upon the environment, the society and the economy. Today, only large Malaysian firms with more than 10 suppliers are engaging in green purchasing and eco-design (Eltayeb & Zailani, 2009). This is true because large firms, especially multinational corporations, often come from developed countries such as the United States and from Europe that have good knowledge about green practices and practicing low carbon initiatives at large.

One important consideration that firms need to understand is that, while transferring technology is sensitive among firms, transferring or sharing low carbon practices among firms will lead to profitability for all the firms involved. Malaysia has become known as a global sourcing hub for electronics parts and materials in the ASEAN region, combining with economic and political stability, low operating costs and cheaper labour cost (compared to developed countries) sharing such knowledge will benefit MNCs sustaining their business as well helping local suppliers to improve their performance according to international standards. If low carbon knowledge is not transferred or learned in Malaysia, firms will pass along the carbon emissions to the end process and the last party will need to take all the blame and corrective actions. Eltayeb et al. (2010) believe that strict regulations for firms are needed ensuring that suppliers follow low carbon regulations will lead to the creation of standards down the supply chain. To make this effort successful, firms in the supply chain must measure and reduce their carbon footprints before moving onwards to the downstream party.

The concept of the low carbon supply chain is clear; each tier in the life cycle of all supply chain members should measure and record their carbon footprint and then reduce the carbon emissions at each tier before sending products to other supply chain members. However, the issue with measuring carbon emissions is obtaining data is difficult. Firms should record their own emission data. Another reason this is because each firm have different capabilities and knowledge of carbon footprint. To solve this problem, firm, suppliers, distributors and government must play their part in measuring carbon footprint according to well-established protocols. Furthermore, recording carbon emissions is very technical, and business managers will find it difficult to either understand or obtain accurate data. Hiring technical staff to monitor and record carbon footprint is extra cost to firms especially in Malaysia where the establishment of SMEs is higher than that of big firms. Thus, until carbon emission recording is made mandatory, measurement data for carbon emission in Malaysia will be difficult to obtain.

CARBON FOOTPRINT AGENDA TOWARDS LOW CARBON SUPPLY CHAIN

Sustainability has received enormous attention in recent years as an effective solution to support the continuous growth and expansion of the manufacturing industry (Yuan, Zhai & Dornfeld, 2012). Drivers for creating green supply chain have been getting stronger in many businesses and in society at large (Seidel, Shahbazpour & Seidel, 2007; Jafartayari, 2010; Millar & Russell, 2011; Vinodh & Joy, 2012; Rosen & Kishawy, 2012). According to Jayal, Badurdeen, Dillon & Jawahir (2010) achieving sustainable manufacturing requires a holistic view spanning the product, manufacturing processes and supply chains, including manufacturing systems across multiple product life cycles. Having said that, firms in a supply chain must learn to work together and use their competitive advantages and capabilities in using green practices to help each other succeed. For example, if a firm can take advantage of renewable energy incentives by the government and suppliers are able to equip green machines at a lower costs because of such incentives, then product costs should be lowered and cost reductions should be shared across supply chain members.

Low carbon emission has become synonymous with green technology and a highly sustainable approach to reducing carbons. Green practices such as using large windows or skylights (Bellona, 2009) or changing incandescent light bulbs to compact fluorescents can be low carbon initiative (green practice). These are some of the small initiatives that manufacturing firm can do. Improving efficiency of vehicles, using hybrid fuel technology (Dey et al., 2011) and maximizing space utilization in shipment or delivery by reducing the empty driving time also can help (Fugate, Davis-Sramek & Goldsby, 2009; Esper & Williams, 2003) and are part of low carbon initiatives from the logistics perspective.

The Malaysian Green Technology Corporation (GreenTech Malaysia), under the purview of Ministry of Energy, Green Technology and Water Malaysia (KeTTHA), has identified four areas for green growth in Malaysia: 1) the Green Malaysia Plan, 2) Green Procurement, 3) Electric Mobility, 4) Mobility Solution and 5) Sustainable Living.

1. **Green Malaysia Plan:** This plan is to designed to promote green technology in the areas of energy, transport buildings, water and waste management. A total of 141 projects have been approved and financed since 2010 equivalent to RM1.9 billion with RM1.5 billion funds left unused. Lim (2014) reported that a total of 2.29 million tonnes of carbon emissions have been saved through these projects. The available funds should be seen as an incentive for potential investors and firms to begin green practices and motivate them to understand the concept of a low carbon supply chain. This is because many firms have already started moving into green industry practices, especially MNCs, for whom SMEs constitute critical parts of their supply chain.

2. **Green Procurement:** The launch of MyHijau, a government-sponsored green products and services labelling program launched in 2012, is an initiative of GreenTech Malaysia to educate consumers, manufacturers and suppliers about green goods and services and to inspire them to use such products. The directory of green products, which included 268 goods and 34 services, is mainly targeted for manufacturing and suppliers. Although Malaysian consumers are not all aware of this green directory and mostly stick to the non-green manufacturers or suppliers of goods and services, the hope is that the directory will inspire future change. The introduction of green initiatives in Malaysia is still relatively new and thus may require more time before useful results are produced.

3. **Electric Mobility:** Electric busses have been tested in several areas in Malaysia to promote green public transportation and to increase public awareness. In the future, more hybrid and electric cars as well as electric-powered public transportation are being targeted to help inspire Malaysian consumers to support low carbon objectives. As of September 2014, the Road Transportation Department of Malaysia reported that 1,024 electric cars and motorcycles had been sold between 2011 and September 2014 (Lim, 2015). Of the total of 627,753 vehicles sold in 2013, 18,967 units were hybrids, which was a 23.5% increase from the 15,355 units sold in 2012 (Mahalingham, 2014). This has been seen as one area in which Malaysia should improve on because as traffic congestion increased the carbon footprint. However, continued growth in this area is threatened because of the relatively high price of hybrids and the expiration of tax exemptions for hybrids (Mahalingham, 2014).

4. **Mobility Solution:** To tackle the culture of Malaysian consumers who prefer pri-

vate over public transportation, GreenTech Malaysia has introduced electric car-sharing programs at several public transport areas such as train and Light Rail Transit (LRT) stations. This program is designed for those commuters using public transport to get to and from their residences or parking areas via sharing and focuses on shifting public transportation preferences and reducing traffic congestion.

5. **Sustainable Living:** The objective of this program, which teams with local municipals councils, is to develop low carbon cities in Malaysia through reducing energy usage and creating indices for measuring energy use in offices, buildings and resident homes. The Green Building Index (GBI) comprises metrics that measure energy usage in structures, materials and resources. The index has six criteria including energy efficiency, indoor environmental quality, sustainable management, material and resources, water efficiency and innovation to ensure buildings are sustainable and eco-friendly. Builders can achieve Gold, Silver, and Platinum status (Green Building Index, n.d.).

The Second National Communication, National Hydraulics Research Institute and Malaysian Meteorological Service (2013) showed that 26.2% of GHG emitters come from industries, 16% from transport energy usage and 11.7% from energy used in manufacturing (Tan, 2013). These three areas are the top GHG emitters in Malaysia, and this is the reason that to reduce carbon emissions in Malaysia, supply chain and manufacturing should be a main priority. The role of energy resources as a contributory factor to sustainability (Halldórsson & Svanberg, 2013) or low carbon footprint needs to be worked into the supply-chain framework. McKinnon (2011) stated that a great deal of carbon emissions derive logistics, and Stock, Boyer & Harmon, (2010) noted that further research on energy is needed in the supply chain context.

Malaysia is currently developing a framework to identify and reduce the carbon footprint among firms in Malaysia, and an agenda geared towards developing a low carbon supply chain includes looking at the role of firms, the supply chain, and the government.

Firms should:

- Measure carbon footprint at organization and manufacturing level. Firms need to start measuring their carbon footprint across organizational structures.
- Eliminate or reduce waste in their manufacturing process.
- Make a checklist of each carbon emitters in manufacturing and find a substitute strategy with lower carbon emitters.
- Join voluntarily a government-sponsored pilot test for determining the level of carbon footprints in firms across Malaysia to enable regulation benefitting all firms.

Supply Chain:

- Distribute knowledge in carbon reduction among members.
- Follow GHGs protocol and ISO in determining carbon emissions and improve processes.
- Increase flexibility to respond to customer requirements while reducing inventory upkeep.
- Adopt similar systems and measurement methods for carbon footprint with other members.

Government:

- Promote Research and Development (R&D) carbon reduction incentives for MNCs so that foreign firms will setup R&D in Malaysia.

- Enable joint projects undertaken for energy efficiency, renewable energy, carbon reduction and carbon credits.
- Introduce incentives for local SMEs to collaborate or undertake projects with MNCs.
- Make reports transparent to stakeholders to learn and benchmark their carbon footprint with other firms. Make reports up to date so that firms will be motivated to be responsive in comparing and reducing their carbon footprints.

LOW CARBON FOOTPRINT IN MALAYSIA MANUFACTURING FIRMS

The Malaysian government has addressed the issue of lowering the country's carbon footprint in several initiatives. In April of 2009, Malaysia embarked upon a green technology plan to address sustainable development issues, creating the Ministry of Energy, Green Technology and Water (KeTTHA) as the body responsible for addressing this issue. Furthermore, on 16th December 2009, in response to the United Nations Framework Convention on Climate Change (UNFCCC) report, Malaysian Prime Minister announced that Malaysia would undertake up to a 40% voluntary reduction of emissions by the year 2020. Later, in June of 2010, the 10th Malaysia Plan highlighted energy efficiency and equipment as well as minimum energy performance standards as the focus for lowering the carbon footprint in manufacturing industries. The 11th Malaysia Plan, announced in June of 2015 focuses on introducing green technology and green research and development (R&D). Malaysia long-term goals are to improve its environmental rating and reducing national energy usage. To further these objectives, the Renewable Energy Act was enacted in 2011. Establishing greener cities and multiplying green industries especially in Small Medium Enterprises (SMEs) are part of its long-term plan. Sadly, as the Kyoto Protocol is nearing its expiry, national and world nations are unable to reduce carbon emission significantly or make progress towards this objective. In order to reach its 2020 objective, Malaysia will need assistance from foreign countries. This is because the Malaysia economy is driven largely by investments from MNCs in the manufacturing industry.

Malaysia's government incentives to promote green practices in Malaysia are still evolving and often revolve around taxes especially for firms generating, using or purchasing renewable energy. Increasing energy efficiencies and providing consultancy on energy conservation are also eligible for tax incentives such as income tax exemptions. Those using imported machinery, equipment, materials, spare parts and consumables that help to reduce energy used also can claim import duty and sales tax exemptions.

Malaysia's understanding of what is needed is in accordance with Ngai et al.'s (2012) statement that identifying, controlling and reducing unnecessary energy and utility consumption can reduce GHGs. Malaysia is trying to make firms understand and change their energy policies to become energy efficiency, adopt renewable energy or invest in green technology. Malaysia 11th (mid-term) and 12th (long-term) plans supporting green causes show that green technology and innovation are considered important with respect to Tier II.

Malaysia has two options to move to Tier III: a) quickly improving and fulfilling renewable energy objectives for manufacturing and businesses or b) restructuring the strategy by promoting incentives for green innovation and technology such as those for green Research and Development (R&D). These incentives are beneficial for firms adopting or selling renewable energy products or services and well-established or heavily invested industries. Other industries have been slow to take up the offer. That is the reason why the manufacturing industry in Malaysia still not has entirely grasped low carbon practices. The Malaysian government has tried to make low carbon manufacturing attractive to other industry by lowering operation costs through the promotion of low carbon initiatives. However, attention should be given to manufacturing industry where it contributes 24.2% of Malaysia Gross Domestic Product (GDP) (Department of Statistics Malaysia, 2012).

Government incentives are designed to help firms prosper by using renewable energy sources or using energy more efficiently. Sometimes, these incentives do not work well. One case is that of hybrid cars. This is because the demand for electric and hybrid cars in Malaysia is poor. As a result, automakers are not too interested in producing electric cars locally. Given that the Malaysian government wants to protect national automakers from losing out to foreign automakers, current incentives are insufficient for promoting low-carbon practices in Malaysia. Thus, in terms of automakers, additional incentives are needed to promote increased use of electric and hybrid cars because Malaysian citizens and firms mostly use private transportation such as cars. Thus, in terms of private transportation, which is widely used in Malaysia and produces high carbon emissions, a need exists to improve. Doing so, might help raise Malaysia's ranking in the United Nations' World Happiness Report, upon which Malaysia ranks 56 out of 156 countries benchmarked (Helliwell, Layard, & Sachs (2013).

CURRENT ISSUES OF CARBON FOOTPRINT IN MANUFACTURING SECTOR IN MALAYSIA

Awareness of green aspects is still not widely diffused throughout Malaysia currently (Eltayeb et al., 2010). To help alleviate this situation, the Ministry of Natural Resources and Environment Malaysia introduced the National Carbon Reporting Program (My Carbon) in December 2013 to encourage Malaysian firms to report voluntarily their carbon footprint. So far in August 2014 only 26 pilot organizations have joined this program, such as Fuji Xerox (photographic), TM (telecommunication), PBA (water management), Maybank (financial institution), Air Asia (flight service) and several local manufacturing firms such as Spritzer (beverage), SKS (furniture) and Asta Chemical (chemicals). Of the 26 firms reporting, 14 firms submitted their carbon footprint using the MyCarbon template, which uses the WRI protocol (refer to Table 2). However, after comparing Malaysia's largest firms with leading firms practicing sustainability, a 2012 Pricewaterhouse Coopers (PWC) reported showed that Malaysian firms did not adequately monitor or measure their carbon footprints.

Malaysia objectives for manufacturing at this moment are mostly directed towards energy used, renewable energy and material or equipment used. Two reasons exist for why Malaysia is focusing on the energy sector: 1) carbon emissions from industry are dominated by production of goods in steel, cement, plastic, paper, and aluminium, and 2) urban development. First, because demand for steel, cement, plastic, paper and aluminium is anticipated to at least double by 2050 (Allwood, Cullen & Milford, 2010), these industries need more attention than others do. As manufacturing is part of the supply chain, a need exists to categorize issues according to stages from supplier to manufacturing to distribution. Second, according to Malaysia Institute of Planners urban developments emitted 50% of the country's carbon emissions, making these areas a priority for more governmental attention.

1. **Energy Sector:** Tenaga Nasional Berhad (TNB), an energy provider in Malaysia, has come out with a plan to promote the use of green technology by consumers and firms. TNB allows users who have installed solar panels that generate electricity to sell excess energy output to TNB at a guaranteed rate. This program is called "Feed-in Tariff (FiT)". However, the cost of installing solar panel in Malaysia is still expensive. As a result, talks are ongoing to allow potential users

Figure 1. Malaysian firms reporting climate change compare to sustainability firms

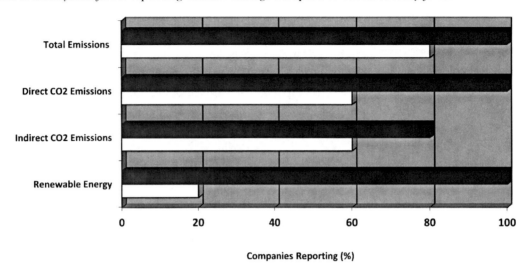

to obtain bank loans, thus encouraging the public to harvest more renewable energy. This program, however, is still its infancy and has not widely adopted by the public. To ensure that this program is ultimately successful, the government should increase green awareness. Even though the rising price of electricity has sparked a temporary public outrage, the public still prefers electricity delivered by traditional power plants. Perhaps educating the public about renewable energy and finding solutions to make renewable energy available at a cheaper prices will encourage more users to adopt green energy. As this system was just introduced, no data or update has been made.

2. **Green Suppliers:** Reportedly, a joint venture between TNB and a United States' firm is developing a 50-Megawatt solar farm project in Kedah to generate electric. This would be the first such project in Malaysia and a great step forward for energy generation because this project will give exposure and experience to TNB in generating electric from renewable energy sources. In the long run, such project will help lower the price of electricity. With the advantage of plentiful sunlight, Malaysia should be able to reduce its carbon footprint.

In past year, public outrage arose when the Malaysian government announced plants to consider nuclear power, perhaps spurred by the Fukushima Daiichi disaster in Japan. Certainly, nuclear plants would provide more energy output and lower the price of electricity but nuclear power has become both an ethical and safety issue. Malaysia is currently researching this energy area slowly because data is scarce and experts are mostly from Western countries.

Malaysia has been a popular manufacturing destination for foreign manufacturers. However, rising land prices have made investors reluctant to shift production to Malaysia even though Malaysia has a stable economy, a good financial system, and a skilled workforce. Malaysia is also absent from any major natural disaster such as earthquakes, volcanic eruptions, tornados and floods. As the prices of real estate and living costs in Malaysia have become increasingly more expensive, Indonesia, Thailand and Vietnam now are becoming hotspots for manufacturers. To counter this development, Malaysia has begun focusing on green trade zone and Free Trade Zone (FTZ) developments. The FTZ is a hub for supply chain members to operate business closely, and a green zone has been developed in Johor to attract foreign investors, especially those from Singapore. As this green trade zone is still in development process, demand remains unknown. What Malaysia needs at this moment is real estate prices for manufacturing fixed at rates competitive with those of neighbouring countries. Then, manufacturing firms might relocate their green suppliers to green zones in Malaysia, which would help Malaysian firms engage in green practices.

3. **Manufacturing:** Issues in manufacturing firms in Malaysia regarding carbon footprint seem mostly about adopting green technology at a lower cost, improving processes though green practices, adhering to international standards for international trade and introducing low carbon initiatives to SMEs. Manufacturers in Malaysia face challenges in retaining talent, an unattractive economic situation and low salary packages for employees. For many small-to-medium size enterprises (SME), financing environmental friendly technology is expensive, even with government incentives. In 2005, Anbumozhi and Kanda found that SME's often did not accept voluntary low carbon initiatives. Because Malaysia SMEs have not embraced low carbon initiatives, reaching its 40% carbon emission reduction in 2020 will be more difficult for Malaysia.

Furthermore, when the economic situation worsens all over the world, manufacturing firms in Malaysia are facing the choice of keeping talented employees happy and productive or of providing costly environmental solutions. Thus, firms must balance improving their financial bottom line while simultaneously trying to meet national and international environmental standards. This is a difficult task, often requiring knowledge and expertise that Malaysia does not possess.

At this moment, the local municipal council in Penang, Malaysia is planning to apply a policy requiring industrial companies to pay for all wastes they produce. However, this policy remains in the planning stage and the local council and state government need feedback from manufacturing firms. Government understands that waste from one firm might be useful for another firm that but they must gather data on industrial waste to ensure the best use of such waste before it is sent to a dumpsite. Once government identifies which firms need the waste materials and put them together with waste producers, there will be less overall waste to worry about and less land needed for dumpsites. This is good practice because once the study is done, government can reinvent the waste management system.

The Electric and Electronics (E&E), crude petroleum, automotive, liquefied natural gas, chemical products and palm oil are main industries in Malaysia. These industries are involved in manufacturing and exporting in the international market. Malaysia was chosen as a place for many companies to setup E&E manufacturing plants because of governmental incentives and low-wage labour. As a result, Malaysia has developed into a global sourcing country thanks to the concentration of many available suppliers and manufacturers. However, E&E is facing several difficulties because of the brain drain of local talent and unattractive salary packages. The Malaysian oil and gas industry has one big player, Petronas, and offers high-wages that attract talented employees. The Malaysian government has worked diligently to ensure the survivability of national automakers Proton and Perodua. The automotive industry not only emits carbon in large quantities during the manufacturing process but also during its entire life cycle. To address this issue, the 10th Malaysian plan encouraged green equipment and energy efficiency.

The other tool that Malaysia has used is its Clean Development Mechanism (CDM). E&E industries in Malaysia are usually part of vast Multinational Corporations (MNCs) that undertake CDM projects to increase their carbon limits but some MNCs voluntary reduce their carbon footprint by promoting green practices. Several examples of manufacturing firms either adopting behavioural changes or through the use of green technology exist. For example, Motorola in Penang offers Six Sigma green belt certification for eliminating waste and time utilization by firms, employees and suppliers. If CDM proves too expensive, perhaps following Intel Malaysia practices for saving energy could serve as a an example, The company managed to save energy consumption from 24 million kilowatt hours (kWh) to 27 million kilowatt hours and reduced water usage from 119,000 cubic meters to 120,000 cubic meters from year 2011 to 2012. These savings were achieved by means of more than 50 energy conservation and water management projects. One such project was adopting lighter aerodynamic cooling blades for central cooling fans at its local plants.

4. **Distribution:** According to Malaysian Institute of Planners (2011), 19% of GHGs came from industrial companies, which is the second largest source compared to urban development 50%. Some argue that that transport, which comprises 13%, should actually be included under included urban development. Others believe that transportation should be included in the industrial total because supply chain is essential to businesses. Whatever the case, transportation is a very visible factor and a critical element when firms use land transportation, especially when they have adopted Just-In-Time (JIT) industrial inventory practices and have numerous suppliers. For example, Toyota Malaysia supplies materials and parts via trailers daily to both the Perodua (Malaysia's second largest automaker) manufacturing plant in Rawang and the Toyota manufacturing plant in Shah Alam through several sub-contractors. Toyota encourages sub-contractors to keep transportation vehicles and warehouses clean and tidy through green continuous quality improvement programs. Although Toyota and Perodua encourage their sub-contractors to meet carbon emissions standards and engage in proper maintenance of their transports, these sub-contractors are often unaware that these practices reduce the carbon footprint because knowledge and awareness regarding this matter remain low in Malaysia. Thus, MNCs could play a major role in sharing green knowledge with local SMEs.

SOLUTIONS AND RECOMMENDATIONS

Creating a low carbon supply chain is a critical issue for manufacturing firms in developing countries especially when buyers for their products are from developed countries. Often times strict regulations in developed countries only allow products with environmental certification and labelling. In the future, Malaysian manufacturing firms will likely lose potential markets if they do not meet these standards, which will adversely affect business performance and reduce business sustainability. Such concerns mean that manufacturing firms must measure their own carbon footprints across their entire supply chains. Frequent measurement of the carbon footprint in their supply chain will assist these firms in reducing costs due to their ability to adhere to environmental standards and take advantage of governmental green incentives.

- **Defects and Wastage:** Firms create products that generate waste and defects during both production and their lifecycle. Recycling electronics wastage requires specific green technology and processes. The extra processes involved in recycling defective products and wastage emits more carbon into environment. Firm should reduce the production of defective products. Turning wastes into profit is tricky but many firms can turn wastage into products. Normally, reverse logistics can be an alternative solution

- **Cycle Times:** Firms should reduce waiting times and lower batch processing times in manufacturing because continuous use of machines and idle time produces carbon emissions. Two ways that firms can reduce carbon emissions are through changing the behaviour of employees and encouraging them to adopt green practices and becoming more efficient. Even though behavioural change is less significant than changing to renewable energy or energy efficiency, behavioural change is the cheapest method and easiest method to adopt. Today, Malaysian manufacturing firms manage their production flow in accordance with current practices, which neither are totally sustainable nor low carbon emitting. By revising production flow to the latest low carbon methods, will firms not only will gain a public relations advantage of being seen as a green firm but also reduce operating costs due to long-run benefits.

- **Inventory:** One of the highest emitters of carbon in manufacturing is inventory upkeep. Firms usually keep some inventory and some inventory requires special conditions or an environment that suits that product (e.g., put in freezer or under spotlight). These processes should be controlled because they emit a high level of carbon gasses. Even firms that do not keep inventory at all or require their suppliers or distributors to keep an inventory also will emit high carbon gasses because suppliers and distributors must frequently transport materials or products to their manufacturing plants. The Just-In-Time (JIT) inventory was introduced is to eliminate waste and inventory costs, however, JIT can also lead to practices that are inefficient with respect to carbon emissions. For example, if a laptop manufacturer needs hard drives from a supplier, the manufacturer will order a certain amount of hard drives and have only a small inventory just to be safe (safety stock). If this firm is using the JIT system, inventory should be kept at a minimum at all times. Nevertheless, practically speaking, this is not always the case. Many firms will stock higher amounts of inventory so that they will save reordering and waiting time. In order to reduce cycle time, management should educate employees in understanding the concept of JIT or "only take what you need" because such education will make the business more sustainable and efficient. Measuring carbon emissions from human activities to storage emissions such as cooling system and electricity is inaccurate today. Carbon emissions produced by storage due to maintenance costs is much higher than human activities at production line.

- **Productivity of Employees:** Managers must utilize employees so that idle time and unnecessary tasks are minimized because idle time and unnecessary tasks will lead to excessive working hour, extra corrective actions and incur extra costs that increase carbon emissions.

- **Flexibility:** Big firms usually are inflexible because they cannot cater to customer specific requirements while SMEs are more nimble and react actively and positively to customer needs. Increased flexibility often means increased costs because production processes must be changed. Thus, having a sufficient range of products that reduces changeover costs is advisable.
- **Uniqueness of Malaysia:** The availability of rainwater and plentiful sunlight are potentially good for green practices that reduce the carbon footprint. For example, Wisma Rehda, collects approximately two million litres of rainwater per year that is being used for landscape irrigation and cleaning purposes. The plenitude of available sunlight enables Wisma Rehda to design its buildings to use more natural light. To meet its 2020 target of carbon emission reduction of 40%, Malaysia has begun giving scholarships and introducing research incentives for universities to undertake green research and carbon reduction. These activities are not new in developed countries but represents a huge step because Malaysia is well known for its emphasis on education.

DISCUSSION AND FUTURE RESEARCH DIRECTIONS

Regulation is an important driver in having firms and consumers adopt green practices. As a result, a carbon tax, which Malaysian government could impose upon firms, can help to reduce carbon emissions and generate earnings for the government. However, this plan could be unpopular to the public because the result will be increased costs passed along to consumers. Nonetheless, it is ideal for Malaysian government to introduce this tax plan so that firms will start to calculate their carbon emissions. This will help the cause of MyCarbon, which is the Malaysian governmental initiative to introduce the notion of carbon footprint to firms so that Malaysia can achieve a carbon emission reduction of 40% by 2020. The Malaysian government can make measuring carbon footprints mandatory for all firms in Malaysia but especially for SMEs. Once all firms begin to calculate their carbon emissions, data for research can be obtained more easily, leading to better policies and improvements to current practices. Then, the government could remove this tax plan, having achieved the main objective of education. This noble cause will then help firms and consumers see that green products are not expensive over the long term and that green practices will benefit them both.

Another point to ponder is the design and layout of manufacturing, retail outlets and offices in Malaysia that do not take advantage of Malaysia's climate and rainwater. There is no debate about the fact that firms must invest in cooling systems at their production sites, but the design of buildings could be improved to reduce energy consumption. Malaysia is rich in natural sunlight but firms are not utilizing that sunlight by installing solar panels or having "solar-friendly" building layouts. Although many real estate developers are creating green projects anticipating the trend from conventional buildings to green buildings, the cost of creating green buildings is still very high in Malaysia. Despite these obstacles, Malaysian real estate developers should focus on designing buildings to enable more sunlight in the buildings and create rainwater system that trap and store rainwater for watering and non-consumable activities.

SMEs in Malaysia have problem related to high operating costs that make it not feasible for them to invest in green technology over the short term. Because green practices currently are still relatively low, training in and awareness of firms in adopting green practices are options. Incentives are largely unavailable for SMEs, thus, green practices mostly are being practiced by MNCs. However, in Malaysia, more than 90% of the firms are SMEs so the Malaysian government must help SMEs obtain green knowledge and green technology.

Then again, MNCs also have an important role to play in knowledge and technology sharing. Often MNCs do not share the latest and green technology with SMEs because MNCs wish to protect their businesses and trade secrets. Because of this, SMEs are paying the price by missing out in competitive market places. For example, import duties on solar panel hurt SMEs in supplying solar panel as they have less power to influence the market compared to MNCs. Thus, the issue of "SMEs vs MNCs" should be looked at from a countrywide perspective. Having said that, more emphasis on SMEs survivability and ability to obtain green knowledge and green technology are needed in Malaysia. After all, to reduce carbon emissions, Malaysia as a country must reduce carbon activities by all firms and consumers. By educating and managing SMEs, green objectives can be achieved.

Having international certifications and meeting international standards is useful for international trade, which might boost the country's economy. However, not all firms can afford to obtain or follow international standards. Furthermore, in the Malaysian context, developing the country's own standards is critical because SMEs can more easily adhere to them. International standards can be in the form of voluntary compliance while national standards could be the mandatory standards that all firms operating in Malaysian must follow. This is important because national standards can help firms learn while exposing them to current acceptable practices around the world. Once firms get used to these, they can upgrade to international standards, which sometimes have more and tougher procedures. As each country and industry is unique and different, adopting practices to local culture and work etiquette are important to ensure that firms are able to adhere to international standards. For example, data and research with respect to carbon footprints are already established in the United States and European countries but in developing countries such as Malaysia, Indonesia or Thailand, the issue and knowledge are still in their infancy. As a result, firms operating in these developing countries are having difficult moments and incur more costs to export to developed countries.

Developing countries, which are often rich in natural resources, should be able to stabilize their economy as suppliers of resources. Because knowledge about green practices is still low in developing countries, and many firms are unable to meet the strict requirement of international standards. Often foreign firms in the form of Foreign Direct Investment (FDI) come to developing countries and extract resources and export them to developed countries. As a result, many local firms, which mostly are SMEs, are losing out. Nevertheless, nowadays more local SMEs are working with MNCs as supply chain partners. That is why, it is important for the Malaysian government to focus on equipping SMEs with knowledge of green practices and having green incentives available to SMEs.

According to Iakovou, Karagiannidis, Vlachos, Toka and Malamakis (2010) and An, Wilhelm and Searcy. (2011), bioenergy has been overlooked in supply chain management. Malaysia is one of the largest palm oil exporters (44% of world exports) and producer (39% of palm oil production) (Malaysia Palm Oil Industry, 2012). Thus, Malaysia should venture into biomass research. Biomass is a fuel created from living organisms and, interestingly, the palm oil tree contains 10% oil and 90% biomass (Salman, 2014). Biomass is considered to be a renewable energy source but apparently, the 168 million tonnes of biomass produce yearly is not being used as source of energy even though the Malaysian government has identified it as a promising option to replace fossil fuel.

This inclusion of renewal energy into the Five Fuel strategy of the 8th Malaysian Plan was designed to spur the contribution of biomass is to at least 5% in 2005 to the Malaysia energy sector. However, until now the dependency upon fossil fuel for energy remains large in Malaysia. Perhaps by conducting more research or encouraging Malaysian palm oil producers to conduct pilot tests using biomass from palm oil tree as source of energy will increase awareness and contribute to shifting from over-dependency of fossil fuel to biomass energy. As most Malaysian firms are SMEs, changing to palm oil of which Malaysia has abundant resources should be cheaper than using current fossil fuel in operation and production.

In the Malaysia context, research on suggestions and recommendations to Malaysian government in developing policies also is another area upon which researchers should focus. This is because carbon footprint knowledge and awareness are low in developing countries such as Malaysia, Indonesia and Thailand. In addition, firms operating in these countries might have been using green or low carbons practices but are unaware of them. Researchers might discover unique or new green practices that can contribute both theoretically and practically.

Shaw et al. (2013) stated that only limited attempts have been made to model the carbon footprint in supply chains. This is an important area upon which researchers should focus as supply chain management has become critical. Many areas exist in which green practices have not yet covered in supply chain literature including energy efficiency, bioenergy, trade credits, low carbon logistics and reverse logistics. For Annex 1 countries, the trade credit relationship with the supply chain could be an interesting topic while for developing countries, energy and bioenergy are important because developing countries are still at the infancy stage of determining energy efficiency and renewable energy for firm and consumer use.

Lastly, barriers to entry and successful drivers from developing countries firms, especially from the Malaysian point of view, are also interesting areas that researchers can look into. This is because developing countries mostly have no financial power or little green knowledge as well as experts compared to developed countries. Thus, firms in developing countries tend to create practices that are geared more towards behavioural change rather than towards technology-based that need large financial investments. These practices can be included in current knowledge in carbon footprint literature.

CONCLUSION

Looking at the situation in 2014, most likely Malaysia will still focus on attracting firms to voluntarily submit their carbon emissions levels and adopting renewable energy because of the uncertainty in petroleum prices and unattractive economic factors. However, the Malaysian government should revise its incentives or promote low carbon footprint practices at large throughout the country through SMEs training and knowledge transfer. Firms should begin to measure carbon emissions and invest in renewable energy as soon as possible by taking advantage the green incentives from the government. For low carbon supply chains to be successful in Malaysia, knowledge transfer between firms about green practices is a must. The president of the Malaysia Biomass Industries Confederation has lauded collaboration between SMEs and MNCs. Supply chain members also should focus on reducing carbon emission and redesign current process to account for the latest low carbon process. Lastly, manufacturing firms, but especially SMEs, must voluntary participate in governmental efforts to promote low carbon practices because this participation will benefit firms when policymakers start to enact environmental policy that is suitable for industrial firms.

Below are reasons why a low carbon footprint is important for manufacturing firms:

1. **Consumer Perspective:** Green and Responsible Association. The oldest argument is that a business exists because of market demand. Horbach (2008) found that

a great demand existed for eco-innovation products, processes and management systems. KIA Motors, a Korean car manufacturer that had stressed design, now has shifted its focus on eco-driver or lower carbon emissions to attract ever-growing green consumers. This trend can be seen elsewhere in manufacturing industries such as energy, telecommunication and computers. Other than in the United States and European countries, green initiatives are considered to be new. Thus, the carbon footprint, which is considered to be an advanced version of green practices remains mostly non-existence in developing countries at this moment. Even though adopting green technology and green practices will increase costs in the short run, the return on investment over the long term will lead to gaining competitive advantage, lower costs and increased market share. Jansson, Marell and Nordlund (2011) stated that customer perception is an important indicator for innovation adoption. Such research shows that a firm's innovation is linked greatly to the marketing of its green products or services to the customer. In other words, green practise helps a firm promote its brand name.

2. **Business Perspective:** Understanding Carbon Footprint Cost. A firm's main objective is to maximize shareholders' wealth. Without understanding the reason why green practices are important, firms will view green practices such as lowering carbon footprint as corporate social responsibility (CSR) and part of regulation obligations. Therefore, a need exists to translate carbon footprint data into meaningful information that top business management can understand to make analytical decisions and strategies. A quick web search on carbon footprint as a keyword shows that there are 59,975 journal publications listed in Science Direct, a reputable research database. However, defining carbon footprint and developing measurement tools is still at an elementary stage. Clearly, helping businesses understand and calculate the costs of their carbon footprint is vital. When firms understand the costs of carbon footprints to their businesses, firms will be more motivated in lowering their carbon footprint using more sound cleaning techniques, green technology, and green policies, hiring experts and changing to green suppliers when other firms are already moving into that direction, even though doing so may be expensive. After all, Multinational Corporations (MNCs) venture to developing countries to lower their cost of production. Thus, implementing a low carbon footprint at the in production makes sense.

3. **Supply Chain Perspective:** Low Cost and Higher Profits. Supply chain refers to firms creating value for customers to reduce costs of products, better service delivery and improved delivery time to customers (Christopher, 2005; Ulaga & Eggert, 2006). As consumers become more aware of green initiatives and responsibilities, the expectation is that a firm will produce goods in a manner that will protect and preserve the environment. As a result, manufacturing firms need to make sure that they have a low carbon footprint along the supply chain. Lowering a carbon footprint is difficult enough but involving all supply chain members will make doing so even harder to coordinate. There can be a moment where the product carbon footprint is low during production but increases during distribution and delivery to a customer. Therefore, to achieve low cost but higher profits while adhering to low carbon footprints, better coordination among members is needed. For that reason, a carbon footprint can be a metric for measuring supply chain performance and cost-profit estimation for a firm. Iskandar Malaysia, the objective of

which is to develop a strong and sustainable metropolis with green concept, is seen as a platform for supply chain members from local and international to operate businesses under a low carbon theme.

4. **Government Perspective:** Robust Environmental Regulations. Regulations are powerful factors that shape the way of firms doing business. For example, Crest Ultrasonic, a firm producing ultrasonic cleaning machine using ultrasonic only began to profitable once United States and Europe regulations banned using alcohol as a cleaning material for machines. That example shows that adhering to regulations and communicating effectively with legislative bodies is important In addition, the example confirms that promoting green practices can be profitable. When firms are able to follow regulations and promote low carbon footprints, government will be able to determine the best incentives and mechanisms to help design robust environmental regulations. Furthermore, international trade requires nations to adhere to multiple regulations and stringent environmental regulations, which can be found in the United States and European countries. As a result, a need exists for manufacturing firms exporting to those countries to adopt stringent production processes. However, government faces a difficult task due to multiple carbon footprint standards definitions, measurement tools and insufficient data.

5. **Academic Perspective:** Contributions to Existing Knowledge. Lastly, gathering mass data from industry will help academicians to improve carbon footprint knowledge, but research in this area currently is still new in Malaysia. Theoretically, Malaysia carbon footprint knowledge is still low, and this topic needs more research and attention from academic researchers in Malaysia. At this moment, MyCarbon, the Malaysian government organization promoting and working on carbon footprint, is the only key player (other than non-governmental organizations) working closely with government to do more research on carbon footprint. Malaysia is currently at the stage of measuring carbon footprints through voluntary means of local firms in records their carbon footprint according to measurement protocol adopted from WRI (see Table 2). This pilot test involves several firms from across various industries that represent supply chain members operating in the country. Once firms understand and undertake carbon footprint measure, successful drivers that help make for low carbon footprints in a supply chain can be identified.

ACKNOWLEDGMENT

The authors convey their appreciation to the Division of Research & Innovation, Universiti Sains Malaysia for funding the research (short-term grant no: 304/PPAMC/6313108).

REFERENCES

Allwood, J. M., Cullen, J. M., & Milford, R. L. (2010). Options for achieving a 50% cut in industrial carbon emissions by 2050. *Environmental Science & Technology*, *44*(6), 1888–1894. doi:10.1021/es902909k PMID:20121181

Amran, A., Zainuddin, Z., & Zailani, S. H. M. (2013). Carbon trading in Malaysia: Review of policies and practices. *Sustainable Development*, *21*(3), 183–192. doi:10.1002/sd.1549

An, H., Wilhelm, W. E., & Searcy, S. W. (2011). Biofuel and petroleum-based fuel supply chain research: A literature review. *Biomass and Bioenergy*, *35*(9), 3763–3774.

Anbumozhi, V., & Kanda, Y. (2005). *Greening the production and supply chains in Asia: Is there a role for voluntary initiatives*. IGES Kansai Research Centre KRC.

Bellona, S. (2009). Justifying energy efficiency as oil prices tumble. *The Chronicle of Higher Education, 55*(22), A14.

Benjaafar, S., Li, Y., & Daskin, M. (2013). Carbon footprint and the management of supply chains: Insights from simple models. *IEEE Transactions on Automation Science and Engineering, 10*(1), 99–116. doi:10.1109/TASE.2012.2203304

Carbon Trust. (2012). Retrieved from www.carbontrust.com/resources/guides/carbon-footprinting-and-reporting/carbon-footprinting

Christopher, M. (2005). Logistics and supply chain management: Creating value-adding networks. *Industry Week, 267*(1), 60.

Department of Statistics Malaysia. (2012). *Key statistics of manufacturing sector, 2012*. Retrieved from http://www.statistics.gov.my/index.php?r=column/cthemeByCat&cat=92&bul_id=enZUVDZxcHI5c1NFYzNSeUtxYWh3QT09&menu_id=SjgwNXdiM0JlT3Q2TDBlWXdKdUVldz09

Dey, A., LaGuardia, P., & Srinivasan, M. (2011). Building sustainability in logistics operations: A research agenda. *Management Research Review, 34*(11), 1237–1259. doi:10.1108/01409171111178774

Eltayeb, T. K., & Zailani, S. (2009). Going green through green supply chain initiatives towards environmental sustainability. *Operations and Supply Chain Management, 2*(2), 93-110

ElTayeb, T. K., Zailani, S., & Jayaraman, K. (2010). The examination on the drivers for green purchasing adoption among EMS 14001 certified companies in Malaysia. *Journal of Manufacturing Technology Management, 21*(2), 206–225. doi:10.1108/17410381011014378

Esper, T. L., & Williams, L. R. (2003). The value of collaborative transportation management (CTM): Its relationship to CPFR and information technology. *Transportation Journal*, 55-65.

Fugate, B. S., Davis-Sramek, B., & Goldsby, T. J. (2009). Operational collaboration between shippers and carriers in the transportation industry. *The International Journal of Logistics Management, 20*(3), 425–447. doi:10.1108/09574090911002850

Green Building Index. (n.d.). *What is green building?* Retrieved from http://www.greenbuildingindex.org/why-green-buildings.html

Halldórsson, Á., & Kovács, G. (2010). The sustainable agenda and energy efficiency: Logistics solutions and supply chains in times of climate change. *International Journal of Physical Distribution & Logistics Management, 40*(1/2), 5–13. doi:10.1108/09600031011018019

Halldórsson, Á., & Svanberg, M. (2013). Energy resources: Trajectories for supply chain management. *Supply Chain Management: An International Journal, 18*(1), 66–73. doi:10.1108/13598541311293186

Helliwell, J., Layard, R., & Sachs, J. (2013). *World happiness report*. United Nations (UN). Retrieved from http://unsdsn.org/wp-content/uploads/2014/02/WorldHappinessReport2013_online.pdf

Horbach, J. (2008). Determinants of environmental innovation—New evidence from German panel data sources. *Research Policy, 37*(1), 163–173. doi:10.1016/j.respol.2007.08.006

Huang, Y. A., Lenzen, M., Weber, C. L., Murray, J., & Matthews, H. S. (2009a). The role of input–output analysis for the screening of corporate carbon footprints. *Economic Systems Research, 21*(3), 217–242. doi:10.1080/09535310903541348

Huang, Y. A., Weber, C. L., & Matthews, H. S. (2009b). Categorization of scope 3 emissions for streamlined enterprise carbon footprinting. *Environmental Science & Technology*, *43*(22), 8509–8515. doi:10.1021/es901643a PMID:20028044

Iakovou, E., Karagiannidis, A., Vlachos, D., Toka, A., & Malamakis, A. (2010). Waste biomass-to-energy supply chain management: A critical synthesis. *Waste Management (New York, N.Y.)*, *30*(10), 1860–1870. doi:10.1016/j.wasman.2010.02.030 PMID:20231084

IPCC. (2006). *2006 IPCC guidelines for national greenhouse gas inventories*. Retrieved from http://www.ipcc-nggip.iges.or.jp./public/2006gl/pdf/0_Overview/V0_1_Overview.pdf

ISO. I. (2006a). 14064-1: 2006-greenhouse gases-Part 1: Specification with guidance at the organization level for quantification and reporting of greenhouse gas emissions and removals. International Organization for Standardization.

ISO. I. (2006b). 14064-2: 2006. Greenhouse gases—Part 2. Specification with guidance at the project level for quantification, monitoring and reporting of greenhouse gas emission reductions or removal enhancements. International Organization for Standardization, Geneva, Switzerland

Jafartayari, S. (2010). *Awareness of sustainable manufacturing practices in Malaysian manufacturers.* (Doctoral dissertation). Universiti Teknologi Malaysia, Faculty of Mechanical Engineering.

Jansson, J., Marell, A., & Nordlund, A. (2011). Exploring consumer adoption of a high involvement eco-innovation using value-belief-norm theory. *Journal of Consumer Behaviour*, *10*(1), 51–60. doi:10.1002/cb.346

Jayal, A. D., Badurdeen, F., Dillon, O. W. Jr, & Jawahir, I. S. (2010). Sustainable manufacturing: Modeling and optimization challenges at the product, process and system levels. *CIRP Journal of Manufacturing Science and Technology*, *2*(3), 144–152. doi:10.1016/j.cirpj.2010.03.006

KeTTHA. (2011). *Low carbon cities framework & assessment system* [PowerPoint Slides]. Retrieved from http://esci-ksp.org/wp/wp-content/uploads/2012/04/Low-Carbon-Cities-Framework-and-Assessment-System.pdf

Kitzes, J., Wackernagel, M., Loh, J., Peller, A., Goldfinger, S., Cheng, D., & Tea, K. (2008). Shrink and share: Humanity's present and future ecological footprint. *Philosophical Transactions of the Royal Society of London. Series B, Biological Sciences*, *363*(1491), 467–475. doi:10.1098/rstb.2007.2164 PMID:17652075

Lenzen, M. (2000). Errors in conventional and input-output—based life—cycle inventories. *Journal of Industrial Ecology*, *4*(4), 127–148. doi:10.1162/10881980052541981

Leonardi, J., & Browne, M. (2010). A method for assessing the carbon footprint of maritime freight transport: European case study and results. *International Journal of Logistics: Research and Applications*, *13*(5), 349-358

Lim, C. Y. (2014, July). Catalyzing green growth: Promoting a sustainable future through four key areas. *Ecowatch: The Star Publication*, 8.

Malaysia Palm Oil Industry. (2012). Retrieved from http://www.mpoc.org.my/Malaysian_Palm_Oil_Industry.aspx

Malaysian Institute of Planners. (2011). *Low carbon cities framework & assessment system* [PowerPoint slides]. Retrieved from http://www.townplan.gov.my/download/lowcarboncities-_paper_3b988287694.pdf

Matthews, H. S., Hendrickson, C. T., & Weber, C. L. (2008). The importance of carbon footprint estimation boundaries. *Environmental Science & Technology, 42*(16), 5839–5842. doi:10.1021/es703112w PMID:18767634

McKinnon, A. C. (2010). Product-level carbon auditing of supply chains: Environmental imperative or wasteful distraction? *International Journal of Physical Distribution & Logistics Management, 40*(1/2), 42–60. doi:10.1108/09600031011018037

McKinnon, A. C. (2011). Developing a decarburization strategy for logistics. In *Proceedings of the 16th Annual Logistics Research Network Conference, Smarter Logistics: Innovation for Efficiency Performance and Austerity*. University of Southampton.

Michael Groppi, P. E., & Burin, J. (2007). *Meeting the carbon challenge, the role of commercial real estate owners, users & managers*. Grubb & Ellis Company.

Millar, H. H., & Russell, S. N. (2011). The adoption of sustainable manufacturing practices in the Caribbean. *Business Strategy and the Environment, 20*(8), 512–526. doi:10.1002/bse.707

Minx, J. C., Wiedmann, T., Wood, R., Peters, G. P., Lenzen, M., Owen, A., & Ackerman, F. et al. (2009). Input-output analysis and carbon footprinting: A review of applications. *Economic Systems Research, 21*(3), 187–216. doi:10.1080/09535310903541298

Ngai, E. W. T., To, C. K., Ching, V. S., Chan, L. K., Lee, M. C., Choi, Y. S., & Chai, P. Y. F. (2012). Development of the conceptual model of energy and utility management in textile processing: A soft systems approach. *International Journal of Production Economics, 135*(2), 607–617. doi:10.1016/j.ijpe.2011.05.016

Pandey, D., Agrawal, M., & Pandey, J. S. (2011). Carbon footprint: Current methods of estimation. *Environmental Monitoring and Assessment, 178*(1-4), 135–160. doi:10.1007/s10661-010-1678-y PMID:20848311

Pas, B. (2008). *2050:2008 specification for the assessment of the life cycle greenhouse gas emissions of goods and services*. British Standards Institution.

Patel, J. (2006). Green sky thinking. *Environment Business*, (122), 32.

Pricewaterhouse Coopers. (2012). *PwC analysis, sustainability reports of analysis sample 2009/2010*. Retrieved from www.pwc.com/my/en/publications/pwcalert97-sncc.jhtml

Rosen, M. A., & Kishawy, H. A. (2012). Sustainable manufacturing and design: Concepts, practices and needs. *Sustainability, 4*(2), 154–174. doi:10.3390/su4020154

Salman, Z. (2014). *Bioenergy development in Malaysia*. BioEnergy Consult. Retrieved from http://www.bioenergyconsult.com/bioenergy-developments-malaysia/

Seidel, R. H. A., Shahbazpour, M., & Seidel, M. C. (2007, February). Establishing sustainable manufacturing practices in SMEs. In *Proceedings of the Second International Conference on Sustainability Engineering and Science*. Auckland, New Zealand: Academic Press.

Shaw, K., Shankar, R., Yadav, S. S., & Thakur, L. S. (2013). Modeling a low-carbon garment supply chain. *Production Planning and Control, 24*(8-9), 851–865. doi:10.1080/09537287.2012.666878

Stern, N. H. (2006). *Stern review: The economics of climate change* (vol. 30). London: HM Treasury.

Stock, J. R., Boyer, S. L., & Harmon, T. (2010). Research opportunities in supply chain management. *Journal of the Academy of Marketing Science, 38*(1), 32–41. doi:10.1007/s11747-009-0136-2

Tan, C. L. (2013, December). A warming trend. *Ecowatch: The Star Publication*, 5.

Ulaga, W., & Eggert, A. (2006). Value-based differentiation in business relationships: Gaining and sustaining key supplier status. *Journal of Marketing, 70*(1), 119–136. doi:10.1509/jmkg.2006.70.1.119

UNFCCC. (2009). *United Nations climate change conference*. Retrieved from https://unfccc.int/files/meetings/durban_nov_2011/statements/application/pdf/111207_cop17_hls_malaysia.pdf

Vinodh, S., & Joy, D. (2012). Structural equation modeling of sustainable manufacturing practices. *Clean Technologies and Environmental Policy, 14*(1), 79–84. doi:10.1007/s10098-011-0379-8

Wackernagel, M., & Rees, W. (1998). *Our ecological footprint: Reducing human impact on the earth (No. 9)*. New Society Publishers.

Wakeley, H. L., Hendrickson, C. T., Griffin, W. M., & Matthews, H. S. (2009). Economic and environmental transportation effects of large-scale ethanol production and distribution in the United States. *Environmental Science & Technology, 43*(7), 2228–2233. doi:10.1021/es8015827 PMID:19452867

West Tennessee Solar Farm. (n.d.). *Homepage*. Retrieved from http://solarfarm.tennessee.edu/

Wiedmann, T. (2009). *Editorial: Carbon footprint and input–output analysis–An introduction*. Academic Press.

Wiedmann, T., & Minx, J. (2008). A definition of 'carbon footprint'. *Ecological Economics Research Trends, 1*, 1-11.

Wiedmann, T. O., Lenzen, M., & Barrett, J. R. (2009). Companies on the scale. *Journal of Industrial Ecology, 13*(3), 361–383. doi:10.1111/j.1530-9290.2009.00125.x

Williams, E., & Tagami, T. (2002). Energy use in sales and distribution via e-commerce and conventional retail: A case study of the Japanese book sector. *Journal of Industrial Ecology, 6*(2), 99–114. doi:10.1162/108819802763471816

WRI/WBCSD. (2004). *The greenhouse gas protocol: A corporate accounting and reporting standard* (revised edition). Geneva: World Business Council for Sustainable Development and World Resource Institute.

WRI/WBCSD. (2005). *The greenhouse gas protocol: Project accounting*. Geneva: World Business Council for Sustainable Development and World Resource Institute.

Yuan, C., Zhai, Q., & Dornfeld, D. (2012). A three dimensional system approach for environmentally sustainable manufacturing. *CIRP Annals-Manufacturing Technology, 61*(1), 39–42. doi:10.1016/j.cirp.2012.03.105

KEY TERMS AND DEFINITIONS

Carbon Footprint: Carbon footprint is a term used to record carbon emissions by individual or a firm in order to see the impact of Greenhouse Gasses (GHSs).

Clean Development Mechanism (CDM): Emission reduction projects undertaken by Annex 1 countries in developing countries in order to earn emission reduction credits.

Economic Input-Output (EIO): EIO relates the production inputs of goods and services in an economy to the production outputs of other sectors.

Free Trade Zone (FTZ): Foreign trade zone or international trade zone, which hosts local and international manufacturers.

Greenhouse Gasses (GHGs): Greenhouse gasses (GHGs), namely, carbon dioxide (CO_2), methane (CH_4), nitrous oxide (N_2O), sulphur hexafluoride, perfluorocarbons and hydroflurocarbons.

Just-in Time (JIT): A system in which materials are delivered just before the production process begin to minimize inventory costs.

Kettha: Ministry of Energy, Green Technology and Water Malaysia.

Life Cycle Assessment (LCA): GHGs are recorded by classifying into Tier 1: direct emission of the firm; Tier 2: energy used and Tier 3: other indirect activities. In the literature, there are some scholars who extend LCA by including Tier 4: whole supply chain.

Low Carbon Supply Chain: Elimination or reduction of waste along the processes in firms and members of supply chain while monitoring and reducing carbon emissions throughout the supply chain.

Mycarbon: Malaysia's corporate GHG reporting program.

Chapter 15
Review of Supply Chain Integration on Green Supply Chain Management (GSCM)

Alia Nadhirah Ahmad Kamal
Universiti Sains Malaysia, Malaysia

Yudi Fernando
Universiti Sains Malaysia, Malaysia

ABSTRACT

The world economy operates on a capitalist market system where more and more natural resources are strained to produce maximum profits on the basis of achieving the efficiency of the economies of scale. As corporations' awareness increases on the jeopardizing impact they have caused to the deteriorating environment, more corporations have established a more eco-friendly operation. Greening the supply chain is one significant example of such moves. Realizing the green supply chain tendency in the industry, this proposed chapter focuses on highlighting the supply chain integration with business partners (suppliers, shippers, distributors, and customers) on Green Supply Chain Management (GSCM) practices. The chapter shows the literature supporting the important integration of GSCM as it enables corporations to gather collective strength, skills, and capabilities in achieving its ecological as well as business objectives. Both practitioners in companies and corporations might find this review useful, as it outlines major lines of research in the field.

INTRODUCTION

The focus of this chapter is to look into one of the most important assets of a company, its supply chain and how the supply chain could be enhanced by integration. Supply chain is the network created amongst different companies that produce, handle or distribute a particular product. It includes all the necessary steps for a specific good or service from the supplier to customer (Gold & Seuring, 2011). Hence, it is crucial to manage the supply chain, as optimized supply chain would result in companies achieving lower costs. It is also important to note the distinction between supply chain and logistics. While logistics look into the distribution process within the company, supply

DOI: 10.4018/978-1-4666-8222-1.ch015

chain includes multiple companies involving suppliers, manufacturers and retailers. Recent corporate practices have observed the consideration of environmental interest in the supply chain, creating the new term green supply chain (GSC). The greed for profit nature of capitalist economic system has caused numerous detrimental impacts towards the environment. Ignoring the principle of sustainability development - achieving a balance between environment, society and economy, capitalist practitioners centered their interest in economic growth by jeopardizing the environmental wellbeing. Thus, the concept of green supply chain management or GSCM has emerged out of this immediate need. It is apparent now that GSC has become more popular in the developed countries and the adoption rate of GSC across the globe has increased significantly (Mathiyazhagan et al., 2013). In other words, corporate organizations have begun to have raised interest on the environmental impacts caused by the operation and have started to modify their business processes to be more eco-friendly.

The data show that as at April 2005, more than 88,800 facilities worldwide had certified their environmental management system (EMS) to ISO 14001, the global EMS standard, while thousands more had adopted uncertified EMSs (Peglau, 2005). EMSs are strategic management approaches consisting internal policies, assessments, plans and implementation actions and as part of EMS's strategy, supply chain process has been rethink and restructured (Darnall, Jolley & Handfield, 2008). Nevertheless very similar to supply chain, this considerably new practice of green supply chain requires extensive monitoring and frequent assessment in order to achieved desired outcome.

Previous researches have suggested that the implementation of green supply chain management (GSCM) could be the best tool for optimizing the ecological, as well as the social and economic performance of the corporation (Rao & Holt, 2005; Mathiyazhagan, Govindan, NoorulHaq & Geng, 2013; Gardas & Narkhede, 2013). However apart from having green internal practices, stakeholder integration should be a priority to achieve corporate green supply chain aspiration (Pedersen, Henriksen, Frier, Søby & Jennings, 2013). Therefore GSCM would require a company to manage and monitor all supply chain members in the process. Nevertheless, integrity has always been a big challenge for GSCM (El-Berishy, Rügge & Scholz-Reiter, 2013) and questions are often raised, are these corporations willing to integrate closely with all supply chain members, sharing information as well business secrets in achieving GSCM ultimate goals?

As previous researches have proven that GSCM has a positive relationship with business economical, environmental, as well as societal performance (Fallah & Ebrahimi, 2014; Ninlawan, Seksan, Tossapol & Pilada, 2010; Zhu et al., 2005; Kumar & Shekhar, 2013), this study would highlight the effect of supply chain integrity to enhance this relationship. Therefore, a content analysis focusing on supply chain integrity was conducted in order to observe the role of integrity in supporting the relationship between GSCM and business performance. This chapter could give descendants insight into conceptualization green supply chain integration for sustainable business performance.

BACKGROUND

Recently there has been a wave, a booming trend of sustainability, where more corporations have publicly announced that they have the best practices that are clean and environmental friendly. In reality, no corporation is able to operate independently and it is misleading to declare a particular company as a green company when the supply chain practice is questionable. For example, it would be unethical for a particular computer producer to state that the computers were produced in a green manner should one of the suppliers throw production toxic waste into its nearby river. In addition, a truly responsible company would be accountable for all the products produced. This is not limited to the sourcing, production and delivery stages but also post product usage, when the products have exhausted their usefulness to the customers. All companies that aim to have greener operation should adopt green supply chain.

The increase in number of academic researches published on green supply chain and firm performance (Zhu & Cote, 2004; Ibrahim, Zailani & Tan, 2012; Mathiyazhagan et al., 2013; Golicic & Smith, 2013) has indicated the effectiveness of GSCM in lowering operation cost (Kamal, 2014). Therefore, GSCM does not only lead to minimization of environmental impact but it also assist in achieving higher efficiency. A research conducted by Kamal (2014) has also managed to identify the major role of focal company in collaborating and integrating with its suppliers to achieve green operations. Information integration was found to be an important factor of GSCM in which all members of the supply chain could only enjoy the benefit of GSCM practices should they agree to share their operational information (Lee, Kim & Choi, 2012; Frohlich & Westbrook, 2001; Stevens, 1989; Van Der Vaart & Van Donk, 2008).

In contrast, some have expressed their cynical view on GSCM, stating that it is a merely idealistic concept that is impossible and costly to achieve (Pagell & Shevchenko, 2014). Following these occurrences, this chapter was enthused in order to find out from the literature, how could integration actually assist in green supply chain management (GSCM).

SUPPLY CHAIN INTEGRATION ON GREEN SUPPLY CHAIN MANAGEMENT (GSCM)

Firstly the concepts that are central to this chapter were explained and discussed followed by the explanation on the methodology used for the content analysis. Subsequently, the analysis was presented and important facts discovered were highlighted. Finally, conclusion was given followed by future research direction.

Environmental Stewardship

In order to develop better understanding on GSCM, sustainability concepts will first be discussed. The International Institute for Sustainable Development has defined sustainable development as the 'development that meets the needs of the present without compromising the ability of future generations to meet their own needs' (United Nations, 1987). One amazing characteristic of ecosystem is it has the natural ability to recover or return to its original stage after being disturbed. However, due to human activities, most of them involving extensive natural resources usage and environmental pollution to serve the capitalist needs of production, this ability has declined globally in the last half century (Chapin, 2009).

Moving forward, services industry such as logistics industry and other environment sensitive industries must adopt GSCM in their operation strategy to ensure that environment is put at the heart of every business process (Seuring & Müller, 2008). Chapin (2009) has reviewed that the collapse of many advanced human societies, including Babylon, the Roman Empire, and the Mayan Civilization, were contributed by environmental degradation. Therefore, it is of vital importance that corporations play their role in ecosystem management and this could be done through environmental stewardship. Environmental stewardship refers to responsible consumption and protection of the natural environment that could be achieved through conservation efforts and sustainable practices (Dorsey, 2003). In this particular research, service industry falls under the doers' category of environmental stewardship. Doers would resolve the problem caused by taking action. As example, if an oil and gas service provider has accidentally caused oil spill in the sea, they would be the volunteers that participate in the cleaning up effort. Next part of this chapter would highlight the difference between conventional supply chain and green supply chain.

Conventional Supply Chain

Supply chain refers to all parties who are involved in satisfying customers' request, which includes the suppliers, transporters, warehouses, retailers as well as the customers themselves (Cox, 1999; Ibrahim, Zailani & Tan, 2012). From the analysis conducted on various supply chain experts, Mentzer et al (2001) has defined supply chain management (SCM) as:

the systemic, strategic coordination of the traditional business functions and the tactics across these business functions within a particular company and across businesses within the supply chain, for the purpose of improving the long-term performance of the individual companies and the supply chain as a whole.

SCM is therefore, highly concerned with operation management, logistics, procurement, information technology and it strives to create an integrated management approach. At first, supply chain is an important element of operational strategy. As the market become more global with shorter life cycle of new product, topped with customers heightened expectations, companies are obliged to focus and improve on their supply chain. In conventional supply chain, raw materials are procured and items are produced at one or more factories, and then shipped to retailers or customers (Fawcett et al., 2007; Christopher, 2012).

However, as SCM is developed purely on capitalist economy, many researchers and environmentalists have viewed this conventional SCM as flawed. SCM disregards the operational impacts on the natural environment while organisational profitability is put at the heart of the process. As the world population increase, more natural resources are consumed and processed globally, resulting in resource scarcity as well as increasing waste production endangering the planet. Moving forward, effective supply chain strategies should enable companies to not just reduce cost and increase service satisfaction levels, but also minimize environmental impacts, through interaction integration at various levels in the supply chain: upstream and downstream. Being aware of the critical need for SCM improvisation, many efforts have been pooled in the creation of a more sustainable approach to SCM. Through this industrial awareness, the green supply chain management (GSCM) was consequently introduced among the practitioners.

GREEN SUPPLY CHAIN MANAGEMENT

Green supply chain management is concerned on ensuring all supply chain deliveries while centralizing the environmental and social concern in every stage of supply chain to reduce product environmental footprint (Kamal, 2014). Hence, integrating the upstream and downstream members in the supply chain are equally important for the focal company, to achieve effective green supply chain management (Lai et al., 2014; Mathiyazhagan, et al., 2013; Ibrahim et al., 2012; Golicic & Smith, 2013). Upstream members would include the first, second, third and fourth tier suppliers of the company that supply products and services to the operation while downstream members refer to the customers.

It was observed that through practicing green supply chain management (GSCM) or also widely referred to as sustainable supply chain management (SSCM), organizations are able to control operational impact on the environment. The direct or indirect impact that production process caused can be monitored from the earliest stage up to the waste disposal procedure. Having the environment and social consideration integrated in key business processes from end-users through original suppliers that provide products, services, and information would add value for customers and other stakeholders (Lee, Kim & Choi, 2012). The research carried out by Lee, Kim and Choi (2012) has also found that after adopting GSCM, business performance is enhanced through the achievement of operational efficiency. Lower resource consumption is achieved via GSCM, while adoption of close-loop operation provides significant benefits (Zhu & Cote, 2004) where a waste produced by a particular production is reused as raw material for a new product and this cycle continues. GSCM ensures a particular corporation meets all conventional supply chain deliveries while putting at heart the environmental and social concern in every stage of supply chain - from the earliest material production to the end customers up to waste management.

Aligning with GSCM's main concern that is to integrate environmental interest into corporate supply chain practices, Rao and Holt (2005) have proposed four possible areas in which environmental initiatives could be integrated. They were inbound logistics, production or the internal supply chain, outbound logistics, and reverse logistics. This was also supported by Ninlawan, Seksan, Tossapol and Pilada (2010). who suggested that green activities in GSCM include green procurement, green manufacturing, green distribution and reverse logistics

Thus, this chapter adapted the theoretical concept proposed by previous research (Ninlawan, Seksan, Tossapol & Pilada, 2010) stating that the components of GSCM were green purchasing, green material handling, green distribution and reverse logistics.

1. Green Purchasing

Green purchasing refers to a purchasing method where environmental and social considerations were taken with equal weight to the price, availability and performance criteria that businesses use to make purchasing decision. Green purchasing was defined by Min and Galle (2001) as an environmentally conscious purchasing practice that reduces sources of waste and promotes recycling and reclamation of purchased materials without affecting the required performance of such materials.

2. Green Material Handling

Green material handling is concerned on using all the input material in the most optimal way that includes pollution prevention and product stewardship (Ninlawan, Seksan, Tossapol & Pilada, 2010).

3. Green Distribution

While traditional distribution concept aims that all produced materials should reach customers in a timely manner, green distribution adds another objective to the system that is to minimise total environmental impact in the logistics process (Rao & Holt, 2005).

4. Reverse Logistics

Reverse logistics closes the supply chain loop in which it reconnects the customers' role to focal company through product after use transfer. It refers to the focal company's responsibility of taking care of produced products once they were no longer desired or can no longer be used by their users (Soleimani & Govindan, 2014).

Next section discussed the element of integration within supply chain from different views.

SUPPLY CHAIN INTEGRATION

Supply chain integration (SCI) is one of the most important aspects in supply chain management and its enablers and outcomes have been extensively studied by researchers (Frohlich & Westbrook, 2001; Narasimhan & Kim, 2002; Vickery et al.,

2003; Droge et al., 2004; Swink et al., 2007; Zhao et al., 2008, 2011; Braunscheidel & Suresh, 2009; Flynn et al., 2010). Integration in this context refers to the extent to which various supply chain activities and processes work together. SCI is defined as the degree to which a firm can strategically collaborate with its supply chain partners and cooperatively manage intra-organizational and inter-organizational processes to achieve effective and efficient flows of products, services, information, financial, and decisions to provide the maximum value to the final customer with low costs and high speed (Frohlich & Westbrook, 2001; Stevens, 1989; Van Der Vaart & Van Donk, 2008). Besides, Pagell (2004) defines SCI as a process of interaction and collaboration in between the companies in a supply chain and they work together in a cooperative manner to achieve the mutual agreement. SCI is a key to the success for the companies and their supply chains.

According to Kumar (2001), SCI is a philosophy for integrating all the activities among and between supply chain partners in the life of a product or a service from the earliest source of raw materials to the ultimate customer. In order to achieve the success of supply chain integration, it is highly dependent on the match between the requirements and offerings that deliver the services. The separation between the requirements and the actors who satisfy these requirements has been recognized as an essence of supply chain integration for dynamic assignment of resources to requests (Mowshowitz, 1997).

Over the past decades, the increasing competition has driven organizations to integrate their suppliers and customers into the overall value chain processes. SCI has been a growing consensus on the strategic importance of integrating suppliers, manufacturers and customers (Barratt, 2004) and organizations are integrating their activities both internally and externally (Wu et al., 2006). Huin et al., (2002) describe internal supply chain as a work aimed at breaking down the barriers between functions within organizations. A framework is needed to describe the key functions of a typical internal supply chain to ensure that all activities are associated with the design and management. Internal integration can be viewed as two dimensional process; technology integration and activity integration. Technology integration is reflected in the level of technology alignment with its supply chain partners, while activity integration is conceptualized as the extent to which a organization coordinates its strategic activities such as planning and forecasting with its supply chain partners (Wu et al., 2006). Technology integration with supply chain partners does not guarantee that the supply chain activities are automatically integrated after the deployment of such technology because its need a higher degree of activity integration. Organizations need to fundamentally shift their ways of doing business with its suppliers from discrete transactions to continuous and consistent transactions in order to achieve activity integration (Wu et al., 2006).

External supply chain integration is a key strategy to achieve competitive advantages in an uncertain environment. Supply chain strategies and operational resources should be used to support business strategies and help the firm achieve competitive advantage (Narasimhan, 1997). According to Golicic and Smith (2013), supplier integration and customer integration are two basic categories of external integration scope. Some researchers argue that besides integration scope, the strategy of integration is also important. Strategic and operational integration can be considered as two major strategies of external integration (Kim, 2009). Operational integration focuses on integration of interdependent processes and information flows that provide ways for partners to improve efficiency and effectiveness while strategic integration focuses on collaborative closeness of a company's relationships with both outside customers and suppliers, including partnerships and strategic alliances (He

& Lai, 2012). In comparison, external integration of supply chain is more powerful than internal integration (Vickery et al., 2003).

The increasing sophistication of information technology (IT) has helped organizations make great strides in resource planning for spanning corporate and national boundaries (Hayes & Pisano, 1994; Barney, 1999; Matusik & Hill, 1998; Wu et al., 2006). Information technology supports key processes in supply chains, including sourcing, procurement, and order fulfillment (Swaminathan & Tayur, 2003) and it plays a central role in supply chain management. IT allows organizations to increase the volume and complexity of information, provide real-time supply chain information, including inventory level, delivery status, and production planning and scheduling which enables organizations to manage and control its supply chain activities, and facilitates the alignment of forecasting and scheduling of operations, allowing better inter-firms coordination. Studies have shown that effective IT connection improves the integration between supply chain partners in terms of material flows (Soliman & Youssef, 2001).

Zou and Cavusgil (2002) noted that one of the seven dimensions of global marketing strategy; coordination and integration of business activities across borders deliver the most significant contribution to an organization's strategic and financial performance. IT supports the sharing of just in time (JIT) schedules and establishes information links that significantly lowered shipment discrepancies in the automotive industry (Srinivasan et al., 1994; Swafford et al., 2008) and its provides the mechanism to effectively gather, store, access, share, and analyze data. Zou and Cavusgil (2002) suggest that IT integration contains three elements: information flow integration, physical flow integration, and financial flow integration (Rai et al., 2006; Swafford et al., 2008). As Frohlich and Westbrook (2001) suggested, the material flow from upstream to the downstream supply chain entities must be supported by the information flow from downstream to upstream while Sheu et al. (2006) found that better IT capabilities and communication contribute to a better platform for both parties to engage in coordination, participation, and problem-solving activities.

While the information integration is important, it is the quantity and the quality of information shared that really matters. Large investments in IT could fail to produce expected benefits if it is not supported by willingness to share needed information (Fawcett et al., 2007). Information sharing requires organizations to exchange strategic supply chain information such as transactional data, materials or product orders. For example, point of sale history helps suppliers to forecast demand that subsequently improves service level and efficiency to their customers. Similarly, real-time inventory position helps suppliers to plan their replenishment and delivery schedules; improve service levels and reduce inventory costs. Such level of information sharing requires frequent and intense communication between firms and suppliers. The intensity of communication constitutes high levels of cooperative behavior between supply chain partners which leads to high degree and symmetry of strategic-information flows between them (Klein et al., 2007) and provides leverages to the supply chain partner for making strategic decision in their operations (Li et.al., 2006).

The other group of studies (Yu et al., 2001; Narasimhan & Nair, 2005; Carr & Kaynak, 2007; Zhou & Benton, 2007; Li & Zhang, 2008; Sezen, 2008) focused on the importance of information sharing and communication between firms and suppliers. Over reliance on technology without willingness to share the critical information pertaining to supply chain will not make the firms meaningfully connected; thus, failing to produce logistics integration. Only firms that are capable of building both the technical and social aspects of information integration will benefits from logistics integration. A number of studies have demonstrated various logistics benefits of having information sharing with supply chain partners concerning inventory management (Lee et al.,

2000; Zhao et al., 2002), agility and flexibility (Swafford etal., 2008), and the bullwhip effect. The bullwhip effect is the uncertainty caused by information flowing in the supply chain where the forecasts of demand become less reliable and it is caused by lack of coordination between organizations and their suppliers (Ravichandran, 2008). For example, Vendor-Managed Inventory (VMI) integration with suppliers has been shown to reduce the bullwhip effect (Disney & Towill, 2003). Thus, both information technology and information sharing are important in integrating the information.

Logistics integration allows organizations to have a smooth production process with well-coordinated flow of materials from suppliers (Frohlich &Westbrook, 2001) and the coordination produces a seamless connection between organizations and suppliers in such a way that the boundary of activities between the two parties is getting blurred (Stock et al., 2000). Lee et al. (1997) argue that having solid logistics integration will reduce various problems, such as the bullwhip effect (Geary et al., 2006). Integrated logistics allow organizations to adopt lean production systems by reliable order cycles and inventory reduction (Cagliano et al., 2006; Schonberger, 2007). Logistics integration also allows organizations and their supply chain partners to act as a single entity which would result in improved performance throughout the chain (Prajogo & Olhager, 2010) and yields a number of operational benefits, including reduction in costs (Nooteboom, 1992; Prajogo & Olhager, 2012), lead time (Liu et al., 2005), and risks (Clemons et al.,1993) as well as improvement in sales, distribution, customer services, and service levels (Prajogo & Olhager, 2012) and customer satisfaction (Kim, 2009).

The majority of empirical surveys on supply chain integration report a positive relationship between integration and performance (Van der Vaart & van Donk, 2008). Higher level of logistical interaction shows that an organizations performing better (De Toni & Nassimbeni, 1999; Prajogo & Olhager, 2012) and in higher levels of collaboration will result operational efficiency (Sheu et.al., 2006). Frohlich and Westbrook (2001) found that the widest arcs of integration had the strongest association with performance improvement while Li et al. (2009) found that supply chain integration is significantly related to supply chain performance. With a proper integration of green supply chain facilities and resources, automotive industry can resolve the environmental issues at the suppliers, distributional, reverse logistic and organization end.

SUPPLY CHAIN INTEGRITY ROLE IN ENHANCING GREEN SUPPLY CHAIN MANAGEMENT

Supply chain integrity plays an important role in the success of any organization or project. In the Oxford Advanced Learner's Dictionary (2006), integrity is defined by two different meanings. The first definition is the quality of being honest and having strong moral principle while the second definition of integrity is the state of being whole and not divided. For the purpose of this particular study, integrity would embrace the second definition which describes the integration of the GSCM members, stakeholders and as well as the process involved in GSCM. Integration would only occur when mutual understanding is achieved within involving parties.

It is apparent that in order for GSCM to succeed, a corporation has to first establish a coherent set of thinking and sustainability goal in the mind of its workers and this is known as internal integration. As stated by Darnall, Jolley and Handfield (2008), internal integration would lead to continued environmental performance of an organization. Nevertheless, the integration of all members of supply chain is equally important as only through total collaboration could GSCM objectives be achieved (Lambert and Cooper, 2000). Ofori (2000) has named integrity as the basic principle of GSCM, noting the vitality of this element in the whole process, but what are exactly the roles of integrity in GSCM? Therefore, in this chapter a content analysis was conducted and the methodology was explained in the next section.

Methods

The purpose of this chapter was to observe the effect of supply chain integrity in enhancing green supply chain (GSCM) leading to escalate level of business performance. Therefore, content analysis method was used to study the literature materials produced in the field of GSCM as well as those that are in relation to supply chain integrity. Content analysis method was adopted as a method allows valid inferences generation from text through a set of procedures (Weber, 1990). Neuendorf (2002) defined content analysis as the systematic, objective, quantitative analysis of message characteristics and it applies to various areas of research. This method is an approach of document analysis (Elo & Kyngas, 2007) that can be used to build a model to describe a phenomenon in a conceptual form.

This particular study adopted the four steps model of content analysis developed by Mayring (2003) and Srivastava (2007). The model is further explained as below:

1. The research journals to be analyzed are sorted and the unit of analysis is defined. This is the phase of material collection.
2. Significant theme of the research journals are evaluated and determined which will provide background for content analysis. This process is named as descriptive analysis.
3. Structural dimensions and related analytic categories are selected, which correlates to the collected journals. This is the category selection process.
4. Based on the categories, the research journals are analyzed according to the analytic dimension. This is the phase of material evaluation.

The structural organization of content analysis process is a vital characteristic that allows tracing and data verification (Gold & Seuring, 2011). It was also quoted that content analysis has a broad application field and reliability and validity of this method are proven (Weber, 1990; Neuendorf, 2002; Elo & Kyngas, 2007; Gold & Seuring, 2011).

This study also incorporated systematic review method which addresses a specific question, uses explicit and transparent methods to perform a deep search on literature and critical appraisal on individual studies, consequently draws conclusions about the existing knowledge and the unexplored area of the question (Briner & Denyer, 2012). The question that was addressed throughout this study was "What role does supply chain integrity have in enhancing green supply chain (GSCM) leading to higher business performance?". The key components of the systematic review adopted from Briner and Denyer (2012) and Tranfield, Denyer and Smart (2003) were listed as follows:

1. Review question identification;
2. Relevant academic journals selection;
3. Academic journals critical reviewing;
4. Review findings reporting and dissemination.

Systematic review was the best analysis approach for this particular study as it was accepted as the most simple, straightforward and logical approach that could be applied across many fields. The analysis of systematic view conducted was presented in the next section of the chapter.

Systematic Review Analysis

As mentioned, the main review questions of this chapter was "What role does supply chain integrity have in enhancing green supply chain (GSCM) leading to higher business performance?" and therefore, some important research articles were presented in Table 1, Table 2 and Table 3 in order to highlight its role in GSCM. The articles were categorized according to dimension, author, year of published, area of research and one column dedicated for supply chain integration role in GSCM.

It was found that over 90% of the articles analyzed have highlighted that integration among supply chain members is the most important element in ensuring the success of GSCM leading to companies performance. Analysis displayed that the integration among supply chain upstream and downstream members with the focal company is central to ensure the success of GSCM (Ibrahim, Zailani & Tan, 2012; Mathiyazhagan et al., 2013; Golicic & Smith, 2013). Through supply chain members integration, the environmental, economical social and operational favorable outcome of GSCM could be enjoyed throughout the whole supply chain, benefitting a larger group (Wilhelm, 2011; Darnall, Jolley & Handfield, 2008; Zhu & Cote, 2004) compared to if the GSCM is practiced independently by the focal company (Mathiyazhagan, Govindan, NoorulHaq & Geng, 2013; Zailani, Jeyaraman, Vengadasan, & Premkumar, 2012).

Integration of Environmental Practices and Technology

Technology may be the best solution to improve GSCM performance. Through integrating sustainable technology, closed loop production is achieved by Guitang Group where the by products and waste of other operations and industries could actually be utilized as input for new production (Zhu & Cote, 2004). To support the importance of technology integration in GSCM, Ibrahim, Zailani and Tan (2012) in their research has found that technological integration with primary suppliers and major customers was positively linked to environmental monitoring and collaboration.

Integration in the Supply Chain Process

Modification in the supply chain process to be more integrated would assist a particular company to be more efficient and effective in resource utilization. As found by Walton, Handfield and Melnyk (1998) in their study, the process that should integrate environmental thinking include material usage, sustainable product design, improvement in suppliers' production process, supplier evaluation and inbound logistics process. On the other hand, Zhu and Cote (2004) has reported the method used by Guitang Group in integrating their supply chain process among different industries in the group as well as with smaller companies operating within China. The group displays the best example of how ecological industry could exist. It has successfully eliminates emission by utilizing its downstream waste and purchasing the by products of other smaller refineries as input material for their products. The integration of technology and industrial symbiosis has established Guitang as the most successful corporation in practicing closed loop production system.

Coopetition Concept

Although some academicians and practitioner may disagree having competitors as an entity influencing GSCM, competitors do affect many factors in the industry which influences the decision making of the focal company. As competing companies operating in the same industry are threatened with the same environmental risks, it is of increasing important that these competing companies pooled their strength and capabilities in developing the best approach to minimize the usage of natural resources and eliminate environmental impacts

Table 1. Supply chain integration role in green supply chain management (GSCM)

Dimension	Author	Year	Journal	Green Supply Chain Area of Research	Supply Chain Integration Role in GSCM
Supply chain members integration	Mathiyazhagan, Govindan, NoorulHaq & Geng	2013	*Journal of Cleaner Production.*	Barrier analysis in implementing green supply chain management	Integration of suppliers in participative decision making process that helps to achieve GSCM. It promotes environmental innovation and leads to reduction in energy consumption as well as cost of materials purchase.
Coopetition and supply chain members integration	Cheng, Yeh & Tu	2008	*Supply Chain Management: An International Journal*	Trust and knowledge sharing in green supply chains.	The research was conducted on Taiwanese companies. This research uses trust as the mediating factor to reflect the level of competition and cooperation (coopetition) relationship between supply chain members from the perspective of inter-organizational knowledge sharing. GSCM would only be functioning well should the firm is confidence on the reliability and integrity of its partners, especially when the process involve knowledge sharing. GSCM requires the integration of all supply chain members in order to establish inter-organizational collaboration that would lead to increasing performance of all supply chain members.
Supply chain members and supply chain process integration	Seuring & Müller	2008	*Journal of cleaner production*	From a literature review to a conceptual framework for sustainable supply chain management	GSCM could be established only with the integration among companies along the supply chain and the integration of environmental and social issues in SC. The supply chain process has to be integrated from raw materials to final customers to achieve performance of operational process
Supply chain members integration	Golicic & Smith	2013	*Journal of Supply Chain Management*	A Meta-Analysis of Environmentally Sustainable Supply Chain Management Practices and Firm Performance	Significant performance results from upstream supplier facing practices as well as design practices. Firm will obtain positive financial results from their environmental supply chain efforts. The skepticism that exists on the outcome of GSCM may hinder the adoption.
Supply chain members integration	Gardas & Narkhede	2013	*International Journal of Application or Innovation in Engineering & Management (IJAIEM)*	Exploring the Green Supply Chain management: A Technical Review.	The research has suggested that for a corporation to successfully implement GSCM, integrating suppliers and distributors in its efforts is needed.
Supply chain members integration	Hutchins & Sutherland	2008	*Journal of Cleaner Production*	An exploration of measures of social sustainability and their application to supply chain decisions	A company seeking to operate in accord with the principles of sustainability or taking an ethical or citizenship approach to corporate social responsibility must consider its entire supply chain, ''not just those links which belong to its own sphere of legal responsibility''.

Table 2. Supply chain integration role in green supply chain management (GSCM)

Dimension	Author	Year	Journal	Green Supply Chain Area of Research	Supply Chain Integration Role in GSCM
Supply chain members integration	Cox	1999	*Supply Chain Management*	Power, value and supply chain management.	Among the eight characteristics of lean approach is to develop close, collaborative, reciprocal and trusting (win-win), rather than arms- length and adversarial (win-lose), relationships with suppliers. Another characteristics related to integration is lean approach is to create a network of suppliers to build common understanding and learning about waste reduction and operational efficiency in the delivery of existing products and services.
Supply chain members and supply chain process integration	Walton, Handfield, & Melnyk	1998	*Journal of Supply Chain Management,*	The green supply chain: integrating suppliers into environmental management processes	Using accepted qualitative research methods for case-based research, several primary areas for change to increase purchasing's impact on environmental results are identified: 1. Materials used in product design for the environment 2. Product design processes 3. Supplier process improvement 4. Supplier evaluation 5. Inbound logistics processes These areas require supply chain members and supply chain process integration
Supply chain members and process integration, integration of environmental practices and technology	Zhu & Cote	2004	*Journal of Cleaner Production*	Integrating green supply chain management into an embryonic eco-industrial development: a case study of the Guitang Group.	Guitang Group is the best example of how ecological industry could exist. The company eliminates emission by utilizing its downstream waste and purchasing the by products of other smaller refineries as input material for their products. The integration of technology and industrial symbiosis have established Guitang as the most successful company in practicing closed loop production system.
Supply chain members integration and integration of environmental practices and technology	Ibrahim, Zailani & Tan	2012	*International Journal of Operations & Production Management*	Extending Green Practices Across the Supply Chain: The Impact of Upstream and Downstream Integration	Technological integration with primary suppliers and major customers was positively linked to environmental monitoring and collaboration. For logistical integration, a linkage was found only with environmental monitoring of suppliers. Finally, as the supply base was reduced, the extent of environmental collaboration with primary suppliers increased.
Supply chain members integration	Darnall, Jolley & Handfield	2008	*Business Strategy and the Environment*	Environmental management systems and green supply chain management: complements for sustainability?	Suppliers integration in product design has minimized environmental impact as well as assist in positioning corporation as leader with high credibility among industry peers and regulators.

Table 3. Supply chain integration role in green supply chain management (GSCM)

Dimension	Author	Year	Journal	Green Supply Chain Area of Research	Supply Chain Integration Role in GSCM
Coopetition and supply chain members integration	Wilhelm	2011	*Journal of Operations Management,*	Managing coopetition through horizontal supply chain relations: Linking dyadic and network levels of analysis	In the automobile industry, integrity is an important factor that assists the development of the next generation drive system which is green. Among the integration observed are: • 1. Formation of development alliances among carmakers (e.g. General Motors, Daimler and BMW) • 2. Formation of development alliances among carmakers and suppliers (e.g. Daimler, BMW and Continental) • 3. Formation of development alliances between suppliers (e.g. ZF Friedrichshafen and Continental)
Supply chain members integration	Testa & Iraldo	2010	*Journal of Cleaner Production*	Shadows and lights of GSCM (Green Supply Chain Management): determinants and effects of these practices based on a multi-national study.	From the research findings, the more a company is able to integrate its business partners in the development of the co-operative environment plans, the more it is able to achieve the expected results and improve its performance. If a company does not make efforts to stimulate and involve its suppliers and partners operating in other phases of product life cycle, the "environmental quality" of a product cannot be guaranteed.
Supply chain members integration	Zailani, Jeyaraman, Vengadasan, & Premkumar	2012	*International Journal of Production Economics*	Sustainable supply chain management (SSCM) in Malaysia: A survey	This study found that from the perspective of transaction cost, strong vertical (the production to marketing processes) integration exist between GSCM practices and the four types of outcome (environmental, economical, social, and operational) concerning GSC performance. It concludes that firms need to integrate across the supply chain in pursuing GSCM practices as a route for firm commercial success rather than as a moral obligation

caused by unsustainable operation (Wilhelm, 2011; Cheng, The and Yu, 2008). Coopetition has also been justified ad more superior compared to competition with the arousing issues of incomplete information, immoral conduct and risk interdependency surrounding the supply chain (Bakshi and Kleindorfer, 2009)

SOLUTIONS AND RECOMMENDATIONS

As previously mentioned, integration has always been a big challenge for GSCM as it involves mutual trust and high credibility. Sharing business information would also mean sharing business secret thus could put the whole corporation at risk. Trust and credibility hence may support or hinder GSCM from achieving its objectives.

The Issues of Trust and Credibility

Realizing the conflict of trust that exists in supply chain, Haghpanah and desJardins (2010) has developed a trust model in decision making for supply chain management. The model incorporates numerous trust factors that are relevant to SCM, and uses both direct and reported observations. In addition, the proposed model was viewed as complete as it consists of several layers in supply network including suppliers, producers, distributors, or retailers. Therefore, this model is recommended for companies having issues with trust in their green supply chain.

The Success Factor of GSCM

Reducing cost of operations, improving inventory, lead times and customer satisfaction, increasing flexibility and cross-functional communication, and remaining competitive appear to be the most important objectives to implement GSCM strategies. These factors mentioned could certainly be achieved through strong integration and collaboration among supply chain members. A research conducted by Tummala, Phillips and Johnson, (2006) has indicated that not enough resources were allocated to implement and support SCM initiatives in general. Therefore, more resources should be directed to improve in the areas of better information systems, greater commitment, setting clear-cut goals, increased training, more personnel, and aligning SCM initiatives with current priorities and resource commitments. Consequently, this would provide greater understanding of strategic and operational issues that support GSCM framework and implementing GSCM strategies to reduce supply chain-wide costs and meeting customer service levels. All these are in addition to integrating all supply chain members

FUTURE RESEARCH DIRECTIONS

While assessing the academic journals, the coopetition concept in green supply chain was discovered to be a very interesting concept to explore into as many researches highlighted the complexity of this relationship (Keating, Quazi, Kriz & Coltman, 2008; Wilhelm, 2011). Coopetition derives from the combination of two words, cooperation and competition. This concept was introduced to describe the gains that a corporation could have through collaborative synergy while at the same time fighting for a larger share of the gain. In this globalization era where competition between firms are getting much tougher, it is vital for business corporations to establish a unique winning strategy to be on top. Coopetition is one of the strategic approaches that GSCM practitioners could adopt where one supplier could cooperate with their rivals in eco raw material development, but consequently compete for market share (Wilhelm, 2011). Therefore, future research may look into the impact of coopetition in GSCM.

It is also fascinating to discover a research highlighting integration of environmental costs of products throughout their life cycles into the market distribution mechanism in order to reduce product harm to the environment. The study conducted by Chen and Sheu (2009) argues that a great functioning green supply chain may not be realized should there be no incentives and public policies. In addition to cost integration, another noteworthy finding found is related to social and ethical integration in the supply chain. Since the benefits of integrating these two elements are viewed as less tangible, they often received less attention in GSCM (Hutchins & Sutherland, 2008). This should not be the case as ethics and societal rights should be put as the core of GSCM. All in, moving forward, corporations have no choice but to integrate, be it with the members of supply chain or even with competitors as the industry challenges are becoming much tougher with the exhaustive natural resources and apparent degradation of Mother Nature. In future, more effort should be pooled into companies' performance post integration.

CONCLUSION

This study has highlighted the role of supply chain integrity in green supply chain management (GSCM) leading to business performance. It offers a conceptualization based on content analysis of academic journals (displayed in Table 1, Table 2 and Table 3. The study found that supply chain integration are required in four major areas, namely integration in the supply chain members and process, integration of environmental practices and technology as well as competitors integration (coopetition). Most of the articles highlighted that integration among supply chain members are the most important factor in GSCM, while competing companies could mutually progress in GSCM should they agree to collaborate. Nevertheless, trust has always been a major factor that hinders integration, especially if the process requires sharing of knowledge on core competencies (Cheng, Yeh & Tu, 2008).

A firm's trust in their supply chain partner is highly associated with both parties' specific asset investments and social exchange theory (Kwon & Suh, 2005). Information sharing has a primary impact on reducing (improving a partner's uncertainty behavior that, in turn, would improve the level of trust. Therefore, the level of integration is strongly related to the level of trust. It was also noted that many companies are still skeptical on GSCM and its capabilities as a catalyst to organizational performance with the absence of credible performance measures (Green, Morten & New, 1998). Thus, adoption of supply chain integration is considerably quite slow especially in developing countries which value competition more than coopetition even along the vertical chain. This should be improved as GSCM could only function and assist in achieving sustainable business performance, should the firm is confidence on the reliability and integrity of its partners (Cheng, Yeh & Tu, 2008).

REFERENCES

Bakshi, N., & Kleindorfer, P. (2009). Co-opetition and investment for supply-chain resilience. *Production and Operations Management*, *18*(6), 583–603. doi:10.1111/j.1937-5956.2009.01031.x

Barratt, M. (2004). Understanding the meaning of collaboration in the supply chain. *Supply Chain Management: An International Journal*, *9*(1), 30–42. doi:10.1108/13598540410517566

Braunscheidel, M. J., & Suresh, N. C. (2009). The organizational antecedents of a firm's supply chain agility for risk mitigation and response. *Journal of Operations Management*, *27*(2), 119–140. doi:10.1016/j.jom.2008.09.006

Briner, R. B., & Denyer, D. (2012). Systematic review and evidence synthesis as a practice and scholarship tool. In Handbook of evidence-based management: Companies, classrooms and research (pp. 112-129). Academic Press.

Cagliano, R., Caniato, F., & Spina, G. (2006). The linkage between supply chain integration and manufacturing improvement programmes. *International Journal of Operations & Production Management*, *26*(3), 282–299. doi:10.1108/01443570610646201

Cheng, J. H., Yeh, C. H., & Tu, C. W. (2008). Trust and knowledge sharing in green supply chains. *Supply Chain Management: An International Journal*, *13*(4), 283–295. doi:10.1108/13598540810882170

Clemons, E. K., Reddi, S. P., & Row, M. C. (1993). The impact of information technology on the organization of economic activity: The 'move to the middle' hypothesis. *Journal of Management Information Systems*, *10*(2), 9–35.

Cox, A. (1999). Power, value and supply chain management. *Supply Chain Management: An International Journal*, *4*(4), 167–175. doi:10.1108/13598549910284480

Darnall, N., Jolley, G. J., & Handfield, R. (2008). Environmental management systems and green supply chain management: Complements for sustainability? *Business Strategy and the Environment*, *17*(1), 30–45. doi:10.1002/bse.557

De Toni, A., & Nassimbeni, G. (1999). Buyer-supplier operational practices, sourcing policies and plant performances: Results of an empirical research. *International Journal of Production Research*, *37*(3), 597–619. doi:10.1080/002075499191698

Disney, S. M., & Towill, D. R. (2003). On bullwhip in supply chains—historical review, present practice and expected future impact. *International Journal of Production Economics*, *101*(1), 2–18.

Droge, C., Jayaram, J., & Vickery, S. K. (2004). The effects of internal versus external integration practices on time-based performance and overall firm performance. *Journal of Operations Management*, *22*(6), 557–573. doi:10.1016/j.jom.2004.08.001

El-Berishy, N., RÃ1/4gge, I., & Scholz-Reiter, B. (2013). The interrelation between sustainability and green logistics. Management and Control of Production and Logistics, 6 (1), 527-531.

Elo, S., & Kyngäs, H. (2008). The qualitative content analysis process. *Journal of Advanced Nursing*, *62*(1), 107–115. doi:10.1111/j.1365-2648.2007.04569.x PMID:18352969

Fallah, M., & Ebrahimi, M. (2014). A study on the effect of green marketing on consumers' purchasing intention. *Management Science Letters*, *4*(3), 421–424. doi:10.5267/j.msl.2014.1.030

Flynn, B. B., Huo, B., & Zhao, X. (2010). The impact of supply chain integration on performance: A contingency and configuration approach. *Journal of Operations Management*, *28*(1), 58–71. doi:10.1016/j.jom.2009.06.001

Frohlich, M. T., & Westbrook, R. (2001). Arcs of integration: An international study of supply chain strategies. *Journal of Operations Management*, *19*(2), 185–200. doi:10.1016/S0272-6963(00)00055-3

Gardas, B. B., & Narkhede, B. E. (2013). Exploring the green supply chain management: A technical review. *International Journal of Application or Innovation in Engineering & Management*, *2*(5), 441–450.

Geary, S., Disney, S. M., & Towill, D. R. (2006). On bullwhip in supply chains—Historical review, present practice and expected future impact. *International Journal of Production Economics*, *101*(1), 2–18. doi:10.1016/j.ijpe.2005.05.009

Gold, S., & Seuring, S. (2011). Supply chain and logistics issues of bio-energy production. *Journal of Cleaner Production*, *19*(1), 32–42. doi:10.1016/j.jclepro.2010.08.009

Golicic, S. L., & Smith, C. D. (2013). A meta-analysis of environmentally sustainable supply chain management practices and firm performance. *Journal of Supply Chain Management*, *49*(2), 78–95. doi:10.1111/jscm.12006

Haghpanah, Y., & desJardins, M. (2010, July). Using a trust model in decision making for supply chain management. In *Proceedings of Workshops at the Twenty-Fourth AAAI Conference on Artificial Intelligence*. AAAI.

Hayes, R. H., & Pisano, G. P. (1994). Beyond world class: The new manufacturing strategy. *Harvard Business Review*, 77–86.

He, Y., & Lai, K. K. (2012). Supply chain integration and service oriented transformation: Evidence Chinese equipment manufacturers. *International Journal of Production Economics*, *135*(2), 291–299. doi:10.1016/j.ijpe.2011.10.013

Huin, S. F., Luong, L. H. S., & Abhary, K. (2002). Internal supply chain planning determinants in small and medium-sized manufacturers. *International Journal of Physical Distribution & Logistics Management*, *32*(9), 771–782. doi:10.1108/09600030210452440

Hutchins, M. J., & Sutherland, J. W. (2008). An exploration of measures of social sustainability and their application to supply chain decisions. *Journal of Cleaner Production*, *16*(15), 1688–1698. doi:10.1016/j.jclepro.2008.06.001

Ibrahim, A., Zailani, T., & Tan, A. C. (2012). Extending green practices across the supply chain: The impact of upstream and downstream integration. *International Journal of Operations & Production Management*, *26*(7), 795–821.

Kamal, A. (2014). *Green supply chain management on sustainable business performance: The green airport suppliers' perspective.* (Unpublished master's thesis). Universiti Sains Malaysia, Penang, Malaysia.

Keating, B., Quazi, A., Kriz, A., & Coltman, T. (2008). In pursuit of a sustainable supply chain: Insights from Westpac Banking Corporation. *Supply Chain Management: An International Journal*, *13*(3), 175–179. doi:10.1108/13598540810871217

Kim, S. W. (2009). An investigation on the direct and indirect effect of supply chain integration on firm performance. *International Journal of Production Economics*, *119*(2), 328–346. doi:10.1016/j.ijpe.2009.03.007

Kumar, K. (2001). Technology for supporting supply chain management: Introduction. *Communications of the ACM*, *44*(6), 58–61. doi:10.1145/376134.376165

Kumar, R., & Shekhar, S. (2013). Green supply chain management – A review. *International Journal of Research in Aeronautical and Mechanical Engineering*, *1*(7), 245–255.

Kwon, I. W. G., & Suh, T. (2005). Trust, commitment and relationships in supply chain management: A path analysis. *Supply Chain Management: An International Journal*, *10*(1), 26–33. doi:10.1108/13598540510578351

Lambert, D. M., & Cooper, M. C. (2000). Issues in supply chain management. *Industrial Marketing Management*, *29*(1), 65–83. doi:10.1016/S0019-8501(99)00113-3

Lee, S. M., Kim, S. T., & Choi, D. (2012). Green supply chain management and organizational performance. *Industrial Management & Data Systems*, *112*(8), 1148–1180. doi:10.1108/02635571211264609

Mathiyazhagan, K., Govindan, K., NoorulHaq, A., & Geng, Y. (2013). An ISM approach for the barrier analysis in implementing green supply chain management. *Journal of Cleaner Production*, *47*, 283–297. doi:10.1016/j.jclepro.2012.10.042

Matusik, S. F., & Hill, C. W. L. (1998). The utilization of contingent work, knowledge creation, and competitive advantage. *Academy of Management Review*, *23*(4), 680–698. doi:10.2307/259057

Mentzer, J. T., DeWitt, W., Keebler, J. S., Min, S., Nix, N. W., Smith, C. D., & Zacharia, Z. G. (2001). Defining supply chain management. *Journal of Business Logistics*, *22*(2), 1–25. doi:10.1002/j.2158-1592.2001.tb00001.x

Mowshowitz, A. (1997). Virtual organization. *Communications of the ACM*, *40*(9), 30–37. doi:10.1145/260750.260759

Narasimhan, R. (1997). Strategic supply management: A total quality management imperative. *Advances in the Management of Organizational Quality*, *2*, 39–86.

Narasimhan, R., & Kim, S. W. (2002). Effect of supply chain integration on the relationship between diversification and performance: Evidence from Japanese and Korean firms. *Journal of Operations Management*, *20*(3), 303–323. doi:10.1016/S0272-6963(02)00008-6

Neuendorf, K. A. (2002). *The content analysis guidebook* (Vol. 300). Thousand Oaks, CA: Sage Publications.

Ninlawan, C., Seksan, P., Tossapol, K., & Pilada, W. (2010). The implementation of green supply chain management practices in electronics industry. In *Proceedings of the International Multiconference of Engineers and Computer Scientists* (vol. 3, pp. 17-19). Academic Press.

Nooteboom, B. (1992). Information technology, transaction costs and the decision to 'make or buy'. *Technology Analysis and Strategic Management, 4*(4), 339–350. doi:10.1080/09537329208524105

Ofori, G. (2000). Greening the construction supply chain in Singapore. *European Journal of Purchasing & Supply Management, 6*(3), 195–206. doi:10.1016/S0969-7012(00)00015-0

Pagell, M. (2004). Understanding the factors that enable and inhibit the integration of operations, purchasing and logistics. *Journal of Operations Management, 22*(5), 459–487. doi:10.1016/j.jom.2004.05.008

Pagell, M., & Shevchenko, A. (2014). Why research in sustainable supply chain management should have no future. *Journal of Supply Chain Management, 50*(1), 44–55. doi:10.1111/jscm.12037

Pedersen, E. R. G., Henriksen, M. H., Frier, C., Søby, J., & Jennings, V. (2013). Stakeholder thinking in sustainability management: The case of Novozymes. *Social Responsibility Journal, 9*(4), 500–515. doi:10.1108/SRJ-08-2012-0101

Peglau, R. (2005). *ISO 14001 certification of the world*. Berlin: Federal Environmental Agency.

Prajogo, D., & Olhager, J. (2012). Supply chain integration and performance: The effects of long term relationships, information technology and sharing, and logistics integration. *International Journal of Production Economics, 135*(1), 514–522. doi:10.1016/j.ijpe.2011.09.001

Rao, P., & Holt, D. (2005). Do green supply chains lead to competitiveness and economic performance? *International Journal of Operations & Production Management, 25*(9), 898–916. doi:10.1108/01443570510613956

Ravichandran, N. (2008). Managing the bullwhip effect: Two case studies. *Journal of Advances in Management Research, 5*(2), 77–87. doi:10.1108/09727980810949151

Schonberger, R. J. (2007). Japanese production management: An evolution with mixed success. *Journal of Operations Management, 25*(2), 403–419. doi:10.1016/j.jom.2006.04.003

Seuring, S., & Müller, M. (2008). From a literature review to a conceptual framework for sustainable supply chain management. *Journal of Cleaner Production, 16*(15), 1699–1710. doi:10.1016/j.jclepro.2008.04.020

Sheu, C., Yen, H. R., & Chae, B. (2006). Determinants of supplier–retailer collaboration: Evidence from an international study. *International Journal of Operations & Production Management, 26*(1), 24–49. doi:10.1108/01443570610637003

Soleimani, H., & Govindan, K. (2014). Reverse logistics network design and planning utilizing conditional value at risk. *European Journal of Operational Research, 237*(2), 487–497. doi:10.1016/j.ejor.2014.02.030

Soliman, F., & Youssef, M. (2001). The impact of some recent developments in e-business on the management of next generation manufacturing. *International Journal of Operations & Production Management, 21*(5-6), 538–564. doi:10.1108/01443570110390327

Srinivasan, K., Kekre, S., & Mukhopadhyay, T. (1994). Impact of electronic data inter change technology on JIT shipments. *Management Science, 40*(10), 1291–1305. doi:10.1287/mnsc.40.10.1291

Srivastava, S. K. (2007). Green supply-chain management: A state-of-the-art literature review. *International Journal of Management Reviews*, *9*(1), 53–80. doi:10.1111/j.1468-2370.2007.00202.x

Stevens, G. C. (1989). Integrating the supply chain. *International Journal of Physical Distribution & Materials Management*, *19*(8), 3–8. doi:10.1108/EUM0000000000329

Stock, G. N., Greis, N. P., & Kasarda, J. D. (2000). Enterprise logistics and supply chain structure: The role of fit. *Journal of Operations Management*, *18*(5), 531–547. doi:10.1016/S0272-6963(00)00035-8

Swafford, P. M., Ghosh, S., & Murthy, N. (2008). Achieving supply chain agility through IT integration and flexibility. *International Journal of Production Economics*, *116*(2), 288–297. doi:10.1016/j.ijpe.2008.09.002

Swaminathan, J. M., & Tayur, S. R. (2003). Models for supply chains in e-business. *Management Science*, *49*(10), 1387–1406. doi:10.1287/mnsc.49.10.1387.17309

Testa, F., & Iraldo, F. (2010). Shadows and lights of GSCM (green supply chain management): Determinants and effects of these practices based on a multi-national study. *Journal of Cleaner Production*, *18*(10), 953–962. doi:10.1016/j.jclepro.2010.03.005

Tranfield, D., Denyer, D., & Smart, P. (2003). Towards a methodology for developing evidence-informed management knowledge by means of systematic review. *British Journal of Management*, *14*(3), 207–222. doi:10.1111/1467-8551.00375

Tummala, V. R., Phillips, C. L., & Johnson, M. (2006). Assessing supply chain management success factors: A case study. *Supply Chain Management: An International Journal*, *11*(2), 179–192. doi:10.1108/13598540610652573

Van der Vaart, T., & Van Donk, D. P. (2008). A critical review of survey-based research in supply chain integration. *International Journal of Production Economics*, *111*(1), 42–55. doi:10.1016/j.ijpe.2006.10.011

Vickery, S. K., Jayaram, J., Droge, C., & Calantone, R. (2003). The effects of an integrative supply chain strategy on customer service and financial performance: An analysis of direct versus indirect relationships. *Journal of Operations Management*, *21*(5), 523–539. doi:10.1016/j.jom.2003.02.002

Walton, S. V., Handfield, R. B., & Melnyk, S. A. (1998). The green supply chain: Integrating suppliers into environmental management processes. *Journal of Supply Chain Management*, *34*(2), 2–11.

Weber, R. (1990). *Basic content analysis* (2nd ed.). Newbury Park, CA: Sage.

Wilhelm, M. M. (2011). Managing coopetition through horizontal supply chain relations: Linking dyadic and network levels of analysis. *Journal of Operations Management*, *29*(7), 663–676. doi:10.1016/j.jom.2011.03.003

Wu, F., Yeniyurt, S., Kim, D., & Cavusgil, S. T. (2006). The impact of information technology on supply chain capabilities and firm performance: A resource-based view. *Industrial Marketing Management*, *35*(4), 493–504. doi:10.1016/j.indmarman.2005.05.003

Zailani, S., Jeyaraman, K., Vengadasan, G., & Premkumar, R. (2012). Sustainable supply chain management (SSCM) in Malaysia: A survey. *International Journal of Production Economics*, *140*(1), 330–340. doi:10.1016/j.ijpe.2012.02.008

Zhao, X., Huo, B., Flynn, B. B., & Yeung, J. (2008). The impact of power and relationship commitment on the integration between manufacturers and customers in a supply chain. *Journal of Operations Management*, *26*(3), 368–388. doi:10.1016/j.jom.2007.08.002

Zhu, Q., & Cote, R. P. (2004). Integrating green supply chain management into an embryonic eco-industrial development: A case study of the Guitang Group. *Journal of Cleaner Production, 12*(8), 1025–1035. doi:10.1016/j.jclepro.2004.02.030

Zhu, Q., Sarkis, J., & Geng, Y. (2005). Green supply chain management in China: Pressures, practices and performance. *International Journal of Operations & Production Management, 25*(5), 449–468. doi:10.1108/01443570510593148

Zou, S., & Cavusgil, S. T. (2002). The GMS: A broad conceptualization of global marketing strategy and its effect on firm performance. *Journal of Marketing, 66*(4), 40–56. doi:10.1509/jmkg.66.4.40.18519

ADDITIONAL READING

Ambrose, E., Marshall, D., & Lynch, D. (2010). Buyer supplier perspectives on supply chain relationships. *International Journal of Operations & Production Management, 30*(12), 1269–1290. doi:10.1108/01443571011094262

Bowersox, D. J., Closs, D. J., & Cooper, M. B. (2002). *Supply chain logistics management* (2nd ed.). New York: McGraw-Hill.

Cagliano, R., Caniato, F., & Spina, G. (2005). E-business strategy: How companies are shaping their supply chain through the internet. *International Journal of Operations & Production Management, 25*(12), 1309–1327. doi:10.1108/01443570510633675

Capgemini, C. (2012). Green Airports [Brochure]. Retrieved on April 28, 2013, from http://www.capgemini.com/m/en/tl/Green_Airports.pdf

Davis, T. (1993). Effective supply chain management. *Sloan Management Review, 34*, 35–35.

DeToni, A., & Tonchia, S. (1998). Manufacturing flexibility: A literature review. *International Journal of Production Research, 36*(6), 1587–1617. doi:10.1080/002075498193183

Doherty, S., & Hoyle, S. (2009). *Supply Chain Decarbonization*. World Economic Forum: Geneva.

Economic Report 2013/2014: GDP to grow 5%-5.5% in 2014. (2014). The Star Online. Retrieved on February 5, 2014 from http://www.thestar.com.my/News/Nation/2013/10/25/Budget-2014-Economic-report.aspx/

Ellinger, A., Shin, H., Northington, W. M., & Adams, F. G. (2011). The influence of supply chain management competency on customer satisfaction and shareholder value. *Supply Chain Management: An International Journal, 17*(3), 249–262. doi:10.1108/13598541211227090

Green, K., Morton, B., & New, S. (1998). Green purchasing and supply policies: Do they improve companies' environmental performance? *Supply Chain Management: An International Journal, 3*(2), 89–95. doi:10.1108/13598549810215405

Haghpanah, Y. (2011). A trust model for supply chain management. *The 10th International Conference on Autonomous Agents and Multiagent Systems 3*, 1375-1376. International Foundation for Autonomous Agents and Multiagent Systems.

Hezri, A. A. (2004). Sustainability indicator system and policy processes in Malaysia: A framework for utilisation and learning. *Journal of Environmental Management, 73*(4), 357–371. doi:10.1016/j.jenvman.2004.07.010 PMID:15531393

Liu, J., Zhang, S., & Hu, J. (2005). A case study of an inter-enterprise workflow supported supply chain management system. *Information & Management, 42*(3), 441–454. doi:10.1016/j.im.2004.01.010

Malaysia's GDP grew 5.1% in Q4 2013. (2014). The Star Online. Retrieved on February 5, 2014 from http://www.thestar.com.my/Business/Business-News/2014/02/12/Malaysia-GDP-grew-5pt1pct-in-Q4-2013/

Najam, A., Runnalls, D., & Halle, M. (2007). *Environment and globalization: five propositions.* International Institute for Sustainable Development.

NAP 2014: Policy would promote Malaysia as production hub. (2014). New Straits Times Online. Retrieved on February 5, 2014 from http://www.nst.com.my/latest/font-color-red-nap-2014-font-policy-would-promote-malaysia-as-production-hub-1.466848

Rai, A., Patnayakuni, R., & Seth, N. (2006). Firm performance impacts of digitally enabled supply chain integration capabilities. *Management Information Systems Quarterly, 30*(2), 225–246.

Swafford, P., Ghosh, S., & Murthy, N. (2006). A frame work for assessing value chain agility. *International Journal of Operations & Production Management, 26*(2), 170–188. doi:10.1016/j.jom.2005.05.002

Vachon, S., & Klassen, R. D. (2006). Extending green practices across the supply chain: The impact of upstream and downstream integration. *International Journal of Operations & Production Management, 26*(7), 795–821. doi:10.1108/01443570610672248

KEY TERMS AND DEFINITIONS

Environmental Management System (EMS): Strategic management approaches consisting internal policies, assessments, plans and implementation actions taken to minimize operational impacts to the environment.

Environmental Stewardship: Human responsible consumption, protection of the natural environment or corrective activities that could be achieved through conservation efforts and sustainable practices.

Green Supply Chain Management: A supply chain management system that is concerned on ensuring all conventional supply chain deliveries while centralizing the environmental and social concern in every stage of supply chain to reduce product environmental footprint.

Integration: The state of being whole and not divided. Integration would only occur when mutual understanding is achieved within involving parties.

Reverse Logistics: Closes the supply chain loop in which it reconnects the customers' role to focal company through product after use transfer. It refers to the focal company's responsibility of taking care of produced products once they were no longer desired or can no longer be used by their users.

Supply Chain Integration: The degree to which a firm can strategically collaborate with its supply chain partners and cooperatively manage intra-organizational and inter-organizational processes to achieve effective and efficient flows of products, services, information, financial, and decisions to provide the maximum value to the final customer with low costs and high speed.

Supply Chain Management: The system used by an organization to improve logistics and planning efficiency, as well as material and information control internally and externally among the upstream and downstream members.

Sustainability: A key for any corporation to develop a sustained business operation as it suggests that businesses should grow and meet the needs of the present without compromising the ability of the future generations to meet their own needs.

Sustainable Business Performance: Achieved when green supply chain management practices were embedded in the business model of an organization.

Compilation of References

Abu-Taha, R. (2011). Multi-criteria applications in renewable energy analysis: A literature review. In *Proceedings of Technology Management in the Energy Smart World (PICMET '11)*. Portland, OR: IEEE.

Acosta, O., & Gonzalez, J. (2010). A thermodynamic approach for emergence of globalization. In K. Deng (Ed.), *Globalization: Today, tomorrow* (pp. 38–49). Sciyo. doi:10.5772/10223

Akinmulegun, S. O. (2011). *Globalization, FDI and economic growth in Nigeria 1986-2009*. (Unpublished doctoral thesis). Adekunle Ajasin University.

Alliance, R. (n.d.). *Standards for sustainable agriculture*. Retrieved from http://www.rainforest-alliance.org/agriculture.cfm?id=san

Allwood, J. M., Cullen, J. M., & Milford, R. L. (2010). Options for achieving a 50% cut in industrial carbon emissions by 2050. *Environmental Science & Technology*, *44*(6), 1888–1894. doi:10.1021/es902909k PMID:20121181

Alonso Aguilar, P. (1997). *Instalación integral de ensayos para la evaluación energética de sistemas solares térmicos de baja temperatura*. Proyecto Fin de Carrera, Universidad de Sevilla.

American Institute of Chemical Engineers Center for Chemical Process Safety. (1989). Guidelines for process equipment reliability data, with data tables. Author.

Amran, A., Zainuddin, Z., & Zailani, S. H. M. (2013). Carbon trading in Malaysia: Review of policies and practices. *Sustainable Development*, *21*(3), 183–192. doi:10.1002/sd.1549

Anbumozhi, V., & Kanda, Y. (2005). *Greening the production and supply chains in Asia: Is there a role for voluntary initiatives*. IGES Kansai Research Centre KRC.

An, H., Wilhelm, W. E., & Searcy, S. W. (2011). Biofuel and petroleum-based fuel supply chain research: A literature review. *Biomass and Bioenergy*, *35*(9), 3763–3774.

Anon. (1890). Mr. Brush's windmill dynamo. Scientific American, 63(25), 54.

Anton, D. K. (2013). Treaty congestion in contemporary international environmental law. In S. Alam, J. Bhuiyan, T. Chowdhury, & E. Techera (Eds.), *Routledge handbook of international environmental law* (pp. 651–665). London: Routledge.

AnyLogic. (n.d.). Retrieved April 20, 2014, from http://www.anylogic.ru

Ardıc, N. (2009). Friend or foe? Globalization and Turkey at the turn of the 21st century. *Journal of Economic and Social Research*, *11*(1), 17–42.

Argan, G. (1993). *Storia dell'arte come storia della città (IT)*. Roma, Italy: Riuniti.

Arvanitis, S., Ley, M., Soltmann, C., Stucki, T., Wörter, M., & Bolli, T. (2011). *Potenziale für Cleantech im Industrie- und Dienstleistungsbereich in der Schweiz, KOF Studien, 27*. Zurich: KOF Swiss Economic Institute.

Ascher, H., & Feingold, H. (1984). *Repairable systems reliability: Modeling, inference, misconceptions and their causes*. New York, NY: Marcel Dekker.

Austin, G. (2009). Cash crops and freedom: Export agriculture and the decline of slavery in colonial West Africa. *International Review of Social History*, *54*(1), 1–37. doi:10.1017/S0020859009000017

Babaev, V. M., Malyarenko, V. A., & Orlova, N. O. (2012). Development and implementation of increase the efficiency of communal energetics. *Energy Saving - Energetics - Energy Audit, 4*(98), 9-22.

Bakshi, N., & Kleindorfer, P. (2009). Co-opetition and investment for supply-chain resilience. *Production and Operations Management, 18*(6), 583–603. doi:10.1111/j.1937-5956.2009.01031.x

Bañuelos-Ruedas, F., Angeles-Camacho, C., & Rios-Marcuello, S. (2011). Methodologies used in the extrapolation of wind speed data at different heights and its impact in the wind energy resource assessment in a region. In Wind farm - Technical regulations, potential estimation and siting assessment. Academic Press.

Barberá, L., González-Prida, V., Moreu, P., & Crespo, A. (2010). *Revisión de herramientas software para el análisis de la fiabilidad, disponibilidad, mantenibilidad y seguridad (RAMS) de equipos industrials: Revista Ingeniería y Gestión de Mantenimiento. Vol. Abril/Mayo/Junio 2010*. España.

Bar-Gill, O., & Parchomovsky, G. (2003). *The value of giving away secrets*. Harvard Law and Econ Discussion Paper No. 417. Cambridge, MA: Harvard Law School.

Barratt, M. (2004). Understanding the meaning of collaboration in the supply chain. *Supply Chain Management: An International Journal, 9*(1), 30–42. doi:10.1108/13598540410517566

Barton, J. H. (2007a). *IP and access to clean energy technologies in developing countries: ICTSD biofuel and wind technologies issues paper*. Geneva: ICTSD. Retrieved from http://www.wipo.int/wipo_magazine/en/2008/01/article_0003.html

Barton, J. H. (2007b). *Intellectual property and access to clean energy technologies in developing countries: An analysis of solar photovoltaic, biofuel and wind technologies*. Geneva: ICTSD. doi:10.7215/GP_IP_20071201

Barton, J. H. (2007c). *New trends in technology transfer*. Geneva: ICTSD.

Beamon, B. M. (1999). Designing the green supply chain. *Logistics Information Management, 12*(4), 332–342. doi:10.1108/09576059910284159

Bebbington, M., Chin-Diew, L., & Zitikis, R. (2009). Balancing burn-in and mission times in environments with catastrophic and repairable failures. *Reliability Engineering & System Safety, 94*(8), 1314–1321. doi:10.1016/j.ress.2009.02.015

Bellona, S. (2009). Justifying energy efficiency as oil prices tumble. *The Chronicle of Higher Education, 55*(22), A14.

Benjaafar, S., Li, Y., & Daskin, M. (2013). Carbon footprint and the management of supply chains: Insights from simple models. *IEEE Transactions on Automation Science and Engineering, 10*(1), 99–116. doi:10.1109/TASE.2012.2203304

Bently Nevada. (1997). *Machinery diagnostics course*. Warrington, UK: Bently Nevada.

Bettencourt, L. M. A., Trancik, J. E., & Kaur, J. (2013). Determinants of the pace of global innovation in energy technologies. *PLoS ONE, 8*(10), e67864. doi:10.1371/journal.pone.0067864 PMID:24155867

Biera, J., Nieto, F. J., Iturriza, I., Viñolas, J., & Goikoetxea, P. (1994a). Characterisation of the behaviour of a CBN wheel when grinding at high speed. In *Proceedigns of the X Congress de Investigatión*. San Sebastián, Spain: Diseño y Utilizatión de Máquinas-Herramientas.

Biera, J., Nieto, F. J., Viñolas, J., & Goikoetxea, P. (1994b). *Proccedings of the XI National Congress of Mechanical Engineering*. Valencia, Spain: Academic Press.

Biera, J., Viñolas, J., & Nieto, F. J. (1997). Time-domain dynamic modelling of the external plunge grinding process. *International Journal of Machine Tools & Manufacture, 37*(11), 1555–1572. doi:10.1016/S0890-6955(97)00024-2

Birnie, P., Boyle, A., & Redgwell, C. (2009). *International law and the environment* (3rd ed.). Oxford, UK: Oxford University Press.

Bivand, R., & Neteler, M. (2000, August). Open source geocomputation: Using the R data analysis language integrated with GRASS GIS and PostgreSQL data base systems. In *Proceedings of the 5th International Conference on GeoComputation*. Academic Press.

Bjornskov, C., & Potrafke, N. (2011). Politics and privatization in Central and Eastern Europe: A panel data analysis. *Economics of Transition*, *19*(2), 201–230. doi:10.1111/j.1468-0351.2010.00404.x

Bouzarour-Amokrane, Y., Tchangani, A., & Pérès, F. (2012). *Definition and measure of risk and opportunity in the bocr analysis*. Paper presented at the 10th International Conference of Modeling and Simulation - MOSIM'12, Bordeaux, France.

Bouzarour-Amokrane, Y., Tchangani, A., & Pérès, F. (2013a). *Evaluation process in end-of-life systems management using BOCR analysis*. Paper presented at the IFAC Conference on Manufacturing modelling, Management and Control, Saint Petersburg, Russia.

Bouzarour-Amokrane, Y., Tchangani, A., & Pérès, F. (2013b). *Résolution des problèmes de décision de groupe par analyse bipolaire*. Paper presented at the 5th Doctoral Days (JDJN) GDRMACS, Strasbourg, France.

Braunscheidel, M. J., & Suresh, N. C. (2009). The organizational antecedents of a firm's supply chain agility for risk mitigation and response. *Journal of Operations Management*, *27*(2), 119–140. doi:10.1016/j.jom.2008.09.006

Briner, R. B., & Denyer, D. (2012). Systematic review and evidence synthesis as a practice and scholarship tool. In Handbook of evidence-based management: Companies, classrooms and research (pp. 112-129). Academic Press.

Brinksmeier, E., Aurich, J. C., Govekar, E., Heinzel, C., Hoffmeister, H. W., Klocke, F., . . . Wittmann, M. (2006). Advances in modelling and simulation of grinding processes. *CIRP Annals - Manufacturing Technology*, *55*(2), 667-696.

Brown, R. (1996). Profile. *Brüel & Kraer*, *4*(3), 10–12.

Cagliano, R., Caniato, F., & Spina, G. (2006). The linkage between supply chain integration and manufacturing improvement programmes. *International Journal of Operations & Production Management*, *26*(3), 282–299. doi:10.1108/01443570610646201

Carbon Trust. (2012). Retrieved from www.carbontrust.com/resources/guides/carbon-footprinting-and-reporting/carbon-footprinting

Carnero, M. C. (2005). Selection of diagnostic techniques and instrumentation in a predictive maintenance program: A case study. *Decision Support Systems*, *38*(4), 539–555. doi:10.1016/j.dss.2003.09.003

Carnero, M. C. (2012). *Programas de mantenimiento predictivo: Análisis de lubricantes y vibraciones*. Saarbrücken, Germany: LAP LAMBERT Academic Publishing GmbH & Co.

Carnero, M. C. (2013). Mantenimiento predictivo en pequeña y mediana empresa. *DYNA Management*, *1*(1). doi:10.6036/MN5790

Carnero, M. C., Gónzalez-Palma, R., Almorza, D., Mayorga, P., & López-Escobar, C. (2010). Statistical quality control through overall vibration analysis. *Mechanical Systems and Signal Processing*, *24*(4), 1138–1160. doi:10.1016/j.ymssp.2009.09.007

Carter, C. (2014, January 11). Wind turbine destroyed by high winds. *The Telegraph*.

Chaly, V. V. (2013). Problemy obespechenia energeticheskoy bezopasnosti v Ukraine. *Upravlinnya rozvytkom*, *22*, 144 – 146.

Chang, C. P., & Berdiev, A. N. (2011). The political economy of energy regulation in OECD countries. *Energy Economics*, *33*(5), 816–825. doi:10.1016/j.eneco.2011.06.001

Charray, C. (2000). Mantenimiento predictivo: Una técnica que reduce o elimina averías inesperadas. *DYNA*, *75*, 28–34.

Chaves Repiso, V. M. (1999). Instalaciones de energía solar térmica con acumulación distribuida. Proyecto Fin de Carrera, University of Seville.

Cheng, J. H., Yeh, C. H., & Tu, C. W. (2008). Trust and knowledge sharing in green supply chains. *Supply Chain Management: An International Journal*, *13*(4), 283–295. doi:10.1108/13598540810882170

Chesbrough, H. (2006). *Open business models: How to thrive in the new innovation landscape*. Boston, MA: Harvard Business School Press.

Choi, S. (2010). Beyond Kantian liberalism: Peace through globalization. *Conflict Management and Peace Science*, *27*(3), 272–295. doi:10.1177/0738894210366513

Choi, T. J., Subrahmanya, N., Li, H., & Shin, Y. C. (2008). Generalized practical models of cylindrical plunge grinding processes. *International Journal of Machine Tools & Manufacture*, *48*(1), 61–72. doi:10.1016/j.ijmachtools.2007.07.010

Choi, T., & Shin, Y. C. (2007). Generalized intelligent grinding advisory system. *International Journal of Production Research*, *45*(8), 1899–1932. doi:10.1080/00207540600562025

Chowdhury, S., Zhang, J., Messac, A., & Castillo, L. (2013). Optimizing the arrangement and the selection of turbines for wind farms subject to varyipreng wind conditions. *Renewable Energy*, *52*, 273–282. doi:10.1016/j.renene.2012.10.017

Chramcov, B., Dostal, P., & Balate, J. (2009, June). Forecast model of heat demand. In *Proceedings of the 28th International Symposium on Forecasting*. Hong Kong, China: Academic Press.

Chramcov, B. (2010). Heat demand forecasting for concrete district heating system. *International Journal of Mathematical Models and Methods in Applied Sciences*, *4*(4), 231–239.

Chramcov, B., Dostál, P., & Baláte, J. (2003). Prediction of the heat supply daily diagram via artificial neural network. In *Proceedings of the 4th International Carpathian Control Conference*. Zittau, Germany: Academic Press.

Christopher, M. (2005). Logistics and supply chain management: Creating value-adding networks. *Industry Week*, *267*(1), 60.

Clemons, E. K., Reddi, S. P., & Row, M. C. (1993). The impact of information technology on the organization of economic activity: The 'move to the middle' hypothesis. *Journal of Management Information Systems*, *10*(2), 9–35.

Collegium of the Ministry Energy and Coal Industry of Ukraine. (2012). *Updates energy strategy of Ukraine till 2030*. Retrieved from http://energetyka.com.ua/normatyvna-baza/384-energetichna-strategiya-ukrajinina-period-do-2030-roku

Colville, F. (2014). *Small-scale solar PV market trends in 2013*. Retrieved January 2, 2014, from http://www.solarbuzz.com/resources/articles-and-presentations/small-scale-solar-pv-market-trends-in-2013

Cormas: Natural Resources and Agent-Based Simulations. (n.d.). Retrieved from: http://cormas.cirad.fr

Cox, A. (1999). Power, value and supply chain management. *Supply Chain Management: An International Journal*, *4*(4), 167–175. doi:10.1108/13598549910284480

Dabiri, J. O. (2011). Potential order-of-magnitude enhancement of wind farm power density via counter-rotating vertical-axis wind turbine arrays. *Journal of Renewable and Sustainable Energy*, *3*(4), 043104. doi:10.1063/1.3608170

Dagdouguia, H., Minciardia, R., Ouammia, A., Robbaa, M., & Sacilea, R. (2010, July). A dynamic optimization model for smart micro-grid: integration of a mix of renewable resources for a green building. In *Proceedings of the 5th Biennial International Congress on Environmental Modelling and Software Society: Modelling for Environment's Sake (ICEMSS'10)*. Academic Press.

Darnall, N., Jolley, G. J., & Handfield, R. (2008). Environmental management systems and green supply chain management: Complements for sustainability? *Business Strategy and the Environment*, *17*(1), 30–45. doi:10.1002/bse.557

De Toni, A., & Nassimbeni, G. (1999). Buyer-supplier operational practices, sourcing policies and plant performances: Results of an empirical research. *International Journal of Production Research*, *37*(3), 597–619. doi:10.1080/002075499191698

Department of Statistics Malaysia. (2012). *Key statistics of manufacturing sector, 2012*. Retrieved from http://www.statistics.gov.my/index.php?r=column/cthemeByCat&cat=92&bul_id=enZUVDZxcHI5c1NFYzNSeUtxYWh3QT09&menu_id=SjgwNXdiM0JlT3Q2TDBlWXdKdUVldz09

DesMarais, C. (2013). *Intel, Microsoft and Kohl's top EPA's green power list*. Retrieved December 1, from http://www.greenbiz.com/blog/2013/07/24/intel-microsoft-kohls-top-epa-lists-green-power

De, U. K., & Pal, M. (2011). Dimensions of globalization and their effects on economic growth and human development index. *Asian Economic and Financial Review*, *1*(1), 1–13.

Dey, A., LaGuardia, P., & Srinivasan, M. (2011). Building sustainability in logistics operations: A research agenda. *Management Research Review*, *34*(11), 1237–1259. doi:10.1108/01409171111178774

Disney, S. M., & Towill, D. R. (2003). On bullwhip in supply chains—historical review, present practice and expected future impact. *International Journal of Production Economics*, *101*(1), 2–18.

Dreher, A., Sturm, J., & Vreeland, J. R. (2009). Development aid and international politics: Does membership on the UN Security Council influence World Bank decisions? *Journal of Development Economics*, *88*(1), 1–18. doi:10.1016/j.jdeveco.2008.02.003

Droge, C., Jayaram, J., & Vickery, S. K. (2004). The effects of internal versus external integration practices on time-based performance and overall firm performance. *Journal of Operations Management*, *22*(6), 557–573. doi:10.1016/j.jom.2004.08.001

Duane, J. T. (1964). Learning curve approach to reliability monitoring. *IEEE Transactions on Aerospace*, *2*(2), 563–566. doi:10.1109/TA.1964.4319640

Duffie, J. A., & Beckman, W. A. (1980). Solar engineering of thermal processes (Vol. 3). New York: Wiley.

El-Berishy, N., RÃ1/4gge, I., & Scholz-Reiter, B. (2013). The interrelation between sustainability and green logistics. Management and Control of Production and Logistics, 6 (1), 527-531.

Elo, S., & Kyngäs, H. (2008). The qualitative content analysis process. *Journal of Advanced Nursing*, *62*(1), 107–115. doi:10.1111/j.1365-2648.2007.04569.x PMID:18352969

Eltayeb, T. K., & Zailani, S. (2009). Going green through green supply chain initiatives towards environmental sustainability. *Operations and Supply Chain Management, 2*(2), 93-110

ElTayeb, T. K., Zailani, S., & Jayaraman, K. (2010). The examination on the drivers for green purchasing adoption among EMS 14001 certified companies in Malaysia. *Journal of Manufacturing Technology Management*, *21*(2), 206–225. doi:10.1108/17410381011014378

Environmental Transport Association. (n.d.). *Air travel's impact on climate change*. Retrieved from http://www.eta.co.uk/env_info/air_travel_climate_change

EPIA. (2011). *Global market outlook for PV until 2015*. Brussels, Belgium: European Photovoltaic Industry Association. Retrieved March 10, 2014, from http://www.epia.org/fileadmin/EPIA_docs/public/ Global_Market_Outlook_for_Photovoltaics_until_2015.pdf

EPIA-Greenpeace. (2011). *Solar generation: Solar photovoltaic electricity empowering the world*. Brussels, Belgium: European Photovoltaic Industry Association. Retrieved March 10, 2014, from web site http://www.epia.org/EPIA_docs/documents/SG6/Solar_Generation_6_2011_Full_report_Final.pdf

Ernst & Young. (2012). *Conflict minerals: What you need to know about the new disclosure and reporting requirements and how Ernst & Young can help*. Retrieved January 23, 2014, from http://www.ey.com/Publication/vwLUAssets/Conflict_minerals/$FILE/Conflict_Minerals_US.pdf

Ernst & Young. (2014). *Renewable energy country attractiveness index*. Ernst & Young Global Limited.

Esper, T. L., & Williams, L. R. (2003). The value of collaborative transportation management (CTM): Its relationship to CPFR and information technology. *Transportation Journal*, 55-65.

EU (2010). Directive 2010/31/EU on the energy performance of buildings. *Official Journal of the European Union*, L153/14 -L153/34.

EUROBSERV'ER. (2012). *Photovoltaic barometer*. Retrieved March 10, 2014, from http://www.eurobserv-er.org/pdf/photovoltaic_2012.pdf

European Commission. (2011). *Roadmap for moving to a competitive low-carbon economy by 2050* [Policy paper]. European Commission.

Fallah, M., & Ebrahimi, M. (2014). A study on the effect of green marketing on consumers' purchasing intention. *Management Science Letters*, *4*(3), 421–424. doi:10.5267/j.msl.2014.1.030

Farkas, K., Frontin, I. F., Maturi, L., Roecker, C., & Scognamiglio, A. (2013,). *Designing architectural integration of solar systems: criteria & guidelines for product developers*. Retrieved August 18, 2014, from web site: http://infoscience.epfl.ch/record/197098/files/task41A3-2-Designing-Photovoltaic-Systems-for-Architectural-Integration.pdf?version=1

Felix, R. (2008). *Multicriteria decision making (MCDM): Management of aggregation complexity through fuzzy interactions between goals or criteria*. Paper presented at the International Conference on Information Processing and Management of Uncertainty in Knowledge-Based Systems (IPMU), Malaga, Span.

Felix, R. (1994). Relationships between goals in multiple attribute decision making. *Fuzzy Sets and Systems, 67*(1), 47–52. doi:10.1016/0165-0114(94)90207-0

Fisch, N. (1995). *Manuskript zur Vorlesung Solartechnik I*. University of Stuttgart.

Fleurey, F., & Solberg, A. (2009). A domain specific modeling language supporting specification, simulation and execution of dynamic adaptive systems. In *Proceedings of Model Driven Engineering Languages and Systems*. Berlin: Springer. doi:10.1007/978-3-642-04425-0_47

Flynn, B. B., Huo, B., & Zhao, X. (2010). The impact of supply chain integration on performance: A contingency and configuration approach. *Journal of Operations Management, 28*(1), 58–71. doi:10.1016/j.jom.2009.06.001

Frohlich, M. T., & Westbrook, R. (2001). Arcs of integration: An international study of supply chain strategies. *Journal of Operations Management, 19*(2), 185–200. doi:10.1016/S0272-6963(00)00055-3

Fthenakis, V., & Bulawka, A. (2004). Environmental impact of photovoltaics. Encyclopedia of Energy, 5, 61-69.

Fugate, B. S., Davis-Sramek, B., & Goldsby, T. J. (2009). Operational collaboration between shippers and carriers in the transportation industry. *The International Journal of Logistics Management, 20*(3), 425–447. doi:10.1108/09574090911002850

Fuglsang, P., & Thomsen, K. (2001). Site-specific design optimization of 1.5–2.0 mw wind turbines. *Journal of Solar Energy Engineering, 123*(4), 296–303. doi:10.1115/1.1404433

Gardas, B. B., & Narkhede, B. E. (2013). Exploring the green supply chain management: A technical review. *International Journal of Application or Innovation in Engineering & Management, 2*(5), 441–450.

Gartner. (2013). *Worldwide semiconductor revenue declined 2.6 percent in 2012*. Retrieved January 23, 2014, from http://www.gartner.com/newsroom/id/2405215

Geary, S., Disney, S. M., & Towill, D. R. (2006). On bullwhip in supply chains—Historical review, present practice and expected future impact. *International Journal of Production Economics, 101*(1), 2–18. doi:10.1016/j.ijpe.2005.05.009

GEIA-STD-0010. (2008, October). *Standard best practices for system safety program development and execution*. Academic Press.

Geletukha G. G., & Zheliezna T. A. (2013). Barriers to bioenergy development in Ukraine. *Promislova teplotekhnika, 35*(5), 43-47.

Geletukha, G. G., Zheliezna, T. A., & Oliinyk, Ye. M. (2013). *Opportunity of heat generation from biomass in Ukraine*. Retrieved from http://www.uabio.org/img/files/docs/position-paper-uabio-6-ua.pdf

Geletukha, G. G., Zheliezna, T. A., Kucheruk, P. P., & Oliinyk, Y. M. (2014). *State of the art and prospects for bioenergy development in Ukraine*. Retrieved from http://www.uabio.org/img/files/docs/position-paper-uabio-9-en.pdf

Georgilakis, P. S. (2006). State-of-the-art of decision support systems for the choice of renewable energy sources for energy supply in isolated regions. *International Journal of Distributed Energy Resources, 2*(2), 129–150.

GeoServer. (n.d.). Retrieved April 20, 2014, from http://geoserver.org/

Giorgini, P., Mylopoulos, J., Nicchiarelli, E., & Sebastiani, R. (2002). Reasoning with goal models. In *Proceedings of the 21st International Conference on Conceptual Modeling*. London, UK: Springer.

Glover, J. l., Champion, D., Daniels, K. J., & Dainty, A. J. D. (2013). An Institutional Theory perspective on sustainable practices across the dairy supply chain. *International Journal of Production Economics, 10*(5), 220–241.

Goeminne, G., & Paredis, E. (2010). The concept of ecological debt: Some steps towards an enriched sustainability paradigm. *Environment, Development and Sustainability, 12*(5), 691–712. doi:10.1007/s10668-009-9219-y

Goetzberger, A., & Hoffmann, V. U. (2005). *Photovoltaic solar energy generation*. Heidelberg, Germany: Springer.

Goldberg, K. (2013, October 30). Clean energy patent boom sets stage for IP wars. In *Law360*. New York, NY: Portfolio Media, Inc.

Goldman, P., & Muszynska, A. (1999). Application of full spectrum to rotating machinery diagnostics. *Orbit (Amsterdam, Netherlands)*, *20*(1), 17–21. PMID:12048694

Gold, S., & Seuring, S. (2011). Supply chain and logistics issues of bio-energy production. *Journal of Cleaner Production*, *19*(1), 32–42. doi:10.1016/j.jclepro.2010.08.009

Golicic, S. L., & Smith, C. D. (2013). A meta-analysis of environmentally sustainable supply chain management practices and firm performance. *Journal of Supply Chain Management*, *49*(2), 78–95. doi:10.1111/jscm.12006

Gonzales-Baixauli, B., Prado Leite, J. C. S., & Mylopoulos, J. (2004). Visual variability analysis for goal models. In *Proceedings of the 12th IEEE International Requirements Engineering Conference (RE'04)*. Kyoto, Japan: IEEE.

González-Prida, V., & Crespo, A. (2012). A reference framework for the warranty management in industrial assets. *Computers in Industry*, *63*, 960–971.

González-Prida, V., Barberá, L., & Crespo, A. (2010). *Practical application of a RAMS analysis for the improvement of the warranty management*. Paper presented at the 1st IFAC workshop on Advanced Maintenance Engineering Services and Technology, Lisbon, Portugal.

González-Prida, V. (2002). *Influencia de la normativa europea en el procedimiento de ensayos de sistemas solares térmicos prefabricados: Propuesta de adaptación de la instalación de ensayos del Instituto Andaluz de Energías Renovables*. Proyecto Fin de Carrera, Universidad de Sevilla.

González-Prida, V., Crespo, A., Pérès, F., De Minicis, M., & Tronci, M. (2012). *Logistic support for the improvement of the warranty management*. Advances in Safety, Reliability and Risk Management. Taylor & Francis Group.

Gonzalez-Prida, V., Parra, C., Gómez, J., Crespo, A., & Moreu, P. (2009). Availability and reliability assessment of industrial complex systems: a practical view applied on a bioethanol plant simulation. In *Safety, reliability and risk analysis: Theory, methods and applications* (pp. 687–695). Taylor & Francis.

Govekar, E., Baus, A., Gradišek, J., Klocke, F., & Grabec, I. (2002). A new method for chatter detection in grinding. *CIRP Annals - Manufacturing Technology*, *51*(1), 267-270.

Grabisch, M., Greco, S., & Pirlot, M. (2008). Bipolar and bivariate models in multicriteria decision analysis: Descriptive and constructive approaches. *International Journal of Intelligent Systems*, *23*(9), 930–969. doi:10.1002/int.20301

Gradišek, J., Baus, A., Govekar, E., Klocke, F., & Grabec, I. (2003). Automatic chatter detection in grinding. *International Journal of Machine Tools & Manufacture*, *43*(14), 1397–1403. doi:10.1016/S0890-6955(03)00184-6

Green Building Index. (n.d.). *What is green building?* Retrieved from http://www.greenbuildingindex.org/why-green-buildings.html

Greenpeace. (2008). *Implementing the energy [R] evolution*. Retrieved March 10, 2014, from http://www.greenpeace.org/international/PageFiles/25109/energyrevolutionreport.pdf

Gruber, E. (1984). *Energieeinsparung und Solarenergienutzung in Eigenheimen: Forschungberich T84-287 des Bundesministeriums für Forschung und Technologie*. German Ministry of Research and Technology.

Grzenda, M. (2012). Consumer-oriented heat consumption prediction. *The Journal of Control and Cybernetics*, *41*, 213–240.

gvSIG. (n.d.). Retrieved April 20, 2014, from: http://www.gvsig.com/

Haghpanah, Y., & desJardins, M. (2010, July). Using a trust model in decision making for supply chain management. In *Proceedings of Workshops at the Twenty-Fourth AAAI Conference on Artificial Intelligence*. AAAI.

Hall, B. H., & Helmers, C. (2010). *The role of patent protection in (clean/green) technology transfer*. NBER Working Papers 16323. Cambridge, MA: National Bureau of Economic Research.

Hall, B. H., & Helmers, C. (2011). Innovation and diffusion of clean/green technology: Can patent commons help? *Journal of Environmental Economics and Management*, *66*(1), 33–51. doi:10.1016/j.jeem.2012.12.008

Halldórsson, Á., & Kovács, G. (2010). The sustainable agenda and energy efficiency: Logistics solutions and supply chains in times of climate change. *International Journal of Physical Distribution & Logistics Management*, *40*(1/2), 5–13. doi:10.1108/09600031011018019

Halldórsson, Á., & Svanberg, M. (2013). Energy resources: Trajectories for supply chain management. *Supply Chain Management: An International Journal*, *18*(1), 66–73. doi:10.1108/13598541311293186

Han, J., Mol, A. P. J., Lu, Y., & Zhang, L. (2009). Onshore wind power development in China: Challenges behind a successful story. *Energy Policy*, *37*(8), 2941–2951. doi:10.1016/j.enpol.2009.03.021

Harris, C. M. (1995). *Shock and vibration handbook*. New York, NY: McGraw-Hill.

Hasan, M. (2013). Sustainable supply chain management practices and operational performance. *American Journal of Industrial and Business Management*, *3*(01), 42–48. doi:10.4236/ajibm.2013.31006

Hase, Y. (2007). *Handbook of power system engineering*. John Wiley & Sons. doi:10.1002/9780470033678

Hayes, R. H., & Pisano, G. P. (1994). Beyond world class: The new manufacturing strategy. *Harvard Business Review*, 77–86.

Health and Safety Concerns of Photovoltaic Solar Panels. (n.d.). Retrieved from http://www.oregon.gov/odot/hwy/oipp/docs/life-cyclehealthandsafetyconcerns.pdf

HELAPCO. (2013). *PV grid: Initial project report*. Retrieved March 10, 2014, from http://www.helapco.gr/pv-grid/pv-grid-initial-project-report/

Helliwell, J., Layard, R., & Sachs, J. (2013). *World happiness report*. United Nations (UN). Retrieved from http://unsdsn.org/wp-content/uploads/2014/02/WorldHappinessReport2013_online.pdf

Henley, E., & Kumamoto, H. (1992). *Probabilistic risk assessment, reliability engineering, design and analysis*. IEEE Press.

Herrera, F., Herrera-Viedma, E., & Verdegay, J. L. (1996). A model of consensus in group decision making under linguistic assessments. *Fuzzy Sets and Systems*, *78*(1), 73–87. doi:10.1016/0165-0114(95)00107-7

Herrera-Viedma, E., Alonso, S., Chiclana, F., & Herrera, F. (2007). A consensus model for group decision making with incomplete fuzzy preference relations. *IEEE Transactions on Fuzzy Systems*, *15*(5), 863–877. doi:10.1109/TFUZZ.2006.889952

Herrera-Viedma, E., Herrera, F., & Chiclana, F. (2002). A consensus model for multiperson decision making with different preference structures. *IEEE Transactions on Systems, Man, and Cybernetics. Part A, Systems and Humans*, *32*(3), 394–402. doi:10.1109/TSMCA.2002.802821

He, Y., & Lai, K. K. (2012). Supply chain integration and service oriented transformation: Evidence Chinese equipment manufacturers. *International Journal of Production Economics*, *135*(2), 291–299. doi:10.1016/j.ijpe.2011.10.013

Horbach, J. (2008). Determinants of environmental innovation—New evidence from German panel data sources. *Research Policy*, *37*(1), 163–173. doi:10.1016/j.respol.2007.08.006

Huang, Y. A., Lenzen, M., Weber, C. L., Murray, J., & Matthews, H. S. (2009a). The role of input–output analysis for the screening of corporate carbon footprints. *Economic Systems Research*, *21*(3), 217–242. doi:10.1080/09535310903541348

Huang, Y. A., Weber, C. L., & Matthews, H. S. (2009b). Categorization of scope 3 emissions for streamlined enterprise carbon footprinting. *Environmental Science & Technology*, *43*(22), 8509–8515. doi:10.1021/es901643a PMID:20028044

Huin, S. F., Luong, L. H. S., & Abhary, K. (2002). Internal supply chain planning determinants in small and medium-sized manufacturers. *International Journal of Physical Distribution & Logistics Management*, *32*(9), 771–782. doi:10.1108/09600030210452440

Hutchins, M. J., & Sutherland, J. W. (2008). An exploration of measures of social sustainability and their application to supply chain decisions. *Journal of Cleaner Production*, *16*(15), 1688–1698. doi:10.1016/j.jclepro.2008.06.001

Iakovou, E., Karagiannidis, A., Vlachos, D., Toka, A., & Malamakis, A. (2010). Waste biomass-to-energy supply chain management: A critical synthesis. *Waste Management (New York, N.Y.)*, *30*(10), 1860–1870. doi:10.1016/j.wasman.2010.02.030 PMID:20231084

Ibrahim, A., Zailani, T., & Tan, A. C. (2012). Extending green practices across the supply chain: The impact of upstream and downstream integration. *International Journal of Operations & Production Management, 26*(7), 795–821.

IEA PVPS. (2003). Energy from the desert, feasibility of very large scale photovoltaic power generation (VLS-PV) systems. London: James & James (Science Publishers).

IEA SHC. (2008). *Task 40-ECBCS annex 52, towards net zero energy solar buildings.* Paris, France: International Energy Agency. Retrieved May 22, 2012, from http://www.iea-shc.org/task40/

IEA. (2010). *JRC, PV status report.* Luxembourg: Office for Official Publications of the European Union. Retrieved March 10, 2014, from http://re.jrc.ec.europa.eu/refsys/pdf/PV%20reports/ PV%20Report%202010.pdf

Imoussaten, A., Montmain, J., Trousset, F., & Labreuche, C. (2011). Multi-criteria improvement of options. In *Proceedings of the 7th conference of the European Society for Fuzzy Logic and Technology (EUSFLAT-2011) and LFA-2011*. Aix-Les-Bains, France: Atlantis Press.

Intel News Release. (2008). *Intel becomes largest purchaser of green power in the U.S.* Retrieved November 24, 2013 from http://www.intel.com/pressroom/kits/green/rec/?wapkw=green+energy

Intel. (2010). *Intel water policy.* Retrieved November 24, 2013 from http://www.intel.com/content/www/us/en/policy/policy-water.html

Intel. (2012). *ESG corporate social responsibility.* Retrieved November 24, 2013 from www.intel.com/content/www/us/en/corporate-responsibility/inspire-the-next-generation.html

Intel. (2013). *2012 Annual report.* Retrieved November 24, 2013 from http://www.intc.com/intel-annual-report/2012/

Intel. (2013). *Conflict minerals whitepaper.* Retrieved November 25, 2013 from http://www.intel.com/content/www/us/en/policy/policy-conflict-minerals.html

Intellectual Property Office UK. (2009). *UK green inventions to get fast-tracked through patent system* [Press release]. Retrieved September 10, 2014, from www.ipo.gov.uk/about/press/press-release/press-release-2009/pressrelease-20090512.htm

International Chamber of Commerce (ICC). (2012). *ICC intellectual property roadmap: Current and emerging issues for business and policymakers 2012.* Paris: ICC.

International Energy Agency (IEA). (2008). *Energy technology perspectives 2008.* Paris: IEA. Retrieved from http://www.iea.org/Textbase/npsum/ETP2008SUM.pdf

International Energy Agency. (2011). *Prospect of limiting the global increase in temperature to 2°C is getting bleaker.* Retrieved from http://www.iea.org/newsroomandevents/news/2011/may/name,19839,en.html

International Energy Agency (IAE). (2012). *Executive summary energy technology perspectives: Pathways to a clean energy system.* Retrieved from http://www.iea.org/Textbase/npsum/ETP2012SUM.pdf

International Tanker Owners Pollution Federation. (n.d.). *Oil tanker statistics 2014.* Retrieved from http://www.itopf.com/information-services/data-and-statistics/statistics/

International Tropical Timber Organization. (2011). *Survey of world's embattled tropical forests reports 50% increase in areas under sustainable management since 2005.* Retrieved from https://www.google.co.nz/url?sa=t&rct=j&q=&esrc=s&source=web&cd=1&cad=rja&uact=8&ved=0CCAQFjAA&url=http%3A%2F%2Fwww.itto.int%2Fdirect%2Ftopics%2Ftopics_pdf_download%2Ftopics_id%3D2642%26no%3D4%26disp%3Dinline&ei=eA7lVMzMDuH2mQWHyYHoBw&usg=AFQjCNEKAP9h8pxnH3l7FKiJS9w1qVmcNA&sig2=r3Pu4JB2vnvNU0V0ghJl1A

Inter P. S. S. Community. (n.d.). Retrieved from: http://www.interpss.org/

IPCC. (2006). *2006 IPCC guidelines for national greenhouse gas inventories.* Retrieved from http://www.ipcc-nggip.iges.or.jp/public/2006gl/pdf/0_Overview/V0_1_Overview.pdf

ISO 9459-2. (2013). *Solar heating. Domestic water heating systems. Part 2: Outdoor tests methods for system performance characterization and yearly performance prediction of solar-only systems.* International Standard.

ISO. I. (2006a). 14064-1: 2006-greenhouse gases-Part 1: Specification with guidance at the organization level for quantification and reporting of greenhouse gas emissions and removals. International Organization for Standardization.

Jafartayari, S. (2010). *Awareness of sustainable manufacturing practices in Malaysian manufacturers*. (Doctoral dissertation). Universiti Teknologi Malaysia, Faculty of Mechanical Engineering.

Jaja, J. M. (2010). Globalization or Americanization: Implications for sub-Saharan Africa. In K. Deng (Ed.), *Globalization: Today, tomorrow* (pp. 138–140). Sciyo.

Jamison, A. (1997), Public participation and sustainable development: Comparing European experiences. In PESTO papers 1. Aalborg University Press.

Jansson, J., Marell, A., & Nordlund, A. (2011). Exploring consumer adoption of a high involvement eco-innovation using value-belief-norm theory. *Journal of Consumer Behaviour*, *10*(1), 51–60. doi:10.1002/cb.346

Jardine, A., Lin, D., & Banjevic, D. (2006). A review on machinery diagnostics and prognostics implementing condition based maintenance. *Mechanical Systems and Signal Processing*, *20*(7), 1483–1510. doi:10.1016/j.ymssp.2005.09.012

Java Agent Development Framework (JADE). (n.d.). Retrieved from http://jade.tilab.com/

Jayal, A. D., Badurdeen, F., Dillon, O. W. Jr, & Jawahir, I. S. (2010). Sustainable manufacturing: Modeling and optimization challenges at the product, process and system levels. *CIRP Journal of Manufacturing Science and Technology*, *2*(3), 144–152. doi:10.1016/j.cirpj.2010.03.006

Jelle, B., & Breivikb, C. (2012). The path to the building integrated photovoltaics of tomorrow. *Energy Procedia*, *20*, 78–87. doi:10.1016/j.egypro.2012.03.010

Joselin-Herbert, G. M., Iniyan, S., Sreevalsan, E., & Rajapandian, S. (2007). A review of wind energy technologies. *Renewable & Sustainable Energy Reviews*, *11*(6), 1117–1145. doi:10.1016/j.rser.2005.08.004

Jungbluth, N., Bauer, C., Dones, R., & Frischknecht, R. (2005). Life cycle assessment for emerging technologies: Case studies for photovoltaic and wind power. *International Journal of Life Cycle Assessment*, *10*(1), 24–34. doi:10.1065/lca2004.11.181.3

Kalenska, S., Rahmetov, D., Kalenskiy, V., Yinik, A., & Kachura, I. (2012). Bioresource potential of Ukraine in settling of production and energy security. *International Scientific Electronic Journal Earth Bioresources and Life Quality*, *1*. Retrieved from http://gchera-ejournal.nubip.edu.ua/index.php/ebql/article/view/20/pdf

Kamal, A. (2014). *Green supply chain management on sustainable business performance: The green airport suppliers' perspective*. (Unpublished master's thesis). Universiti Sains Malaysia, Penang, Malaysia.

Keating, B., Quazi, A., Kriz, A., & Coltman, T. (2008). In pursuit of a sustainable supply chain: Insights from Westpac Banking Corporation. *Supply Chain Management: An International Journal*, *13*(3), 175–179. doi:10.1108/13598540810871217

Kefela, G. T. (2011). Driving forces of globalization in emerging market economies developing countries. *Asian Economic and Financial Review*, *1*(2), 83–94.

KeTTHA. (2011). *Low carbon cities framework & assessment system* [PowerPoint Slides]. Retrieved from http://esci-ksp.org/wp/wp-content/uploads/2012/04/Low-Carbon-Cities-Framework-and-Assessment-System.pdf

Khalifa, O. O., Densibali, A., & Faris, W. (2006). Image processing for chatter identification in machining processes. *International Journal of Advanced Manufacturing Technology*, *31*(5-6), 443–444. doi:10.1007/s00170-005-0233-4

Khorshid, S. (2010). Soft consensus model based on coincidence between positive and negative ideal degrees of agreement under a group decision-making fuzzy environment. *Expert Systems with Applications*, *37*(5), 3977–3985. doi:10.1016/j.eswa.2009.11.018

Kijima, M., & Sumita, N. (1986). A useful generalization of renewal theory: Counting process governed by non-negative markovian increments. *Journal of Applied Probability*, *23*(1), 71–88. doi:10.2307/3214117

Kim, S. W. (2009). An investigation on the direct and indirect effect of supply chain integration on firm performance. *International Journal of Production Economics*, *119*(2), 328–346. doi:10.1016/j.ijpe.2009.03.007

Kiss, P., & Jánosi, I. M. (2007). Comprehensive empirical analysis of ERA-40 surface wind distribution over. *Europe*.

Kitzes, J., Wackernagel, M., Loh, J., Peller, A., Goldfinger, S., Cheng, D., & Tea, K. (2008). Shrink and share: Humanity's present and future ecological footprint. *Philosophical Transactions of the Royal Society of London. Series B, Biological Sciences*, *363*(1491), 467–475. doi:10.1098/rstb.2007.2164 PMID:17652075

Konechenkov, A. (2013). *Renewable energy. Focusing: Ukraine vision 2050*. Retrieved from http://www.inforse.org/europe/pdfs/VisionUA_ppt.pdf

Kosmo. (n.d.). Retrieved April 20, 2014, from http://www.opengis.es/

Krauter, S. (2006). *Solar electric power generation - Photovoltaic energy systems*. Heidelberg, Germany: Springer.

Krivtsov, V. (2000). *Monte Carlo approach to modeling and estimation of the generalized renewal process in repairable system reliability analysis*. (Dissertation for the degree of Doctor of Philosophy). University of Maryland, College Park, MD.

Kumar, K. (2001). Technology for supporting supply chain management: Introduction. *Communications of the ACM*, *44*(6), 58–61. doi:10.1145/376134.376165

Kumar, R., & Shekhar, S. (2013). Green supply chain management – A review. *International Journal of Research in Aeronautical and Mechanical Engineering*, *1*(7), 245–255.

Kuznetsova, A., & Kutsenko, K. (2010). Biogas and "green" tariffs in Ukraine – A profitable investment? *German–Ukrainian Policy Dialogue in Agriculture. Institute for Economic Research and Policy Consulting*, *26*, 1–38.

Kwon, I. W. G., & Suh, T. (2005). Trust, commitment and relationships in supply chain management: A path analysis. *Supply Chain Management: An International Journal*, *10*(1), 26–33. doi:10.1108/13598540510578351

Lambert, D. M., & Cooper, M. C. (2000). Issues in supply chain management. *Industrial Marketing Management*, *29*(1), 65–83. doi:10.1016/S0019-8501(99)00113-3

Lamla, M. (2009). Long-run determinants of pollution: A robustness analysis. *Ecological Economics*, *69*(1), 135–144. doi:10.1016/j.ecolecon.2009.08.002

Lane, E. (2009). Clean tech reality check: Nine international green technology transfer deals unhindered by intellectual property rights. *Santa Clara Computer and High-Technology Law Journal*, *26*, 533.

Lane, E. (2011). *Clean tech intellectual property: Ecomarks, green patents, and green innovation*. Oxford, UK: Oxford University Press.

Law of Ukraine on Electricity. (n.d.). Retrieved April 20, 2014, from http://zakon2.rada.gov.ua/laws/show/575/97-%D0%B2%D1%80

Layton, J. (2013). Do wind turbines cause health problems?. *HowStuffWorks*.

Lazarou, S., Oikonomou, D. S., & Ekonomou, L. (2012). A platform for planning and evaluating distributed generation connected to the hellenic electric distribution grid. In *Proceedings of the 11th WSEAS International Conference on Circuits, Systems, Electronics, Control & Signal Processing*. Montreux, Switzerland: WSEAS Press.

Lee, A. H. I., Chen, H., & Kang, H. (2009). Multi-criteria decision making on strategic selection of wind farms. *Renewable Energy*, *34*(1), 120–126. doi:10.1016/j.renene.2008.04.013

Lee, S. M., Kim, S. T., & Choi, D. (2012). Green supply chain management and organizational performance. *Industrial Management & Data Systems*, *112*(8), 1148–1180. doi:10.1108/02635571211264609

Leitao, N. C. (2011). Foreign direct investment: Localization and institutional determinants. *Management Research and Practice*, *3*(2), 1–6.

Lenzen, M. (2000). Errors in conventional and input-output—based life—cycle inventories. *Journal of Industrial Ecology*, *4*(4), 127–148. doi:10.1162/10881980052541981

Leonardi, J., & Browne, M. (2010). A method for assessing the carbon footprint of maritime freight transport: European case study and results. *International Journal of Logistics: Research and Applications, 13*(5), 349-358

Lepawsky, J., & McNabb, C. (2010). Mapping international flows of electronic waste. *Canadian Geographer, 54*(2), 177–195. doi:10.1111/j.1541-0064.2009.00279.x

Levitt, J. (2003). *The complete guide to preventive and predictive maintenance*. New York, NY: Industrial Press Inc.

Lewis, A. (2011). *Europe breaking electronic waste export ban*. Retrieved from http://www.bbc.co.uk/news/world-europe-10846395

Lezhnev, S., & Hellmuth, A. (2012). *Taking conflict out of consumer gadgets*. Retrieved January 23, 2014 from http://www.enoughproject.org/files/CorporateRankings2012.pdf

Li, H., & Shin, Y. C. (2006a). Wheel regenerative chatter of surface grinding. Transactions of ASME. *Journal of Manufacturing Science and Engineering, 128*(2), 393–403. doi:10.1115/1.2137752

Li, H., & Shin, Y. C. (2006b). A time-domain dynamic model for chatter prediction of cylindrical plunge grinding processes. *Journal of Manufacturing Science and Engineering, 128*(2), 404–416. doi:10.1115/1.2118748

Li, H., & Shin, Y. C. (2007). A study on chatter boundaries of cylindrical plunge grinding with process condition-dependent dynamics. *International Journal of Machine Tools & Manufacture, 47*(10), 1563–1572. doi:10.1016/j.ijmachtools.2006.11.009

Lim, C. Y. (2014, July). Catalyzing green growth: Promoting a sustainable future through four key areas. *Ecowatch: The Star Publication*, 8.

Lim, S. S., Vos, T., Flaxman, A. D., Danaei, G., Shibuya, K., Adair-Rohani, H., & Blyth, F. et al. (2012). A comparative risk assessment of burden of disease and injury attributable to 67 risk factors and risk factor clusters in 21 regions, 1990–2010: A systematic analysis for the global burden of disease study 2010. *Lancet, 380*(9859), 2224–2260. doi:10.1016/S0140-6736(12)61766-8 PMID:23245609

Linacre, E. (1992). *Climate data and resources*. London: Routledge.

López-Escobar, C., González-Palma, R., Almorza, D., Mayorga, P., & Carnero, M. C. (2012). Statistical quality control through process self-induced vibration spectrum analysis. *International Journal of Advanced Manufacturing Technology, 58*(9-12), 1243–1259. doi:10.1007/s00170-011-3462-8

Luque, A., & Hegedus, S. (2003). *Handbook of photovoltaic science and engineering* (1st ed.). West Sussex, UK: John Wiley & Sons. doi:10.1002/0470014008

Maboye, B. (1995). *Technology transfer overlooked in GEF solar project*. Stockholm: Stockholm Environmental Institute.

Makarovsky, E.L. (2004). Energetichesky potential netradicionnyh i vozobnovlyaemyx istochnikov Ukrainy. *Integrovani technologii ta energozberezennya, 3*, 75-83.

Malaysia Palm Oil Industry. (2012). Retrieved from http://www.mpoc.org.my/Malaysian_Palm_Oil_Industry.aspx

Malaysian Institute of Planners. (2011). *Low carbon cities framework & assessment system* [PowerPoint slides]. Retrieved from http://www.townplan.gov.my/download/lowcarboncities-_paper_3b988287694.pdf

Malhi, Y., Aragão, L., Galbraith, D., Huntingford, C., Fisher, R., Zelazowski, P., ... Meir, P. (2009). Exploring the likelihood and mechanism of a climate-change-induced dieback of the Amazon rainforest. *Proceedings of the National Academy of Sciences of the United States of America, 106*(49), 20610-20615. doi:10.1073/pnas.0804619106

MapGuide Project Home. (n.d.). Retrieved April 20, 2014, from http://mapguide.osgeo.org/

Massel, L. V., & Bakhvalov, K. S. (2012). Open integrated environment InterPSS as a basis of smart grid it-infrastructure. *Vestnik ISTU, 7*, 10–15.

Mathiyazhagan, K., Govindan, K., NoorulHaq, A., & Geng, Y. (2013). An ISM approach for the barrier analysis in implementing green supply chain management. *Journal of Cleaner Production, 47*, 283–297. doi:10.1016/j.jclepro.2012.10.042

MatLAB. (n.d.). Retrieved April 20, 2014, from http://uk.mathworks.com/

Matthews, H. S., Hendrickson, C. T., & Weber, C. L. (2008). The importance of carbon footprint estimation boundaries. *Environmental Science & Technology*, *42*(16), 5839–5842. doi:10.1021/es703112w PMID:18767634

Matusik, S. F., & Hill, C. W. L. (1998). The utilization of contingent work, knowledge creation, and competitive advantage. *Academy of Management Review*, *23*(4), 680–698. doi:10.2307/259057

McCarthy, M. (2010). *The big question: How big is the problem of electronic waste, and can it be tackled?* Retrieved from http://www.independent.co.uk/environment/green-living/the-big-question-how-big-is-the-problem-of-electronic-waste-and-can-it-be-tackled-1908335

McKinnon, A. C. (2011). Developing a decarburization strategy for logistics. In *Proceedings of the 16th Annual Logistics Research Network Conference, Smarter Logistics: Innovation for Efficiency Performance and Austerity*. University of Southampton.

McKinnon, A. C. (2010). Product-level carbon auditing of supply chains: Environmental imperative or wasteful distraction? *International Journal of Physical Distribution & Logistics Management*, *40*(1/2), 42–60. doi:10.1108/09600031011018037

Mentzer, J. T., DeWitt, W., Keebler, J. S., Min, S., Nix, N. W., Smith, C. D., & Zacharia, Z. G. (2001). Defining supply chain management. *Journal of Business Logistics*, *22*(2), 1–25. doi:10.1002/j.2158-1592.2001.tb00001.x

Michael Groppi, P. E., & Burin, J. (2007). *Meeting the carbon challenge, the role of commercial real estate owners, users & managers*. Grubb & Ellis Company.

Millar, H. H., & Russell, S. N. (2011). The adoption of sustainable manufacturing practices in the Caribbean. *Business Strategy and the Environment*, *20*(8), 512–526. doi:10.1002/bse.707

Miller, J. (2012). *Intel, Apple praised for clean mineral efforts*. Retrieved November 17, 2013 from http://www.cbsnews.com/8301-501465_162-57494957-501465/intel-apple-praised-for-clean-mineral-efforts

Millien, R., & Laurie, R. (2007). *A summary of established & emerging IP business models*. Paper presented at the Sedona Conference, Phoenix, AZ.

Millien, R., & Laurie, R. (2008). *Meet the middlemen*. London: Globe White Page Ltd.

Mimiko, N. O. (2010). *Swimming against the tide: Development challenge for the long-disadvantaged in a fundamentally skewed global system: An inaugural lecture delivered at the Oduduwa Hall, Obefemi Awolowo University, Ile-Ife, Nigeria, on Tuesday October 12, 2010*. Ile-Ife, Nigeria: Obafemi Awolowo University Press.

Ministry of Energy and Coal Industry of Ukraine. (n.d.). Retrieved from http://mpe.kmu.gov.ua

Minx, J. C., Wiedmann, T., Wood, R., Peters, G. P., Lenzen, M., Owen, A., & Ackerman, F. et al. (2009). Input-output analysis and carbon footprinting: A review of applications. *Economic Systems Research*, *21*(3), 187–216. doi:10.1080/09535310903541298

Mobley, R. K. (2001a). *An introduction to predictive maintenance*. Woburn, MA: Butterworth-Heinemann.

Mobley, R. K. (2001b). *Plant engineer's handbook*. Woburn, MA: Butterworth-Heinemann.

Molony, T., & Smith, J. (2010). Biofuels, food security, and Africa. *African Affairs*, *109*(436), 489–498. doi:10.1093/afraf/adq019

Monti, A., Ponci, F., Smith, A., & Liu, R. (2009). A design approach for digital controllers using reconfigurable network-based measurement. In *Proceedings of the International Instrumentation and Measurement Technology Conference 2009*. Piscataway, NJ: IEEE. doi:10.1109/IMTC.2009.5168415

Montoya, F. G., Manzano-Agugliaro, F., López-Márquez, S., Hernández-Escobedo, Q., & Gil, C. (2014). Wind turbine selection for wind farm layout using multi-objective evolutionary algorithms. *Expert Systems with Applications*, *41*(15), 6585–6595. doi:10.1016/j.eswa.2014.04.044

Mowshowitz, A. (1997). Virtual organization. *Communications of the ACM*, *40*(9), 30–37. doi:10.1145/260750.260759

Muslih, I. M., & Abdellatif, Y. (2011). Hybrid micropower energy station: Design and optimization by using HOMER modeling software. In *Proceedings of the 2011 International Conference on Modeling, Simulation and Visualization Methods*. Pittsburgh, PA: ACTAPRESS.

Muszynska, A. (1992). Vibrational diagnostics of rotating machinery malfunctions. Course Rotor Dynamics and Vibration in Turbomachinery, Belgium.

Mytelka, L. (2007). *Technology transfer issues in environmental goods and services*. ICTSD Trade and Environment Papers No. 6. Geneva: ICTSD. Retrieved from http://www.ictsd.org/sites/default/files/research/2008/04/2007-04-lmytelka.pdf

Narasimhan, R. (1997). Strategic supply management: A total quality management imperative. *Advances in the Management of Organizational Quality, 2*, 39–86.

Narasimhan, R., & Kim, S. W. (2002). Effect of supply chain integration on the relationship between diversification and performance: Evidence from Japanese and Korean firms. *Journal of Operations Management, 20*(3), 303–323. doi:10.1016/S0272-6963(02)00008-6

National Aeronautics and Space Administration. (2000). *Reliability centered maintenance guide for facilities and collateral equipment*. Washington, DC: National Aeronautics and Space Administration.

National Renewable Energy Lab. (2011). *Best research-cell efficiencies, rev. 12-2011*. Retrieved May 15, 2014, from http://www.nrel.gov/ncpv/images/efficiency_chart.jpg

National Report on the Man-Made and Natural Security in 2012. (2012). Retrieved from http://www.mns.gov.ua/files/prognoz/report/2012/3_4_2012.pdf

NEPLAN Desktop Overview. (n.d.). Retrieved April 20, 2014, from http://www.neplan.ch

Network Simulation (OPNET Modeler Suite). (n.d.). Retrieved April 20, 2014, from http://www.riverbed.com/products/performance-management-control/network-performance-management/network-simulation.html

Neuendorf, K. A. (2002). *The content analysis guidebook* (Vol. 300). Thousand Oaks, CA: Sage Publications.

Newell, R. G. (2010). International climate technology strategies. In J. Aldy & R. Stavins (Eds.), *Post-Kyoto international climate policy: Implementing architectures for agreement*. Cambridge, UK: Cambridge University Press.

Ngai, E. W. T., To, C. K., Ching, V. S., Chan, L. K., Lee, M. C., Choi, Y. S., & Chai, P. Y. F. (2012). Development of the conceptual model of energy and utility management in textile processing: A soft systems approach. *International Journal of Production Economics, 135*(2), 607–617. doi:10.1016/j.ijpe.2011.05.016

Ninlawan, C., Seksan, P., Tossapol, K., & Pilada, W. (2010). The implementation of green supply chain management practices in electronics industry. In *Proceedings of the International Multiconference of Engineers and Computer Scientists* (vol. 3, pp. 17-19). Academic Press.

Nissen, M. E. (2001). Agent-based supply chain integration. *Journal of Information Technology Management, 2*(3), 289–312. doi:10.1023/A:1011449109160

NIST/SEMATECH. (2014). *e-Handbook of statistical methods*. Retrieved June 6, 2014. http://www.itl.nist.gov/div898/handbook/

Nooteboom, B. (1992). Information technology, transaction costs and the decision to 'make or buy'. *Technology Analysis and Strategic Management, 4*(4), 339–350. doi:10.1080/09537329208524105

Obadan, M. I. (2010). *Globalization: Concepts, opportunities and challenges for Nigeria development*. Paper presented at the Training Programme on Economic Reform.

OECD. (2004). *Patents and innovation: Trends and policy challenges*. Paris: OECD Publishing.

OECD. (2006a). *OECD science, technology and industry outlook 2006*. Paris: OECD Publishing.

OECD. (2006b). *Biotechnology update*. Paris: OECD.

OECD. (2006c). *Guidelines for the licensing of genetic inventions* [Policy guideline]. Retrieved from http://www.oecd.org/sti/biotechnology/licensing

OECD. (2013). *Helping companies source minerals responsibly*. Retrieved November 17, 2013 from http://www.oecd.org/daf/inv/mne/mining.htm

Ofori, G. (2000). Greening the construction supply chain in Singapore. *European Journal of Purchasing & Supply Management, 6*(3), 195–206. doi:10.1016/S0969-7012(00)00015-0

On the Basic Parameters of the Boiler and Heating Networks in Ukraine by 2013. (2014). *The statistical bulletin* (Pub. No. 03.3-5/178-14). Kyiv: The State Statistics Committee of Ukraine.

OpenMap. (n.d.). Retrieved April 20, 2014, from http://openmap.bbn.com

Orynski, F., & Pawlowski, W. (2004). Simulation and experimental research of the grinder's wheelhead dynamics. *Journal of Vibration and Control*, *10*(6), 915–930. doi:10.1177/1077546304041369

Pagell, M. (2004). Understanding the factors that enable and inhibit the integration of operations, purchasing and logistics. *Journal of Operations Management*, *22*(5), 459–487. doi:10.1016/j.jom.2004.05.008

Pagell, M., & Shevchenko, A. (2014). Why research in sustainable supply chain management should have no future. *Journal of Supply Chain Management*, *50*(1), 44–55. doi:10.1111/jscm.12037

Paish, O. (2002). Small hydro power: Technology and current status. *Renewable & Sustainable Energy Reviews*, *6*(6), 537–556. doi:10.1016/S1364-0321(02)00006-0

Pandey, D., Agrawal, M., & Pandey, J. S. (2011). Carbon footprint: Current methods of estimation. *Environmental Monitoring and Assessment*, *178*(1-4), 135–160. doi:10.1007/s10661-010-1678-y PMID:20848311

Papadopoulou, E. (2012). *Energy management in buildings using photovoltaics*. London: Springer. doi:10.1007/978-1-4471-2383-5

Paredis, E., Goeminne, G., Vanhove, W., Maes, F., & Lambrecht, J. (2009). *The concept of ecological debt: Its meaning and applicability in international policy*. Gent, Belgium: Academia Press.

Parfenenko, Yu. V., Okopnyi, R. P., & Nenja, V. G. (2012). Creation of control instruments for buildings' heating. *The Journal of the National Technical University NTU KPI. Series Information Systems and Networks*, *34*, 93–97.

Parfenenko, Yu. V., Shendryk, V. V., Nenja, V. G., & Okopnyi, R. P. (2013). Information-analysis system for monitoring and prediction of heat-supply buildings. *Visnyk of East Ukrainian National University named after V. Dahl*, *743*, 38–43.

Park, J. H., Park, I. Y., Kwun, Y. C., & Tan, X. (2011). Extension of the TOPSIS method for decision making problems under interval-valued intuitionistic fuzzy environment. *Applied Mathematical Modelling*, *35*(5), 2544–2556. doi:10.1016/j.apm.2010.11.025

Parra, C., Crespo, A., Cortés, P., & Fygueroa, S. (2006). On the consideration of reliability in the life cycle cost analysis (LCCA). In *A review of basic models: Safety and reliability for managing risk* (pp. 2203–2214). Taylor & Francis.

Pas, B. (2008). *2050:2008 specification for the assessment of the life cycle greenhouse gas emissions of goods and services*. British Standards Institution.

Patel, J. (2006). Green sky thinking. *Environment Business*, (122), 32.

Pavliuk, S. (2011). *The efficient national energy efficiency policy – Basis modernization of housing and communal services*. Retrieved from http://www.eu.prostir.ua/files/1341335344039/energy%20effeciencyPP_engl.pdf

Pedersen, E. R. G., Henriksen, M. H., Frier, C., Søby, J., & Jennings, V. (2013). Stakeholder thinking in sustainability management: The case of Novozymes. *Social Responsibility Journal*, *9*(4), 500–515. doi:10.1108/SRJ-08-2012-0101

Peglau, R. (2005). *ISO 14001 certification of the world*. Berlin: Federal Environmental Agency.

Pestana-Barros, C., & Sequeira-Antunes, O. (2011). Performance assessment of Portuguese wind farms: Ownership and managerial efficiency. *Energy Policy*, *39*(6), 3055–3063. doi:10.1016/j.enpol.2011.01.060

Peters, G., Minx, J., Weber, C., & Edenhofer, O. (2011). Growth in emission transfers via international trade from 1990 to 2008. *Proceedings of the National Academy of Sciences of the United States of America*, *108*(21), 8903–8908. doi:10.1073/pnas.1006388108 PMID:21518879

Pinson, P. (2013). Wind energy: Forecasting challenges for its operational management. *Statistical Science*, *28*(4), 564–585. doi:10.1214/13-STS445

Popp, A. (2000). Swamped in information but starved of data' information and intermediaries in clothing supply chains. *Supply Chain Management*, *5*(3), 151–161. doi:10.1108/13598540010338910

PostGIS. (n.d.). Retrieved April 20, 2014, from http://postgis.refractions.net

Prajogo, D., & Olhager, J. (2012). Supply chain integration and performance: The effects of long term relationships, information technology and sharing, and logistics integration. *International Journal of Production Economics*, *135*(1), 514–522. doi:10.1016/j.ijpe.2011.09.001

Preston, B. J. (2012). Benefits of judicial specialization in environmental law: The land and environment court of New South Wales as a case study. *Pace Environmental Law Review*, *29*(2), 396–440.

Price, T. J. (2005). James Blyth - Britain's first modern wind power engineer. *Wind Engineering*, *29*(3), 191–200. doi:10.1260/030952405774354921

Pricewaterhouse Coopers. (2012). *PwC analysis, sustainability reports of analysis sample 2009/2010*. Retrieved from www.pwc.com/my/en/publications/pwcalert97-sncc.jhtml

Pring, G., & Pring, C. (2009). *Greening justice: Creating and Improving environmental courts and tribunals*. Washington, DC: The Access Initiative.

Product Suite, P. S. S. (n.d.). Retrieved April 20, 2014, from http://www.simtec.cc/sites_en/sincal.asp/

Pusavec, F., Krajnik, P., & Kopac, J. (2010). Transitioning to sustainable production – Part I: Application on machining technologies. *Journal of Cleaner Production*, *18*(2), 174–184. doi:10.1016/j.jclepro.2009.08.010

Quantum, G. I. S. (n.d.). Retrieved April 20, 2014, from http://qgis.org/ru/site/

Quantz, L. (1924). *Wasserkraftmaschinen - Eine Einführung in Wesen, Bau und Berechnung neuzeitlicher Wasserkraftmaschinen und Wasserkraftanlagen*. Berlin: Verlag von Julius Springer.

Rahbari, O., Vafaeipour, M., Fazelpour, F., Feidt, M., & Rosen, M. A. (2014). Towards realistic designs of wind farm layouts: Application of a novel placement selector approach. *Energy Conversion and Management*, *81*, 242–254. doi:10.1016/j.enconman.2014.02.010

Ramachandra, T. V. (2009). RIEP: Regional integrated energy plan. *Renewable & Sustainable Energy Reviews*, *13*(2), 285–317. doi:10.1016/j.rser.2007.10.004

Ramirez-Rosado, I., Garcia-Garrido, E., Fernandez-Jimenez, L., Zorzano-Santamaria, P., Monteiro, C., & Miranda, V. (2008). Promotion of new wind farms based on a decision support system. *Renewable Energy*, *33*(4), 558–566. doi:10.1016/j.renene.2007.03.028

Rao, B. B., Tamazian, A., & Vadlamannati, K. C. (2011). Growth effects of a comprehensive measure of globalization with country specific time series data. *Applied Economics*, *43*(4-6), 551–568. doi:10.1080/00036840802534476

Rao, B. B., & Vadlamannati, K. C. (2011). Globalization and growth in the low income African countries with the extreme bounds analysis. *Economic Modelling*, *28*(3), 795–805. doi:10.1016/j.econmod.2010.10.009

Rao, B. K. N. (1996). *Handbook of condition monitoring*. Oxford, UK: Elsevier.

Rao, P., & Holt, D. (2005). Do green supply chains lead to competitiveness and economic performance? *International Journal of Operations & Production Management*, *25*(9), 898–916. doi:10.1108/01443570510613956

Ravichandran, N. (2008). Managing the bullwhip effect: Two case studies. *Journal of Advances in Management Research*, *5*(2), 77–87. doi:10.1108/09727980810949151

RelPro®, Reliability and Production, Analysis and Simulation. (2014). *RelPro SpA*. Retrieved April 10, 2014. http://www.relpro.pro

Renewable Energy Institute NAS of Ukraine. (n.d.). Retrieved April 20, 2014, from http://www.ive.org.ua/

Renewables 2013: Global status report. (2013). *REN 21 steering committee*. Retrieved from http://www.ren21.net/Portals/0/documents/Resources/GSR/2013/GSR2013_lowres.pdf

Renssen, S. (2011). Driving technology transfer. *Nature Climate Change*, *1*(6), 289–290. doi:10.1038/nclimate1193

Report, S. (2008, March). *Literature search on the potential health impacts associated with wind-to-energy turbine operations*. Retrieved from http://www.healthy.ohio.gov/~/media/ODH/ASSETS/Files/eh/HAS/windturbinehealthimpactreport.ashx

Reuters. (2012). *Intel, Apple leading charge against conflict minerals -report*. Retrieved November 23, 2013 from http://in.reuters.com/article/2012/08/16/conflictminerals-enoughproject-idINL2E8JF7HR20120816

Rigby, P., Fillon, B., Gombert, A., Ruenda, J., Kiel, E., Mellikov, E., . . . Warren, P. (2011). *Strategic energy technology plan: Photovoltaic technology, scientific assessment in support of the materials roadmap enabling low carbon energy technologies*. Retrieved March 10, 2014, from http://setis.ec.europa.eu/system/files/Scientific_Assessment_PV.pdf

Rimmer, M. (2011). *Intellectual property and climate change: Inventing clean technologies*. Cheltenham, UK: Edgar Elgar Publishing. doi:10.4337/9780857935885

Rockett, A. (2010). The future of energyn-photovoltaics. *Current Opinion in Solid State and Materials Science*, *14*(6), 117–122. doi:10.1016/j.cossms.2010.09.003

Roncero, J. R. (2008). Integration is key to smart grid management. *Smart Grids for Distribution, CIRED Seminar*, *9*, 1–4.

Rosenberger, C. (2012). Die policy of Ukraine in the power industry field (KAS Policy Paper 18). Retrieved from http://www.kas.de/wf/doc/kas_33444-1522-1-30.pdf?130206154511

Rosen, M. A., & Kishawy, H. A. (2012). Sustainable manufacturing and design: Concepts, practices and needs. *Sustainability*, *4*(2), 154–174. doi:10.3390/su4020154

Rossi, A. (2011). *L'architettura della città (IT)*. Macerata, Italy: Quodlibet.

Ryasnoy, D. (2010). *Energy becomes green: The introduction of «green» tariffs gave rise the development of alternative power generation in Ukraine*. Retrieved from http://www.ges-ukraine.com/maininfo_15-9.html

Safe Mode the Work of DEC Ukraine. (n.d.). Retrieved April 20, 2014, from http://www.ukrenergo.energy.gov.ua/ukrenergo/control/publish/category

Safe Setbacks: How Far Should Wind Turbines be from Homes ?. (2008, July 30). [Web log comment]. Retrieved from http://kirbymtn.blogspot.com/2008/07/safe-setbacks-how-far-should-wind.html

Salman, Z. (2014). *Bioenergy development in Malaysia*. BioEnergy Consult. Retrieved from http://www.bioenergyconsult.com/bioenergy-developments-malaysia/

Satterthwaite, D. (2008). Cities' contribution to global warming: Notes on the allocation of greenhouse gas emissions. *Environment and Urbanization*, *20*(2), 539–549. doi:10.1177/0956247808096127

Saunders, J. (2009). *The war over eco-labels*. Retrieved from http://www.greenbiz.com/blog/2009/04/28/war-over-eco-labels

Scheffer, C., & Girdhar, P. (2004). *Machinery vibration analysis & predictive maintenance*. Oxford, MA: Elsevier.

Schonberger, R. J. (2007). Japanese production management: An evolution with mixed success. *Journal of Operations Management*, *25*(2), 403–419. doi:10.1016/j.jom.2006.04.003

Scognamiglio, A. R. H., & Røstvik, H. N. (2013). Photovoltaics and zero energy buildings: A new opportunity and challenge for design. *Progress in Photovoltaics: Research and Applications*, *21*(6), 1319–1336. doi:10.1002/pip.2286

Scognamiglio, A., Ossenbrink, H., & Annunziato, M. (2011). Forms of cities for energy self-sufficiency., In *Proceedings UIA 2011, The XXIV World Congress of Architecture*. Tokyo: Academic Press.

Seidel, R. H. A., Shahbazpour, M., & Seidel, M. C. (2007, February). Establishing sustainable manufacturing practices in SMEs. In *Proceedings of the Second International Conference on Sustainability Engineering and Science*. Auckland, New Zealand: Academic Press.

Sen, Z. (2008). *Solar energy fundamentals and modeling techniques* (1st ed.). London: Springer.

Seuring, S., & Müller, M. (2008). From a literature review to a conceptual framework for sustainable supply chain management. *Journal of Cleaner Production*, *16*(15), 1699–1710. doi:10.1016/j.jclepro.2008.04.020

Shaw, K., Shankar, R., Yadav, S. S., & Thakur, L. S. (2013). Modeling a low-carbon garment supply chain. *Production Planning and Control*, *24*(8-9), 851–865. doi:10.1080/09537287.2012.666878

Sheu, C., Yen, H. R., & Chae, B. (2006). Determinants of supplier–retailer collaboration: Evidence from an international study. *International Journal of Operations & Production Management, 26*(1), 24–49. doi:10.1108/01443570610637003

Shevtsov, A. I., Barannik, V. O., Zemlyaniy, M. G., & Ryauzova, T. V. (2010). *State and perspectives of reforming the heat supply system in Ukraine*. Retrieved from http://www.db.niss.gov.ua/docs/energy/Teplozabezpechennya.pdf

Shuhong, L., Shangfeng, W., Michihiro, N., & Yulin, W. (2008). *Flow simulation and Performance prediction of a Kaplan Turbine*. Paper presented at the 4th international Symposium on Fluid Machinery and Fluid Engineering, Beijing, China.

Sinha, S., Mahesh, P., Donders, E., & Van Breusegem, W. (2011). *Waste electrical and electronic equipment: The EU and India: Sharing best practices*. Available: http://eeas.europa.eu/delegations/india/documents/eu_india/final_e_waste_book_en.pdf

Skjoett-Larsen, T. (2000). European logistics beyond 2000. *International Journal of Physical Distribution & Logistics Management, 30*(5), 377–387. doi:10.1108/09600030010336144

SMA. (2012). *SMA technology brochure 7, OptiTrac*. Retrieved March 10, 2014, from http://files.sma.de/dl/3491/TECHOPTITRAC-AEN082412.pdf

SMA. (2014). *Sunny portal*. Retrieved July 10, 2014, from http://www.sunnyportal.com/Templates/PublicPagesPlantList.aspx

SmartGrids Strategic Deployment Document Finalized. (n.d.). Retrieved April 20, 2014, from http://www.smartgrids.eu

Smith, D. (2001). *Reliability, maintainability, and risk: practical methods for engineers*. Newnes.

Smithers, J., & Blay-Palmer, A. (2001). Technology innovation as a strategy for climate adaptation in agriculture. *Applied Geography (Sevenoaks, England), 21*(2), 175–197. doi:10.1016/S0143-6228(01)00004-2

Snoeys, R. (1971). Principal parameters of grinding machine stability. *Ingenieursblad, 40*(4), 87–95.

Snoeys, R., & Brown, D. (1969). Dominating parameters in grinding wheel, and workpiece regenerative chatter. In *Proceedings of the 10th International MTDR Conference, Advance in Machine Tool Design and Research* (pp. 325-348). Manchester, UK: Academic Press.

Snoeys, R., & Wang, I. C. (1968). Analysis of the static and dynamic stiffnesses of the grinding wheel surface. In *Proceedings of the 9th MTDR Conference* (pp. 1133-1148). Birmingham, UK: MTDR.

Soleimani, H., & Govindan, K. (2014). Reverse logistics network design and planning utilizing conditional value at risk. *European Journal of Operational Research, 237*(2), 487–497. doi:10.1016/j.ejor.2014.02.030

Soliman, F., & Youssef, M. (2001). The impact of some recent developments in e-business on the management of next generation manufacturing. *International Journal of Operations & Production Management, 21*(5-6), 538–564. doi:10.1108/01443570110390327

Source Intelligence. (2012). *What are conflict minerals?* Retrieved November 19, 2013 from http://www.sourceintelligence.com/what-are-conflict-minerals

Srinivasan, K., Kekre, S., & Mukhopadhyay, T. (1994). Impact of electronic data inter change technology on JIT shipments. *Management Science, 40*(10), 1291–1305. doi:10.1287/mnsc.40.10.1291

Srivastava, S. K. (2007). Green supply-chain management: A state-of-the-art literature review. *International Journal of Management Reviews, 9*(1), 53–80. doi:10.1111/j.1468-2370.2007.00202.x

St. Gallen, U., & Fraunhofer, M. O. E. Z. (2012). *Creating a financial marketplace for IPR in Europe: Final report for European Commission to EU tender No 3/PP/ENT/CIP/10/A/NO2S003*. European Commission.

State Agency on Energy Efficiency & Energy Saving of Ukraine. (n.d.). Retrieved April 20, 2014, from http://saee.gov.ua/en/

Steger, M. (2009). *Globalization*. New York, NY: Sterling.

Stern, N. H. (2006). *Stern review: The economics of climate change* (vol. 30). London: HM Treasury.

Stevens, G. C. (1989). Integrating the supply chain. *International Journal of Physical Distribution & Materials Management*, *19*(8), 3–8. doi:10.1108/EUM0000000000329

Stigler, G. J. (1951). The division of labor is limited by the extent of the market. *Journal of Political Economy*, *59*(3), 185–193. doi:10.1086/257075

Stirling, W. C. (2003). *Satisficing games and decision making: With applications to engineering and computer science*. Cambridge, UK: Cambridge University Press. doi:10.1017/CBO9780511543456

Stock, G. N., Greis, N. P., & Kasarda, J. D. (2000). Enterprise logistics and supply chain structure: The role of fit. *Journal of Operations Management*, *18*(5), 531–547. doi:10.1016/S0272-6963(00)00035-8

Stock, J. R., Boyer, S. L., & Harmon, T. (2010). Research opportunities in supply chain management. *Journal of the Academy of Marketing Science*, *38*(1), 32–41. doi:10.1007/s11747-009-0136-2

Strategic Approach to International Chemicals Management. (2009). *Background information in relation to the emerging policy issue of electronic waste*. Paper presented at the SAICM/ICCM.2/INF36 Disseminated at the International Conference on Chemicals Management, Geneva, Switzerland.

Strategic Environmental Review. Renewable Energy Scenarios. Interconnection and Transmission Considerations in Renewable Energy Development in Ukraine. (2013). Retrieved from http://www.uself.com.ua/fileadmin/uself-ser-en/3/E%20-%20Transmission.pdf

Strong, S. (2010). *Building integrated photovoltaics (BIPV), whole building design guide*. Retrieved November 11, 2014, from http://www.wbdg.org/resources/bipv.php

Šúri, M., Huld, T., Dunlop, E., & Ossenbrink, H. (2007). Potential of solar electricity generation in the European Union member states and candidate countries. *Solar Energy*, *81*(10), 1295–1305. doi:10.1016/j.solener.2006.12.007

Sustainable Energy for All (SE4ALL). (2013). Retrieved April 20, 2014, from http://data.worldbank.org/data-catalog/sustainable-energy-for-all

Swafford, P. M., Ghosh, S., & Murthy, N. (2008). Achieving supply chain agility through IT integration and flexibility. *International Journal of Production Economics*, *116*(2), 288–297. doi:10.1016/j.ijpe.2008.09.002

Swaminathan, J. M., & Tayur, S. R. (2003). Models for supply chains in e-business. *Management Science*, *49*(10), 1387–1406. doi:10.1287/mnsc.49.10.1387.17309

Takai, M., Martin, J., & Bagrodia, R. (2001). Effects of wireless physical layer modeling in mobile ad hoc networks. In *Proceedings of the 2nd ACM International Symposium on Mobile Ad Hoc Networking & Computing*. New York, NY: ACM. doi:10.1145/501426.501429

Tan, C. L. (2013, December). A warming trend. *Ecowatch: The Star Publication*, 5.

Tapia-Garcí, A., Del Moral, J. M., Martí, M. J., Nez, M. A., & Herrera-Viedma, E. (2012). A consensus model for group decision-making problems with interval fuzzy preference relations. *International Journal of Information Technology & Decision Making*, *11*(04), 709–725. doi:10.1142/S0219622012500174

Tchangani, A. (2013). Bipolarity in decision analysis: A way to cope with human judgment. In Exploring innovative and successful applications of soft computing. Granada, Spain: IGI Global.

Tchangani, A., Bouzarour-Amokrane, Y., & Pérès, F. (2012). Evaluation model in decision analysis: Bipolar approach. *Informatica*, *23*(3), 461–485.

Testa, F., & Iraldo, F. (2010). Shadows and lights of GSCM (green supply chain management): Determinants and effects of these practices based on a multi-national study. *Journal of Cleaner Production*, *18*(10), 953–962. doi:10.1016/j.jclepro.2010.03.005

The NS-3 Consortium. (n.d.). Retrieved April 20, 2014, from https://www.nsnam.org/

The Official Website Company Bioresurce Ukraine. (n.d.). Retrieved April 20, 2014, from http://bioresource.pro/index.php/uslugi

The Official Website Consulting Company NIKO Analytical Section. (2013). Retrieved from http://www.kua.niko.ua/wp-content/uploads/2010/07/Sector_Review_Energy_2011_full-v_ENG.pdf

The QualNet Communications Simulation Platform. (n.d.). Retrieved April 20, 2014, from http://web.scalable-networks.com/content/qualnet

Thomsen, K., Schepers, G., & Fuglsang, P. (2001). Potentials for site-specific design of mw sized wind turbines. *Journal of Solar Energy Engineering*, *123*(4), 304–309. doi:10.1115/1.1408611

Thurlow, L. C. (1997). Needed a new system of IP rights. *Harvard Business Review*, *75*(5), 94–103. PMID:10170334

Tiba, C., Candeias, A. L. B., Fraidenraich, N., Barbosa, E. M. S., & de Carvalho Neto, P. B. (2010). A GIS-based decision support tool for renewable energy management and planning in semi-arid rural environments of northeast of Brazil. *Renewable Energy*, *35*(12), 2921–2932. doi:10.1016/j.renene.2010.05.009

Tietze, F., & Herstatt, C. (2010). *Technology market intermediaries and innovation.* Paper presented at DRUID Summer Conference, London, UK.

Tisue, S., & Wilensky, U. (2004). NetLogo: A simple environment for modeling complexity. In *Proceedings of the International Conference on Complex Systems.* Boston: NECSI Knowledge Press.

Tonshoff, H. K., Peters, J., Inasaki, I., & Paul, T. (1992). Modelling and simulation of grinding processes. *CIRP Annals - Manufacturing Technology*, *41*(2), 677–688.

Tranfield, D., Denyer, D., & Smart, P. (2003). Towards a methodology for developing evidence-informed management knowledge by means of systematic review. *British Journal of Management*, *14*(3), 207–222. doi:10.1111/1467-8551.00375

Tsai, M. (2007). Does globalization affect human well-being? *Social Indicators Research*, *81*(1), 103–126. doi:10.1007/s11205-006-0017-8

Tummala, V. R., Phillips, C. L., & Johnson, M. (2006). Assessing supply chain management success factors: A case study. *Supply Chain Management: An International Journal*, *11*(2), 179–192. doi:10.1108/13598540610652573

Tysiac, K. (2012). *Build a strong team to comply with conflict minerals rule.* Retrieved November 23, 2013 from http://www.cgma.org/magazine/news/pages/20126875.aspx

Tysiac, K. (2012). *Highly scrutinized SEC conflict mineral regs. include new audit requirement.* Retrieved November 23, 2013 from http://www.journalofaccountancy.com/News/20126307.htm

Tysiac, K. (2013). *Conflict minerals rule poses compliance challenge.* Retrieved November 23, 2013 from http://www.journalofaccountancy.com/Issues/2013/Apr/20127083.htm

Ulaga, W., & Eggert, A. (2006). Value-based differentiation in business relationships: Gaining and sustaining key supplier status. *Journal of Marketing*, *70*(1), 119–136. doi:10.1509/jmkg.2006.70.1.119

Ullmer, E. (1995). *Theoretische Validierung einer Prüfmethode für solare Brauchwasserwärmungsanlage.* University of Stuttgart.

UN. (2009b). *UN-DESA policy brief no. 25.* United Nations Department of Economic and Social Affairs. Accessed on November 1, 2014 at http://www.un.org/esa/ analysis/policy briefs/policy brief 25.pdf

UNE-EN 1151. (1999). *Bombas. Bombas rotodinámicas. Bombas de circulación cuyo consumo de energía no excede de 200 W, destinadas a la calefacción central y a la distribución de agua caliente sanitaria doméstica. Requisitos, ensayos, marcado.* International Standard.

UNE-EN 12976-1. (2006). *Sistemas solares térmicos y sus componentes. Sistemas prefabricados. Parte 1: Requisitos generales.* International Standard.

UNE-EN 1489. (2001). *Válvulas para la edificación. Válvulas de seguridad. Ensayos y requisitos.* International Standard.

UNE-EN 1490. (2001). *Válvulas para la edificación. Válvulas de alivio de presión y temperatura. Ensayos y requisitos.* International Standard.

UNE-EN 1717. (2001). *Protección contra la contaminación de agua potable en las instalaciones de agua y requisistos generales de los dispositivos para evitar la contaminación por reflujo.* International Standard.

UNE-EN 307. (1999). *Intercambiadores de calor. Directrices para elaborar las instrucciones de instalación. Funcionamiento y mantenimiento, necesarias para mantener el rendimiento de cada uno de los tipos de intercambiadores de calor.* International Standard.

UNE-EN 60335-1/A15. (2001). *Seguridad de los aparatos electrodomésticos y análogos. Parte 1: Requisitos generales*. International Standard.

UNE-EN 60335-2-21/A1. (2001). *Seguridad de los aparatos electrodomésticos y análogos. Parte 2-21: Requisitos particulares para los termos eléctricos*. International Standard.

UNE-EN 806-1. (2001). *Especificaciones para instalaciones de conducción de agua destinada al consumo humano en el interior de edificios. Parte 1: Generalidades*. International Standard.

UNE-EN 809. (1999). *Bombas y grupos motobombas para líquidos. Requisitos comunes de seguridad*. International Standard.

UNE-ENV 1991-2-3. (1998). *Eurocódigo 1: Bases de proyecto y acciones en estructuras. Parte 2-3: Acciones en estructuras. Cargas de nieve*. International Standard.

UNEP. (2009c). Making Tourism More Sustainable: A Guide for Policy Makers.

UNEP. (2011b). *A guidebook to the green economy issue 1: Green economy, green growth, and low-carbon development – History, definitions and a guide to recent publications division for sustainable development*. UN-DESA. Retrieved from https://sustainabledevelopment.un.org/content/.../GE%20Guidebook.pdf

UNEP. (2011b). *Keeping Track of our Changing Environment: From Rio to Rio+20*. UNEP.

UNFCCC. (2009). *United Nations climate change conference*. Retrieved from https://unfccc.int/files/meetings/durban_nov_2011/statements/application/pdf/111207_cop17_hls_malaysia.pdf

United Nations Department of Economic and Social Affairs. (2009b). *UN-DESA policy brief no. 25: The challenges of adapting to a warmer planet for urban growth and development*. Retrieved from http://www.un.org/en/development/desa/policy/publications/policy_briefs/policybrief25.pdf

United Nations Development Programme. (2011). *Human development report 2010: The real wealth of nations: Pathways to human development: Summary*. Retrieved from http://hdr.undp.org/sites/default/files/hdr_2010_en_summary.pdf

United Nations Environment Programme Paris UNCTAD. (2012c). *Investment policy framework for sustainable development*. Retrieved from 12 November 2014 from http://unctad.org/en/PublicationsLibrary/webdiaepcb2012d6_en.pdf

United Nations Environment Programme. (2009a). World population ageing 2009. New York, NY: Author.

United Nations Environment Programme. (2011a). Keeping track of our changing environment from Rio to Rio + 20 (1992-2012). New York, NY: Author.

United Nations Environment Programme. (2013). *Global chemicals outlook: Towards sound management of chemicals: Trends and changes*. Retrieved from http://www.unep.org/chemicalsandwaste/Portals/9/Mainstreaming/GCO/Rapport_GCO_calibri_greendot_20131211_web.pdf

Valverde, A. (1994). *Análisis de la disponibilidad de los equipos dinámicos y su incidencia en el mantenimiento en plantas industriales*. (Thesis). UNED, Madrid, Spain.

Van der Vaart, T., & Van Donk, D. P. (2008). A critical review of survey-based research in supply chain integration. *International Journal of Production Economics*, *111*(1), 42–55. doi:10.1016/j.ijpe.2006.10.011

Vanalle, R. M., Lucato, W. C., & Santos, L. B. (2011). Environment requirements in the automotive supply chain: An evaluation of a first tier company in the Brazilian auto industry. *Procedia Environmental Science*, *10*, 337–343. doi:10.1016/j.proenv.2011.09.055

Vance, A. (2012, August 9). Patents - Startups' new creed: Patent first, prototype later. *Bloomberg Businessweek*. Retrieved from http://www.bloomberg.com/bw/articles/2012-08-09/startups-new-creed-patent-first-prototype-later

Varga, A., & Hornig, R. (2008). An overview of the OMNET++ simulation environment. In *Proceedings of the 1st International Conference on Simulation Tools and Techniques for Communications, Networks and Systems*. Marseille, France: ACM Digital Library. doi:10.4108/ICST.SIMUTOOLS2008.3027

Veber, B., Nagode, M., & Fajdiga, M. (2008). Generalized renewal process for repairable systems based on finite Weibull mixture. *Reliability Engineering & System Safety*, *93*(10), 1461–1472. doi:10.1016/j.ress.2007.10.003

Ventures, O. (2011). *Understanding e-waste*. Retrieved from: http://lasepa.org/E-waste%20Day%201/understanding%20e waste%20final%20of%20all.pdf

Ventur, R., Scott Brown, D., & Izenour, S. (1977). *Learning from Las Vegas: The forgotten symbolism of architectural form*. Cambridge, MA: MIT Press.

Veugelers, R. (2011). Europe's clean technology investment challenge. *Bruegel Policy Contribution*. Retrieved from http://www.bruegel.org/download/parent/561-europes-clean-technology-investment-challenge/file/1408-europes-clean-technology-investment-challenge/

Vickery, S. K., Jayaram, J., Droge, C., & Calantone, R. (2003). The effects of an integrative supply chain strategy on customer service and financial performance: An analysis of direct versus indirect relationships. *Journal of Operations Management*, *21*(5), 523–539. doi:10.1016/j.jom.2003.02.002

Vinodh, S., & Joy, D. (2012). Structural equation modeling of sustainable manufacturing practices. *Clean Technologies and Environmental Policy*, *14*(1), 79–84. doi:10.1007/s10098-011-0379-8

Viñolas, J., Biera, J., Nieto, J., Llorente, J. I., & Vigneau, J. (1997). The use of an efficient and intuitive tool for the dynamic modelling of grinding processes. *CIRP Annals - Manufacturing Technology*, *46*(1), 239-252.

Vintr, M. (2007). *Reliability assessment for components of complex mechanisms and machines*. Paper presented at 12th IFToMM World Congress, Besançon, France.

Voss, K., & Musal, E. (2012). Net zero energy buildings - International projects on carbon neutrality in buildings (2nd ed.). Institut für Internationale Architektur-Dokumentation.

Voss, K., Kiefer, K., Reise, C., & Meyer, T. (2003). Building energy concepts with photovoltaics—concept and examples from Germany. *Advances in Solar Energy*, *15*, 235–259.

Wackernagel, M., & Rees, W. (1998). *Our ecological footprint: Reducing human impact on the earth (No. 9)*. New Society Publishers.

Wakeley, H. L., Hendrickson, C. T., Griffin, W. M., & Matthews, H. S. (2009). Economic and environmental transportation effects of large-scale ethanol production and distribution in the United States. *Environmental Science & Technology*, *43*(7), 2228–2233. doi:10.1021/es8015827 PMID:19452867

Walby, S. (2009). *Globalization & inequalities: Complexity and contested modernities*. London: Sage.

Walton, S. V., Handfield, R. B., & Melnyk, S. A. (1998). The green supply chain: Integrating suppliers into environmental management processes. *Journal of Supply Chain Management*, *34*(2), 2–11.

Weber, R. (1990). *Basic content analysis* (2nd ed.). Newbury Park, CA: Sage.

West Tennessee Solar Farm. (n.d.). *Homepage*. Retrieved from http://solarfarm.tennessee.edu/

Wiedmann, T. (2009). *Editorial: Carbon footprint and input–output analysis–An introduction*. Academic Press.

Wiedmann, T., & Minx, J. (2008). A definition of 'carbon footprint'. *Ecological Economics Research Trends*, *1*, 1-11.

Wiedmann, T. O., Lenzen, M., & Barrett, J. R. (2009). Companies on the scale. *Journal of Industrial Ecology*, *13*(3), 361–383. doi:10.1111/j.1530-9290.2009.00125.x

Wijnmalen, D. J. D. (2007). Analysis of benefits, opportunities, costs, and risks (BOCR) with the AHP-ANP: A critical validation. *Mathematical and Computer Modelling*, *46*(7-8), 892–905. doi:10.1016/j.mcm.2007.03.020

Wilhelm, M. M. (2011). Managing coopetition through horizontal supply chain relations: Linking dyadic and network levels of analysis. *Journal of Operations Management*, *29*(7), 663–676. doi:10.1016/j.jom.2011.03.003

Wilhelm, S. (1988). Galvanic corrosion caused by corrosion products. In H. Hack (Ed.), *Galvanic corrosion*. Philadelphia: ASTM. doi:10.1520/STP26189S

Wilkof, N. (2012, August 21). *So which is it for a start-up: A patent or a proto-type?* [Blog post]. Retrieved from http://ipfinance.blogspot.de/2012/08/so-which-is-it-for-start-up-patent-or.html

Williams, E., & Tagami, T. (2002). Energy use in sales and distribution via e-commerce and conventional retail: A case study of the Japanese book sector. *Journal of Industrial Ecology*, *6*(2), 99–114. doi:10.1162/108819802763471816

World Wildlife Fund. (2010). *Living planet report 2010: Biodiversity, biocapacity and development*. Gland, Switzerland: Author.

World Wind Energy Association (WWEA). (2012). *Half-year technical report*. Retrieved from http://www.wwindea.org/webimages/Half-year report 2012.pdf

WRI/WBCSD. (2004). *The greenhouse gas protocol: A corporate accounting and reporting standard* (revised edition). Geneva: World Business Council for Sustainable Development and World Resource Institute.

WRI/WBCSD. (2005). *The greenhouse gas protocol: Project accounting*. Geneva: World Business Council for Sustainable Development and World Resource Institute.

Wright, B., & Shih, T. (2010). *Agricultural innovation*. NBER Working Paper 15793. Cambridge, MA: National Bureau of Economic Research.

WTO-UNEP. (2009). *Trade and climate change*. WTO-UNEP Report. Available at: http://unfccc.int/files/adaptation/adverse_effects_and_response_measures_art_48/application/pdf/ part_iv_trade_and_climate_change_report.pdf

Wu, F., Yeniyurt, S., Kim, D., & Cavusgil, S. T. (2006). The impact of information technology on supply chain capabilities and firm performance: A resource-based view. *Industrial Marketing Management*, *35*(4), 493–504. doi:10.1016/j.indmarman.2005.05.003

Wu, Y., Geng, S., Xu, H., & Zhang, H. (2014). Study of decision framework of wind farm project plan selection under intuitionistic fuzzy set and fuzzy measure environment. *Energy Conversion and Management*, *87*, 274–284. doi:10.1016/j.enconman.2014.07.001

Wu, Y., Li, Y., Ba, X., & Wang, H. (2013). Post-evaluation indicator framework for wind farm planning in China. *Renewable & Sustainable Energy Reviews*, *17*, 26–34. doi:10.1016/j.rser.2012.09.013

Yanaglsawa, T., & Guellee, D. (2009). *The emerging patent marketplace*. OECD Science, Technology and Industry Working Paper, 2009/9. Paris: OECD Publishing.

Yanez, M., Joglar, F., & Modarres, M. (2002). Generalized renewal process for analysis of repairable system with limited failure experience. *Reliability Engineering & System Safety*, *77*(2), 167–180. doi:10.1016/S0951-8320(02)00044-3

Yassin, A., Yebesi, F., & Tingle, R. (2005). *Occupational exposure to crystalline silica dust in the United States: 1988–2003*. Academic Press.

Yi, Ch.-Y., Lee, J.-H., & Shim, M.-P. (2010). Site location analysis for small hydropower using geo-spatial information system. *Renewable Energy*, *35*(4), 852–861. doi:10.1016/j.renene.2009.08.003

Yuan, C., Zhai, Q., & Dornfeld, D. (2012). A three dimensional system approach for environmentally sustainable manufacturing. *CIRP Annals-Manufacturing Technology*, *61*(1), 39–42. doi:10.1016/j.cirp.2012.03.105

Zahedi, A. (2006). Solar photovoltaic (PV) energy latest developments in the building integrated and hybrid PV systems. *Renewable Energy*, *31*(5), 711–718. doi:10.1016/j.renene.2005.08.007

Zailani, S., Jeyaraman, K., Vengadasan, G., & Premkumar, R. (2012). Sustainable supply chain management (SSCM) in Malaysia: A survey. *International Journal of Production Economics*, *140*(1), 330–340. doi:10.1016/j.ijpe.2012.02.008

Zerkalov, D. V. (2012). *Energy saving in Ukraine*. Kyiv: Osnova.

Zhao, X., Huo, B., Flynn, B. B., & Yeung, J. (2008). The impact of power and relationship commitment on the integration between manufacturers and customers in a supply chain. *Journal of Operations Management*, *26*(3), 368–388. doi:10.1016/j.jom.2007.08.002

Zhu, Q., & Cote, R. P. (2004). Integrating green supply chain management into an embryonic eco-industrial development: A case study of the Guitang Group. *Journal of Cleaner Production*, *12*(8), 1025–1035. doi:10.1016/j.jclepro.2004.02.030

Zhu, Q., Geng, Y., Fujita, T., & Hashimoto, S. (2010). Green supply chain management in leading manufacturers: Case studies in Japanese large companies. *Management Research Review*, *33*(4), 380–392. doi:10.1108/01409171011030471

Zhu, Q., Sarkis, J., & Geng, Y. (2005). Green supply chain management in China: Pressures, practices and performance. *International Journal of Operations & Production Management*, *25*(5), 449–468. doi:10.1108/01443570510593148

Zhu, Q., Sarkis, J., & Lai, K.-. (2014). Green supply chain management: Pressures, practices and performance within the Chinese automobile industry. *Journal of Cleaner Production*, *15*(11-12), 1041–1052. doi:10.1016/j.jclepro.2006.05.021

Zou, S., & Cavusgil, S. T. (2002). The GMS: A broad conceptualization of global marketing strategy and its effect on firm performance. *Journal of Marketing*, *66*(4), 40–56. doi:10.1509/jmkg.66.4.40.18519

Zsidisin, G. A., & Siferd, S. P. (2001). Environmental purchasing: A framework of theory development. *European Journal of Purchasing and Supply Management*, *7*(1), 61–73. doi:10.1016/S0969-7012(00)00007-1

Zucatto, L. C. (2008). *Inovações em processos como uma forma de estruturar uma cadeia de suprimentos sustentável: são possíveis?* Rio de Janeiro: XXVIII Encontro Nacional de Engenharia de Produção.

About the Contributors

Vicente González-Prida Díaz is a PhD in Industrial Engineering by the University of Seville, and Executive MBA by the Chamber of Commerce. He currently works as Program Manager in the company General Dynamics – European Land Systems. He shares his professional performance with the development of research projects in the Department of Industrial Organization and Management at the University of Seville. He has written multitude of articles for conferences and publications. His main interest is related to industrial asset management, specifically the reliability (design specification, data analysis), maintenance (outsourcing, e-maintenance), and sales management (cost analysis, logistics, organization). Recently, he has published the book *After-Sales Service of Engineering Industrial Assets* from Springer-Verlag.

Anthony Raman is an Academic Staff Member at NTEC Tertiary Group, New Zealand. Has extensive experience within multicultural, bicultural, international and indigenous environments focusing in education & training and international business development. In the past, he held managerial and senior positions in two large public educational institutions in New Zealand and undertaken international business consultancy. He has an interest in transdisciplinary approach in the areas of management, marketing and international business. His qualifications are a postgraduate qualification in Marketing, Associateship of the Chartered Institute of Arbitrators, Member of the Chartered Institute of Marketing and a postgraduate level Graduate Certificate in Research Methods (Australia). Apart from being a Chartered Marketer with the Chartered Institute of Marketing (UK), he is an elected Council Member and a Registered Professional Marketer with the Canadian Institute of Marketing.

* * *

Alia Nadhirah Ahmad Kamal is a final year student in Universiti Sains Malaysia (USM) pursuing Master of Business Administration (MBA) with a concentration in Sustainable Development. She received her Bachelor's Degree in Applied Language Studies (English for Professional Communication) with a minor in Business Management from Universiti Teknologi MARA, Malaysia. In year 2011, Alia was awarded with 'Tan Sri Arshad Ayub' Excellent Award by the 13th Yang di-Pertuan Agong Tuanku Mizan Zainal Abidin, for her contribution in the society and excellent academic achievement. Being highly involved in the Graduate School of Business Club is an advantage when Alia was chosen as The Most Outstanding Student Award by Graduate School of Business, USM. She was also among the two Malaysian students selected for One More Step, Erasmus Mundus Student Exchange Scholarship Program which enabled her to experience studying in Europe for six months.

David Almorza was born in Cádiz (Spain). He obtained the Mathematic Degree at the University of Sevilla (Spain) and the phd at the University of Cádiz. Since 1991. Dr. Almorza currently works as a lecturer of the Statistical Department at the University of Cádiz. In 2003. He was appointed as Vice-Rector of the University of Cádiz and still continues to develop that role for the University. Dr. Almorza is a researcher focused on engineering statistic and genetics' statistic. His main collaborations in these areas involve the Wessex Institute of Technology (UK) and Genetic Institute Ewald Favret (Argentina). On the engineering statistics side, He is co-author of several papers published, among others, in *Mechanical Systems and Signal Processing, International Journal of Advanced Manufacturing Technology*, and ESAIM: P&S (*European Series in Applied and Industrial Mathematics: Probability and Statistics*). On the genetics' statistics area, Dr. Almorza is co-author of several papers published, among others, in *BAG: Journal of Basic & Applied Genetics, Neotropical Entomology, and Human Heredity*.

Yasmina Bouzarour-Amokrane received engineer degree from Ecole Nationale Polytechnique d'Alge, Algeria, a master of science degree in industrial engineering from Ecole Natinale Supérieure des Mines de Saint Etienne, France, and PhD degree from Toulouse Université, France. Her researches revolve around decision analysis and uncertainty modeling. She is actually temporarily associated to teaching and research at Institue Nationale Polytechnique de Toulouse.

María Carmen Carnero is an associate professor (with tenure) in the Technical School of Industrial Engineering of the University of Castilla-La Mancha (Spain). She received her PhD degree from the University of Castilla-La Mancha in 2001. She has published research articles in *Omega, Decision Support System, European Journal of Operational Research, Reliability Engineering and System Safety, Mechanical Systems and Signal Processing, Journal of Manufacturing Systems, Production Planning and Control,* etc., and has participated in some research projects supported by European Union and National and Regional Administration. Her research interests are multiple criteria decision making, predictive maintenance and performance evaluation applied to maintenance.

Adolfo Crespo Márquez is currently Full Professor at the School of Engineering of the University of Seville, and Head of the Department of Industrial Management. He holds a PhD in Industrial Engineering from this same University. His research works have been published in journals such as the *International Journal of Production Research, International Journal of Production Economics, European Journal of Operations Research, Journal of Purchasing and Supply Management, International Journal of Agile Manufacturing, Omega, Journal of Quality in Maintenance Engineering, Decision Support Systems, Computers in Industry, Reliability Engineering and System Safety,* and *International Journal of Simulation and Process Modeling*, among others. Prof. Crespo is the author of seven books, the last four with Springer-Verlag in 2007, 2010, 2012, and 2014 about maintenance, warranty, and supply chain management. Prof. Crespo leads the Spanish Research Network on Dependability Management and the Spanish Committee for Maintenance Standardization (1995-2003). He also leads a research team related to maintenance and dependability management currently with 5 PhD students and 4 researchers. He has extensively participated in many engineering and consulting projects for different companies, for the Spanish Departments of Defense, Science and Education as well as for the European Commission (IPTS). He is the President of INGEMAN (a National Association for the Development of Maintenance Engineering in Spain) since 2002.

About the Contributors

Yudi Fernando is a Senior Lecturer in Operations Management at the Graduate School of Business, Universiti Sains Malaysia (USM). He received his PhD from the School of Management at USM. Dr. Yudi also worked for several years in the electronics industry. His research focus is in the area of sustainability supply chain, green logistics, green production and service management. Dr. Yudi currently teaches operations strategy: sustainability and supply chain management, business research methodology, advance business statistics subjects at MBA/DBA levels. He has also been supervising several PhD/DBA students with research topics on operations and supply chain management. He has been a member of editorial boards for International Journal of Operations Research and Information Systems (IJORIS). Dr. Yudi has actively participated as an expert reviewer for international journals and conferences. Thus far, he has published in several journals such *International Journal of Services and Operations Management, International Journal of Applied Logistics, Food Control, International Journal of Information Management, International Journal Logistics Systems and Management, International Journal of Operations and Supply Chain Management, International Journal of Productivity and Quality Management, International Journal of Value Chain Management*, and so on.

Rafael González-Palma, Professor in Mechanical Engineer 1987, Dr. in Science Physics 1994, Naval technical engineer1966. 27 years of experience as engineer in companies. Thirty years of experience as professor and investigator, over 35 JCR publication in reviews, taking part in many congress.

Shahul Hameed has been a lecturer for last two decades in India/Malaysia/New Zealand. Currently tutoring social work in Social Work with Te Wananga o Aotearoa (TWoA), Hamilton, New Zealand. His research interest lies in multi-disciplinary subjects in Social Sciences. He is an Immigrant to New Zealand and settled with his spouse and four children.

George Ch. Ioannidis was born in Athens, Greece, in 1970. He received the M.Sc. degree and the Ph.D. degree in electrical engineering both from National Technical University of Athens in 1993 and 1998 respectively. He is currently Assistant Professor at the Piraeus University of Applied Sciences/Faculty of Technological Applications/Dept. of Electrical Engineering. His main research interests include dc to dc converters, ac to dc converters, robust control of power converters, converters for fuel cells, photovoltaic systems and power quality.

Stavros D. Kaminaris received the Diploma and Ph.D. in Electrical Engineering from NTUA in 1989 and 1999 respectively. He is currently Associate Professor at the Piraeus University of Applied Sciences/Dept. of Electrical Engineering. His research interests include applications of AI techniques/intelligent based systems/fuzzy logic & expert systems applications on power systems, RES applications, smart electrical installations, intelligent control and decision making, energy saving & management systems and energy storage.

Kurtar Kaur is a currently doing her MBA in Universiti Sains Malaysia. She recieved her BBA from Open Universiti Malaysia. She has 32+ years working in the semiconductor manufacturing industry. She currently resides in Kedah, Malaysia with her husband and 2 daughters. Her areas of interest are sustainability and conservation.

Fredy Kristjanpoller, Researcher and Academic at Technical University Federico Santa María, Chile, has been active in national and international research, both for important journals and conferences. He has developed projects in important Chilean and International industries. Consultant specialist in the area of Reliability, Asset Management, System Modeling, and Evaluation of Engineering Projects. He is Industrial Engineer and Master in Asset Management and Maintenance.

Carlos López-Escobar in an engineer with working experience in large multinational organizations including Delphi automotive, Valeo automotive, and Alcoa. He holds a consolidated international management experience with exposure to different businesses and processes, including automotive, aerospace, food packaging, primary metals, rolling products, engineered products, and others. Most of his background is on Operational Excellence (OpEx) which includes strategy deployment, leading change, maintenance, production, operations, quality, process/quality management and other aspects related to operations. Dr. López Escobar has been exposed to demanding customers, including Toyota, VW group, PSA group, Nissan, the Coca-Cola company, Pepsi, and others. He is an experienced OpEx internal auditor and consultant. Dr. López-Escobar has lead Operational Excellence assessments mainly in Europe, Asia and Australia. He is Co-Author of articles published in technical magazines and a experienced OpEx trainer. His quality management experience has consisted in roles as Alcoa CSI regional Quality manager for Europe and North Africa operations. In addition, he held positions as quality Director in Valeo automotive systems. Previous positions in Delphi automotive included maintenance management, process engineering and production supervisory roles.

Lutz Maicher is head of the technology transfer research group at the computer science department of Jena University, Germany. He is computer scientist (PhD) by training with a strong background in the field of innovation, technologies and intellectual property. Besides his position at Jena University, he is head of the research group "Competitive Intelligence" at the Fraunhofer MOEZ, Leipzig. His research area is the usage of data analytics techniques (data and text mining; as well visual analytics) for competitive intelligence applications. Before joining Fraunhofer he was head of the Topic Maps Lab at University of Leipzig, which was dedicated to flexible data integration and analysis.

Pedro Mayorga is a researcher with a background in Engineering and experience in both industry and academia. He has been working as researcher in energy for more than 15 years, participating in or coordinating more than forty research projects. As entrepreneur he joined others to establish EnerOcean, a pioneer marine energy services company.

Warren Naylor has over 38 years of system engineering, software engineering, field engineering, and system safety engineering experience. Mr. Naylor started his career in system safety engineering in 1993 while at BAE Systems where he became the North America Lead System Safety Engineer. In 2004 Mr. Naylor moved to Northrop Grumman Electronic Systems in Baltimore serving currently as the System Safety Senior Consultant Engineer. In this capacity, Mr. Naylor is technically responsible for all NGES System Safety Programs is centered out of corporate headquarters in Baltimore. Mr. Naylor is known for his innovative solutions to the emerging issues facing system safety and has published in excess of over 15 foundation system safety technical papers on subjects including Commercial-Off-the-Shelf (COTS), programmatic risk, humans and their impact on safety, software safety and reliability, cost

About the Contributors

and schedules negative impact on system safety among others. He has also collaborated in the writing of many Government Papers and Guidance including being a contributing author of RTCA DO-278 and ANSI-STD-0010. Mr. Naylor is Fellow Member and a Past President of the International System Safety Society (ISSS) a member of INCOSE, and an Associate Editor for RAMS. Mr. Naylor also instructs several system safety classes. EDUCATION Bachelor of Science Information Systems Management, Potomac College, Washington, DC.

Bartosz M. Olszański, M. Sc., Eng. Graduated from Warsaw University of Technology in 2013, majoring in aerospace engineering. Currently a PhD student at the Department of Aerodynamics, IAAT WUT. His professional experience includes sixteen-month work on A350XWB project while performing a static and fatigue stress analysis for metalic and composite elements of secondary wing structure and optimization & experimental studies of a new type of laminar airfoil. His main scientific interests are missile aerodynamics, experimental fluid mechanics, hydrodynamics, macro and micro-economic analysis. In his free time, he is a host of a radio broadcast.

Yuliia Parfenenko is a researcher of computer science at the Sumy State University, Sumy, Ukraine. She has a Bachelor of Computer science from Sumy State University in 2007. She received her Master's Degree of Information Technology of Design (Diploma with distinction) in 2008. From November 2008 till November 2011 she was a PhD student at the Sumy State University. Research direction is creation information technologies for energy management in district heating. Additional information you can find on the personal webpage: http://itp.elit.sumdu.edu.ua/index.php/home/staff/38-parfenenko-yuliya-viktorivna.

François Pérès is a French engineer graduated from the ENIT (Ecole Nationale d'Ingénieurs de Tarbes). After obtaining a PhD degree in Bordeaux University, he entered the Ecole Centrale of Paris as assistant professor where he spent five years working in the Industrial Engineering Laboratory. Head of the research group dealing with robust design and dependability activities he was also, at that time, director of a technology transfer center specialized in prototyping. Following a research supervisor accreditation he joined the Institut National Polytechnique and is currently full professor at Toulouse University, working inside the Laboratoire Génie de Production as leader of the Uncertainty Risk and Decision group. His research deals with risk management applied mainly to the design and maintenance strategies of manufactured products. Lately some extensions of his work have concerned human factor-based risk assessment approach. François Peres has been author or co-author of about 80 articles, 30 of them in international journals indexed in Isiweb or Scopus and director or co-director of 11 PhD students. He is involved in various national and international projects, member of several scientific committees and affiliated to international research communities like IFAC, IFIP, or IMDR.

Constantinos S. Psomopoulos was born in Athens, Greece, in 1973. He received the M.Sc. degree and the Ph.D. degree in electrical engineering both from National Technical University of Athens in 1997 and 2002 respectively. He is currently Associate Professor at the Piraeus University of Applied Sciences/Faculty of Technological Applications/Dept. of Electrical Engineering. His main research interests include ecodesign of ErPs, electromechanical installations, high voltage engineering and high field effects, power generation, renewable energy sources, energy efficiency and GHG emissions, photovoltaic systems and power quality.

Muhammad Shabir Shaharudin received his MBA in International Business from Universiti Sains Malaysia and Bachelor Degree in International Business from Limkokwing University, Malaysia. He is currently pursuing PhD at the Graduate School of Business, Universiti Sains Malaysia. He has two years working experience in two SMEs whose core activities are transport & logistics as well as retail & distribution. His current research is in Carbon Footprint of supply chain and his area of interest includes business activities within an Islamic environment & economic system, eco-innovation, sustainability development, and supply chain & operations.

Vira Shendryk started her academic career in Sept. 2002 at the Sumy State University, Ukraine. She received her PhD degree from the Sumy State University in 2003. Currently she is an Associate Professor in Computer Science Department at the Sumy State University. She was a visiting scholar at McMaster University, Canada in 2012 and was a visiting research fellow at the Department of Computer Science of the Faculty of Technology and Society, the Malmo University, Sweden in 2013. Her research interest is focused on the field of Information Systems and Decision Science particularly in decision making under uncertainty. She has written over 100 journal articles and conference papers and presentations. She has also been a Member of the Editorial Review Board of international multi-disciplinary quarterly journal *Information Technology and Economics*, which focuses on the intersection of Information Technology and Economics and was Member of Programme Committees the International Conference on Information and Software Technologies (ICIST 2014, Lithuania) and Advanced Information Systems and Technologies (AIST 2012-2014, Ukraine).

Olha Shulyma received the degree of MSc from Sumy State University, Ukraine in 2012. Since 2012, she is a PhD Student in Sumy State University (Sumy, Ukraine). Her research interests centre on using IT in power system based on RES. In 2014, she was enrolled in a PhD Sandwich in Computer Science in Malmö University, Sweden.

René Tapia, Industrial Engineer graduated of the Technical University Federico Santa Maria, Chile, works as a freelance consultant for projects related to Asset Management and Maintenance in the industrial sector, currently in Chile, Peru and Spain; has several scientific publications in the area, participating actively in international conferences. Designer and expert modeler in RelPro Software. Also has experience in Software Development Projects, with over 13 years of programming experience. Specialties: Analysis of Production Systems in terms of Reliability, Availability, Maintenance and Production; Sizing of Production Lines; Implementation of Reliability Engineering and Performance Metrics; Simulation of Production Processes; Development of models for optimal maintenance policies; Technical and financial studies for optimal replacement of equipment and components; Simulation and design of transport systems; Estimated production and economic evaluation in production lines with maintenance activities, using discrete and continuous simulation.

About the Contributors

Ayeley Philippe Tchangani received Ingénieur degree (1995) from Ecole Centrale de Lille, France, MSc degree (1995) and a PhD degree (1999) from Université des Sciences et Technologies de Lille, France in control and automation. After a post-doctoral fellowship at French South Africa Technical Institute in electronics (Pretoria, South Africa), he joins Université Toulouse III – IUT de Tarbes in 2001 where he is currently an associate professor. He holds a research position at Laboratoire Génie de Production (LGP) of Ecole Nationale d'Ingenieurs de Tarbes (ENIT) since 2003. Dr. Tchangani's current research interests are in decision analysis, uncertainty modeling and risk assessment and management. Results of his researches appeared in a number of international refereed journals; a list of his publications is available at http://publicationslist.org/tchangani. He regularly serves as Internal Program Committee (IPC) member of international conferences. Dr. Tchangani is a member of IEEE and MCDM societies.

Liina Tonisson is a researcher at Fraunhofer MOEZ Competitive Intelligence group. She studied economics and business administration in Stockholm School of Economics in Riga and holds a master degree in Sustainable Development from University of Leipzig and Hiroshima University. In 2011 September she joined Fraunhofer MOEZ after working with Intellectual Property related Venture Capital/ Private Equity fund. She focuses on IPR related services and clean technology transfer processes.

Pablo Viveros, Researcher and Academic at Technical University Federico Santa María, Chile, has been active in national and international research, both for important journals and conferences. Also he has developed projects in the Chilean industry. Consultant specialist in the area of Reliability, Asset Management, System Modeling and Evaluation of Engineering Projects. He is Industrial Engineer and Master in Asset Management and Maintenance.

Ika Sari Wahyuni-TD obtained her Bachelor's degree in Accounting from Andalas University, Indonesia. She has previously completed additional education of the Accounting profession from Andalas University under certified of The Indonesian Institute of Accountants. She then received Master's degree of Accounting in Kulliyah of Economics & Management Sciences, International Islamic University Malaysia (IIUM), followed by several years of experience working as an assistant lecturer and lecturer in public and private institutions. She has been actively involved in accounting discussions and forums for more than two years.

Jan H. Wiśniewski has a M.Sc. (Eng.) in Sustainable Energy Planning and Management at Aalborg University and M.Sc. in Environmental Protection at the Warsaw University of Technology and is currently in his final year of PhD studies in Fluid Mechanics at the Warsaw University of Technology. His main interests are wind turbine aerodynamics, developing as well as calculating economic feasibility of electricity generation systems, noise and electromagnetic field measurements and fencing.

Index

A

Absorbance (α) 61-62, 89
Analysis of Variance 229, 233-234, 243

B

Balance of System (BOS) 131, 159
Bio-Diversity 260
Bipolar Approach 2, 8-9, 21
BOCR Analysis 5, 11, 21
Building Adapted Photovoltaics 160
Building Integrated Photovoltaics 160
Business Sustainability 264, 323, 336

C

Carbon Footprint 133, 324-329, 331-333, 336-338, 340, 346
Cash Crops 250, 261
Chaffing 195, 203
Chattering 211, 215, 219, 229, 232-234, 237, 243
Clean Development Mechanism (CDM) 328, 336, 347
Clean Technology 263-276, 279
Collector Circuit 75, 122
Collector performance 122
Consensus 3, 7-10, 14, 17-18, 21, 254, 261, 266, 353
Corrosion 92, 105-106, 195, 203, 213

D

Decision Support System 1, 162, 179, 181-183, 185-186, 192
decreasing wear 22
Degradation 102, 246, 248, 251, 295, 307-308, 350, 361
Distributed Generation 129, 165, 181
District Heating 162, 165-166, 173, 186, 192

E

Ecological Debt 248-250, 261
Ecological Footprints 250, 261
Economic Input-Output (EIO) 328, 347
Energy-Carrier Fluid 122
Energy Management 163, 179, 183, 185-186
Environmental Management System (EMS) 349, 368
Environmental Stewardship 350, 368
E-Waste 261

F

Failure Rate 289, 291-293, 296-297, 300-301, 308
Flat Collector 122
Francis turbine 37, 44-45, 48, 50
Free Trade Zone (FTZ) 335, 347

G

gauge readings 35, 38, 43-44, 48, 50
Geographic Information System 180, 192
Globalization 244-248, 251-253, 256-257, 261, 361
Green Global Business 323
Greenhouse Gasses (GHGs) 325, 347
Green Innovations 323
green supply chain 311-312, 320-321, 323, 325, 328-329, 348-351, 355-357, 361, 368
Green Supply Chain Management 312, 323, 325, 348-351, 355, 361, 368
Green Tariff 160, 176
grid-connected systems 127, 130
Gross Domestic Product (GDP) 41, 246, 251, 261, 274, 325, 333
Group Decision Making Problem (GDMP) 1, 4, 21

H

Hazardous Material 203

HeatCAM System 185, 192
H-Type Vertical Axis Wind Turbines 22, 34
Hydroelectric power plant 35, 37, 39, 52

I

Imperfect Maintenance 309
Incidence Angle (δ) 61-62, 89
Industrial Property 264-265, 267, 269-273, 276, 280
Industrial property management 269, 271
Integration 17, 134, 136, 160, 162-163, 181, 186, 236, 244-245, 247-248, 256, 261, 312, 321, 348-355, 357, 360-362, 368
Intellectual Property 244, 253-254, 263-276, 280-281
intellectual property management 271, 275
Intellectual Property Rights 244, 253-254, 263-268, 270-274, 276, 280-281
Intellectual Property Services 269, 271-273, 275, 280
Irradiation 81, 122, 125-126, 133, 141, 160

J

Just-in Time (JIT) 347

K

Kaplan turbine 36-40, 44-45, 49-50
Kettha 325, 330, 332, 347
K-Omega SST Model 34

L

Life Cycle Assessment (LCA) 327, 347
Low Carbon Supply Chain 324-325, 328-329, 331, 336, 347

M

manufacturing firm 330
Mitigation 1, 203, 255, 265
Multi-Criteria Problem 21
Multilateral 244, 247, 251, 253, 255-256, 261
Multilateral Environmental Agreements 251, 253, 261
Mycarbon 333, 338, 347

N

Nation States 244-245, 247, 261
Nominal Capacity 38-39, 52
Nominal Conditions 52
Non Repairable System 309

O

Operating Point 132, 160
Overall Vibration Value 204, 209, 214, 229, 231-234, 236, 238, 243
Overall vibration value analysis 229, 236, 238

P

Parameterization 309
Parasitic Energy (QPAR) 82, 122
Peak Sun Hours 160
Permanent Regime 122
photovoltaic generators 155
Photovoltaic (PV) Cell 160
Photovoltaic (PV) System 160
Physical Environment 244-246, 248, 253, 261
Power Output 39, 53, 132
Pyranometer 89, 94, 101
Pyrheliometer 89, 94, 101

R

Radiometer 89
Reflectance (θ) 89
Repairable System 289, 309
Residual Risk 195, 203
Reverse Logistics 312, 340, 352, 368
roof-top installations 134, 155

S

Self-induced vibrations 210-211, 237-238
Simulation 24-25, 27, 181-183, 210-211, 219, 287-289, 295, 299, 302, 305, 307
small-scale photovoltaics 124
Small Scale PV Installations 133, 140, 161
Smart Grid 181-182, 192
Solar Azimuth (ϒs) 89
Solar Collector 61-64, 73, 75, 105, 122, 183
Solar Contribution 122
Solar Fraction (f) 81, 122
Spalart-Allmaras Model 34
Spectral Identity 233-234, 237-238, 243
Spectral Intercomparison 209, 233, 237-238, 243
Supply Chain Integration 321, 348-350, 352-353, 355, 357, 362, 368
Supply Chain Management 312, 323, 325, 339-340, 348-352, 354-355, 360-361, 368
Supply Chain Sustainable Development 323
Surrounding Air Speed 89
Sustainable Business Performance 349, 362, 368
System Safety 194-195, 197, 199, 201, 203

T

technology management 255, 271
Terminal 184, 192
Thermal Load 122
Tolerance Level 305, 309
Tracking Array 161
Trade Liberalization 245, 247, 262
Transmittance (τ) 61-62, 89

U

UNDESA 247, 251, 262
UNDP 253, 255, 262
UNEP 246, 251-253, 262
urban areas 124, 130, 133, 155, 251

V

Vertical Axis Wind Turbine (VAWT) 22-24, 26, 31

W

Waterfalls 215, 233, 235-236, 243
Water Gauge 35, 38, 53
Wind Farm Installation 1-3, 21
Wind Turbine 3, 21-28, 30-31, 34, 36, 170, 173, 182-183, 195, 197-198, 200-201, 203, 272
World Radiometric Reference (WRR) 89

X

XFOIL 24, 34

CPSIA information can be obtained at www.ICGtesting.com
Printed in the USA
BVOW07*0421070515
399330BV00009B/159/P